Accession no.
01060245

KU-301-589

WITHDRAWN

LIBRARY

TEL. 01244 375444 EXT. 3301

This book is to be returned on or
before the last date stamped below.

**UNIVERSITY
COLLEGE
CHESTER**

Handbook of
Experimental Pharmacology

Volume 136

Editorial Board

G.V.R. Born, London
P. Cuatrecasas, Ann Arbor, MI
D. Ganten, Berlin
H. Herken, Berlin
K. Starke, Freiburg i. Br.
P. Taylor, La Jolla, CA

. 1307101

CHESTER COLLEGE

ACC. No.	DEPT.
O 10 60245	

CLASS No.
ε 572.6 LAT

LIBRARY

572.6

Springer
Berlin
Heidelberg
New York
Barcelona
Hong Kong
London
Milan
Paris
Singapore
Tokyo

CHESTER COLLEGE

LIBRARY

Stress Proteins

Contributors

H. Abe, D.M. Altmann, A.-P. Arrigo, M. Bachelet, L. Battistini,
R. Bomprezzi, G. Borsellino, I.R. Brown, C. Buttinelli, R. Carroll,
R.S. Gupta, J.B. Harrub, K. Himeno, A.D. Johnson, D.E. Kovacs,
D.S. Latchman, G.C. Li, C. Mattei, E. Mimnaugh, N.F. Mivechi,
C. Montesperelli, M. Morange, R.I. Morimoto, G. Multhoff,
K. Nagata, L. Neckers, S.G. Newton, J. Nishizawa, T.S. Nowak, Jr.,
C.M. Pickart, B.S. Polla, C. Pozzilli, X. Préville, G. Ristori,
M. Salvetti, T.W. Schulte, F.R. Sharpe, Y. Shi, B.J. Soltys,
P.K. Srivastava, A. Stephanou, W.J. Valentine, W. van Eden,
M. Vignola, D.M. Yellon, Q. Zhou

Editor

David S. Latchman

 Springer

Professor
David S. LATCHMAN, M.A. Ph.D., D.Sc.
Head, Department of Molecular Pathology
Director, Windeyer Institute of Medical Sciences
University College London
Cleveland Street
London W1P 6DB
United Kingdom

With 49 Figures and 7 Tables

ISBN 3-540-65017-2 Springer-Verlag Berlin Heidelberg New York

Library of Congress Cataloging-in-Publication Data
Stress proteins / contributors, H. Abe ... [et al.]; editor, David S. Latchman.
     p.   cm. – (Handbook of experimental pharmacology; v. 136)
     Includes bibliographical references and index.
     ISBN 3-540-65017-2 (alk. paper)
     1. Heat shock proteins.   2. Stress (Physiology)   I. Abe, H.   II. Latchman, David S.   III. Series.
QP905.H3   vol. 136
[QP552.H43]
615′.1s – dc21   98-42186
[572′.6]   CIP

This work is subject to copyright. All rights are reserved, whether the whole or part of the material is concerned, specifically the rights of translation, reprinting, reuse of illustrations, recitation, broadcasting, reproduction on microfilm or in any other way, and storage in data banks. Duplication of this publication or parts thereof is permitted only under the provisions of the German Copyright Law of September 9, 1965, in its current version, and permission for use must always be obtained from Springer-Verlag. Violations are liable for prosecution under the German Copyright Law.

© Springer-Verlag Berlin Heidelberg 1999
Printed in Germany

The use of general descriptive names, registered names, trademarks, etc. in this publication does not imply, even in the absence of a specific statement, that such names are exempt from the relevant protective laws and regulations and therefore free for general use.

Product liability: The publishers cannot guarantee the accuracy of any information about dosage and application contained in this book. In every individual case the user must check such information by consulting the relevant literature.

Cover design: *design & production* GmbH, Heidelberg

Typesetting: Best-set Typesetter Ltd., Hong Kong

Production Editor: Angélique Gcouta

SPIN: 10628216   27/3020 – 5 4 3 2 1 0 – Printed on acid-free paper

Preface

This work is concerned with a group of proteins which were originally considered to be an esoteric phenomenon but which have now been shown to play critical roles both in normal and stressed cells as well as being involved in a variety of human diseases. It is the purpose of this work to give a comprehensive view of these proteins and their various aspects.

After an introductory chapter providing an overview of these proteins, the work is divided into four main sections each of which deals with one important aspect of these proteins. Thus, the first section contains a series of chapters which describe individual stress proteins and their roles in particular biological phenomena. Evidently, the induction of these proteins by elevated temperature or other stresses is their defining feature and the second section of this book therefore considers the regulation of stress protein gene expression both by stressful stimuli such as elevated temperature or ischaemia and by non-stressful stimuli such as cytokines.

This dual approach to the stress proteins involving their study in both non-stressed and stressed cells is continued in the third section, which deals with the role of these proteins. Thus specific chapters in this section deal with their role in normal cellular processes such as embryonic development and the immune response, whilst others deal with their ability to protect specific tissues such as the brain and the heart from the damaging effects of stressful stimuli. Finally, the last section of the book deals with the role of stress proteins in specific human diseases, illustrating once again their ability to play a protective role as well as acting as a target for damaging immune responses in autoimmune disease.

Overall, therefore, it is hoped that this book will provide a comprehensive review of the many facets of these proteins and their dual role in normal and stressed cells. I am most grateful to the editorial board of the *Handbook of Experimental Pharmacology* for deciding to include a volume on stress proteins in this series and for inviting me to edit it, as well as to Doris Walker and the staff at Springer-Verlag for their efficiency in the production of this book.

London, November 1998   DAVID S. LATCHMAN

List of Contributors

ABE, H., Department of Neurology, University of Tennessee,
855 Monroe Avenue, Link 415, Memphis, TN 38163, USA

ALTMANN, D.M., Transplantation Biology Group, Clinical Sciences Centre,
Imperial College School of Medicine, Hammersmith Hospital,
Du Cane Road, London W12 ONN, UK

ARRIGO, A.-P., Laboratoire du Stress Cellulaire, Centre de Génétique
Moléculaire et Cellulaire, CNRS UMR-5534, Université Claude
Bernard, Lyon-1, 69622 Villeurbanne, France

BACHELET, M., Laboratoire de Physiologie Respiratoire, UFR Cochin
Port-Royal, 24 Rue du Faubourg Saint Jacques, 75014 Paris, France

BATTISTINI, L., IRCCS S. Lucia, Rome, Italy

BOMPREZZI, R., Department of Neurological Science,
University 'la Sapienza v.le dell, Universita 30–00185 Rome, Italy

BORSELLINO, G., Department of Neurological Science,
University 'la Sapienza v.le dell, Universita 30–00185 Rome, Italy

BROWN, I.R., Division of Life Sciences, University of Toronto,
Scarborough Campus, Toronto, Ontario M1C1A4, Canada

BUTTINELLI, C., Department of Neurological Science,
University 'la Sapienza v.le dell, Universita 30–00185 Rome, Italy

CARROLL, R., Hatter Institute for Cardiovascular Studies, Department of
Academic Cardiology, University College London Hospitals and
Medical School, Gower Street, London WC1 E6AU, UK

GUPTA, R.S., Department of Biochemistry, McMaster University, Hamilton,
Ontario, Canada L8 N 3Z5

HARRUB, J.B., Department of Neurology, University of Tennessee,
855 Monroe Avenue, Link 415, Memphis, TN 38163, USA

HIMENO, K., Department of Parasitology and Immunology, University of
Tokushima School of Medicine, Tokushima 770, Japan

JOHNSON, A.D., Department of Biology, Wake Forest University,
Box 7325 Reynolda Station, Winston-Salem, NC 27109, USA

KOVACS, D.E., Department of Neurological Science,
University 'la Sapienza v.le dell, Universita 30–00185 Rome, Italy

LATCHMAN, D.S., Department of Molecular Pathology, The Windeyer
Institute of Medical Sciences, University College London,
The Windeyer Building, 46 Cleveland Street, London W1P 6DB, UK

LI, G.C., Department of Medical Physics and Department of Radiation
Oncology, Memorial Sloan Kettering Cancer Center, 1275 York Avenue,
New York, NY 10021, USA

MATTEI, C., Department of Neurological Science,
University 'la Sapienza v.le dell, Universita 30–00185 Rome, Italy

MIMNAUGH, E., National Cancer Institute, National Institute of Health,
Key West Facility, 9610 Medical Center Drive, Rockville, MD 20850,
USA

MIVECHI, N.F., Institute of Molecular Medicine and Genetics,
MCG CB2803, 1120 15th St, Augusta, GA 30912, USA

MONTESPERELLI, C., Department of Neurological Science,
University 'la Sapienza v.le dell, Universita 30–00185 Rome, Italy

MORANGE, M., Unité de Génétique Moléculaire, ENS, 46 rue d'Ulm,
75230 Paris, Cedex 05, France

MORIMOTO, R.I., Department of Biochemistry, Molecular Biology
and Cell Biology, Rice Institute for Biomedical Research,
Northwestern University, Evanston, IL 60208-3500, USA

MULTHOFF, G., Institute for Clinical Haematology, GSF Munich, Germany

NAGATA, K., Department of Molecular and Cellular Biology, Institute for
Frontier Medical Sciences, Kyoto University, CREST, J. ST. Sakyo-ku
Kyoto 606-8397, Japan

NECKERS, L., Medicine Branch, National Cancer Institute, National Institute
of Health, Key West Facility, 9610 Medical Centre Drive, Rockville,
MD 20850, USA

NEWTON, S.G., Transplantation Biology Group, Clinical Sciences Centre,
Imperial College School of Medicine, Hammersmith Hospital,
Du Cane Road, London W12 ONN, UK

NISHIZAWA, J., Department of Cardiovascular Surgery, Takeda Hospital,
Nishinotoin Shiokojidori Shimogyo-ku Kyoto 600-8558, Japan

NOWAK JR, T.S., Department of Neurology, University of Tennessee,
855 Monroe Avenue, Link 415, Memphis, TN 38163, USA

PICKART, C.M., Department of Biochemistry, School of Public Health, Johns Hopkins University, 615 North Wolfe Street, Baltimore, MD 21205–2179, USA

POLLA, B.S., Laboratoire de Physiologie Respiratoire, UFR Cohin Port-Royal, 24 rue du Faubourg St Jacques, 75014 Paris, France

POZZILLI, C., Department of Neurological Science, University 'la Sapienza v.le dell, Universita 30–00185 Rome, Italy

PREVILLE, X., Laboratoire du Stress Cellulaire, Centre de Genetique Moleculaire et Cellulaire, CNRS UMR-5534, Université Claude Bernard, Lyon-1, 69622 Villeurbanne, France

RISTORI, G., Department of Neurological Science, University 'la Sapienza v.le dell, Universita 30–00185 Rome, Italy

SALVETTI, M., Department of Neurological Science, University 'la Sapienza v.le dell, Universita 30–00185 Rome, Italy

SCHULTE, T.W., National Cancer Institute, National Institute of Health, Key West Facility, 9610 Medical Center Drive, Rockville, MD 20850, USA

SHARP, F.R., Department of Neurology, University of California at San Francisco and Department of Veterans Affairs Medical Centre, 4150 Clements Street, San Francisco, CA 94121, USA

SHI, Y., Department of Biochemistry, Molecular Biology and Cell Biology, Rice Institute for Biomedical Research, Northwestern University, Evanston, IL 60208-3500, USA

SOLTYS, B.J., Department of Biochemistry, McMaster University, Hamilton Ontario, Canada L8 N 3Z5

SRIVASTAVA, P.K., Centre for Immunotherapy of Cancer and Infectious Diseases, MC 1601, University of Connecticut Health Centre, Farmington, CT 06030, USA

STEPHANOU, A., Department of Molecular Pathology, The Windeyer Institute of Medical Sciences, University College London, The Windeyer Building, 46 Cleveland Street, London W1P 6DB, UK

VALENTINE, W.J., Department of Neurology, University of Tennessee, 855 Monroe Avenue, Link 415, Memphis, TN 38163, USA

VAN EDEN, W., Institute of Infectious Diseases and Immunology, Utrecht University, Yalelaan 1, 3584 CL Utrecht, Netherlands

VIGNOLA, M., Italian National Research Council, Institute of Lung Pathophysiology, Palermo, Italy

YELLON, D.M., Hatter Institute for Cardiovascular Studies, Department of
    Academic Cardiology, University College Hospital, Gower Street,
    London WC1 E6AU, UK

ZHOU, Q., Department of Neurology, University of Tennessee,
    855 Monroe Avenue, Link 415, Memphis, TN 38163, USA

Contents

CHAPTER 3

Heat Shock Protein 70

CHAPTER 4

**Mitochondrial Molecular Chaperones hsp60 and
mhsp70: Are Their Roles Restricted to Mitochondria?**
B.J. Soltys and R.S. Gupta. With 4 Figures 69

CHAPTER 5

Role of Hsp27 and Related Proteins
A.-P. Arrigo and X. Préville. With 4 Figures 101

CHAPTER 9

**Regulation of Heat Shock Transcription Factors by Hypoxia or
Ischemia/Reperfusion in the Heart and Brain**

CHAPTER 10

Autoregulation of the Heat Shock Response

CHAPTER 11

The Cellular Stress Gene Response in Brain

CHAPTER 12

Heat Stress Proteins and Their Relationship to Myocardial Protection
R. CARROLL and D.M. YELLON   265

CHAPTER 13

Heat Shock Proteins in Inflammation and Immunity
M. BACHELET, G. MULTHOFF, M. VIGNOLA, K. HIMENO and B.S. POLLA
With 4 Figures   281

CHAPTER 14

Heat Shock Proteins in Embryonic Development
M. MORANGE   305

CHAPTER 15

Heat Shock Proteins in Rheumatoid Arthritis
W. VAN EDEN   329

CHAPTER 16

Heat Shock Protein 60 and Type I Diabetes

CHAPTER 17

Heat Shock Proteins and Multiple Sclerosis

CHAPTER 18

Heat Shock Proteins in Atherosclerosis

CHAPTER 19

Heat Shock Protein-Peptide Interaction: Basis for a New Generation of Vaccines Against Cancers and Intracellular Infections

CHAPTER 1

Stress Proteins: An Overview

D.S. LATCHMAN

A. Introduction

When, in 1962, Ritossa reported the appearance of a new puffing pattern in the salivary gland polytene chromosomes of *Drosophila busckii* following exposure to elevated temperature, he can hardly have foreseen that this phenomenon would ultimately be of interest to those concerned with processes as diverse as protein folding and human autoimmune diseases. Yet, this finding represented the first step in the study of a group of proteins whose synthesis is induced by exposure of cells to elevated temperature or other stresses and which are therefore known as the heat shock or stress proteins. Studies over the past 35 years have revealed the importance of these proteins in fundamental cellular processes such as protein folding as well as elucidating their role in a variety of different human diseases. It is the purpose of this book to discuss these proteins and detailed reviews of their various functions and their involvement in specific diseases are presented in the individual chapters which follow. The purpose of this introductory chapter is to provide an overview of these proteins and the most important themes which emerge from their study.

B. The Stress Proteins

Following the initial report of Ritossa, subsequent studies revealed that similar elevated synthesis of a few proteins following exposure to heat or other stresses occurs in all organisms studied ranging from prokaryotic bacteria such as *E. coli* to mammals including man. Initially, the major stress or heat-inducible proteins were identified simply by exposing cells to such stresses in the presence of radiolabelled amino acids and identifying newly labelled proteins by autoradiography of polyacrylamide gels. Subsequently, following the cloning of the genes encoding stress proteins identified in this manner, other genes encoding related stress proteins were identified on the basis of their homology. It then became clear that many stress proteins are encoded by multi-gene families encoding proteins with similar but distinct features. In addition, other proteins which were identified in other situations were also subsequently shown to be induced by specific stresses when they were analyzed in detail.

In this manner, a number of different stress-inducible proteins have been defined. The major stress proteins which are normally referred to as hsp N where N is the molecular weight in kilo-daltons are listed in Table 1 together with brief details concerning their features. Further details of the proteins and

Table 1. Major eukaryotic stress proteins

Family	Members	Prokaryotic homologues	Functional role	Comments
Hsp100	Hsp104, Hsp100	ClpA, ClpB	Protein turnover	Appear to be involved in tolerance to extreme temperature. Have ATPase activity
Hsp90	Hsp90, Grp94	C62.5	Maintenance of proteins such as steroid receptors in an inactive form until appropriate	*Drosophila* and yeast proteins known as hsp83
Hsp70	Grp78 (=Bip), Hsp70, Hsc70, Hsx70	dnaK	Protein folding and unfolding, assembly of multi-protein complexes	Hsx70 only in primates
Hsp60	Hsp60	gro EL, 65 kDa antigen	Protein folding and mycobacterial unfolding, organelle translocation	Major antigen of many bacteria and parasites which infect man
Hsp56	Hsp56 (also known as FKBP59)	None	Protein folding, associated with hsp90 and hsp70 in steroid receptor complex	Have peptidyl prolyl isomerase activity, target of immuno-suppressive drugs
Hsp47	Hsp47	None	Protein folding of collagen and possibly other proteins	Has homology to protease inhibitors
Hsp27	Hsp27, Hsp26, etc.	Mycobacterial 18 kDa antigen	Protein folding, actin binding proteins	Very variable in size (12–40 kDa) and number in different organisms
Ubiquitin	Ubiquitin	None	Protein degradation	Also found conjugated to histone H2A in the nucleus

their functional roles and involvement in specific human diseases are discussed extensively in the appropriate chapters of this book.

C. Functions of Stress Proteins

Although the stress proteins were evidently initially defined on the basis of their enhanced synthesis in response to exposure to heat or other stresses, many of these proteins are also expressed in normal unstressed cells. Thus, for example, hsp90 constitutes approximately 1% of the total protein in unstressed cells and is induced 2–5 fold following exposure to elevated temperature. Indeed, even where a specific stress protein is absent in unstressed cells, a homologous protein exists in normal unstressed cells. This is seen in the case of the hsp70 family where the major stress inducible protein (known either as hsp70 or hsp72) is absent in unstressed cells which contain however, large amounts of another member of the family (known as hsc70 or hsp73) which is only mildly induced by exposure to stress.

This indicates therefore the function of the hsps is likely to be one which is required in normal cells but is of even greater importance in stressed cells. Moreover, this function is likely to be of great importance in all cells ranging from bacteria to man since it has been highly conserved in evolution. Thus, not only does exposure to stress induce the synthesis of specific proteins in all organisms, but these proteins themselves have been highly conserved in evolution as indicated in Table 1 which shows that prokaryotic homologues of most eukaryotic hsps can be identified in bacteria.

Numerous studies have indicated that the primary role of the majority of hsps lies in producing proper folding of other proteins so that they can fulfil their appropriate functions. This is evidently required in all normal cells but will be of greater importance in stressed cells where exposure to elevated temperature or other noxious stimuli will result in the production of aberrant proteins which are improperly folded. Some of the roles of the hsps in proper protein folding are illustrated in Fig. 1. As discussed in subsequent chapters, these include preventing the inappropriate associations of other proteins with each other, the correct intracellular transport of other proteins and the maintenance of specific proteins in an inactive form until the correct signal for their activation is received. These functions are evidently of importance in the normal cell, whilst in the stressed cell, abnormal interactions which are highly undesirable will evidently increase and must be dealt with by the increased concentrations of hsps produced by the stress.

Evidently, in some situations the damaged proteins produced by the stress will be too abnormal to be refolded properly. For this reason, some stress-inducible proteins such as the small hsp ubiquitin are involved in protein degradation and therefore assist in the destruction of damaged proteins in the stressed cell whilst also playing a role in normal protein turnover in unstressed cells.

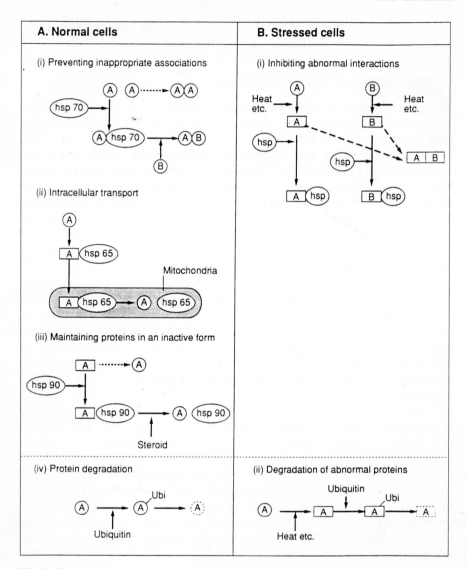

Fig. 1. Functions of the stress proteins

D. Hsp Expression and Regulation

The dual role of the stress proteins in both normal and stressed cells, evidently requires the existence of complex regulatory processes which ensure that the correct expression pattern is produced. Indeed, such processes must be operative at the very earliest stages of embryonic development since the genes encoding stress proteins such as hsp70 and hsp90 have been shown to be amongst the first genes which are transcribed from the embryonic genome.

A number of studies have shown that a specific transcription factor known as heat shock factor 1 (HSF1) is responsible for the induction of the stress protein genes following exposure to heat or other stresses. Thus, this factor exists in an inactive monomeric form in unstressed cells. Following exposure to heat or other stresses, it forms a trimeric form which is able to bind to appropriate DNA sequences in the promoters of the stress protein genes known as heat shock elements (HSE). Subsequently HSF1 becomes phosphorylated and then activates transcription of these genes.

Much less is known, about the regulation of these genes in response to non stressful stimuli or in normal cells. It has been shown that a second form of HSF, known as HSF2 which also binds to the HSE mediates the induction of the hsp70 gene which occurs when an erythroleukaemia cell line is induced to differentiate following exposure to hemin. However, it is now becoming clear that transcription factors which bind to sites within the stress protein gene promoters that are distinct from HSE sequences, also play a role in the regulation of these genes in normal cells and in response to specific non stressful stimuli. Thus, for example, it has been shown that the enhanced synthesis of hsp70 and hsp90 following exposure of cells to cytokines such as interleukin-6 (IL-6) is mediated by transcription factors such as NF-IL6 and STAT-3 which bind to sites in their gene promoters adjacent to HSE sequences. Hence, the expression of the stress proteins in normal and stressed cells is parallelled by their regulation by transcription factors which are activated by non stressful and stressful stimuli respectively.

E. Stress Proteins and Protection

The important role of the stress proteins in processes such as protein folding and their expression in normal embryonic development suggests that they are likely to be essential for the survival of cells and organisms. In agreement with this idea, no "knock-out" mice lacking specific hsp genes have been reported in mammals suggesting that such mice might not be viable and indeed, inactivation of the hsp90 gene has been shown to be lethal in both *Drosophila* and in yeast. Thus, although the inactivation of an individual stress protein gene may not be lethal in all cases since another member of the family might substitute for its functions, the complete elimination of the function of a specific family of stress proteins is likely to be lethal in the vast majority of cases.

As always, in discussing the stress proteins it is necessary to consider not only their function in normal cells but their function in stressed cells. A very wide variety of studies have left no doubt that induction of the stress proteins by a specific stress does have a protective effect and enhances the ability of the organism to deal with that stress. Such a conclusion was initially reached on the basis of experiments showing that a mild stress sufficient to induce the hsps was able to protect cells against a subsequent more severe stress. Subsequently, such studies were supplemented by the over expression of individual hsps in cultured cells which showed that this made the cells more resistant to a subsequent stress. Conversely the elimination of a particular stress protein by micro-injection of an antibody to it or the use of an anti-sense approach resulted in a reduced tolerance to stress. These experiments were brought to their logical conclusion by the finding that transgenic animals overexpressing hsp70 showed enhanced resistance to cardiac ischaemia.

These experiments are of obvious therapeutic relevance in terms of human diseases which involve damage caused by such stressful stimuli particularly cerebral ischaemia (stroke) or cardiac ischaemia. For this reason, great efforts are being made to identify drugs which could induce the synthesis of stress proteins in a non stressful manner as well the potential development of means of delivering stress protein genes to damaged organs in vivo using gene therapy procedures.

F. Stress Proteins and Human Disease

As described above, the stress proteins play a critical role in the functioning of normal cells and particularly in their stress resistance. These effects are obviously of great benefit to the organism. In addition however, the stress proteins have also been implicated in specific human diseases. Somewhat surprisingly, unlike the situation with most other essential proteins, such alterations in the stress proteins in human diseases do not generally involve the reduction in their expression or the abolition of their function.

Rather, the enhanced expression of stress proteins has been reported in a number of different human diseases. Evidently, many of such reports will simply arise from the fact that particular cells in a specific disease are exposed to stress due to the effects of the disease and therefore respond by inducing the synthesis of the stress proteins. In other cases however, the specific synthesis of an individual stress protein has been reported with no effects on other stress proteins suggesting that the effect is a specific one which may be related to the nature of the disease. Thus, for example, hsp90 has been reported to be specifically overexpressed in a subset of patients with the autoimmune disease systemic lupus erythematosus (SLE) where the expression of other stress proteins is unaffected.

Probably the greatest significance of such general or specific over expression of stress proteins in particular diseases, is that they can provoke an autoimmune response leading to the production of antibodies or T cells which recognise these proteins and can produce a damaging immune response. Such immune responses to individual stress proteins, particular hsp60, have been reported in diseases as diverse as rheumatoid arthritis, multiple sclerosis and type 1 diabetes. Similar autoimmune responses to endogenous mammalian stress proteins can be provoked by exposure of animals to bacterial pathogens such as the Mycobacteria whose major antigen is homologous to hsp60.

Although such autoimmune responses appear to play a critical role, for example, in animal models of rheumatoid arthritis involving Mycobacterial infection, it is at present unclear whether they play a similar role in the pathogenesis of human rheumatoid arthritis and other autoimmune diseases such as type 1 diabetes or multiple sclerosis. This is discussed further in the appropriate chapters of this book.

G. Conclusion

As will be seen from this brief introduction and more particularly from the detailed examples described in this book, the stress proteins are unique in terms of their dual role in both normal and stressed cells. This critical role, is paralleled by the existence of regulatory processes which ensure the appropriate patterns of expression of these stress proteins in normal and stressed cells. Similarly, therapeutic interventions which involve the stress proteins may involve both the enhancement of their expression by pharmacological means or gene therapy procedures to enhance resistance to stress as well as interventions targeted at diseases which involve enhanced stress protein expression with a consequent autoimmune response. We have indeed come a long way from the salivary glands of an insect to the realisation that these proteins play key roles in the normal development of humans and other animals as well as in specific human diseases. The next 35 years is likely to see further progress as well as the development of interventions aimed at manipulating the expression of the stress proteins and/or the immune response to them for therapeutic benefit.

CHAPTER 2
The Hsp90 Chaperone Family

L. NECKERS, E. MIMNAUGH, and T.W. SCHULTE

A. General Aspects

Heat shock protein 90 (Hsp90) is one of the most abundant proteins in eukaryotic cells, comprising 1–2% of total cellular protein even under non-stress conditions. Hsp90 is highly evolutionarily conserved, with homologs in such divergent species as bacteria, yeast, *Drosophila* and humans (for review, see PRATT and TOFT 1997). Mutational analysis in yeast has demonstrated an absolute Hsp90 requirement for cell survival (PARSELL and LINDQUIST 1993) (although it is not essential in prokaryotes), and recent studies have shown that Hsp90 participates in multiple signal transduction pathways. The development of early eukaryotic cells from prokaryotic progenitors appears to have been accompanied by gene duplication of both Hsp70 and Hsp90 (GUPTA and GOLDING 1996). In mammalian cells, there are two Hsp90 isomers in the cytosol (Hsp90α and Hsp90β in humans (HICKEY et al. 1989), Hsp86 and Hsp84 in mice (PERDEW et al. 1993)), while a third homolog, glucose regulated protein 94 (Grp94), is localized primarily to the endoplasmic reticulum (LITTLE et al. 1994; WEARSCH and NICCHITTA 1996). Two recent reports have identified an additional, truncated, cytosolic member of the family, designated Hsp75 (CHEN et al. 1996a; SONG et al. 1995). Hsp90 exists as a dimer, and while homodimers appear to be more common (MINAMI et al. 1991, 1994), heterodimerization of Hsp90α and Hsp90β can occur (PERDEW et al. 1993). Although primarily cytoplasmic in unstressed cells, Hsp90 rapidly accumulates in cell nuclei following various stresses including heat shock (AKNER et al. 1992; GASC et al. 1990).

Hsp90, in concert with other chaperones and co-chaperones, plays an important role in the folding of certain newly synthesized proteins and in the stabilization and refolding of proteins after thermal stress (BUCHNER 1996; HARTL 1996). Hsp90 interacts with a distinct, though expanding, subset of signalling proteins (see Table 1), and has been shown to be necessary for their function and stability. Thus, Hsp90 is required for full activity of several "ligand-dependent" transcription factors including members of the steroid receptor family, the aryl hydrocarbon receptor and the retinoid receptor. "Ligand-independent" transcription factors which bind to Hsp90 include MyoD, the heat shock factor Hsf-1, mutated p53 and hypoxia inducible factor 1-α. Hsp90 or Grp94 have also been demonstrated to be necessary for proper

Table 1. Hsp90 client proteins

Client protein	Reference
Transcription factors	
Steroid hormone receptors	
Glucocorticoid receptor	Sanchez et al. 1985
Progesterone receptor	Catelli et al. 1985; Schuh et al. 1985
Estrogen receptor	Redeuilh et al. 1987
Androgen receptor	Veldscholte et al. 1992
Mineralocorticoid receptor	Rafestin-Oblin et al. 1989
Aryl hydrocarbon receptor	Denis et al. 1988; Perdew 1988
v-erbA	Privalsky 1991
Retinoid receptor	Holley and Yamamoto 1995
Sim	Shue and Kohtz 1994
Myo D1	Shaknovich et al. 1992
Heat shock factor 1	Nadeau et al. 1993
Mutated p53	Blagosklonny et al. 1995
Protein kinases	
Tyrosine kinases	
$p60^{v\text{-src}}$	Brugge et al. 1981; Opperman et al. 1981
$p60^{c\text{-src}}$	Hutchison et al. 1992a
Src family kinases	Hartson and Matts 1994
Wee1	Aligue et al. 1994
Sevenless	Cutforth and Rubin 1994
$p185^{erbB2}$	Chavany et al. 1996[a]
Fps/Fes	Lipsich et al. 1982; Nair et al. 1996
Serine/threonine kinases	
Raf-1/v-Raf	Stancato et al. 1993; Wartmann and Davis 1994
B-Raf	Jaiswal et al. 1996
MEK	Stancato et al. 1997
Heme regulated eIF-2α kinase	Matts and Hurst 1989
eEF-2 kinase	Palmquist et al. 1994
Casein kinase 2	Miyata and Yahara 1992
CDK4	Stepanova et al. 1996
Other proteins	
Cytoskeletal proteins	
Actin	Koyasu et al. 1986
Tubulin	Sanchez et al. 1988
Intermediate filaments	Fostinis et al. 1992
Calmodulin	Minami et al. 1993
G protein $\beta\gamma$-subunit	Inanobe et al. 1994
Proteasome	Tsubuki et al. 1994; Wagner and Margolis 1995
Hepatitis B virus reverse transcriptase	Hu and Seeger 1996; Hu et al. 1997

[a] $p185^{erbB2}$ interacts with Grp94, not Hsp90.

function and correct cellular localization of a wide variety of tyrosine and serine/threonine kinases, including members of the Src family, oncogenic EGF receptor-related tyrosine kinase $p185^{erbB-2}$, cyclin-dependent kinase Cdk4, cell cycle-associated kinase Wee1, and serine-threonine kinase Raf-1, a member of the MAP kinase pathway.

Hsp90 comprises the core of several multimolecular chaperone complexes – best characterized in their association with steroid receptors – which interact with proteins at different stages of their maturation, perhaps with different functional consequences. The ability of Hsp90 to participate in the assembly of multiple higher order chaperone complexes certainly must contribute to its involvement in diverse cellular activities, although factors which regulate its participation in these multiple signalling pathways are only now being characterized.

B. The Early Protein Folding Complex

As proteins are synthesized on polyribosomes, many, if not all, initially interact with what has been termed "the early protein folding complex" (see Fig. 1). This contains Hsp70/Hsc70 (DnaK in bacteria, reviewed in GEORGOPOULOS et al. 1994) as the central chaperone protein. Hsp70 binds to an unfolded polypeptide which requires binding of ATP to its N-terminal ATPase domain. Stable binding of Hsp70 to the polypeptide requires hydrolysis of Hsp70-bound ATP, a process stimulated by the co-chaperone Hsp40 (human homolog named Hdj for human DnaJ-like protein; Ydj1 in yeast) and the binding of p48[Hip] (JOHNSON and CRAIG 1997; ZIEGELHOFFER et al. 1996). This early protein folding-chaperone complex leads to partially folded intermediates which require further interactions with chaperonins (CCT/TriC in human, GroEl/ES in bacteria; FENTON and HORWICH 1997; HARTL 1996) to achieve a mature conformation. In the case of proteins such as steroid receptors, the mature conformation is achieved by interactions with a series of Hsp90-containing multiprotein complexes (Fig. 1).

C. Hsp90-Containing Multimolecular Complexes

I. The Steroid Receptor-Associated Hsp90-Containing Intermediate Folding Complex

Maturation of the progesterone receptor (PR) requires the co-chaperone p60[Hop] (SMITH et al. 1993, Sti1 in yeast). Since p60[Hop] binds to both Hsp90 and Hsp70 it is thought to serve as an adaptor protein which assembles an Hsp70/Hsp90/p60[Hop]-PR intermediate complex (CHEN et al. 1996c; LÄSSLE et al. 1997; SMITH et al. 1993). This complex is transient and does not convey hormone binding capability to the receptor (JOHNSON et al. 1996). The benzoquinone ansamycin, geldanamycin, which inhibits many important functions of Hsp90 (see below), stabilizes this intermediate complex and prevents formation of the mature complex (JOHNSON and TOFT 1995; SMITH et al. 1995).

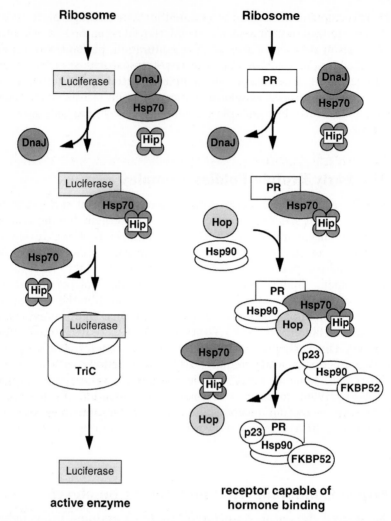

Fig. 1. Model for the sequential action of various chaperone-protein complexes. Folding of newly synthesized luciferase or progesterone receptor (PR) illustrates the TriC and Hsp90 folding pathways. See text for further details

II. The Steroid Receptor-Associated Hsp90-Containing Mature Folding Complex

As a general rule, PR bound to the intermediate Hsp90 complex is not capable of binding ligand. The ligand binding-competent conformation is stabilized by a second Hsp90-containing complex which exists in dynamic equilibrium with the intermediate complex. This multiprotein complex is stabilized by molybdate and contains the co-chaperone p23 and an immunophilin, such as FKBP52 (formerly referred to as Hsp56, p56, p59, FKBP59, or HBI), FKBP51 or CyP40 (Cpr6, Cpr7 in yeast; DITTMAR et al. 1997; JOHNSON et al. 1994, 1996; JOHNSON and TOFT 1994; RENOIR et al. 1990; SULLIVAN et al. 1997). Upon hormone binding, the Hsp90 complex dissociates from the PR (KOST et al. 1989). Certain types of steroid receptors may also require Hsp70 and a DnaJ family member to be present for the hormone-bound receptor to achieve its activated conformation (BOHEN et al. 1995; CAPLAN et al. 1995; KIMURA et al. 1995).

D. Individual Chaperone and Co-chaperone Proteins Found in Hsp90 Complexes

Below is a brief overview of the individual members of Hsp90-containing multimolecular chaperone complexes. Space does not permit a more detailed description of these proteins, and the interested reader is referred to the listed references for more information.

I. Hsp70

The Hsp70 family of proteins includes the constitutively expressed cytosolic chaperone Hsc70, the heat stress-induced Hsp70, and the endoplasmic reticulum-localized Grp78 (BiP; GETHING and SAMBROOK 1992; HARTL 1996; HENDRICK and HARTL 1993). Hsp70 family members bind to a large variety of newly synthesized proteins and other hydrophobic peptides. Even though Hsp70 has high protein folding and refolding activity on its own, it cannot fold PR to a hormone-binding conformation without the participation of Hsp90 (JOHNSON et al. 1996).

II. p48[Hip]

The recently cloned co-chaperone p48[Hip] binds to Hsp70's ATPase domain and stabilizes its ADP-bound state (HOHFELD et al. 1995; PRAPAPANICH et al. 1996a). Functional analysis of p48[Hip] has revealed an N-terminal homo-oligomerization domain, and three tetratricopeptide repeat elements flanked by a charged region. Tetratricopeptide repeats (TPRs) are degenerate sequences of 34 amino acids that often occur in tandem and appear to mediate protein-protein interaction (SIKORSKI et al. 1990). The TPR motif in p48[Hip] is important for Hsp70 binding. p48[Hip] also contains a C-terminal domain that shares homology with another co-chaperone, p60[Hop] (IRMER and HOHFELD 1997; PRAPAPANICH et al. 1996b).

III. p60[Hop]

p60[Hop] binds to Hsp70 via a N-terminal TPR motif and to Hsp90 via a second TPR motif which is closer to the C-terminus (CHEN et al. 1996c; LÄSSLE et al.

1997; SMITH et al. 1993). p60[Hop] by itself does not have refolding activity and therefore should not be classified as a true chaperone (BOSE et al. 1996; FREEMAN et al. 1996). When incubated with purified Hsp70 and Hsp90, p60[Hop] forms a tripartite complex with the two other proteins, thus creating the intermediate Hsp90 folding complex (see above).

IV. p23

p23 is a highly acidic phosphoprotein that is a component of the mature PR complex (DITTMAR et al. 1997; JOHNSON et al. 1994, 1996; JOHNSON and TOFT 1994; SMITH et al. 1990). p23 is ubiquitously expressed in tissues and is highly conserved across species (homologs are found in plants, yeast and humans; OWENS et al. 1996b). p23 is frequently found in association with Hsp90 and an immunophilin, even in the absence of PR (JOHNSON and TOFT 1994). p23-Hsp90 complexes are stabilized by exogenously added molybdate and are disrupted by geldanamycin (see below; JOHNSON et al. 1996). Although p23 does not possess refolding activity, it binds to early unfolding intermediates of proteins in a manner similar to Hsp90 and slows their inactivation following heat shock (BOSE et al. 1996; FREEMAN et al. 1996). In addition to participating in steroid receptor-associated Hsp90 complexes, p23 can also be found in Hsp90 complexes with hepatitis B virus reverse transcriptase, mutated p53, aryl hydrocarbon receptor, and heat shock transcription factor Hsf-1 (BLAGOSKLONNY et al. 1996; HU et al. 1997; NAIR et al. 1996).

V. Immunophilins

Immunophilins are ubiquitously expressed, highly conserved proteins which possess peptidylprolyl isomerase (PPIase) activity. The immunosuppressant drugs FK506, cyclosporin A, and rapamycin bind to the immunophilins and inhibit their PPIase activity (PEATTIE et al. 1992; SCHMID 1993; SCHREIBER 1991; WALSH et al. 1992). Immunophilins are subdivided into those which bind either FK506 (FKBPs) or cyclosporin A (CyPs). FKBP52 and CyP40 have also been shown to display chaperone activity which is not affected by immunosuppressants (BOSE et al. 1996; FREEMAN et al. 1996). Immunophilins bind to Hsp90 via their TPR motifs and replace p60[Hop] in the multichaperone complex (OWENS et al. 1996a). Binding of purified FKBP52 to Hsp90 was found to be ATP-independent and unaffected by either FK506 or rapamycin (RADANYI et al. 1994). The same binding site on Hsp90 can also associate with the TPR domains of FKBP51, CyP40, protein phosphatase 5 or the yeast outer mitochondrial membrane protein Mas70p (also called Tom70p; see below; OWENS et al. 1996a; SILVERSTEIN et al. 1997).

    Immunophilins may have a transport function since FKBP52 has been found associated with both microtubules and nuclear structures, and it is an important component of the hypothesized "transportosome" (CZAR et al. 1994; PERROT et al. 1995; PRATT 1993; RUFF et al. 1992). FKBP52 contains a sequence of eight amino acids with six negatively charged residues that is electrostatically complementary to the positively charged nuclear localization signal of the GR (LEBEAU et al. 1992; PEATTIE et al. 1992; SCHMITT et al. 1993). Intracellular injection of an antibody against this sequence inhibits the dexamethasone-mediated shift of GR into the nucleus (CZAR et al. 1995). Geldanamycin, which inhibits formation of the mature FKBP52-containing GR-Hsp90 complex, effectively blocks nuclear translocation of hormone-bound GR (CZAR et al. 1997), further supporting a role for the FKBP52-containing complex in protein translocation.

VI. Other TPR-Containing Proteins

Recently, two other proteins containing TPR motifs have been shown to bind to Hsp90. PP5 is a serine/threonine phosphatase that contains 4 TPR motifs in its N-terminal domain (BECKER et al. 1994; CHEN et al. 1994, 1996b). In L cell cytosol, 35% of GR-Hsp90 complexes were bound to PP5 while 50% were bound to FKBP52 (SILVERSTEIN et al. 1997). PP5 is characterized by low affinity binding to FK506, and its C-terminal phosphatase domain shares 50% amino acid homology with the PPIase domain of FKBP52, with which it also shares an identical pattern of subcellular localization (CHEN et al. 1994; SILVERSTEIN et al. 1997). Although the role of PP5 in the GR-Hsp90 complex is not known, its enzymatic activity suggests that it participates in regulating the phosphorylation status of one or more members of the complex.

Mas70p is a component of the protein import machinery in the outer mitochondrial membrane which contains 7 TPR motifs (LITHGOW et al. 1995), and it also binds to Hsp90 (OWENS et al. 1996a). Although proteins that might be targeted for intra-mitochondrial import by this mechanism remain to be identified, the existence of a Mas70p-Hsp90 complex suggests that Hsp90 may participate in protein movement into mitochondria.

VII. p50^{Cdc37}

p50 was initially identified as a phosphoprotein which could be co-precipitated with v-Src and Hsp90 (BRUGGE et al. 1981; OPPERMAN et al. 1981), and Cdc37 was first characterized in S. cerevisiae through a temperature-sensitive muta-tion that causes G1 arrest (REED 1980). When the mammalian homolog of Cdc37 was identified, it was found to be identical to the previously identified p50 (DAI et al. 1996; GRAMMATIKAKIS et al. 1995; STEPANOVA et al. 1996). Presently known as p50^{Cdc37}, this chaperone protein replaces immunophilin in Hsp90 complexes with kinases such as Cdk4, v-Src and Raf-1 (BRUGGE 1986; OWENS et al. 1996a; STANCATO et al. 1994; WARTMANN and DAVIS 1994). p50^{Cdc37} does not possess TPR motifs but instead binds to a site close to Hsp90's TPR binding domain (PRATT and TOFT 1997).

The function of p50^{Cdc37} remains unclear. Interestingly, the kinases with which it associates either must be transported to the plasma membrane instead of the nucleus in order to accomplish their function (i.e., Raf-1 and v-Src), or they must be sequestered in an inactive form in the cytoplasm bound to an Hsp90/p50^{Cdc37} complex (Cdk4). Activation of the sequestered kinase results in its dissociation from the chaperone complex followed by nuclear transport (Stepanova et al. 1996). Whether p50^{Cdc37} plays a role in cytoplasmic localization of these kinases remains to be determined.

p50^{Cdc37} stabilizes Cdk4 by helping it to associate with Hsp90, and thus has been proposed to be a "kinase targeting" subunit of Hsp90 (Stepanova et al. 1996). However, recent work in vitro has shown that p50^{Cdc37} also exerts chaperone activity toward β-galactosidase and casein kinase II independently from Hsp90 (Kimura et al. 1997). Further, in yeast, Cdc37 is able to compensate for Hsp90 deficiency in maintaining the activity of p60$^{v\text{-}Src}$, but not GR (Kimura et al. 1997). Thus, evidence exists that Cdc37 may sometimes function as an independent chaperone rather than merely a targeting subunit of Hsp90.

E. Refolding of Denatured Proteins

Although the molecular interactions between Hsp90-containing chaperone complexes and steroid hormone receptors have been most intensively studied to date, a number of observations clearly indicate that Hsp90 also plays an active role in the refolding of proteins following chemical or thermal denaturation. Although Hsp90 acting alone does not appear to be capable of correctly folding polypeptides, when included in near stoichiometric quantities, highly-purified Hsp90 has the ability to bind to and suppress the aggregation and at least partially restore the enzymatic activities of chemically- or thermally-denatured enzymes in vitro, including casein kinase II, firefly luciferase, citrate synthase, β-galactosidase and dihydrofolate reductase. This property of Hsp90 seems to stem from its ability to bind to hydrophobic, normally unexposed surfaces of partially unfolded proteins, thereby diminishing those inter- and intra-molecular interactions that eventually cause denatured proteins to aggregate in an aqueous environment. Importantly, Hsp90 stabilizes proteins at early or intermediate stages of unfolding, before they undergo massive aggregation, and interestingly, in contrast to Hsp70, Hsp90 performs its anti-aggregation function in an ATP-independent fashion (Jakob et al. 1995; Wiech et al. 1992). Although the interaction of guanidine-denatured β-galactosidase with purified Hsp90 maintains the enzyme in a folding-competent state, both Hsp70 (or Hsc70) and Hdj-1, in addition to Hsp90, are required for renaturation of denatured β-galactosidase to full enzymatic activity (Freeman and Morimoto 1996).

The participation of chaperones in the renaturation of thermally-denatured enzymes has frequently been studied using rabbit reticulocyte lysate, which contains abundant heat shock proteins in addition to the enzymatic machinery for protein synthesis. Dissection of this system reveals that Hsp90,

Hsp70, the chaperonin TriC and Mg-ATP are all necessary for the renaturation of guanidinium-inactivated firefly luciferase (NIMMESGERN and HARTL 1993). THULASIRAMAN and MATTS (1996) have reported that the renaturation of thermally denatured luciferase in rabbit reticulocyte lysate is inhibited by geldanamycin, a natural product that tightly and specifically binds to Hsp90 and inhibits its chaperone function (WHITESELL et al. 1994), indicating that Hsp90 is critical to the refolding process. However, Hsp90, Hsc70, p60^{Hop}, p48^{Hip} and p23 were co-precipitated from the lysate with luciferase. Geldanamycin increased the amount of luciferase that could be isolated with the Hsp90-Hsp70-p60^{Hop} chaperone complex and blocked the co-precipitation of the p23-containing chaperone complex with luciferase (THULASIRAMAN and MATTS 1996). The same Hsp90-containing chaperone complex that binds the steroid hormone aporeceptors thus appears to be the major chaperone complex that refolds denatured luciferase. Recent data from David Toft's laboratory demonstrates that luciferase is most effectively renatured in vitro in the presence of Hsp90, p60^{Hop}, Hsp70 and Ydj1 (JOHNSON et al., in press). Apparently Ydj1 binding to Hsp70 stimulates the hydrolysis of Hsp70-bound ATP, potentiating the binding of p60^{Hop}. Since Hop also binds to an ADP-dependent conformation of Hsp90 (see below), p60^{Hop} might serve as a scaffold for the assembly of an Hsp70/Hsp90 complex.

SCHNEIDER et al. (1996) have obtained similar results demonstrating that Hsp90, in cooperation with Hsp70 and p60^{Hop}, mediates the ATP-dependent refolding of heat-denatured luciferase. They also observed that geldanamycin inhibited the refolding of unfolded luciferase in cell extracts and in intact cells, apparently by preventing the dissociation of luciferase from the Hsp90/p60^{Hop}/Hsp70 complex. Finally, in comparing the refolding activities of rabbit reticulocyte lysate with highly-purified chaperone proteins, SCHUMACHER and coworkers (1996) found that Hsp90, Hsp70 and DnaJ cooperated in the renaturation of thermally denatured luciferase. These investigators found that Hsp70 and DnaJ proteins could renature luciferase in the absence of Hsp90, although less efficiently than when Hsp90 was present.

Geldanamycin might antagonize Hsp90 folding activity in one of two ways (which are not necessarily mutually exclusive). (1) Fully functional refolding requires transfer of the partially folded protein from an Hsp90/p60^{Hop}/Hsp70 complex to an Hsp90/p23-containing complex prior to protein dissociation from the chaperone system, and geldanamycin blocks this by destabilizing the later complex. (2) Functional refolding can be accomplished by an Hsp90/p60^{Hop}/Hsp70 complex, but release of the fully folded protein from this complex requires conversion of Hsp90 to its ATP-bound conformation (thereby stimulating dissociation of p60^{Hop}/Hsp70 from Hsp90 and releasing the folded protein). Geldanamycin inhibits this process by substituting for ADP on Hsp90 and thereby preventing nucleotide exchange (see Figure 2). While evidence exists for the ability of geldanamycin to disrupt Hsp90/p23 complexes (JOHNSON and TOFT 1995; SULLIVAN et al. 1997), there is currently no direct evidence for the later scenario, although several independent observations that geldanamycin causes accumulation of Hsp90/p60[Hop]/Hsp70 complexes in vitro (JOHNSON et al., in press; SMITH et al. 1995) certainly support this idea.

Interestingly, in addition to Hsp90 and Hsp70, two other members of this large complex, FKBP52 (independent of its prolyl isomerase activity) and p23, acting as individual proteins, potently suppress the aggregation of thermally-denatured citrate synthase (BOSE et al. 1996). Similarly, the cyclophilin, CyP40, and p23 have been shown to individually maintain guanidine-denatured β-galactosidase in a non-native, folding-competent conformation (FREEMAN et al. 1996). Thus, these observations suggest that several members of Hsp90-containing multimolecular complexes may function cooperatively to recognize protein abnormalities and to refold partially-denatured proteins.

F. Benzoquinone Ansamycins and Nucleotide Binding to Hsp90

Until recently, yeast in which Hsp90 is either mutated or conditionally suppressed have served as the only means by which to study the many functions of this chaperone in vivo (BOHEN and YAMAMOTO 1993; PICARD et al. 1990; XU and LINDQUIST 1993). Although manipulating the overall level of Hsp90 is a powerful tool, this technique does not allow for analysis of the role of individual Hsp90-containing chaperone complexes in modulating target protein activity. Several years ago, this laboratory reported that a class of drugs known as benzoquinone ansamycins, including herbimycin A and geldanamycin, specifically bind to Hsp90 (WHITESELL et al. 1994) and affect its function (JOHNSON and TOFT 1995; SMITH et al. 1995; SULLIVAN et al. 1997). A study of structure-activity relationships has demonstrated a high correlation between the biologic effects of the benzoquinone ansamycins and their ability to bind to Hsp90 (AN et al. 1997). Thus, these drugs have proven very helpful in unravelling relationships between the various Hsp90 chaperone complexes. In addition, the ansamycins are proving quite useful in identifying previously unrecognized client proteins which depend on Hsp90 for stability and function. The fact that benzoquinone ansamycins have also been shown to possess anti-tumor activity in preclinical models identifies the Hsp90 chaperone family as a novel target for anti-cancer drug development.

Because of the utility of these compounds in studying Hsp90 activity, it was important to characterize the site of interaction of ansamycins with Hsp90. During the spring and summer of 1997, three groups independently identified the Hsp90 binding site for geldanamycin (GRENERT et al. 1997; PRODROMOU et al. 1997; STEBBINS et al. 1997). Unexpectedly, this helped resolve a long-simmering controversy of whether Hsp90 contains a nucleotide binding site. STEBBINS et al. (1997) first published the crystal structure of an amino terminal fragment of human Hsp90 and using co-crystallization techniques, identified a binding pocket for geldanamycin. These authors suggested that the hydrophobic pocket they described was normally a site of interaction with denatured protein substrates. At the same time, BERGERAT et al. (1997) described a novel, glycine-rich, ATP-binding motif characteristic of bacterial DNA gyrase, and they reported strong homology between this motif and a nearly identical sequence present in the amino terminal portion of Hsp90. Comparison of the amino acid sequence identified by Bergerat et al. and the ansamycin binding pocket identified by STEBBINS et al. reveals the ATP-binding motif to be fully contained within the ansamycin binding pocket (see Fig. 2A).

Whether or not Hsp90 binds ATP and possesses ATPase activity has been an area of controversy for several years. Early claims that Hsp90 could bind ATP affinity resins and the photoaffinity analog, 8-azido-ATP, and that Hsp90 contained an ATPase activity (CSERMELY et al. 1993, 1995; NADEAU et al. 1993) have not been reproduced in other laboratories (WEARSCH and NICCHITTA 1997), and a systematic study by JAKOB et al. (1996) failed to detect such activities. On the other hand, the binding of Hsp90 to p23 clearly requires ATP and is inhibited by ADP (JOHNSON and TOFT 1995; SULLIVAN et al. 1997). A solution to these discrepant findings has now emerged. A recent report by SCHEIBEL et al. (1997) used electron spin resonance spectrometry to demonstrate weak (Kd = 400 μM), but specific, ATP binding to Hsp90. These investigators demonstrated that while ribose-modified, spin-labeled ATP analogs were weakly bound by Hsp90, analogs whose adenine moiety was modified (at either the C8 or N6 position) were not accepted by the protein. Interestingly, previous studies demonstrating a lack of ATP binding to Hsp90 utilized either ATP affinity columns in which the nucleotide was coupled to the resin via the C8 position of adenine, or used affinity-labeled derivatives modified at the C8 position.

Coincident with the publication of the crystal structure of the human Hsp90 amino terminus, GRENERT et al. (1997) recently reported biochemical data employing deletion and point mutants prepared from chicken Hsp90. These investigators found that full-length Hsp90 bound to an ATP column if ATP was coupled to the resin via its gamma phosphate group, leaving the ribose and adenine moieties free to interact with the protein. Interestingly, an amino terminal deletion fragment nearly identical to the region crystalized by STEBBINS et al. retained the ability to bind ATP. Full-length Hsp90, in the presence of ATP, bound p23, but when three glycine residues in the ATP binding domain were mutated to valine, p23 binding was lost. Using a geldanamycin column, GRENERT et al. (1997) further demonstrated that the amino terminal fragment of Hsp90 bound to the drug as effectively as the full-length protein. Mutation of the three glycines in the GXXGXG motif which abrogated ATP-dependent p23 binding also eliminated geldanamycin binding. Coincident with these studies, a structural analysis of the amino terminal portion of yeast Hsp90 was published (PRODROMOU et al. 1997). These authors confirmed the original data of STEBBINS et al. (1997) but, in agreement

A

```
         #                                      #  #  #
         *                    *                 *  *  *
90       D                    A  D              V  V  V       140
     IVDTGIGMTKADLVNNLGTIAKSGTKAFMEALQAGADISMIGQFGVGSYSA
```

```
141                                             #              190
                                                *
                                                Q  D          A
     YLVAEKVTVITKHNDDEQYAWESSAGGSFTVRLDNGEPLGRGTKVILHLK
```

surface groove of GA binding site   * impairs GA binding
pocket region of GA binding site   # impairs p23 binding
proposed ATP binding site

B

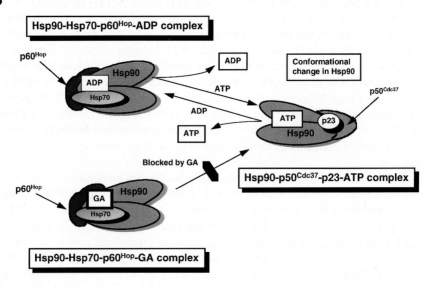

with the study of GRENERT et al. (1997), identified the geldanamycin binding pocket as a nucleotide binding site, capable of binding either ATP or ADP. The importance of this region for Hsp90 function has been documented in a study in *S. cerevisiae*, which showed that a deletion mutant of the *HSP82* gene (an Hsp90 homolog) which lacks the ATP (and geldanamycin) binding motif failed to complement a complete deletion of *HSP82* (LOUVION et al. 1996).

Based on these data, GRENERT et al. (1997) have proposed a model in which ATP binds to Hsp90 in the hydrophobic pocket described by STEBBINS et al. as the geldanamycin binding site. In this model, the more hydrophobic adenine moiety binds within the pocket and the more polar sugar and phosphate groups lie closer to the pocket surface. This is consistent with a space-filling model of adenylic acid, which adopts a "C"-like configuration highly analogous to that reported for Hsp90-bound geldanamycin. Structural analysis of the ATP binding site on the DNA gyrase B protein of *E. coli* has identified a similar, mostly hydrophobic, adenine binding pocket. The existence of a GXXGXG motif near the opening of the pocket (found in the ATP binding motifs of both DNA gyrase B and Hsp90) allows for close contact of the phosphates with the binding site. Although the glycines in this motif are not predicted to be critical for ansamycin binding based on the structural studies, the biochemical data described above clearly demonstrate the importance of these amino acids.

Taken together, the data further suggest that the ATP/geldanamycin binding site acts as a conformational switch region which regulates the assembly of Hsp90-containing multi-chaperone complexes (Fig. 2B). For example, in its ATP-bound state Hsp90 can interact with p23, whereas in its ADP-bound state it instead reacts with p60[Hop]. Recent data from Toft's laboratory confirm

---

Fig. 2. A Homology of the geldanamycin (*GA*) binding pocket of Hsp90 with a unique amino terminal nucleotide binding domain. A partial amino terminal sequence of chicken Hsp90α is shown from amino acids 90–190. Single amino acid mutations that impair GA or p23 binding are indicated by * and #, respectively. The surface groove and pocket regions of the GA binding site as identified by STEBBINS et al. (1997) are indicated, as are the ATP binding homology domains identified by BERGERAT et al. (1997). Amino acid substitutions are shown above the sequence (adapted from GRENERT et al. 1997). **B** Nucleotide regulation of Hsp90 conformation. The Hsp90-Hsp70-p60[Hop] heterocomplex (*top left*) is shown with ADP bound to the Hsp90 dimer. ADP binding to Hsp90 is probably stabilized by p60[Hop], and favors a Hsp90 conformation which allows Hsp90/Hsp70/p60[Hop] interaction. Upon exchange of ATP for ADP, Hsp90 adopts an alternative conformation which no longer favors association of Hsp90 with Hsp70/p60[Hop], but instead supports association with either an immunophilin or p50[Cdc37], as well as p23 (*top right*). Since Hsp90 lacks ATPase activity, the factors which regulate nucleotide exchange in this model are currently unknown, although the two chaperone complexes are probably in dynamic flux. Geldanamycin binds to the nucleotide-binding pocket of Hsp90 and stabilizes the conformation favored by ADP, thus preferentially locking Hsp90 into an Hsp70/p60[Hop]-containing complex (*bottom left*). By restricting Hsp90 to this "ADP-bound" conformation, geldanamycin prevents p23 and p50[Cdc37] (or immunophilin) from binding to Hsp90

that the binding of p23 and p60^{Hop} to Hsp90 are mutually exclusive (JOHNSON et al., in press). By preventing ATP binding, geldanamycin blocks p23 association with Hsp90 and stabilizes p60^{Hop} association, thereby locking the chaperone into the equivalent of its ADP-bound conformation. This hypothesis predicts that Hsp90 binding phenomena which are not ATP-dependent, or which are stimulated by ADP, would not be affected by GA. Indeed, Hsp90 binding to hydrophobic resins, which is inhibited by ATP, is favored by the presence of either ADP or geldanamycin (SULLIVAN et al. 1997). This hydrophobic binding activity, which may identify a site for protein substrates, exists outside of the amino terminal domain. Thus, these findings support the hypothesis that the nucleotide/geldanamycin binding site on Hsp90, perhaps by regulating Hsp90 conformation, is capable of modifying the properties of other regions of the molecule. Based on their structural analysis of a yeast Hsp90 amino terminal fragment co-crystalized with either ATP or ADP, PRODROMOU et al. (1997) have reached a similar conclusion.

G. Hsp90 Client Proteins

I. Transcription Factors

1. Steroid Receptors

The role of Hsp90 in the function of steroid receptors, especially the glucocorticoid receptor (GR) and the progesterone receptor (PR), has been extensively studied. Much of what is known about Hsp90 multimolecular complexes has been learned from the careful study of these receptors. The estrogen receptor (ER), androgen receptor (AR) and mineralocorticoid receptor (MR) share many properties with the GR and PR (BINART et al. 1991; CHAMBRAUD et al. 1990; FANG et al. 1996; SEGNITZ and GEHRING 1997). Subtle differences exist in vivo between the various steroid receptor proteins. One notable example is that GR and MR are primarily cytosolic in the absence of ligand binding, while unliganded PR, ER and AR are found in the nucleus (DEFRANCO et al. 1995). It is not clear what role Hsp90 chaperone complexes play in the nuclear localization of these receptors, although Hsp90 is thought to be required for localization of unliganded GR to the cytoplasm.

Nonetheless, the interaction of Hsp90-containing, higher order complexes with both GR and PR, at least in vitro, is somewhat similar. Both receptor proteins interact with Hsp90 with a stoichiometry of 1:2, thus demonstrating that Hsp90 binds to these receptors as a dimer. Hsp90 dimers associate with the hormone binding domains of both unliganded receptor proteins (DALMAN et al. 1991; SCHOWALTER et al. 1991), and based on studies with Hsp90 deletion mutants, several regions of Hsp90 are important for this interaction. A C-terminal fragment of about 200 amino acids appears to contain both the region that interacts with steroid receptor proteins and the Hsp90 dimerization domain (SULLIVAN and TOFT 1993). The primary interaction with the receptor proteins is ATP- and co-chaperone independent, but amino acids

381–441 of chicken Hsp90 are also necessary for the receptor proteins to attain a ligand binding conformation, and this region might serve as a binding site for other components of an Hsp90-containing multimolecular complex. The N-terminal domain, as described above, serves as a nucleotide binding switch domain which probably regulates p23 and p60^{Hop} binding (GRENERT et al. 1997).

In cells, besides being necessary for ligand binding and nuclear trafficking, Hsp90 is also required for maintenance of GR stability (WHITESELL and COOK 1996). For a more detailed discussion of the role of Hsp90 in steroid receptor function, the reader is referred to two recent, excellent reviews (PRATT 1997; PRATT and TOFT 1997).

2. Aryl Hydrocarbon Receptor

The aryl hydrocarbon receptor (AhR) avidly binds polycyclic benzoflavones, heterocyclic amines and halogenated aromatic hydrocarbons such as dioxin. This signal-transducing protein is a basic helix-loop-helix (bHLH) domain transcription factor which induces expression of cytochrome P4501A1, glutathione S-transferase and other xenobiotic-metabolizing enzymes (for a review see BOCK 1994). The unliganded, latent cytosolic AhR is complexed with Hsp90 dimers, but following ligand binding, the receptor dissociates from Hsp90, dimerizes with its partner protein, ARNT (aryl hydrocarbon receptor nuclear translocator) in the nucleus, then binds to specific DNA response elements (FUKUNAGA et al. 1995; PERDEW 1988; WHITELAW et al. 1993; WILHELMSSON et al. 1990). The binding of the receptor to Hsp90 has been shown to be required for AhR signalling in a yeast expression system with genetically regulatable levels of Hsp90 (CARVER et al. 1994). The AhR must associate, in an ATP-dependent manner, with an Hsp90-containing multichaperone complex for the correct functional folding of both its ligand-binding and its DNA-binding bHLH domains (ANTONSSON et al. 1995). When the ligand binding domain of the receptor is overexpressed in wheat germ extract or in bacteria (two systems deficient in p23 activity), the purified domain is devoid of dioxin binding and cannot be reconstituted by exogenous purified Hsp90 (COUMAILLEAU et al. 1995). These observations strongly suggest that the AhR ligand domain requires interaction with an Hsp90/p23-containing chaperone complex in order to achieve a functionally active conformation. This hypothesis is supported by a recent in vitro study of NAIR et al. (1996) which documented the association of AhR protein translated in reticulocyte lysate with a Hsp90/p23-containing complex.

Hsp90 is also required for the DNA binding activity of other bHLH transcription factors, including MyoD, E12, Sim, and hypoxia inducible factor-1α (GRADIN et al. 1996; McGUIRE et al. 1995; SHUE and KOHTZ 1994), suggesting that this class of transcription factors requires conformational modification by Hsp90 before they become active. Likewise, at least a transient role for Hsp90 in retinoid receptor function was found by HOLLEY and YAMAMOTO

(1995), who demonstrated that retinoid receptors expressed in yeast strains with very low Hsp82 (yeast Hsp90 homolog) levels have impaired signal transduction capability.

3. Mutated p53

We, and others, have recently shown that mutated (but not wild type) forms of the tumor suppressor protein p53 require at least transient interaction with an Hsp90/p23-containing chaperone complex in order to achieve the conformation characteristic of the mutated protein (BLAGOSKLONNY et al. 1996; SEPEHRNIA et al. 1996). If mutated p53 is translated in reticulocyte lysate in the presence of geldanamycin, or in wheat germ extract without drug, the protein is undetectable by mutant conformation-specific anti-p53 antibodies. However, if p23 is added to wheat germ extract, the mutant p53 protein produced is detected by conformation-specific antibodies. Geldanamycin added together with p23 abrogates the restorative properties of this co-chaperone in wheat germ extract because it prevents p23 from binding to Hsp90. In a similar fashion, if another benzoquinone ansamycin, macbecin, is added to yeast which have been transfected with mutant p53, the treated yeast do not produce p53 protein detectable by mutant conformation-specific antibodies; the p53 they do produce is recognized by non-conformation-specific p53 antibodies. A similar result is obtained if mutant p53 is transfected into yeast deficient in Hsp90. Wild type 53 is not affected by either benzoquinone ansamycins, p23 deficiency, or the lack of Hsp90 in vitro or in vivo.

Although wild type p53 has a very short half-life, the mutant form of the protein is quite stable. Because the protein is conformationally flexible and normally undergoes self-association, in cells which express one mutant and one wild type allele, the more abundant mutant protein is frequently able to convert the short-lived wild type protein to a "mutant" conformation, thereby abrogating its activity. By preventing attainment of mutant p53's correct conformation, benzoquinone ansamycins destabilize the mutant protein and reduce its aberrantly prolonged half-life. Thus, in tumor cells heterozygous for mutant p53, these drugs have been shown to restore the transcriptional activity of the wild type protein (BLAGOSKLONNY et al. 1995).

4. Heat Shock Factor

The heat shock transcription factor Hsf-1 has been reported to interact with Hsp90, suggesting a role for Hsp90 in regulating the heat shock response, perhaps by sequestering Hsf-1 in a monomeric state and preventing it from trimerizing to its activated DNA-binding form. Importantly, the binding of Hsf-1 to Hsp90 appears to be temperature-sensitive, in that association of these two proteins is lost at 43°C, coincidentally a temperature that activates the heat shock response (NADEAU et al. 1993; NAIR et al. 1996). The ability of herbimycin A and geldanamcyin to induce the synthesis of both Hsp90 and Hsp70 (HEGDE et al. 1995; MURAKAMI et al. 1994; WHITESELL et al. 1994) is consistent with the ability of these drugs to disrupt Hsp90/p23-containing chaperone complexes with monomeric Hsf-1, thereby freeing the transcription factor to trimerize and attain DNA binding capability (NAIR et al. 1996).

II. Protein Kinases

1. Tyrosine Kinases

a) The Src Family Kinases

The transforming protein of Rous sarcoma virus, p60[v-src], is a tyrosine kinase that associates with the plasma membrane via N-terminal myristylation (JOVE and HANAFUSA 1987). Biochemical analyses in the early 1980's revealed that p60[v-src] is complexed with Hsp90 and a 50 kD protein (BRUGGE et al. 1981; OPPERMAN et al. 1981), recently identified as mammalian p50[Cdc37] (DAI et al. 1996; STEPANOVA et al. 1996). The p60[v-src]-Hsp90-p50[Cdc37] complex forms immediately after translation of the kinase and stays intact during myristylation (BRUGGE et al. 1983; COURTNEIDGE and BISHOP 1982). Complexed p60[v-src] is phosphorylated predominantly on serine residues and is kinase inactive (BRUGGE 1986). Based on experiments with temperature sensitive mutants, the Hsp90 multimolecular complex may serve as a vehicle to transport p60[v-src] to the plasma membrane, keeping the kinase inactive until it reaches its final destination.

The p60[v-src]-Hsp90 multimolecular complex has been studied in vitro using the rabbit reticulocyte lysate model (HUTCHISON et al. 1992a) and shares many biochemical similarities with the steroid aporeceptor-chaperone complex, including stabilization by molybdate (HUTCHISON et al. 1992b). However, as stated above, the p60[v-src] complex contains p50[Cdc37] instead of immunophilin. The C-terminal catalytic domain of p60[v-src] is sufficient to allow binding by the Hsp90 multimolecular complex (JOVE et al. 1986).

Recently, we suggested a possible role for the serine/threonine phosphorylation of Hsp90 in regulating p60[v-src]-Hsp90 association (MIMNAUGH et al. 1995). We observed that the p60[v-src]-Hsp90 complex was destabilized in cells treated with the phosphatase inhibitor okadaic acid, resulting in complex dissociation and a resultant loss of p60[v-src] protein from the treated cells. Concomitant with these events, phosphorylation of Hsp90 serine residues was doubled and the phosphorylation of threonine residues was increased 20-fold. Since the phosphothreonine content of Hsp90 normally is barely detectable, but is so markedly increased upon phosphatase inhibition, threonine phosphorylation of Hsp90 must be tightly controlled, and therefore is a candidate mechanism for triggering release of p60[v-src] protein from the chaperone complex.

Experiments in yeast have supported the importance of the Hsp90 multimolecular complex for the function of p60[v-src] in vivo. Mutations lowering Hsp90 expression lead to a decrease in p60[v-src] levels, a decrease in tyrosine phosphorylated proteins and rescued cell growth by a release from p60[v-src]-

dependent cell cycle arrest (NATHAN and LINDQUIST 1995; XU and LINDQUIST 1993). Mutations in other members of Hsp70- and Hsp90-containing complexes, including Ydj1 (the yeast DnaJ homologue; DEY et al. 1996a; KIMURA et al. 1995), Sti1 (the yeast p60^{Hop} homolog; CHANG et al. 1997), Cpr6 and Cpr7 (yeast CyP40 homologues; DUINA et al. 1996), and p50^{Cdc37} (DEY et al. 1996b) produce similar effects. As stated above, Cdc37 overexpression can compensate for Hsp90 deficiency in yeast, suggesting the possible existence of redundant mechanisms for regulating this kinase (KIMURA et al. 1997).

Interestingly, the benzoquinone ansamycin antibiotics were first identified as inhibitors of p60^{v-src} tyrosine kinase activity (UEHARA et al. 1986, 1988). However, recently it was found that kinase inhibition is indirect. By disrupting the p60^{v-src}-Hsp90 complex, these drugs mimic Hsp90 depletion and cause destabilization and loss of the kinase protein (WHITESELL et al. 1994). Taken together, these data demonstrate that p60^{v-src} depends upon Hsp90 complexes in a way very similar to the steroid receptors.

The role of Hsp90 in the function and stability of normal cellular c-src is less clear. While the p60^{v-src}-Hsp90 complex is abundant and stable, especially in temperature sensitive v-Src mutants, early experiments failed to detect native c-Src-Hsp90 complexes (BRUGGE 1986). Also, c-Src expressed in yeast was not destabilized or inactivated in an Hsp90-deficient mutant (XU and LINDQUIST 1993). However, c-Src-Hsp90 complexes can be assembled in reticulocyte lysate and are disrupted by geldanamycin (HUTCHISON et al. 1992a; WHITESELL et al. 1994). More recently, the normal cellular Src-family kinases Lck and Fgr, which share a high homology with c-Src in their C-terminal regions, have been detected in complexes with Hsp90 (HARTSON et al. 1996; HARTSON and MATTS 1994). In addition to Src family kinases, other non-receptor tyrosine kinases such as Fes and its viral oncogenic homolog Fps have been found to exist in Hsp90-containing complexes (LIPSICH et al. 1982; NAIR et al. 1996).

b) Wee1 Kinase

Wee1 is an *S. pombe* tyrosine kinase that phosphorylates Cdc2 on Y15 and thus controls the length of the G2 phase of the cell cycle. A protein, Swo1, was identified in a screen for mutations which rescue cell cycle progression in Wee1-overexpressing cells. Swo1 is a Hsp90 homolog shown to directly bind to Wee1 by co-immunoprecipitation and to be required for function and stability of the kinase (ALIGUE et al. 1994). Further details about this protein complex remain to be identified.

c) Sevenless Tyrosine Kinase

The sevenless protein-tyrosine kinase is required for the differentiation of the R7 photoreceptor neuron during development of the compound eye in *Drosophila melanogaster*. Signalling through the sevenless receptor pathway is inhibited by mutations in Enhancer of sevenless, E(sev)3 A and E(sev)3B.

E(sev)3 A encodes a Hsp90 homolog, and E(sev)3B a homolog of p50^{Cdc37} (CUTFORTH and RUBIN 1994). These observations support a role for an Hsp90 multimolecular complex in the sevenless pathway, even though a direct inter- action of Hsp90, p50^{Cdc37} and the sevenless kinase remains to be demonstrated.

d) Receptor Tyrosine Kinases – p185^{erbB2}

Several years ago, benzoquinone ansamycins were observed to stimulate the rapid degradation of the p185^{erbB2} receptor tyrosine kinase (MILLER et al. 1994). It was subsequently shown that ansamycins do not interact with the kinase directly, but instead bind to a 100 kDa protein later identified as the Hsp90 homolog Grp94 (CHAVANY et al. 1996). p185^{erbB2} is associated with Grp94, both in the endoplasmic reticulum and perhaps also at the plasma membrane. By disrupting this association, benzoquinone ansamycins cause rapid polyubiquitination of p185^{erbB2} followed by its proteasome-dependent degradation (MIMNAUGH et al. 1996). Other receptor tyrosine kinases, includ- ing the insulin receptor and IGF-1 receptor, are similarly sensitive to ansamycins (SEPP et al. 1995). Whether this sensitivity reveals dependence of these kinases on association with a member of the Hsp90 family remains to be determined.

2. Serine/Threonine Kinases

a) Raf-1 Kinase

The serine/threonine kinase Raf-1 is part of one or more protein kinase cascades that are important for mitogenic signalling, but it has also been described to play a role in differentiation and control of apoptosis (MAGNUSON et al. 1994; MARSHALL 1995). Raf-1 is recruited to the plasma membrane by activated Ras and is subsequently activated by phosphorylation and possibly other biologic events (MORRISON and CUTLER 1997). Like p60^{v-src}, Raf-1 is associated with Hsp90 and p50^{Cdc37}, both as a native complex in cells and when reconstituted in vitro in reticulocyte lysate (STANCATO et al. 1993; WARTMANN and DAVIS 1994). The C-terminal catalytic domain of the protein is sufficient to allow Hsp90 binding (STANCATO et al. 1993). The Raf-1-containing multimo- lecular complex binds [^{3}H]FK506, suggesting that it may also contain an FKBP immunophilin (STANCATO et al. 1994).

The function of the Hsp90/Raf-1 complex in mediating kinase activity is far from clear. When Hsp90 is inhibited with geldanamycin, Raf-1 is destabi- lized (SCHULTE et al. 1995) and degraded rapidly by the proteasome (SCHULTE et al. 1997), in a manner similar to that described for steroid hormone recep- tors and receptor tyrosine kinases (SEPP et al. 1995; WHITESELL and COOK 1996). Thus, association with Hsp90 must be important for Raf-1 stability, even though the exact mechanism by which Raf-1 becomes targeted to the proteasome following geldanamycin remains to be elucidated.

Inhibition of Hsp90 has also been shown to inhibit Raf-1 signalling, in part by preventing newly synthesized Raf-1 from reaching the plasma membrane

(Schulte et al. 1996; Stancato et al. 1997), thus supporting an additional role for Hsp90 as a "transportosome" for this kinase. Further, since the downstream target of Raf-1, MEK1, has also been detected in a complex with both Raf-1 and Hsp90 (Stancato et al. 1997), a third function of Hsp90 might be to provide a scaffold for the assembly of a Raf-1/MEK1 multiprotein signalling unit.

In *Drosophila*, mutations that produce dominant negative Hsp83 inhibit signalling through Raf (van der Straten et al. 1997). Interestingly, a mutated Raf protein which is constitutively targeted to the plasma membrane independently of Ras still requires functional Hsp83 for signalling. These data support the hypothesis that only Hsp90-bound Raf is in a conformation which can be activated by mitotic signals, similar to the Hsp90-steroid receptor model (van der Straten et al. 1997).

b) Casein Kinase II

Casein kinase II (CKII) is a widely expressed serine/threonine kinase consisting of two catalytic alpha and two regulatory beta subunits (Bodenbach et al. 1994). Many details about the function and regulation of CKII remain to be elucidated, even though many possible substrates of CKII have been identified, and the kinase appears to be especially important in proliferating cells. The interaction between CKII and Hsp90 is complex. Hsp90 binds tightly to CKII, protects it from self-aggregation and enhances its kinase activity (Miyata and Yahara 1992). But CKII can also phosphorylate Hsp90α, Hsp90β and Grp94 in vitro (Cala and Jones 1994; Shi et al. 1994) with unknown consequences. In fact, inadvertent co-purification of CKII may have been responsible for the ATPase activity previously associated with Hsp90 preparations reported in earlier publications.

c) Heme-Regulated eIF-2α Kinase (HRI)

The heme-regulated eukaryotic initiation factor 2α (eIF-2α) kinase (also called heme-regulated inhibitor, HRI) regulates protein synthesis in rabbit reticulocytes depending upon heme concentration. During translation, recycling of eIF-2 requires exchange of GDP for GTP catalyzed by eIF-2B. If eIF-2α is phosphorylated by HRI, it binds and sequesters eIF-2B and thus indirectly arrests protein synthesis (Chen and London 1995). HRI is activated by heme deficiency, but also by heat shock, sulfhydryl reagents, oxidants, glucose deficiency and ethanol. HRI binds co-translationally to Hsp90-containing multimolecular complexes that contain Hsc70, FKBP52 and p23, thus closely resembling steroid receptor folding complexes (Matts and Hurst 1989; Xu et al. 1997). Hsp90 association is required to fold nascent HRI into an activatable conformation and to stabilize it prior to activation. When activated by heme deficiency, HRI undergoes either autophosphorylation or phosphorylation by other kinases such as CKII (Mendez and de Haro 1994), dissociates from Hsp90, and acquires an active, stable state. Benzoquinone ansamycins disrupt the HRI-Hsp90 complex, resulting in an incorrectly folded protein which cannot undergo autophosphorylation (UMA et al. 1997).

d) Cdk4/Cdk6

Cdk4/cyclin D complexes play an essential role during progression through the G1 phase of the cell cycle by phosphorylating the retinoblastoma protein (SHERR 1994; WEINBERG 1995). Cdk4 has recently been shown to exist in a complex with Hsp90 and p50^{Cdc37} (DAI et al. 1996; LAMPHERE et al. 1997; STEPANOVA et al. 1996). Interestingly, the complexes Cdk4 forms with Hsp90/p50^{Cdc37} are mutually exclusive to those it forms with cyclin D (STEPANOVA et al. 1996), suggesting that Hsp90 serves to fold and sequester Cdk4 in an inactive but primed conformation, which dissociates from the chaperone complex upon binding to cyclin D. Although Cdk4 is found in both cytosol and nucleus, the active, cyclin D-bound form is found predominately in the nucleus. If Hsp90 function is inhibited by geldanamycin, Cdk4 levels are decreased by post-translational destabilization of the protein, similar to steroid receptors and Raf-1. Since Cdk4 binds Hsp90 only in the presence of p50^{Cdc37}, the later protein has been described as a kinase-targeting subunit of Hsp90 (STEPANOVA et al. 1996).

Cdk6 has also been reported to form a complex with Hsp90 and p50^{Cdc37} (STEPANOVA et al. 1996), and the protein is destabilized upon pharmacologic inhibition of Hsp90 function by herbimycin A (AKAGI et al. 1996). A detailed analysis of the nature of this heterocomplex is still lacking.

III. Other Proteins

1. Cytoskeletal Proteins

Besides transcription factors and kinases, Hsp90 has been shown to complex with cytoskeletal proteins including actin (KOYASU et al. 1986), tubulin (REDMOND et al. 1989; SANCHEZ et al. 1988) and intermediate filaments (CZAR et al. 1996; FOSTINIS et al. 1992).Whether these complexes reflect chaperoning and assembly of the cytoskeletal subunit proteins themselves, or whether Hsp90 links signalling proteins to the cytoskeleton in order to provide docking or intracellular transportation requires further study.

2. Calmodulin

Calmodulin binds Hsp90 (MINAMI et al. 1993) and may regulate certain of its functions, since Ca^{2+}-dependent interaction of calmodulin with Hsp90 has been shown to inhibit binding of the chaperone to actin filaments (MIYATA and YAHARA 1991; NISHIDA et al. 1986). Further, upon treatment with a calmodulin antagonist, GR becomes incapable of ligand binding, even though GR remains bound to Hsp90 (NING and SANCHEZ 1996). Additional studies are needed to understand the biologic function of the Hsp90-calmodulin interaction, and whether calmodulin plays a more general role in assembly/disassembly of Hsp90-containing multiprotein complexes.

3. $\beta\gamma$-Subunits of Trimeric GTP-Binding Proteins

Trimeric GTP-binding proteins (G proteins) consist of α, β and γ-subunits and serve to transmit and amplify signals from membrane receptors to effector proteins (TAUSSIG and GILMAN 1995). Activation of a trimeric G protein requires dissociation of GDP and binding of GTP to the α-subunit, as well as dissociation of the $\beta\gamma$-dimers. Hsp90 specifically binds to the $\beta\gamma$-dimers (INANOBE et al. 1994), and interaction with Hsp90 may be necessary for their formation. $\beta\gamma$-dimers form spontaneously when these subunits are synthesized in reticulocyte lysate (SCHMIDT and NEER 1991), but do not associate when synthesized in bacteria (HIGGINS and CASEY 1994) or wheat germ extract (MENDE et al. 1995), both of which are deficient in p23 activity. Further, Hsp90 was found to associate with $\beta\gamma$-dimers only when they were dissociated from the α-subunit. Although, the exact impact of Hsp90 on trimeric G protein signalling remains to be elucidated, the data suggest that Hsp90 stabilizes $\beta\gamma$-dimers in a conformation primed to trimerize with the α-subunit.

4. Proteasome

While it has been proposed that chaperones associate with denatured proteins and direct them towards either refolding or degradation, an association be-tween Hsp90 and the proteasome multi-catalytic proteinase complex has recently been described (REALINI et al. 1994; WAGNER and MARGOLIS 1995). Hsp90 was found to non-competitively inhibit the benzyloxycarbonyl(Z)-Leu-Leu-Leu-4-methylcoumaryl-7-amide (ZLLL-MCA) degrading activity of the proteasome with a stoichiometry of 1:1 (TSUBUKI et al. 1994; WAGNER and MARGOLIS 1995). In contrast, purified Hsp90 has no inhibitory activity against trypsin, chymotrypsin or calpains at a 10-fold molar excess. Given the great abundance of Hsp90 in cells, and assuming uniform intracellular distribution, one might expect all proteasome complexes in the cell to be saturated with Hsp90 and thus to be inhibited. Although this is clearly not the case, the factors which regulate proteasome activity are unknown and this provocative observation suggests participation of Hsp90 in some aspect of proteasome regulation. Indeed, Hsp90 has been reported to protect the trypsin-like activ-ity of the proteasome from oxidant-induced inactivation (CONCONI and FRIGUET 1997). Interestingly, Hsp90 itself does not appear to be a substrate for proteasomal degradation (WAGNER and MARGOLIS 1995).

5. Hepadnavirus Reverse Transcriptase

Hepatitis B viruses (hepadnaviruses) are a group of small DNA viruses that depend upon reverse transcription for replication. Hsp90 binds to hepadnavirus reverse transcriptase and induces a conformational change that requires the cooperation of Hsp70, Hsp40, p23 and possibly other components (Hu and Seeger 1996; Hu et al. 1997). This conformational change triggers the binding of εRNA, which allows for nucleocapsid assembly and the initiation of viral DNA synthesis.

6. Tumor Necrosis Factor Receptor and Retinoblastoma Protein

Recently, two independent studies described a new member of the Hsp90 family capable of associating with the tumor necrosis factor receptor I and with the retinoblastoma protein (Chen et al. 1996a; Song et al. 1995). Designated Hsp75, neither the function of this chaperone nor the multimolecular complexes in which it participates are known, but it may represent yet another example of the involvement of the Hsp90 family in the regulation of diverse signal transduction pathways.

H. Hsp90 and Drug Development

Although it has been known for many years that Hsp90 is important for the function of steroid receptors and perhaps certain viral oncogenic kinases, the past few years have witnessed an exponential expansion in our knowledge of this chaperone. Although mutational analysis suggests Hsp90 is essential for eukaryotic cell survival (at least in yeast and *Drosophila*), cancer cells seem to be especially sensitive to pharmacologic interference in Hsp90 function. Perhaps this is because so many growth-regulating proteins in tumor cells depend upon either stable or transient interactions with Hsp90 for their activities.

The remarkable sensitivity of cancer cells to pharmacologic disruption of Hsp90 function suggests that the Hsp90 family may be a novel target for anti-cancer drug development. Benzoquinone ansamycins have been identified as first generation antagonists of Hsp90, and their efficacy as anti-tumor drugs will soon be tested in human clinical trials. These drugs may also be of value in certain instances where steroid hormone antagonism is indicated. Their use for intervention in hormone-dependent prostate cancer and breast cancer is currently under consideration.

Further detailed characterization of the ansamycin binding site on Hsp90 should allow for future development of more potent and specific Hsp90 inhibitors. Eventually, it may be possible to target specific sites on the chaperone with small molecules to selectively disrupt the binding of other members of Hsp90-containing multiprotein complexes, thus permitting very selective interference in Hsp90 function. For now, the ansamycins continue to prove useful as pharmacologic probes to investigate both the many activities of Hsp90 and the diverse array of target proteins with which it interacts.

I. Conclusion

Hsp90 is clearly a chaperone protein that has acquired additional functions during evolution. Besides its role in refolding proteins after partial denaturation, Hsp90 interactions are also necessary for several client proteins to gain their mature, functional conformation. At the same time, Hsp90 can prevent premature activation of the client protein, either by steric interference, by sequestering the client protein in a cellular compartment that precludes its binding to its biologic target, or by supporting a conformation that needs further modification for full activity. Hsp90 may also serve as a molecular scaffold to assemble signalling proteins in a multimeric complex with their downstream effectors. Finally, Hsp90 regulates the stability of many of its client proteins in vivo, possibly by favoring certain "protected" conformations, masking residues which are substrates for ubiquitination, or by playing a role in targeting proteins for proteasomal degradation.

References

Akagi T, Ono H, Shimotohno K (1996) Tyrosine kinase inhibitor herbimycin A reduces the stability of cyclin-dependent kinase Cdk6 protein in T-cells. Oncogene 13:399–405

Akner G, Mossberg K, Sundqvist KG, Gustafsson JA, Wikstrom AC (1992) Evidence for reversible, non-microtubule and non-microfilament-dependent nuclear translocation of hsp90 after heat shock in human fibroblasts. Eur J Cell Biol 58:356–364

Aligue R, Akhavan NH, Russell P (1994) A role for Hsp90 in cell cycle control: Wee1 tyrosine kinase activity requires interaction with Hsp90. EMBO J 13:6099–6106

An WG, Schnur RC, Neckers L, Blagosklonny MV (1997) Depletion of p185[erbB2], Raf-1 and mutant p53 proteins by geldanamycin derivatives correlates well with antiproliferative activity. Cancer Chemother Pharmacol 40:60–64

Antonsson C, Whitelaw ML, McGuire J, Gustafsson JA, Poellinger L (1995) Distinct roles of the molecular chaperone hsp90 in modulating dioxin receptor function via the basic helix-loop-helix and PAS domains. Mol Cell Biol 15:756–765

Becker W, Kentrup H, Klumpp S, Schultz JE, Joost HG (1994) Molecular cloning of a serine/threonine phosphatase containing a putative regulatory tetratricopeptide repeat domain. J Biol Chem 269:22586–22592

Bergerat A, de MB, Gadelle D, Varoutas PC, Nicolas A, Forterre P (1997) An atypical topoisomerase II from Archaea with implications for meiotic recombination. Nature 386:414–417

Binart N, Lombes M, Rafestin-Oblin ME, Baulieu EE (1991) Characterization of human mineralocorticosteroid receptor expressed in the baculovirus system. Proc Natl Acad Sci USA 88:10681–10685

Blagosklonny MV, Toretsky J, Bohen S, Neckers L (1996) Mutant conformation of p53 translated in vitro or in vivo requires functional HSP90. Proc Natl Acad Sci USA 93:8379–8383

Blagosklonny MV, Toretsky J, Neckers L (1995) Geldanamycin selectively destabilizes and conformationally alters mutated p53. Oncogene 11:933–939

Bock KW (1994) Aryl hydrocarbon or dioxin receptor: biologic and toxic responses. Rev Physiol Biochem Pharmacol 125:1–42

Bodenbach L, Fauss J, Robitzki A, Krehan A, Lorenz P, Lozeman FJ, Pyerin W (1994) Recombinant human casein kinase II. A study with the complete set of subunits (alpha, alpha' and beta), site-directed autophosphorylation mutants and a bicistronically expressed holoenzyme. Eur J Biochem 220:263–273

Bohen SP, Kralli A, Yamamoto KR (1995) Hold 'em and fold 'em: chaperones and signal transduction. Science 268:1303–1304

Bohen SP, Yamamoto KR (1993) Isolation of Hsp90 mutants by screening for decreased steroid receptor function. Proc Natl Acad Sci USA 90:11424–11428

Bose S, Weikl T, Bugl H, Buchner J (1996) Chaperone function of Hsp90-associated proteins. Science 274:1715–1717

Brugge JS (1986) Interaction of the Rous sarcoma virus protein pp60[src] with cellular proteins pp50 and pp90. In: Compans RW, Cooper M, Koprowski H et al (eds) Current topics in microbiology and immunology, vol 123. Springer, Berlin Heidelberg New York, pp 1–22

Brugge JS, Erikson E, Erilson RL (1981) The specific interaction of the Rous sarcoma virus transforming protein, pp60[v-src], with two cellular proteins. Cell 25:363–372

Brugge JS, Yonemoto W, Darrow D (1983) Interaction between the Rous sarcoma virus transforming protein and two cellular phosphoproteins: analysis of the turnover and distribution of this complex. Mol Cell Biol 3:9–19

Buchner J (1996) Supervising the fold: functional principles of molecular chaperones. FASEB J 10:10–19

Cala SE, Jones LR (1994) Grp94 resides within cardiac sarcoplasmic reticulum vesicles and is phosphorylated by casein kinase II. J Biol Chem 269:5926–5931

Caplan AJ, Langley E, Wilson EM, Vidal J (1995) Hormone-dependent transactivation by the human androgen receptor is regulated by a dnaJ protein. J Biol Chem 270:5251–5257

Carver LA, Jackiw V, Bradfield CA (1994) The 90-kDa heat shock protein is essential for Ah receptor signaling in a yeast expression system. J Biol Chem 269: 30109–30112

Catelli MG, Binart N, Jung-Testas I, Renoir JM, Baulieu EE, Feramisco JR, Welch WJ (1985) The common 90-kd protein component of non-transformed '8S' steroid receptors is a heat shock protein. EMBO J 4:3131–3135

Chambraud B, Berry M, Redeuilh G, Chambon P, Baulieu EE (1990) Several regions of human estrogen receptor are involved in the formation of receptor-heat shock protein 90 complexes. J Biol Chem 265:20686–20691

Chang HC, Nathan DF, Lindquist S (1997) In vivo analysis of the Hsp90 cochaperone Sti1 (p60). Mol Cell Biol 17:318–325

Chavany C, Mimnaugh E, Miller P, Bitton R, Nguyen P, Trepel J, Whitesell L, Schnur R, Moyer J, Neckers L (1996) p185[erbB2] binds to GRP94 in vivo. Dissociation of the p185[erbB2]/GRP94 heterocomplex by benzoquinone ansamycins precedes depletion of p185[erbB2]. J Biol Chem 271:4974–4977

Chen CF, Chen Y, Dai K, Chen PL, Riley DJ, Lee WH (1996a) A new member of the hsp90 family of molecular chaperones interacts with the retinoblastoma protein during mitosis and after heat shock. Mol Cell Biol 16:4691–4699

Chen JJ, London IM (1995) Regulation of protein synthesis by heme-regulated eIF-2 kinase. Trends Biochem Sci 20:105–108

Chen MS, Silverstein AM, Pratt WB, Chinkers M (1996b) The tetratricopeptide repeat domain of protein phosphatase 5 mediates binding to glucocorticoid receptor heterocomplexes and acts as a dominant negative mutant. J Biol Chem 271:32315–32320

Chen MX, McPartlin AE, Brown L, Chen YH, Barker HM, Cohen PT (1994) A novel human protein serine/threonine phosphatase, which possesses four tetratricopeptide repeat motifs and localizes to the nucleus. EMBO J 13:4278–4290

Chen S, Prapapanich V, Rimerman RA, Honore B, Smith DF (1996c) Interactions of p60, a mediator of progesterone receptor assembly, with heat shock proteins hsp90 and hsp70. Mol Endocrinol 10:682–693

Conconi M, Friguet B (1997) Proteasome inactivation upon aging and on oxidation-effect of HSP 90. Mol Biol Rep 24:45–50

Coumailleau P, Poellinger L, Gustafsson JA, Whitelaw ML (1995) Definition of a minimal domain of the dioxin receptor that is associated with Hsp90 and maintains wild type ligand binding affinity and specificity. J Biol Chem 270:25291–25300

Courtneidge S, Bishop JM (1982) Transit of pp60^{v-src} to the plasma membrane. Proc Natl Acad Sci USA 79:7117–7121

Csermely P, Kajtar J, Hollosi M, Jalsovszky G, Holly S, Kahn CR, Gergely PJ, Soti C, Mihaly K, Somogyi J (1993) ATP induces a conformational change of the 90-kDa heat shock protein (hsp90). J Biol Chem 268:1901–1907

Csermely P, Miyata Y, Schnaider T, Yahara I (1995) Autophosphorylation of grp94 (endoplasmin). J Biol Chem 270:6381–6388

Cutforth T, Rubin GM (1994) Mutations in Hsp83 and cdc37 impair signaling by the sevenless receptor tyrosine kinase in Drosophila. Cell 77:1027–1036

Czar MJ, Galigniana MD, Silverstein AM, Pratt WB (1997) Geldanamycin, a heat shock protein 90-binding benzoquinone ansamycin, inhibits steroid-dependent translocation of the glucocorticoid receptor from the cytoplasm to the nucleus. Biochemistry 36:7776–7785

Czar MJ, Lyons RH, Welsh MJ, Renoir JM, Pratt WB (1995) Evidence that the FK506-binding immunophilin heat shock protein 56 is required for trafficking of the glucocorticoid receptor from the cytoplasm to the nucleus. Mol Endocrinol 9:1549–1560

Czar MJ, Owens GJ, Yem AW, Leach KL, Deibel MJ, Welsh MJ, Pratt WB (1994) The hsp56 immunophilin component of untransformed steroid receptor complexes is localized both to microtubules in the cytoplasm and to the same nonrandom regions within the nucleus as the steroid receptor. Mol Endocrinol 8:1731–1741

Czar MJ, Welsh MJ, Pratt WB (1996) Immunofluorescence localization of the 90-kDa heatshock protein to cytoskeleton. Eur J Cell Biol 70:322–330

Dai K, Kobayashi R, Beach D (1996) Physical interaction of mammalian CDC37 with CDK4. J Biol Chem 271:22030–22034

Dalman FC, Scherrer LC, Taylor LP, Akil H, Pratt WB (1991) Localization of the 90-kDa heat shock protein-binding site within the hormone-binding domain of the glucocorticoid receptor by peptide competition. J Biol Chem 266:3482–3490

DeFranco DB, Madan AP, Tang Y, Chandran UR, Xiao N, Yang J (1995) Nucleo-cytoplasmic shuttling of steroid receptors. Vitam Horm 51:315–338

Denis M, Cuthill S, Wikstrom AC, Poellinger L, Gustafsson JA (1988) Association of the dioxin receptor with the Mr 90,000 heat shock protein: a structural kinship with the glucocorticoid receptor. Biochem Biophys Res Commun 155:801–807

Dey B, Caplan AJ, Boschelli F (1996a) The Ydj1 molecular chaperone facilitates formation of active p60^{v-src} in yeast. Mol Biol Cell 7:91–100

Dey B, Lightbody JJ, Boschelli F (1996b) CDC37 is required for p60^{v-src} activity in yeast. Mol Biol Cell 7:1405–1417

Dittmar KD, Demady DR, Stancato LF, Krishna P, Pratt WB (1997) Folding of the glucocorticoid receptor by the heat shock protein (hsp) 90-based chaperone machinery. The role of p23 is to stabilize receptor.hsp90 heterocomplexes formed by hsp90.p60.hsp70. J Biol Chem 272:21213–21220

Duina AA, Chang HC, Marsh JA, Lindquist S, Gaber RF (1996) A cyclophilin function in Hsp90-dependent signal transduction. Science 274:1713–1715

Fang Y, Fliss AE, Robins DM, Caplan AJ (1996) Hsp90 regulates androgen receptor hormone binding affinity in vivo. J Biol Chem 271:28697–28702

Fenton WA, Horwich AL (1997) GroEL-mediated protein folding. Protein Sci 6:743–760

Fostinis Y, Theodoropoulos PA, Gravanis A, Stournaras C (1992) Heat shock protein HSP90 and its association with the cytoskeleton: a morphological study. Biochem Cell Biol 70:779–786

Freeman BC, Morimoto RI (1996) The human cytosolic molecular chaperones hsp90, hsp70 (hsc70) and hdj-1 have distinct roles in recognition of a non-native protein and protein refolding. EMBO J 15:2969–2979

Freeman BC, Toft DO, Morimoto RI (1996) Molecular chaperone machines: chaperone activities of the cyclophilin Cyp-40 and the steroid aporeceptor-associated protein p23. Science 274:1718–1720

Fukunaga BN, Probst MR, Reisz PS, Hankinson O (1995) Identification of functional domains of the aryl hydrocarbon receptor. J Biol Chem 270:29270–29278

Gasc JM, Renoir JM, Faber LE, Delahaye F, Baulieu EE (1990) Nuclear localization of two steroid receptor-associated proteins, hsp90 and p59. Exp Cell Res 186:362–367

Georgopoulos C, Liberek K, Zylicz M, Ang D (1994) Properties of the heat shock proteins of Escherichia coli and the autoregulation of the heat shock response. In: Morimoto R, Tissieres A, Georgopoulos C (eds) The biology of heat shock proteins and molecular chaperones. Cold Spring Harbor Laboratory Press, New York, p 209

Gething MJ, Sambrook J (1992) Protein folding in the cell. Nature 355:33–45

Gradin K, McGuire J, Wenger RH, Kvietikova I, Fhitelaw ML, Toftgard R, Tora L, Gassmann M, Poellinger L (1996) Functional interference between hypoxia and dioxin signal transduction pathways: competition for recruitment of the Arnt transcription factor. Mol Cell Biol 16:5221–5231

Grammatikakis N, Grammatikakis A, Yoneda M, Yu Q, Banerjee SD, Toole BP (1995) A novel glycosaminoglycan-binding protein is the vertebrate homologue of the cell cycle control protein, Cdc37. J Biol Chem 270:16198–16205

Grenert JP, Sullivan WP, Fadden P, Haystead T, Clark J, Mimnaugh E, Krutzsch H, Ochel HJ, Schulte TW, Sausville E, Neckers LM, Toft DO (1997) The amino-terminal domain of heat shock protein 90 (hsp90) that binds geldanamycin is an ATP/ADP switch domain that regulates hsp90 conformation. J Biol Chem 272:23843–23850

Gupta RS, Golding GB (1996) The origin of the eukaryotic cell. Trends Biochem Sci 21:166–171

Hartl FU (1996) Molecular chaperones in cellular protein folding. Nature 381:571–579

Hartson SD, Barrett DJ, Burn P, Matts RL (1996) Hsp90-mediated folding of the lymphoid cell kinase p56[lck]. Biochemistry 35:13451–13459

Hartson SD, Matts RL (1994) Association of Hsp90 with cellular Src-family kinases in a cell-free system correlates with altered kinase structure and function. Biochemistry 33:8912–8920

Hegde RS, Zuo J, Voellmy R, Welch WJ (1995) Short circuiting stress protein expression via a tyrosine kinase inhibitor, herbimycin A. J Cell Physiol 165:186–200

Hendrick JP, Hartl FU (1993) Molecular chaperone functions of heat shock proteins. Annu Rev Biochem 62:349–384

Hickey E, Brandon SE, Smale G, Lloyd D, Weber LA (1989) Sequence and regulation of a gene encoding a human 89-kilodalton heat shock protein. Mol Cell Biol 9:2615–2626

Higgins JB, Casey PJ (1994) In vitro processing of recombinant G protein γ-subunits. Requirements for assembly of an active $\beta\gamma$ complex. J Biol Chem 269:9067–9073

Hohfeld J, Minami Y, Hartl FU (1995) Hip, a novel cochaperone involved in the eukaryotic hsc70/hsp40 reaction cycle. Cell 83:589–598

Holley SJ, Yamamoto KR (1995) A role for Hsp90 in retinoid receptor signal transduction. Mol Biol Cell 6:1833–1842

Hu J, Seeger C (1996) Hsp90 is required for the activity of a hepatitis B virus reverse transcriptase. Proc Natl Acad Sci USA 93:1060–1064

Hu J, Toft DO, Seeger C (1997) Hepadnavirus assembly and reverse transcription require a multi-component chaperone complex which is incorporated into nucleocapsids. EMBO J 16:59–68

Hutchison KA, Brott BK, De LJ, Perdew GH, Jove R, Pratt WB (1992a) Reconstitution of the multiprotein complex of pp60src, hsp90, and p50 in a cell-free system. J Biol Chem 267:2902–2908

Hutchison KA, Stancato LF, Jove R, Pratt WB (1992b) The protein-protein complex between pp60^v-src and hsp90 is stabilized by molybdate, vanadate, tungstate, and an endogenous cytosolic metal. J Biol Chem 267:13952–13957

Inanobe A, Takahashi K, Katada T (1994) Association of the beta gamma subunits of trimeric GTP-binding proteins with 90-kDa heat shock protein, hsp90. J Biochem (Tokyo) 115:486–492

Irmer H, Hohfeld J (1997) Characterization of functional domains of the eukaryotic co-chaperone Hip. J Biol Chem 272:2230–2235

Jaiswal RK, Weissinger E, Kolch W, Landreth GE (1996) Nerve growth factor-mediated activation of the mitogen-activated protein (MAP) kinase cascade involves a signaling complex containing B-Raf and HSP90. J Biol Chem 271:23626–23629

Jakob U, Lilie H, Meyer I, Buchner J (1995) Transient interaction of Hsp90 with early unfolding intermediates of citrate synthase. Implications for heat shock in vivo. J Biol Chem 270:7288–7294

Jakob U, Scheibel T, Bose S, Reinstein J, Buchner J (1996) Assessment of the ATP binding properties of Hsp90. J Biol Chem 271:10035–10041

Johnson BD, Schumacher RJ, Ross ED, Toft DO (1998) Hop modulates hsp70/hsp90 interactions in protein folding. J Biol Chem 273:3679–3686

Johnson J, Corbisier R, Stensgard B, Toft D (1996) The involvement of p23, hsp90, and immunophilins in the assembly of progesterone receptor complexes. J Steroid Biochem Mol Biol 56:31–37

Johnson JL, Beito TG, Krco CJ, Toft DO (1994) Characterization of a novel 23-kilodalton protein of unactive progesterone receptor complexes. Mol Cell Biol 14:1956–1963

Johnson JL, Craig EA (1997) Protein folding in vivo: unraveling complex pathways. Cell 90:201–204

Johnson JL, Toft DO (1994) A novel chaperone complex for steroid receptors involving heat shock proteins, immunophilins, and p23. J Biol Chem 269:24989–24993

Johnson JL, Toft DO (1995) Binding of p23 and hsp90 during assembly with the progesterone receptor. Mol Endocrinol 9:670–678

Jove R, Garber EA, Iba H, Hanafusa H (1986) Biochemical properties of p60^v-src mutants that induce different cell transformation parameters. J Virol 60:849–857

Jove R, Hanafusa H (1987) Cell transformation by the viral src oncogene. Annu Rev Cell Biol 3:31–56

Kimura Y, Rutherford SL, Miyata Y, Yahara I, Freeman BC, Yue L, Morimoto RI, Lindquist S (1997) Cdc37 is a molecular chaperone with specific functions in signal transduction. Genes Dev 11:1775–1785

Kimura Y, Yahara I, Lindquist S (1995) Role of the protein chaperone YDJ1 in establishing Hsp90-mediated signal transduction pathways. Science 268:1362–1365

Kost SL, Smith DF, Sullivan WP, Welch WJ, Toft DO (1989) Binding of heat shock proteins to the avian progesterone receptor. Mol Cell Biol 9:3829–3838

Koyasu S, Nishida E, Kadowaki T, Matsuzaki F, Iida K, Harada F, Kasuga M, Sakai H, Yahara I (1986) Two mammalian heat shock proteins, HSP90 and HSP100, are actin-binding proteins. Proc Natl Acad Sci USA 83:8054–8058

Lamphere L, Fiore F, Xu X, Brizuela L, Keezer S, Sardet C, Draetta GF, Gyuris J (1997) Interaction between Cdc37 and Cdk4 in human cells. Oncogene 14:1999–2004

Lässle M, Blatch GL, Kundra V, Takatori T, Zetter BR (1997) Stress-inducible, murine protein mSTI1. J Biol Chem 272:1876–1884

Lebeau MC, Massol N, Herrick J, Faber LE, Renoir JM, Radanyi C, Baulieu EE (1992) P59, an hsp90-binding protein. Cloning and sequencing of its cDNA and preparation of a peptide-directed polyclonal antibody. J Biol Chem 267:4281–4284

Lipsich LA, Cutt J, Brugge JS (1982) Association of the transforming proteins of Rous, Fujinami and Y73 avian sarcoma viruses with the same two cellular proteins. Mol Cell Biol 2:875–880

Lithgow T, Glick BS, Schatz G (1995) The protein import receptor of mitochondria. Trends Biochem Sci 20:98–101

Little E, Ramakrishnan M, Roy B, Gazit G, Lee AS (1994) The glucose-regulated proteins (GRP78 and GRP94): functions, gene regulation, and applications. Crit Rev Eukaryot Gene Expr 4:1–18

Louvion JF, Warth R, Picard D (1996) Two eukaryote-specific regions of Hsp82 are dispensable for its viability and signal transduction functions in yeast. Proc Natl Acad Sci USA 93:13937–13942

Magnuson NS, Beck T, Vahidi H, Hahn H, Smola U, Rapp UR (1994) The Raf-1 serine/threonine protein kinase. Semin Cancer Biol 5:247–253

Marshall CJ (1995) Specificity of receptor tyrosine kinase signaling: transient versus sustained extracellular signal-regulated kinase activation. Cell 80:179–185

Matts RL, Hurst R (1989) Evidence for the association of the heme-regulated eIF-2 alpha kinase with the 90 kDa heat shock protein in rabbit reticulocyte lysate in situ. J Biol Chem 264:15542–15547

McGuire J, Coumailleau P, Whitelaw ML, Gustafsson JA, Poellinger L (1995) The basic helix-loop-helix/PAS factor Sim is associated with hsp90. Implications for regulation by interaction with partner factors. J Biol Chem 270:31353–31357

Mende U, Schmidt CJ, Yi F, Spring DJ, Neer EJ (1995) The G protein γ subunit. Requirements for dimerization with β subunits. J Biol Chem 270:15892–15898

Mendez R, de Haro C (1994) Casein kinase II is implicated in the regulation of heme-controlled translational inhibitor of reticulocyte lysates. J Biol Chem 269:6170–6176

Miller P, DiOrio C, Moyer M, Schnur RC, Bruskin A, Cullen W, Moyer JD (1994) Depletion of the erbB-2 gene product p185 by benzoquinoid ansamycins. Cancer Res 54:2724–2730

Mimnaugh EG, Chavany C, Neckers L (1996) Polyubiquitination and proteasomal degradation of the p185[c-erbB-2] receptor protein-tyrosine kinase induced by geldanamycin. J Biol Chem 271:22796–22801

Mimnaugh EG, Worland PJ, Whitesell L, Neckers LM (1995) Possible role for serine/threonine phosphorylation in the regulation of the heteroprotein complex between the hsp90 stress protein and the pp60[v-src] tyrosine kinase. J Biol Chem 270:28654–28659

Minami Y, Kawasaki H, Miyata Y, Suzuki K, Yahara I (1991) Analysis of native forms and isoform compositions of the mouse 90-kDa heat shock protein, HSP90. J Biol Chem 266:10099–10103

Minami Y, Kawasaki H, Suzuki K, Yahara I (1993) The calmodulin-binding domain of the mouse 90-kDa heat shock protein. J Biol Chem 268:9604–9610

Minami Y, Kimura Y, Kawasaki H, Suzuki K, Yahara I (1994) The carboxy-terminal region of mammalian HSP90 is required for its dimerization and function in vivo. Mol Cell Biol 14:1459–1464

Miyata Y, Yahara I (1991) Cytoplasmic 8 S glucocorticoid receptor binds to actin filaments through the 90-kDa heat shock protein moiety. J Biol Chem 266:8779–8783

Miyata Y, Yahara I (1992) The 90-kDa heat shock protein, HSP90, binds and protects casein kinase II from self-aggregation and enhances its kinase activity. J Biol Chem 267:7042–7047

Morrison DK, Cutler RE (1997) The complexity of Raf-1 regulation. Curr Opin Cell Biol 9:174–179

Murakami Y, Mizuno S, Uehara Y (1994) Accelerated degradation of 160 kDa epidermal growth factor (EGF) receptor precursor by the tyrosine kinase inhibitor herbimycin A in the endoplasmic reticulum of A431 human epidermoid carcinoma cells. Biochem J 301:63–68

Nadeau K, Das A, Walsh CT (1993) Hsp90 chaperonins possess ATPase activity and bind heat shock transcription factors and peptidyl prolyl isomerases. J Biol Chem 268:1479–1487

Nair SC, Toran EJ, Rimerman RA, Hjermstad S, Smithgall TE, Smith DF (1996) A pathway of multi-chaperone interactions common to diverse regulatory proteins: estrogen receptor, Fes tyrosine kinase, heat shock transcription factor Hsf1, and the aryl hydrocarbon receptor. Cell Stress Chaperones 1:237–250

Nathan DF, Lindquist S (1995) Mutational analysis of Hsp90 function: interactions with a steroid receptor and a protein kinase. Mol Cell Biol 15:3917–3925

Nimmesgern E, Hartl FU (1993) ATP-dependent protein refolding activity in reticulocyte lysate. Evidence for the participation of different chaperone components. FEBS Lett 331:25–30

Ning YM, Sanchez ER (1996) In vivo evidence for the generation of a glucocorticoid receptor-heat shock protein-90 complex incapable of binding hormone by the calmodulin antagonist phenoxybenzamine. Mol Endocrinol 10:14–23

Nishida E, Koyasu S, Sakai H, Yahara I (1986) Calmodulin-regulated binding of the 90-kDa heat shock protein to actin filaments. J Biol Chem 261:16033–16036

Opperman H, Levinson W, Bishop JM (1981) A cellular protein that associates with the transforming protein of the Rous sarcoma virus is also a heat shock protein. Proc Natl Acad Sci USA 78:1067–1071

Owens GJ, Czar MJ, Hutchison KA, Hoffmann K, Perdew GH, Pratt WB (1996a) A model of protein targeting mediated by immunophilins and other proteins that bind to hsp90 via tetratricopeptide repeat domains. J Biol Chem 271:13468–13475

Owens GJ, Stancato LF, Hoffmann K, Pratt WB, Krishna P (1996b) Binding of immunophilins to the 90 kDa heat shock protein (hsp90) via a tetratricopeptide repeat domain is a conserved protein interaction in plants. Biochemistry 35:15249–15255

Palmquist K, Riis B, Nilsson A, Nygard O (1994) Interaction of the calcium and calmodulin regulated eEF-2 kinase with heat shock protein 90. FEBS Lett 349:239–242

Parsell DA, Lindquist S (1993) The function of heat-shock proteins in stress tolerance: degradation and reactivation of damaged proteins. Annu Rev Genet 27:437–496

Peattie DA, Harding MW, Fleming MA, DeCenzo MT, Lippke JA, Livingston DJ, Benasutti M (1992) Expression and characterization of human FKBP52, an immunophilin that associates with the 90-kDa heat shock protein and is a component of steroid receptor complexes. Proc Natl Acad Sci USA 89:10974–10978

Perdew GH (1988) Association of the Ah receptor with the 90-kDa heat shock protein. J Biol Chem 263:13802–13805

Perdew GH, Hord N, Hollenback CE, Welsh MJ (1993) Localization and characterization of the 86- and 84-kDa heat shock proteins in Hepa 1c1c7 cells. Exp Cell Res 209:350–356

Perrot–Applanat M, Cibert C, Geraud G, Renoir JM, Baulieu EE (1995) The 59 kDa FK506-binding protein, a 90 kDa heat shock protein binding immunophilin (FKBP59-HBI), is associated with the nucleus, the cytoskeleton and mitotic apparatus. J Cell Sci 108(Pt 5): 2037–2051

Picard D, Khursheed B, Garabedian MJ, Fortin MG, Lindquist S, Yamamoto KR (1990) Reduced levels of hsp90 compromise steroid receptor action in vivo. Nature 348:166–168

Prapapanich V, Chen S, Nair SC, Rimerman RA, Smith DF (1996a) Molecular cloning of human p48, a transient component of progesterone receptor complexes and an Hsp70-binding protein. Mol Endocrinol 10:420–431

Prapapanich V, Chen S, Toran EJ, Rimerman RA, Smith DF (1996b) Mutational analysis of the hsp70-interacting protein Hip. Mol Cell Biol 16:6200–6207

Pratt W (1993) The role of heat shock proteins in regulating the function, folding, and trafficking of the glucocorticoid receptor. J Biol Chem 268:21455–21458

Pratt WB (1997) The role of the hsp90-based chaperone system in signal transduction by nuclear receptors and receptors signaling via MAP kinase. Annu Rev Pharmacol Toxicol 37:297–326

Pratt WB, Toft DO (1997) Steroid receptor interactions with heat shock protein and immunophilin chaperones. Endocr Rev 18:306–360

Privalsky ML (1991) A subpopulation of the v-erb A oncogene protein, a derivative of the thyroid hormone receptor, associates with heat shock protein 90. J Biol Chem 266:1456–1462

Prodromou C, Roe SM, O'Brien R, Ladbury JE, Piper PW, Pearl LH (1997) Identification and structural characterization of the ATP/ADP-binding site in the Hsp90 molecular chaperone. Cell 90:65–75

Radanyi C, Chambraud B, Baulieu EE (1994) The ability of the immunophilin FKBP59-HBI to interact with the 90-kDa heat shock protein is encoded by its tetratricopeptide repeat domain. Proc Natl Acad Sci USA 91:11197–11201

Rafestin-Oblin ME, B. C, Radanyi C, Lombes M, Baulieu EE (1989) Mineralocorticosteroid receptor of the chick intestine. Oligomeric structure and transformation. J Biol Chem 264:9304–9309

Realini C, Dubiel W, Pratt G, Ferrell K, Rechsteiner M (1994) Molecular cloning and expression of a gamma-interferon-inducible activator of the multicatalytic protease. J Biol Chem 269:20727–20732

Redeuilh G, Moncharmon B, Secco C, Baulieu EE (1987) Subunit composition of the molybdate-stabilized "8–9S" nontransformed estradiol receptor purified from calf uterus. J Biol Chem 262:6969–6975

Redmond T, Sanchez ER, Bresnick EH, Schlesinger MJ, Toft DO, Pratt WB, Welsh MJ (1989) Immunofluorescence colocalization of the 90-kDa heat-shock protein and microtubules in interphase and mitotic mammalian cells. Eur J Cell Biol 50:66–75

Reed SI (1980) The selection of S. cerevisiae mutants defective in the start event of cell division. Genetics 95:561–577

Renoir JM, Radanyi C, Faber LE, Baulieu EE (1990) The non-DNA-binding heterooligomeric form of mammalian steroid hormone receptors contains a hsp90-bound 59-kilodalton protein. J Biol Chem 265:10740–10745

Ruff VA, Yem AW, Munns PL, Adams LD, Reardon IM, Deibel MR, Leach KL (1992) Tissue distribution and cellular localization of hsp56, an FK506-binding protein. Characterization using a highly specific polyclonal antibody. J Biol Chem 267:21285–21288

Sanchez ER, Toft DO, Schlesinger MJ, Pratt WB (1985) Evidence that the 90-kDa phosphoprotein associated with the untransformed L-cell glucocorticoid receptor is a murine heat shock protein. J Biol Chem 260:12398–12401

Sanchez ER, Redmond T, Scherrer LC, Bresnick EH, Welsh MJ, Pratt WB (1988) Evidence that the 90-kilodalton heat shock protein is associated with tubulin-containing complexes in L cell cytosol and in intact PtK cells. Mol Endocrinol 2:756–760

Scheibel T, Neuhofen S, Weikl T, Mayr C, Reinstein J, Vogel PD, Buchner J (1997) ATP-binding properties of human Hsp90. J Biol Chem 272:18608–18613

Schmid FX (1993) Prolyl isomerase: enzymatic catalysis of slow protein folding reactions. Annu Rev Biophys Biomol Struct 22:123–143

Schmidt CJ, Neer EJ (1991) In vitro synthesis of G protein $\beta\gamma$ dimers. J Biol Chem 266:4538–4544

Schmitt J, Pohl J, Stunnenberg HG (1993) Cloning and expression of a mouse cDNA encoding p59, an immunophilin that associates with the glucocorticoid receptor. Gene 132:267–271

Schneider C, Sepp LL, Nimmesgern E, Ouerfelli O, Danishefsky S, Rosen N, Hartl FU (1996) Pharmacologic shifting of a balance between protein refolding and degradation mediated by Hsp90. Proc Natl Acad Sci USA 93:14536–14541

Schowalter DB, Sullivan WP, Maihle NJ, Dobson AD, Conneely OM, O'Malley BW, Toft DO (1991) Characterization of progesterone receptor binding to the 90- and 70-kDa heat shock proteins. J Biol Chem 266:21165–21173

Schreiber SL (1991) Chemistry and biology of the immunophilins and their immunosuppressive ligands. Science 251:283–287

Schuh S, Yonemoto W, Brugge J, Bauer VJ, Riehl RM, Sullivan WP, Toft DO (1985) A 90000-dalton binding protein common to both steroid receptors and the

Rous sarcoma virus transforming protein, pp60^{v-src}. J Biol Chem 260:14292–14296

Schulte TW, An WG, Neckers LM (1997) Geldanamycin-induced destabilization of Raf-1 involves the proteasome. Biochem Biophys Res Commun 239:655–659

Schulte TW, Blagosklonny MV, Ingui C, Neckers L (1995) Disruption of the Raf-1-Hsp90 molecular complex results in destabilization of Raf-1 and loss of Raf-1-Ras association. J Biol Chem 270:24585–24588

Schulte TW, Blagosklonny MV, Romanova L, Mushinski JF, Monia BP, Johnston JF, Nguyen P, Trepel J, Neckers LM (1996) Destabilization of Raf-1 by geldanamycin leads to disruption of the Raf-1-MEK-mitogen-activated protein kinase signalling pathway. Mol Cell Biol 16:5839–5845

Schumacher RJ, Hansen WJ, Freeman BC, Alnemri E, Litwack G, Toft DO (1996) Cooperative action of Hsp70, Hsp90, and DnaJ proteins in protein renaturation. Biochemistry 35:14889–14898

Segnitz B, Gehring U (1997) The function of steroid hormone receptors is inhibited by the hsp90-specific compound geldanamycin. J Biol Chem 272:18694–18701

Sepehrnia B, Paz IB, Dasgupta G, Momand J (1996) Heat shock protein 84 forms a complex with mutant p53 protein predominantly within a cytoplasmic compartment of the cell. J Biol Chem 271:15084–15090

Sepp LL, Ma Z, Lebwohl DE, Vinitsky A, Rosen N (1995) Herbimycin A induces the 20 S proteasome- and ubiquitin-dependent degradation of receptor tyrosine kinases. J Biol Chem 270:16580–16587

Shaknovich R, Shue G, Kohtz DS (1992) Conformational activation of a basic helix-loop-helix protein (MyoD1) by the C-terminal region of murine HSP90 (HSP84). Mol Cell Biol 12:5059–5068

Sherr CJ (1994) Cycling on cue. Cell 75:551–555

Shi Y, Brown ED, Walsh CT (1994) Expression of recombinant human casein kinase II and recombinant heat shock protein 90 in Escherichia coli and characterization of their interactions. Proc Natl Acad Sci USA 91:2767–2771

Shue G, Kohtz DS (1994) Structural and functional aspects of basic helix-loop-helix protein folding by heat-shock protein 90. J Biol Chem 269:2707–2711

Sikorski RS, Boguski MS, Goebl M, Hieter P (1990) A repeating amino acid motif in CDC23 defines a family of proteins and a new relationship among genes required for mitosis and RNA synthesis. Cell 60:307–317

Silverstein AM, Galigniana MD, Chen MS, Owens GJ, Chinkers M, Pratt WB (1997) Protein phosphatase 5 is a major component of glucocorticoid receptor.hsp90 complexes with properties of an FK506-binding immunophilin. J Biol Chem 272:16224–16230

Smith D, Faber LE, Toft DO (1990) Purification of unactivated progesterone receptor and identification of novel receptor-associated proteins. J Biol Chem 265:3996–4003

Smith DF, Sullivan WP, Marion TN, Zaitsu K, Madden B, McCormick DJ, Toft DO (1993) Identification of a 60-kilodalton stress-related protein, p60, which interacts with hsp90 and hsp70. Mol Cell Biol 13:869–876

Smith DF, Whitesell L, Nair SC, Chen S, Prapapanich V, Rimerman RA (1995) Progesterone receptor structure and function altered by geldanamycin, an hsp90-binding agent. Mol Cell Biol 15:6804–6812

Song HY, Dunbar JD, Zhang YX, Guo D, Donner DB (1995) Identification of a protein with homology to hsp90 that binds the type 1 tumor necrosis factor receptor. J Biol Chem 270:3574–3581

Stancato LF, Chow YH, Hutchison KA, Perdew GH, Jove R, Pratt WB (1993) Raf exists in a native heterocomplex with hsp90 and p50 that can be reconstituted in a cell-free system. J Biol Chem 268:21711–21716

Stancato LF, Chow YH, Owens GJ, Yem AW, Deibel MJ, Jove R, Pratt WB (1994) The native v-Raf.hsp90.p50 heterocomplex contains a novel immunophilin of the FK506 binding class. J Biol Chem 269:22157–22161

Stancato LF, Silverstein AM, Owens GJ, Chow YH, Jove R, Pratt WB (1997) The hsp90-binding antibiotic geldanamycin decreases Raf levels and epidermal growth factor signaling without disrupting formation of signaling complexes or reducing the specific enzymatic activity of Raf kinase. J Biol Chem 272:4013–4020

Stebbins CE, Russo AA, Schneider C, Rosen N, Hartl FU, Pavletich NP (1997) Crystal structure of an Hsp90-geldanamycin complex: targeting of a protein chaperone by an antitumor agent. Cell 89:239–250

Stepanova L, Leng X, Parker SB, Harper JW (1996) Mammalian p50^{Cdc37} is a protein kinase-targeting subunit of Hsp90 that binds and stabilizes Cdk4. Genes Dev 10:1491–1502

Sullivan W, Stensgard B, Caucutt G, Bartha B, McMahon N, Alnemri ES, Litwack G, Toft D (1997) Nucleotides and two functional states of hsp90. J Biol Chem 272:8007–8012

Sullivan WP, Toft DO (1993) Mutational analysis of hsp90 binding to the progesterone receptor. J Biol Chem 268:20373–20379

Taussig R, Gilman AG (1995) Mammalian membrane-bound adenylyl cyclases. J Biol Chem 270:1–4

Thulasiraman V, Matts RL (1996) Effect of geldanamycin on the kinetics of chaperone-mediated renaturation of firefly luciferase in rabbit reticulocyte lysate. Biochemistry 35:13443–13450

Tsubuki S, Saito Y, Kawashima S (1994) Purification and characterization of an endogenous inhibitor specific to the Z-Leu-Leu-Leu-MCA degrading activity in proteasome and its identification as heat-shock protein 90. FEBS Lett 344:229–233

Uehara Y, Hori M, Takeuchi T, Umezawa H (1986) Phenotypic change from transformed to normal induced by benzoquinonoid ansamycins accompanies inactivation of p60^{src} in rat kidney cells infected with Rous sarcoma virus. Mol Cell Biol 6:2198–2206

Uehara Y, Murakami Y, Suzukake TK, Moriya Y, Sano H, Shibata K, Omura S (1988) Effects of herbimycin derivatives on src oncogene function in relation to antitumor activity. J Antibiot (Tokyo) 41:831–834

Uma S, Hartson SD, Chen JJ, Matts RL (1997) Hsp90 is obligatory for the heme-regulated eIF-2alpha kinase to acquire and maintain an activable conformation. J Biol Chem 272:11648–11656

van der Straten A, Rommel C, Dickson B, Hafen E (1997) The heat shock protein 83 (Hsp83) is required for Raf-mediated signalling in Drosophila. EMBO J 16:1961–1969

Veldscholte J, Berrevoets CA, Zegers ND, van der Kwast T, Grootegoed JA, Mulder E (1992) Hormone-induced dissociation of the androgen receptor-heat-shock protein complex: use of a new monoclonal antibody to distinguish transformed from nontransformed receptors. Biochemistry 31:7422–7430

Wagner BJ, Margolis JW (1995) Age-dependent association of isolated bovine lens multicatalytic proteinase complex (proteasome) with heat-shock protein 90, an endogenous inhibitor. Arch Biochem Biophys 323:455–462

Walsh CT, Zydowsky LD, McKeon FD (1992) Cyclosporin A, the cyclophilin class of peptidylprolyl isomerases, and blockade of T cell signal transduction. J Biol Chem 267:13115–13118

Wartmann M, Davis RJ (1994) The native structure of the activated Raf protein kinase is a membrane-bound multi-subunit complex. J Biol Chem 269:6695–6701

Wearsch PA, Nicchitta CV (1996) Purification and partial molecular characterization of GRP94, an ER resident chaperone. Protein Expr Purif 7:114–121

Wearsch PA, Nicchitta CV (1997) Interaction of endoplasmic reticulum chaperone GRP94 with peptide substrates is adenine nucleotide-independent. J Biol Chem 272:5152–5156

Weinberg RA (1995) The retinoblastoma protein and cell cycle control. Cell 81:323–330

Whitelaw M, Pongratz I, Wilhelmsson A, Gustafsson JA, Poellinger L (1993) Ligand-dependent recruitment of the Arnt coregulator determines DNA recognition by the dioxin receptor. Mol Cell Biol 13:2504–2514

Whitesell L, Cook P (1996) Stable and specific binding of heat shock protein 90 by geldanamycin disrupts glucocorticoid receptor function in intact cells. Mol Endocrinol 10:705–712

Whitesell L, Mimnaugh EG, De CB, Myers CE, Neckers LM (1994) Inhibition of heat shock protein HSP90-pp60[v-src] heteroprotein complex formation by benzoquinone ansamycins: essential role for stress proteins in oncogenic transformation. Proc Natl Acad Sci USA 91:8324–8328

Wiech H, Buchner J, Zimmermann R, Jakob U (1992) Hsp90 chaperones protein folding in vitro. Nature 358:169–170

Wilhelmsson A, Cuthill S, Denis M, Wikstrom AC, Gustafsson JA, Poellinger L (1990) The specific DNA binding activity of the dioxin receptor is modulated by the 90 kd heat shock protein. EMBO J 9:69–76

Xu Y, Lindquist S (1993) Heat-shock protein hsp90 governs the activity of pp60[v-src] kinase. Proc Natl Acad Sci USA 90:7074–7078

Xu Z, Pal JK, Thulasiraman V, Hahn HP, Chen JJ, Matts RL (1997) The role of the 90-kDa heat-shock protein and its associated cohorts in stabilizing the heme-regulated eIF-2alpha kinase in reticulocyte lysates during heat stress. Eur J Biochem 246:461–470

Ziegelhoffer T, Johnson J, Craig E (1996) Chaperones get Hip. Current Biol 6:272–275

CHAPTER 3

Heat Shock Protein 70

G.C. Li and N.F. Mivechi

A. Introduction

One of the most interesting aspects of thermal biology in the mammalian system is the response of heated cells to subsequent heat challenges. When exposed to a mild, non-lethal heat shock, mammalian cells develop a transient resistance to subsequent thermal stress (thermotolerance). On the molecular level, heat shock activates a specific set of genes, so-called heat shock genes, and results in the preferential synthesis of heat shock proteins (hsps). The heat shock response was originally described as a phenomenon of inducible gene expression, and has rapidly become an extensively studied adaptive response to a diverse array of environmental stress. Heat shock and stress response was ubiquitous, and the genes encoding these heat shock proteins (or stress proteins) were highly conserved at the level of nucleotide sequences. The field of stress protein research, in the last decade, has attracted the attention of a wide spectrum of investigators, ranging from molecular and cell biologists to medical oncologists. Studies on stress proteins now include topics in molecular biology on gene expression via the complex signal transduction pathways that control the regulation of stress genes, in cell biology on the role of stress proteins as molecular chaperones that regulate various aspects of protein folding and transport, and in pathophysiology and medicine on the role of the stress proteins in human disease, as well as in immunobiology and infectious diseases on the involvement of these proteins in the immune response during tissue damage and infection. In this chapter, we have chosen, of necessity, to highlight only selected topics on the expression, function and regulation of the 70-kDa heat shock protein, hsp70. In addition, we focus primarily on studies with mammalian cells.

B. Expression and Function of hsp70

I. Hsp70, Transient Thermotolerance and Permanent Heat Resistance

Mammalian cells when exposed to a non-lethal heat shock, can acquire a transient resistance to one or more subsequent exposures at elevated temperatures. This phenomenon has been termed thermotolerance (Gerner and

Schneider 1975; Henle and Leeper 1976; Henle and Dethlefsen 1978). On the molecular level, heat shock activates a specific set of genes, the so-called heat shock genes, resulting in the preferential synthesis of heat shock proteins (Lindquist 1986; Lindquist and Craig 1988; Morimoto et al. 1990). The mechanism underlying thermotolerance has been studied extensively, and many reports suggest that the heat shock proteins (hsps) are involved in the development of thermotolerance (Landry et al. 1982; Li et al. 1982; Subjeck et al. 1982). In the early 1980s, qualitative evidence for a causal relationship between hsp synthesis and thermotolerance was established using cell survival assay and analysis of radiolabeled proteins separated by polyacrylamide gel electrophoresis (Landry et al. 1982; Li and Werb 1982; Laszlo and Li 1985). Results from these studies can be summarized as follows: (i) heat shock induces transiently enhanced synthesis of hsps that correlates temporally with the development of thermotolerance; (ii) the persistence of thermotolerance correlates well with the stability of hsps; (iii) agents known to induce hsps induce thermotolerance; (iv) conversely, agents known to induce thermotolerance induce hsp (Ritossa 1962, 1963; Li and Hahn 1978; Li 1983; Hahn et al. 1985; Boon-Niermeijer et al. 1986, 1988; Haveman et al. 1986; Henle et al. 1986; Li and Mivechi 1986; Lee and Dewey 1987; Laszlo 1988; Crete and Landry 1990; Kampinga et al. 1992; Amici et al. 1993; Burgman et al. 1993); and (v) stable heat-resistant variant cells express high levels of hsps constitutively (Laszlo and Li 1985). Quantitatively, of the many hsps preferentially synthesized after heat shock, the concentration of the 70-kDa heat shock protein (hsp70) appears to correlate best with heat resistance, either permanent or transient (Li and Werb 1982; Laszlo and Li 1985; Li 1985).

In mammalian cells, several types of experiments (e.g., microinjection of affinity-purified anti-hsp70 antibodies or hsp70 protein, amplification of hsp70 promoter sequence) were performed to vary the intracellular concentration of hsp70 and to correlate this change with cellular response to thermal stress (Johnston and Kucey 1988; Riabowol et al. 1988; Li 1989). More recently, the expression of hsp70 under heterologous promoters has yielded additional insight into its structure and function. Transient expression of *Drosophila* hsp70 in monkey COS cells demonstrated that hsp70 accelerates the recovery of cell nucleoli after heat shock (Munro and Pelham 1984). Similarly, the domains of human hsp70 responsible for nucleolar localization and for ATP-binding were dissected (Milarski and Morimoto 1989). Using a retroviral-mediated gene transfer technique, Li et al. (1992) have generated rat cell lines stably and constitutively expressing a cloned human hsp70 gene. These cell lines provide a direct means of studying the effects of hsp70 expression on cell survival after heat shock.

It is clearly demonstrated (Fig. 1A) that the clonogenic survival of five independent pools of MVH-infected Rat-1 cells (constitutively expressing human hsp70) were approximately 100-fold higher (for a 60 or 75 min, 45°C heat treatment) than that of the parental Rat-1 or the MV6-infected Rat-1 cells (vector control). Similar results were obtained for individually isolated

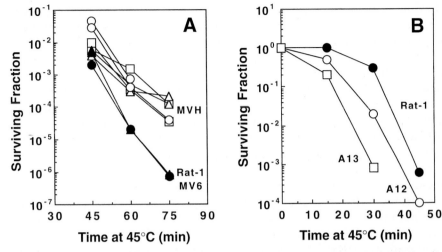

Fig. 1. A Expression of human hsp70 gene confers thermal resistance to Rat-1 cells. Survival at 45°C of pooled MVH-infected (expressing human hsp70), MV6-infected (vector control), and uninfected Rat-1 cells. Survivals from five pooled populations of MVH-infected cells independently derived from separate infection experiments are shown (*open symbols*). Each pool is derived by pooling 200–600 colonies. Survival values after 30-min heating at 45°C are clustered around 10% for all cells and are, therefore, omitted for clarity (redrawn from Li et al. 1992). **B** Cells expressing antisense *hsc70* (designated A12 and A13) show increased thermal sensitivity. Cellular survivals after heat shock at 45°C were determined by the colony formation assay

clones of Rat-1 cells expressing human hsp70 (Li et al. 1992). Cells expressing higher levels of human hsp70 generally survive thermal stress better than cells expressing lower levels, and there appears to be a good correlation between levels of exogenous human hsp70 expressed and the degree of thermal resistance (Li et al. 1992). These data provide direct evidence for a causal relation between expression of a functional form of mammalian hsp70 and survival of cells at elevated temperatures. However, production of hsp70 is only part of the program of protein biosynthesis initiated after heat shock, and other components of this response might also enhance cell survival.

The role of hsp70/hsc70 in thermotolerance development was also examined using a cell line that constitutively expresses antisense hsc70 RNA (Nussenzweig and Li 1993). The antisense RNA, complementary to 712 nucleotides of the coding strand of the rat hsc70 gene, was designed to also target hsp70, which is >75% homologous to hsc70. Similar to the finding of Li and Hightower (Li and Hightower 1995), heat-induced hsp70 expression is significantly reduced and delayed in the transfected cells relative to the parental Rat-1 cells. For example, after a 45°C, 15 min heat shock, increased hsp70 protein level is apparent within 4 h in Rat-1 cells, 2–4 h earlier than in the antisense RNA-transfected cells. In contrast, there were no significantly detectable changes in hsc70 level in these cell lines. This is in agreement with

CHESTER COLLEGE LIBRARY

previous attempts to regulate gene expression by antisense RNA in which a large excess of antisense over sense transcripts is often necessary to produce significant results. Consistent with the above, the antisense transfected cells are more heat-sensitive than Rat-1 cells (Fig. 1B), and show a profoundly reduced thermotolerance level as well as a delay in thermotolerance development.

It is well established that heat shock inhibits RNA and protein synthesis. This inhibition is reversible, and the transcriptional and translational activity recovers gradually when heated cells are returned to 37°C. Thermotolerant cells subjected to heat shock exhibit less translational inhibition (MIZZEN and WELCH 1988; LIU et al. 1992). Furthermore, the time required for the recovery of RNA and protein synthesis upon returning the heated cells to 37°C is also found to be considerably shorter for thermotolerant cells than for control nontolerant cells (BLACK and SUBJECK 1989; LASZLO 1992; LIU et al. 1992). The role of hsp70 in these processes was examined using Rat-1 fibroblasts expressing a cloned human hsp70 gene, designated M21 (LI et al. 1992). The constitutive expression of the human hsp70 gene in Rat-1 cells confers heat resistance as evidenced by the enhanced survival of heat-treated cells; furthermore, overexpression of hsp70 confers resistance against heat-induced translational inhibition (termed translational tolerance) (LIU et al. 1992). In addition, after a 45°C, 25-min heat treatment, the time required for RNA and protein synthesis to recover was considerably shorter in M21 cells which overexpress human hsp70 than in control Rat-1 cells. These data demonstrate that the expression of human hsp70 in Rat-1 cells confers heat resistance and translational tolerance, and enhances the ability of cells to recover from heat-induced translational and transcriptional inhibition.

The ability of heat shock proteins to protect cellular processes and enzymatic activities against heat stress has been demonstrated in vitro and in vivo. In vitro, E. coli hsp70 homologue, DnaK, can protect RNA polymerase from heat-induced inactivation, and thermally inactivated RNA polymerase can be reactivated by DnaK in a process that requires ATP hydrolysis (SKOWYRA et al. 1990). Furthermore, the thermal protection efficiency of DnaK is enhanced by the action of its partner proteins DnaJ and GrpE (HENDRICK and HARTL 1993). Similarly, mitochondrial hsp60 has been shown to have thermal protective functions through its ability to prevent the thermal inactivation of dihydrofolate reductase imported into the mitochondria (MARTIN et al. 1992). It was also demonstrated that in vivo the enzymatic activities of luciferase and β-galactosidase were protected in thermotolerant mouse and Drosophila cells against a heat challenge (NGUYEN et al. 1989). To extend these studies further to test the possibility that the intranuclear microenvironment may affect the stability of proteins under heat shock, MICHELS et al. (1995) compared the in situ thermal stability of a reporter protein localized in the nucleus or in the cytoplasm, using recombinant firefly luciferase targeted to the cytoplasm (cyt-luciferase) or to the nucleus (nuc-luciferase), respectively. In each case, decreased luciferase activity and solubility were found in lysates from heat-shocked cells, indicating thermal denaturation in situ. However, nuc-luciferase was more susceptible suggesting that the microenvironment of an intracellular compartment may modulate the thermal stability of proteins. In thermotolerant cells, the thermal inactivation of the recombinant luciferase by a second heat shock occurred at a slower rate with a lesser effect for the nuc-luciferase (threefold) than for the cyt-luciferase (sevenfold). The heat-inactivated luciferases were partially reactivated during recovery after stress, suggesting the capacity of both the cytoplasmic and nuclear compartments to reassemble proteins from an aggregated state.

To test the hypothesis that hsp70 plays a role in protecting cells from thermal stress, the reporter gene firefly luciferase was constitutively expressed in mammalian Rat-1 cells or M21 cells (Rat-1 cells expressing high level of human hsp70 via stable transfection). The effects of 43°C heat shock on luciferase activity in these cells were determined; in parallel, the solubility of luciferase in heat-shocked cells were monitored by immunoblot analysis of the soluble and insoluble protein fractions using antisera against luciferase, respectively. Our data demonstrate that in cells expressing a high level of intact human hsp70, the heat inactivation of luciferase enzyme activity is attenuated (LI et al. 1992). Similarly, a decrease in the insoluble fraction of luciferase molecules in these cells is observed (LI et al. 1992). These data further support the hypothesis that hsp70 protects cells either by stabilization and prevention of thermal denaturation of normal proteins, and/or by facilitating the dissociation of protein aggregates.

II. Hsp70 and Apoptosis

Heat induces cell death both in terms of loss of clonogenicity and apoptosis. It has been shown that thermotolerance protects against heat-induced apoptosis in a human acute lymphoblastic leukemic T-cells (MOSSER and MARTIN 1992). However, whether heat shock proteins have a direct role in the protection against apoptosis is not clearly defined. We studied heat-induced apoptosis and loss of clonogenicity in Rat-1 fibroblasts, thermotolerant Rat-l (TT Rat-l), Rat-l transfected with the human hsp70 gene (M21) and Rat-l transfected with the human c-myc proto-oncogene (Rat-l:myc). A significant fraction of 44°C heat-treated rat fibroblasts exhibited nuclear chromatin condensation, cytoplasmic contraction, loss of adhesion, dye exclusion, and DNA laddering (Fig. 2A–D). Relative to Rat-l, TT Rat-l and M21 cells are heat-resistant, but Rat-1:myc cells are heat-sensitive, both in terms of apoptosis induction and clonogenic survival (Fig. 2E,F). This study suggests that apoptosis is an important mechanism of heat-induced cell killing in some cell lines, which is suppressed by an overexpression of the human hsp70 gene, but enhanced by c-myc expression. These data also suggest that the effects of these genes on thermal sensitivity is in part due to their modulation of apoptosis; although, other factors as yet unknown may also contribute.

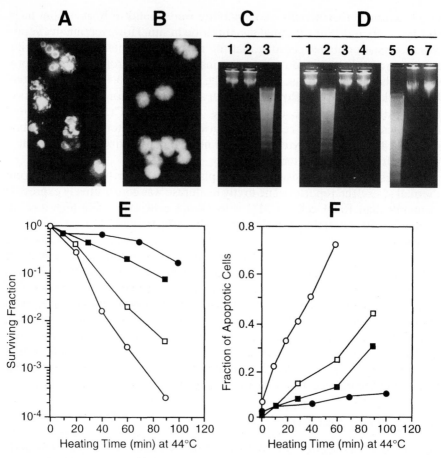

Fig. 2A–F. Induction of apoptosis in Rat-1, thermotolerant Rat-1 (TT Rat-1) and Rat-1 cells transfected with human hsp70 (M21) or c-myc proto-oncogene (Rat-1:myc). In **A** and **B**, cells were heated at 44°C for 30 min and studied 12 h after heating. Cells were stained with 10 μg/ml acridine orange and studied under fluorescence microscope at 200×. **A** Detached cells: most cells are apoptotic as evidenced by chromatin condensation or nuclear fragmentation. **B** Attached cells: most cells are non-apoptotic. **C** DNA gel electrophoresis. *Lane 1,* Rat-1:myc, unheated control; *lane 2,* Rat-1:myc, attached cells, 12 h after heating at 44°C for 30 min; *lane 3,* Rat-1: myc, detached cells, same heat treatment as for lane 2. Equal amount of DNA (3 μg) was loaded per lane. **D** DNA gel electrophoresis of control and heated (24 h after heating at 44°C for 40 min) Rat-1, TT Rat-1 and M21 cells. *Lanes 1, 2,* heated M21 and Rat-1 cells; *lanes 3, 4,* control M21 and Rat-1 cells; *lanes 5, 6,* heated Rat-1 and TT Rat-1 cells; *lane 7,* control TT Rat-1 cells. The attached and detached cells were pooled together for DNA gel electrophoresis analysis (3 μg DNA was loaded per lane). **E** Surviving fraction of Rat-1, TT Rat-1, M21 and Rat-1:myc cells heated for various periods at 44°C. -□-, Rat-1; -●-, TT Rat-1; -■-, M21; -○-, Rat-1:myc. **F** Dose response of heat-induced apoptosis assayed 24 h after heating for Rat-1, TT Rat-1, M21 and Rat-1:myc cells. -□-, Rat-1; -●-, TT Rat-1; -■-, M21; -○-, Rat-1:myc. As in **D**, cells were pooled and used for the analysis

STRASSER and ANDERSON (1995) have shown both *Bcl*-2 gene expression and thermotolerance protect cells from heat-induced apoptosis and promote cell proliferation after thermal stress. Based on these results, the authors suggested that heat shock induces apoptosis by two independent pathways, one inhibited by *Bcl*-2 expression and the other by the heat shock proteins. Our finding that hsp70 overexpression reduces heat-induced apoptosis complements their findings. It has been reported that hsp70 can interact with and alter the conformation of p53 protein (HAINAUT and MILNER 1992), although the biological significance of p53-hsp70 complexes is far from understood. Given the current belief that the induction of apoptosis requires p53 (LOWE et al. 1993a,b), the putative inactivation of p53 by hsp70 may underly the observed decrease in apoptosis in M21 and TT Rat-1 cells.

III. Hsp70 Protects Cells from Oxidative Stress

To investigate whether hsp70 protects cells from oxidative stress, the hydrogen peroxide (H_2O_2)-induced cytotoxicity was examined in Rat-1 cells and Rat-1 cells expressing intact or mutant human hsp70 protein (LI and LEE 1992). Our results show that cells expressing higher levels of the hsp70 protein generally tolerate thermal stress better, whereas cells expressing either of two mutated hsp70-encoding genes, one with a 4-base pair out-of-frame deletion and one with an in-frame deletion of codons 436–618 are heat sensitive. These results provide strong evidence that expression of hsp70 leads directly to thermal tolerance. Interestingly, cells expressing a mutant hsp70 missing codons 120–428 (the ATP-binding domain) are nevertheless heat resistant. This result suggests that ATP-binding appears to be dispensable in the hsp70-mediated protection of cells from thermal stress (LI et al. 1992). In parallel, cells expressing high level of intact human hsp70 are more resistant to H_2O_2, whereas cells expressing the mutant human hsp70 missing codons 436–618, are H_2O_2 sensitive. Suprisingly, cells expressing the mutant hsp70 missing its ATP-binding domain and being heat resistant are also resistant to H_2O_2 (Fig. 3).

Taken together, these results show that the constitutive and stable expression of human hsp70 gene confers heat resistance and resistance to H_2O_2, which again implies a direct link between expression of a functional mammalian hsp70 and cell survival during heat shock and oxidative stress. Furthermore, these data suggest that the ATP-binding may not be necessary for the protective function of hsp70. Perhaps, hsp70 lacking the ATP-binding domain can still bind to cellular proteins and prevent their aggregation upon thermal or oxidative stress. On the other hand, ATP binding and/or hydrolysis may be used to enable hsp70 to dissociate from its substrates or to facilitate the dissociation of aggregated proteins.

IV. Hsp70 Protects Cells from X-Ray Damage

As shown in the previous paragraphs, hsp70 plays an important role in protecting rodent cells from thermal and oxidative stress. To examine the role of

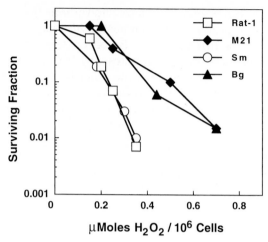

Fig. 3. Expression of human hsp70 gene in Rat-1 cells confers resistance to H_2O_2. Monolayers of Rat-1 cells or Rat-1 cells expressing intact (M21) or mutant human hsp70 (Bg, Sm) were exposed to graded doses of H_2O_2. Survival was determined by cloning forming assay. The data were plotted as a function of μmoles H_2O_2/cell. ─□─, Rat-1 cells; ─◆─, M21, Rat-1 cells expressing high level of intact human hsp70; ─▲─, Bg, Rat-1 cells overexpressing mutant hsp70 missing its ATP-binding domain; ─○─, Sm, Rat-1 cells overexpressing mutant hsp70 missing its substrate binding domain

hsp70 in modulating cellular response to X-ray irradiation, a human hsp70-encoding gene and a neomycin-resistance gene were co-infected into Rat-1 cells using a retroviral expression vector. Selection of infected cells with antibiotic G418- (400 μg/ml) generated colonies that were isolated as either (a) individual colonies or (b) pooled populations from a minimum of 1000 individual colonies. As a control, Rat-1 cells were infected with the helper-free, replication-defective, ecotropic retroviruses bearing only the neomycin-resistance gene, but not the human hsp70 gene. For X-ray survival studies, monolayers of exponentially growing cells were subjected to graded doses of X-ray (0–15 Gy). Survival was assessed with standard colony formation assay. Our data show that X-ray survival of cells overexpressing human hsp70 is, in general, higher than the control cells only expressing neomycin resistance gene (Fig. 4). For example, the survival level at 12 Gy for the pooled population and individual colonies relative to the control was increased 4–7, >10-fold, respectively. The survival level at 15 Gy was also increased for both the pooled population and the individual colonies. Thus, it appears that hsp70 may also play a protective role in rodent cells exposed to X-rays.

V. Hsp70 as Molecular Chaperone

Under stress conditions, the increased levels of hsp70 appear to provide protection against aggregation and to help misfolded proteins to refold (BUKAU

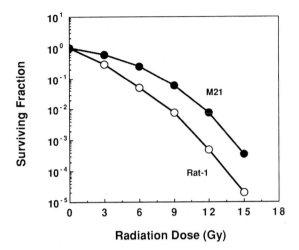

Fig. 4. Expression of human hsp70 in Rat-1 cells confers resistance to ionizing radiation. Monolayers of Rat-1 cells (Rat-1) or Rat-1 cells overexpressing intact human hsp70 (M21) were exposed to graded doses of ionizing radiation. Survival was determined by colony forming assay

and HORWICH 1998) (for review see BUKAU and HORWICH 1998). Under normal physiological growth conditions, hsp70 seems to assist newly synthesized proteins to fold correctly, disassemble oligomeric protein structures and to assist in proteolysis of unstable proteins. So far in limited studies, hsp70 association with transcription factors has also been described (HARTL 1996; SHI et al. 1998). In the cases where hsp70 has been shown to assist folding, one molecule of hsp70 binds to one molecule of substrate, and substrates undergo repeated cycles of binding and dissociation. As part of a multidirectional and dynamic folding network, the substrates released from the chaperone undergo kinetic partitioning between folding to native state, aggregation, rebinding to hsp70, and binding to other chaperones or proteases (SZABO et al. 1994).

Information related to the biochemical and molecular function of hsp70 has been provided by the crystal structure of the carboxyl-terminal substrate binding domain of bacterial homologue, DnaK protein (SCHMID et al. 1994). A consensus motif in polypeptides where DnaK is able to recognize have been identified and it appears to consist of 4–5 hydrophobic residues flanked by basic residues (RUDIGER et al. 1997). The ability of hsp70 to bind to its substrates is driven by ATP that binds to the amino terminal domain of hsp70. This binding of ATP to hsp70 leaves the substrate binding pocket open and therefore hsp70 has a high affinity for its substrates (BUKAU and HORWICH 1998). In its ADP-bound state, hsp70 has its binding pocket closed and therefore, has low affinity for its substrates. How the opening and closing of hsp70 pocket is achieved by ATP hydrolysis is not known. Most likely the co-chaperones such as DnaJ are regulating the ATP hydrolysis of ATP-bound hsp70 (McCARTY et al. 1995). The capacity of hsp70 to submit to cycles of binding to hydrophobic peptides and release from them suggest that hsp70 plays a crucial role in a variety of intracelluar processes.

C. Regulation of hsp70

I. Heat Shock Transcription Factor (HSF), the Transcriptional Regulator of hsp70

The heat shock transcription factors (HSFs) have been cloned from a variety of organisms. Studies suggest that yeast (*Saccharomyces cerevisiae* and *Schizosaccharomyces pombe*) and *Drosophila* each contain one gene (Sorger and Pelham 1988; Wiederrecht et al. 1988; Gallo et al. 1993), while two separate genes have been cloned in mouse cells and three genes each in human and chicken (Rabindran et al. 1991; Sarge et al. 1991; Schuetz et al. 1991; Nakai and Morimoto 1993). Comparisons of HSF protein structure in a variety of organisms indicate the presence of a conserved DNA binding domain and three hydrophobic heptad repeats that constitute the trimerization domain. These domains are located within the amino-terminal region of the protein. The stress-responsive transcriptional activation domain is located in the carboxyl-terminal end of the molecule. Intra-molecular interactions between the amino- and carboxyl-terminal coiled coil domains of HSF keep the protein in an inactive state under nonstress growth conditions (Rabindran et al. 1993).

In eukaryotes, HSF-1 binds to conserved regulatory sequences known as heat shock elements (HSEs) where it controls the expression of heat shock proteins (hsps) in response to stress (Wu 1984; Sorger et al. 1987; Zimarino and Wu 1987; Larson et al. 1988; Abravaya et al. 1992; Mivechi et al. 1992). The HSE consists of multiple inverted repeats of AGAAN located upstream of heat shock genes (Cunnif and Morgan 1993; Kroeger and Morimoto 1994; Wu 1995). In the yeast *S. cerevisiae*, HSF is constitutively bound to HSE, while in most organisms a stress is required for HSF to trimerize and bind to DNA (Sorger et al. 1987; Gardina and Lis 1995; Mivechi and Giaccia 1995; Cotto et al. 1996). In *Drosophila* and mammals, stress causes an increase in trimerization, DNA binding, and phosphorylation of HSF-1 (Sorger et al. 1987; Sarge et al. 1993; Kim et al. 1997). In mammalian cells, HSF-1 is phosphorylated under normal physiological growth conditions and this phosphorylation has been shown to repress the activity of HSF-1 (Chu et al. 1996; Knauf et al. 1996; Kim et al. 1997; Kline and Morimoto 1997).

HSF in yeast appears to be an essential gene. In *Drosophila*, it is dispensable for general cell growth and viability (Sorger and Pelham 1988; Jedlicka et al. 1997), but seems to be required during oogenesis and early larval development (Jedlicka et al. 1997). These data suggest that the function of HSF may not be solely to control transcription of hsps under stress, but it may also control the expression of non-heat shock genes under normal physiological growth conditions (Jedlicka et al. 1997). The presence of multiple HSF genes in higher eukaryotes, in contrast to the single HSF gene present in yeast and *Drosophila*, could indicate that individual proteins may be responsible for different biological functions. In human and mouse, HSF-1 has been shown to activate transcription of various hsps in response to heat shock as well as other environmental stresses (SARGE et al. 1991; SCHUETZ et al. 1991; MIVECHI et al. 1992). HSF-2, on the other hand, does not respond to heat stress but has been shown to activate transcription of hsps in response to hemin in erythroleukemia cells (SISTONEN et al. 1992; MIVECHI et al. 1994). HSF-2 activity is also detected during mouse spermatogenesis (SARGE et al. 1994). The third HSF isoform, HSF-3, is found in chicken and has recently been shown to be activated by c-myb in the absence of cellular stress (NAKAI and MORIMOTO 1993; KANEI-ISHII et al. 1997). Another isoform, HSF-4, has been found in human cells but seems to lack the property of a transcriptional activator (NAKAI et al. 1997). Other levels of regulation described for HSF-1 and HSF-2 in mammalian cells is the presence of alpha and beta isoforms (FIORENZA et al. 1995; GOODSON et al. 1995). These are alternate splice variants of HSF-1 and HSF-2 that recently were shown to be expressed differentially during development (LEPPA et al. 1997).

II. Signal Transduction Leading to Modulation of hsp70 Levels

Heat shock activates many signaling pathways which results in the activation of down stream enzymes such as MAPK/ERKs, JNK/SAPK, pp90rsk, pp70S6 K, PKB/Akt, GSK-3, p38/HOG1 and others (ROUSE et al. 1994; ADLER et al. 1995; MIVECHI and GIACCIA 1995; KONISHI et al. 1996; YANG et al. 1997; HE et al. 1998). The consequences of activation of these signaling cascades by heat shock have only now begun to be investigated. In the following section we will briefly describe the signaling pathways which are known to modulate the activity of HSF-1 and therefore, the levels of hsp70. We will also describe the implications of activation of functionally opposing signaling pathways by heat shock, that is, pathways which are known to regulate cell growth and those that regulate apoptosis.

Activated Ras triggers a kinase cascade involving the serine/threonine kinase Raf-1 (MOODIE et al. 1993; STOKOE et al. 1994), which phosphorylates and activates MEK (MAP/ERK kinase or MAP kinase kinase; CREWS and ERIKSON 1992), which then phosphorylates and activates MAP kinase via tyrosine and threonine phosphorylation (ANDERSON et al. 1990). Therefore, constitutive activation of Ras or Raf-1 permits the cell to escape the regulation of normal growth factor control. In addition to growth factor regulation, MAP kinases have been shown to be induced by cellular exposure to free radical generating agents such as ionizing radiation, hydrogen peroxide, ultraviolet light, low oxygen conditions, and heat shock, indicating that MAP kinases also respond to non-growth factor mediated regulation as well (HIBI et al. 1993; KOONG et al. 1994a,b; STEVENSON et al. 1994). As a genetic means to demonstrate the role of MAP kinases in cell signaling after stress, inhibitory mutants

of ERK1 and ERK2 were used to show that Ets-1/AP1, but not c-Jun activation was a direct consequence of MAP kinase activation (Westwick et al.
1994). These dominant negative MAP kinases seemed to be specific in their
activity as they did not interfere with pathway involved in the activation of the
NF-kB transcription factor, a pathway shown to be independent of MAP
kinase activity (Koong et al. 1994b).

   One of the ways MAP kinases and their downstream effectors may modulate the heat shock response is through the regulation of HSF-1. HSF-1 is
phosphorylated under normal physiological growth conditions and it becomes
highly phosphorylated on serine residues after heat shock. The heat induced
phosphorylation of HSF-1 can increase its apparent molecular weight by as
much as 11 kDa (Baler et al. 1993; Sarge et al. 1993; Westwood and Wu 1993;
Mivechi et al. 1994, 1995). At present, the mechanism and function of HSF-1
phosphorylation is still unclear, but is presumed to be important in modulating
the heat shock response. Although the increased transcriptional activity of
hsps seems to be in large part mediated by HSF-1, the promoter regions of
heat shock genes contain several regulatory sequences that respond to diverse
stimuli such as changes in serum concentrations. For example, the increase in
the level of hsp70 transcription induced by serum is modulated through the
serum response factor (SRF). It is, however, unclear whether these other
elements synergistically interact with HSF to increase the heat shock response
or act as negative regulators of the heat shock response under unstressed
conditions (Wu et al. 1985, 1986).

   Although, the signal transduction pathway leading to HSF-1 activation is
not known at the present time, there is some evidence that MAP kinases may
regulate HSF-1 activity. For example, studies on *Saccharomyces cerevisiae*
suggest that mutations in the Ras-cAMP pathway confer heat sensitivity.
Saccharomyces cerevisiae that possess mutations in the protein kinase SLK1,
a MAP kinase family member, fail to become thermotolerant and rapidly lose
viability after heat shock as compared to their wild-type counterparts
(Costigan and Snyder 1994; Engelberg et al. 1994). Since the activation of
MAP kinases by heat shock closely correlates with HSF-1 activity, we sought
to determine the possible regulation of HSF-1 by MAP kinases, as described
below.

III. Negative Regulatory Effect of ERK1 on hsp70 Gene Expression

To examine the role of MAP kinases (ERK1 and ERK2) in the regulation of
the heat shock response, a hsp70 reporter gene construct with different com-
binations of wild-type ERK1, ERK2 or dominant negative alleles of ERK1
and ERK2 (ERK1KR, ERK2–52R) was co-transfected into *Ha-ras* inducible
NIH3T3 cells (Koong et al. 1994b; Mivechi and Giaccia 1995). The hsp 70
reporter gene contained 188 bp of the hsp 70 promoter from the start site of
transcription (Williams et al. 1989). In Ha-ras uninduced cells, wild-type
ERK1 and ERK2 had little effect on heat inducible hsp70 gene expression

(Fig. 5A). Similar results were obtained when kinase defective alleles of ERK1 (ERK1KR) and ERK2 (ERK2–52R) were transiently co-transfected with the hsp70-luciferase reporter construct in Ha-ras uninduced NIH3T3 cells (Fig. 5B). However, in *Ha-ras* induced cells, wild type ERK1 inhibited hsp70-luciferase expression by 50% after heat shock and kinase dead ERK1KR increased hsp70 luciferase activity by 30-fold after heat shock (Fig. 5B). This inhibitory effect seemed to be specific for kinase defective ERK1 and wild type ERK1, as wild type ERK 2 and the kinase dead allele of ERK 2 had little effect on hsp70-luciferase induction by heat shock. These results indicate that ERK1 acts as a negative regulator of hsp70 gene expression in cells that possess oncogenic Ras mutations.

IV. Mutational Analysis of HSF-1 Phosphorylation by ERK1 Protein Kinase

Sequence analysis suggests that HSF-1 contains multiple potential ERK1 phosphorylation sites. To map the region of HSF-1 phosphorylation by ERK1, immune complex kinase assays were performed with purified wild-type histidine-tagged HSF-1 or a series of deletion mutants as substrates. ERK1 was immunoprecipitated from NIH3T3 cells that had not received any treatment (control), which were heated (43°C for 30 min), or which were pretreated with sodium vanadate (500 μM for 2 h at 37°C). The immunoprecipitated ERK1 protein kinase was then used to phosphorylate the purified wild-type HSF-1 protein (amino acids 1–450) or a mutant of HSF-1 protein that contained deletions between amino acids 203 and 445 (mutant Δ06). This deletion removes the C-terminal portion of HSF-1 that has regulatory and transcriptional activation functions (GREEN et al. 1995; SHI et al. 1995), but leaves the N-terminal DNA binding and trimerization domains intact. The results showed that wild-type HSF-1 was phosphorylated by immunoprecipitated ERK1 (KIM et al. 1997). Consistent with previous results, the level of phosphorylation was increased when ERK1 was precipitated from cells that had been heat treated (43°C 30 min, 7-fold increase) or treated with sodium vanadate, which increases the level of ERK1 activity by inhibiting MAP kinase phosphatase (12-fold increase; MIVECHI and GIACCIA 1995).

When mutant Δ06 was used as a substrate, phosphorylation by ERK1 immunoprecipitated from control cells was reduced. Strikingly, the phosphorylation of mutant Δ06 was almost unaffected by heating or sodium vanadate treatment (1–2 fold). These results suggest that the major region of phosphorylation of HSF-1 by ERK1 protein kinase resides between amino acids 203 and 450 and that there is little phosphorylation of the DNA binding or trimerization domain located between amino acids 1 and 203.

We next performed immune complex kinase assays to analyze phosphorylation of HSF-1 deletion mutants containing mutational deletions located in the regions between amino acid residues 203 and 445. Mutant Δ34 had a

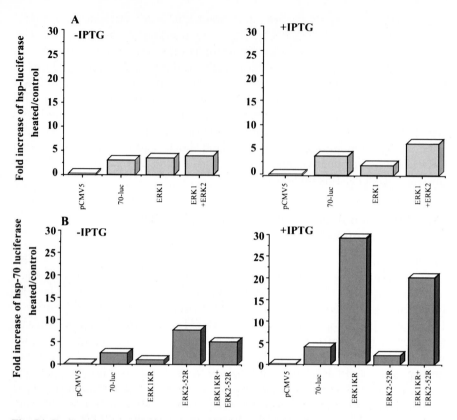

Fig. 5A,B. Dominant negative allele of ERK1 increases hsp70 accumulation after heat shock. NIH3T3 cells stably transfected with a conditionally inducible *Ha-ras* under the control of an IPTG inducible promoter were transiently transfected with the indicated plasmids. Twenty-four hours after transfection, some groups were made to induce *Ha-ras* (+IPTG) for 24 h before heat shock. Cells were heated at 43°C for 20 min, incubated at 37°C for 6 h to allow expression of luciferase, lysed, and a fraction of the cell lysates were used to determine luciferase activity (**A**). Control experiments containing pCMV vector only and pCMVβ Gal were included to determine the background luciferase expression and transfection efficiency, respectively (**B**). ERK1, ERK2 are wild type MAP kinase alleles, and ERK1KR and ERK2–52R are kinase dead alleles of ERK1 and ERK2, respectively (data from Mivechi and Giaccia 1995)

deletion of sequences between amino acids 308 and 370, mutant Δ24 had a deletion of sequences between amino acids 280 and 370, mutant Δ36 had a deletion of sequences between amino acids 308 and 445, and mutant Δ56 had a deletion of sequences between amino acids 420 and 445 (Fig. 6A). As shown in Fig. 6B, ERK1 immunoprecipitated from cells under control conditions, phosphorylates HSF-1 wild-type, mutants Δ34, Δ36 and Δ56 to the same extent. However, there was an approximate 7-fold reduction in phosphorylation of HSF-1 mutant Δ24. These results indicate that the phosphorylation of HSF-1 by ERK1 protein kinase is highly dependent on a sequence between

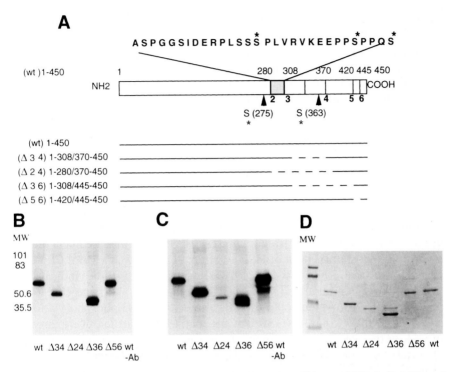

Fig. 6A–D. Immune complex kinase assays using the wild-type HSF-1 (1–450) and individual HSF-1 deletion mutants as substrates. **A** Map of HSF-1 (1–450) (M, 61 000) and HSF-l deletion mutants (Δ34, Δ24, Δ36, and Δ56 with M, 54 000, 50 000, 43 000, 60 000, respectively). The *dotted lines* correspond to the sequences absent from each mutant. The serine residues with an asterisk (*) are potential MAP kinases phosphorylation sites. **B** Immune complex kinase assays of immunoprecipitated ERK1 protein kinase from control NIH3T3 cells. The immunoprecipitated ERK1 was used to phosphorylate 1 μg of wild-type (*wt*) or various deletion mutants indicated in the figure. The position of the molecular weight markers is shown *on the left*. **C** Immune complex kinase assays of immunoprecipitated ERK1 protein kinase from heated (43°C for 30 min) NIH3T3 cells. Reaction contained 1 μg of wild-type (*wt*) or deletion mutants as substrates as indicated in the figure. The lane labeled as *wt, -Ab* in **B** and **C** indicates that no antibody was used during immunoprecipitation reaction. **D** One microgram of the purified wild-type HSF-1 (1–450) (*wt*) or deletion mutants analyzed by SDS-PAGE and stained with Coomassie blue, Molecular weight (*MW*) markers are shown *on the left* (data from KIM et al. 1997)

amino acids 280 and 308. Similar results were obtained using ERK1 immuno-precipitated from heat treated (43°C, 30 min) cells (Fig. 6C). The immune complex kinase assays performed with cells pretreated with sodium vanadate (500 μM) showed similar patterns of phosphorylation, that is, the HSF-1 mutant Δ24 showed a reduction in HSF-1 phosphorylation (data not shown). As shown in Fig. 6A, the amino acids between 280 and 308 in HSF-1 protein contain 3 serine residues with potential ERK1 phosphorylation motifs. These serine residues are located at amino acid residues 292, 303, and 307. Figure 6D shows that all mutants were present in approximately similar amounts, as measured by Coomasie blue staining (Kim et al. 1997).

Consistent with the results described above, phosphopeptide mapping of HSF-1 suggest that HSF-1 is phosphorylated on multiple serine residues and, perhaps, a threonine residue. Constitutive phosphorylation of serines 303 and 307, which are located distal to the transcriptional activation domain, negatively regulates HSF-1 function, since mutations to alanine of either serine 303 alone or of both serines 303 and 307 cause constitutive transcriptional activation of HSF-1 (Shi et al. 1995).

V. Modulation of HSF-1 by Other Protein Kinases

The protein kinases that phosphorylate HSF-1 have not been identified with certainty in vivo. However, as described above, it has been suggested that one or more mitogen activated protein kinases (MAPKs) family members may be involved in phosphorylation of HSF-1 (Mivechi and Giaccia 1995; Chu et al. 1996; Knauf et al. 1996; Kim et al. 1997). Recent in vitro studies indicate that extracellular signal regulated kinases (ERKs) may phosphorylate HSF-1 protein on serine 307, which could then facilitate serine 303 phosphorylation by glycogen synthase kinase-3 (GSK3β; Chu et al. 1996). Further, overexpression of GSK3β that has been shown recently to cause inactivation and nuclear exit of certain activated transcription factors, was recently shown to cause rapid dispersion of HSF-1 from the punctuated granules after heat shock. This function of GSK3β requires the activity of ERK1 protein kinase, as inhibition of ERK1 abolishes the GSK3β effect (He et al. 1998).

Other protein kinases such as p38/RK/Mpk2/CSBP protein kinases, the c-Jun amino terminal kinases/stress activated protein kinases (JNK/SAPK) and PKC have also been suggested to take part in inactivation of HSF-1 (Knauf et al. 1996; He et al. 1998b; Soncin et al. 1998). A recent report also suggests that Casein Kinase II phosphorylation of threonine 142 has a positive effect on HSF-1 transcription (Soncin et al. 1998).

VI. Implication of HSF-1 Regulation by Functionally Opposing Signaling Cascades

As described above, HSF-1 is inactivated by multiple signaling pathways, some of which are required for cell survival (such as ERK and GSK-3) and others induce apoptosis (such as JNK/SAPK). For example, GSK/ERK and JNK may cooperate to repress HSF-1 for two reasons. First, repression of HSF-1 is required for continuation of cell growth and survival as during a heat shock response most intracellular events are severely reduced. Second, a severe heat treatment which could result in sustained JNK activation could cause rapid down regulation of HSF-1 and drive the cells into apoptosis. As with other stresses and growth factors, findings thus far, also suggest the pleiotropic nature of heat-induced intracellular signaling (Xia et al. 1995).

Signals initiating from ERK, GSK, JNK, PKC and p38 trigger a wide range of potential outcomes, some of which, as in the case of promotion (JNK) and suppression (ERK and GSK) of apoptosis, appears to be contradictory. Activation of multiple signaling pathways after heat shock can induce both cellular protection against further stress stimuli (heat shock protein synthesis) or induce apoptosis, but the net effect depends upon the status of anti-apoptotic signals within the cell.

VII. Regulation of Heat Shock Response: Possible Involvement of Ku Autoantigen

Although the importance of the heat shock transcription factor HSF-1 in the regulation of mammalian heat shock genes is well-established (Morimoto et al. 1990), recent data indicate that activation of HSF-1, by itself, is not sufficient for the induction of hsp70 mRNA synthesis, suggesting the existence of additional regulatory factors (Jurivich et al. 1992; Liu et al. 1993; Mathur et al. 1994). We have recently characterized a constitutive heat shock element binding factor (CHBF) in rodent cells that exhibits a tight inverse correlation with the HSE-binding activity of HSF-1 in vitro both upon heat shock and during post-heat shock recovery (Liu et al. 1993; Kim et al. 1995a). Additionally, we and others have found that HSF-1-HSE binding is not sufficient for heat shock gene induction; agents such as sodium salicylate (Jurivich et al. 1992) and arsenite (Liu et al. 1993) elicit considerable HSF-1-HSE binding activity, yet do not result in significant induction of hsp70 mRNA synthesis. Interestingly, unlike heat shock stimuli, these agents do not inactivate the CHBF activity, which remains at the high levels observed in untreated cells (Liu et al. 1993). These findings suggest that CHBF may be involved in the regulation of heat shock gene expression, possibly by acting as a repressor (Liu et al. 1993). To study the role of CHBF in the regulation of heat shock gene expression, we have purified this protein and found that CHBF is identical or closely related to the Ku autoantigen (Kim et al. 1995b). The Ku autoantigen is a heterodimer of 70-kDa (Ku70) and 86-kDa (Ku80) polypeptides (Mimori et al. 1981; Reeves and Sthoeger 1989). In vitro studies have shown that this abundant nuclear protein binds to the termini of double-stranded DNA and DNA ending in stem-loop structures (Mimori and Hardin 1986). By virtue of its DNA binding activity, Ku serves as a regulatory component of the mammalian DNA-dependent protein kinase, DNA-PK, which is activated by DNA ends (Gottlieb and Jackson 1993). The other component of DNA-PK is a 460-kDa catalytic subunit (DNA-PK$_{cs}$). Various cellular roles for Ku have been suggested, including a role in transcription, recombination, replication, and DNA repair (Yaneva and Busch 1986; Amabis et al. 1990; Lees-Miller et al. 1990; Lees-Miller and Anderson 1991; Reeves 1992; Bannister et al. 1993; Dvir et al. 1993; Falzon et al. 1993; Gottlieb and Jackson 1993). However, until recently there has been a paucity of experimental data on the in vivo function of Ku.

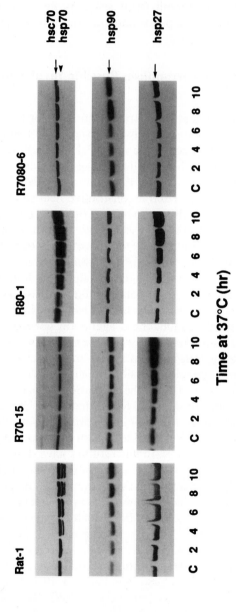

Fig. 7. Western blot analysis of heat shock proteins during post-heat-shock recovery at 37°C. Rat-1, R70-15, R80-1 and R7080-6 cells were exposed to 45°C for 15 min, and returned to 37°C for 2–10 h. The heat-induced expression of hsp90, hsc70, hsp70, hsp60, and hsp27 was analyzed by Western blot. *C,* control unheated cells. *2, 4, 6, 8, 10,* incubation time (h) at 37°C. Hsp90, hsc70, hsp60 and hsp27 are *indicated by arrows,* hsp70 is *indicated by an arrowhead.* The induction of hsp70 is repressed in cells overexpressing Ku-70 (R70-15), both Ku-70 and Ku-80 (R7080-6), but not Ku-80 alone (R80-1). Induction kinetics of the other heat shock proteins, hsp90, hsc70, hsp60, and hsp27, upon heat shock are similar in all cell lines

To test whether the Ku-protein plays a role in the heat shock response, we chose to modulate the intracellular concentration of Ku through the use of retroviral-mediated gene transfer technique, and have established rodent cell lines that stably and constitutively overexpress one or both subunits of the human Ku protein. Our data (Fig. 7) show that coexpression of human Ku70 and Ku80 or expression of only the Ku70 subunit in rodent cells specifically inhibits the induction of hsp70 upon heat shock (LI et al. 1995; YANG et al. 1996). On the other hand, expression of human Ku80 alone does not have this effect. Thermal induction of other heat shock proteins in all of the Ku-overexpressing cell lines appears not to be significantly affected (Fig. 7), nor the state of phosphorylation or the DNA-binding ability of HSF-1 (YANG et al. 1996). While both Ku70 and Ku80 are involved in DNA repair, Ku70 but not Ku80, appears to be involved in the regulation of hsp70 gene expression. Thus, different subunits of Ku or different Ku-containing complexes may be involved in distinct biological processes.

The fact that Ku is involved in gene regulation is not without precedent. Both the 70- and 86-kDa subunits of Ku contain leucine/serine repeats which are reminiscent of similar "zipper" motifs found in a variety of transcription factors. Furthermore, Ku is a component of DNA-PK which phosphorylates several transcription factors, such as SP1 (GOTTLIEB and JACKSON 1993), c-Jun (BANNISTER et al. 1993), p53 (LEES-MILLER et al. 1990), c-Myc, Oct-1, and Oct-2 (LEES-MILLER and ANDERSON 1991) in vitro. Studies also indicate that the Ku/DNA-PK$_{cs}$ complex modulates RNA polymerase I mediated transcription (KNUTH et al. 1990; HOFF and JACOB 1993; KUHN et al. 1993, 1995). Additionally, Ku has been found to be localized on certain transcriptionally active loci of chromosomal DNA (AMABIS et al. 1990; REEVES 1992). These data further support a role for DNA-PK/Ku in transcription; specific genes regulated by DNA-PK/Ku, however, need yet to be identified.

References

Abravaya K, Myers MP, Murphy SP, Morimoto RI (1992) The human heat shock protein hsp70 interacts with HSF, the transcription factor that regulates heat shock gene expression. Genes Dev 6:1153–1164

Adler VJ, Schaffer A, Kim J, Donlan L, Ronai Z (1995) UV irradiation and heat shock mediate JNK activation via alternate pathways. J Biol Chem 270:26071–26077

Amabis JM, Amabis DC, Kahuraki J, Stollar BO (1990) The presence of an antigen reactive with a human autoantibody in Trichosia pubescens (Diptera:sciaridae) and its association with certain transcriptionally active regions of the genome. Chromosoma 99:102–110

Amici C, Palamara T, Santoro MG (1993) Induction of thermotolerance by prostaglandin A in human cells. Exp Cell Res 207:230–234

Anderson NG, Maller JL, Tonks NK, Sturgill TW (1990) Requirement for integration of signals from two distinct phosphorylation pathways for activation of MAP kinase. Nature 343:651–653

Baler R, Dahl G, Voellmy R (1993) Activation of human heat shock gene is accompanied by oligomerization, modification, and rapid translocation of heat shock transcription factor HSF1. Mol Cell Biol 13:2486–2496

Bannister AJ, Gottlieb TM, Kouzarides T, Jackson SP (1993) c-Jun is phosphorylated by the DNA-dependent protein kinase in vitro; definition of the minimal kinase recognition motif. Nucleic Acids Res 21:1289–1295

Black AR, Subjeck JR (1989) Involvement of rRNA synthesis in the enhanced survival and recovery of protein synthesis seen in thermotolerance. J Cell Physiol 138:439–449

Boon-Niermeijer EK, Souren JEM, De Waal AM, Van Wijk R (1988) Thermotolerance induced by heat and ethanol. Int J Hyperthermia 4:211–222

Boon-Niermeijer EK, Tuyl M, Van der Scheur H (1986) Evidence for two states of thermotolerance. Int J Hyperthermia 2:93–105

Bukau B, Horwich AL (1998) The HSP-70 and HSP-60 chaperone machines. Cell 92:351–366

Burgman PWJJ, Kampinga HH, Konings AWT (1993) Possible role of localized protein denaturation in the induction of thermotolerance by heat, sodium-arsenite, and ethanol. Int J Hyperthermia 9:151–162

Chu B, Soncin F, Price BD, Stevenson MA, Calderwood SK (1996) Sequential phosphorylation by mitogen activated protein kinase and glycogen synthase kinase-3 represses transcriptional activation by heat shock factor-1. J Biol Chem 271:30847–30857

Costigan C, Snyder M (1994) SLK1, a yeast homolog of MAP kinase activators, has a ras/cAMP independent role in nutrient sensing. Mol Genet 243:286–296

Cotto JJ, Kline M, Morimoto RI (1996) Activation of heat shock factor 1 DNA binding precedes stress-induced serine phosphorylation. Evidence for a multistep pathway of regulation. J Biol Chem 271:3355–3358

Crete P, Landry J (1990) Induction of hsp27 phosphorylation and thermoresistance in chinese hamster cells by arsenite, cycloheximde, A23187, and EGTA. Radiat Res 121:320–327

Crews CM, Erikson RL (1992) Purification of a murine protein-tyrosine/threonine kinase that phosphorylates and activates the Erk-1 gene product: relationship to the fission yeast byr1 gene product. Proc Natl Acad Sci USA 89:8205–8209

Cunnif NFA, Morgan WD (1993) Analysis of heat shock element recognition by saturation mutagenesis of the human HSP70.1 gene promoter. J Biol Chem 268:8317–8324

Dvir A, Stein LY, Calore BL, Dynan WS (1993) Purification and characterization of a template-associated protein kinase that phosphorylates RNA polymerase II. J Biol Chem 268:10440–10447

Engelberg D, Zandi E, Parker CS, Karin M (1994) The yeast and mammalian Ras pathway control transcription of heat shock genes independently of heat shock transcription factor. Mol Cell Biol 14:4929–4937

Falzon M, Fewell JW, Kuff EL (1993) EBP-80, a transcription factor closely resembling the human autoantigen Ku, recognizes single- to double-strand transitions in DNA. J Biol Chem 268:10546–10552

Fiorenza MT, Farkas T, Dissing M, Kolding D, Zimarino V (1995) Complex expression of murine heat shock transcription factors. Nucleic Acids Res 23:467–474

Gallo GJ, Prentice H, Kingston RE (1993) Heat shock factor is required for growth at normal temperatures in the fission yeast Schizosaccharomyces pombe. Mol Cell Biol 13:749–761

Gardina C, Lis JT (1995) Dynamic protein-DNA architecture of a yeast heat shock promoter. Mol Cell Biol 15:2737–2744

Gerner EW, Schneider MJ (1975) Induced thermal resistance in HeLa cells. Nature 256:500–502

Goodson ML, Park-Sarge O-K, Sarge KD (1995) Tissue-dependent expression of heat shock factor 2 isoforms with distinct transcriptional activities. Mol Cell Biol 15:5288–5293

Gottlieb TM, Jackson SP (1993) The DNA-dependent protein kinase: requirement for DNA ends and association with Ku antigen. Cell 72:131–142

Green M, Schuetz TJ, Sullivan EK, Kingston RE (1995) A heat shock-responsive domain of human HSF1 that regulates transcription activation domain function. Mol Cell Biol 15:3354–3362

Hahn GM, Shiu EC, West B, Goldstein L, Li GC (1985) Mechanistic implications of the induction of thermotolerance in Chinese hamster cells by organic solvents. Cancer Res 45(9):4138–4143

Hainaut P, Milner J (1992) Interaction of heat-shock protein 70 with p53 translated in vitro: evidence for interaction with dimeric p53 and for a role in the regulation of p53 conformation. Eur Mol Biol Org 11:3513–3520

Hartl FU (1996) Molecular chaperones in cellular protein folding. Nature 381:571–580

Haveman J, Li GC, Mak JY, Kipp JB (1986) Chemically induced resistance to heat treatment and stress protein synthesis in cultured mammalian cells. Int J Radiat Biol 50:51–64

He B, Meng Y-H, Mivechi NF (1998) GSK-3b and ERK MAPK inactivate HSF-1 by facilitating the disappearance of transcriptionally active granules after heat shock. Mol Cell Biol (In press)

He B, Meng Y-H, Mivechi NF (1998) GSK-3, ERK MAPK and JNK cooperate to inactivate the heat shock transcription factor-1. Molecular Chaperones and Heat Shock Response, Cold Spring Harbor, p 96

Hendrick JP, Hartl F-U (1993) Molecular chaperone functions of heat-shock proteins. Annu Rev Biochem 62:349–384

Henle KJ, Dethlefsen LA (1978) Heat fractionation and thermotolerance: a review. Cancer Res 38:1843–1851

Henle KJ, Leeper DB (1976) Interaction of hyperthermia and radiation in CHO cells: recovery kinetics. Radiat Res 66:505–518

Henle KJ, Moss AJ, Nagle WA (1986) Temperature-dependent induction of thermotolerance by ethanol. Radiat Res 66:505–518

Hibi M, Lin A, Smeal T, Minden A, Karin M (1993) Identification of an oncoprotein- and uv-responsive protein kinase that binds and potentiates the c-jun activation domain. Genes Dev 7:2135–2148

Hoff CM, Jacob ST (1993) Characterization of the factor EIBF from a rat hepatoma that modulates ribosomal RNA gene transcription and its relationship to the human Ku autoantigen. Biochem Biophys Res Commun 190:747–753

Jedlicka P, Mortin MA, Wu C (1997) Multiple functions of Drosophila heat shock transcription factor in vivo. EMBO J 16:2452–2462

Johnston RN, Kucey BL (1988) Competitive inhibition of hsp70 gene expression causes thermosensitivity. Science 242:1551–1554

Jurivich DA, Sistonen L, Kroes RA, Morimoto RI (1992) Effect of sodium salicylate on the human heat shock response. Science 255:1243–1245

Kampinga HH, Brunsting JF, Konings AWT (1992) Acquisition of thermotolerance induced by heat and arsenite in HeLa cells: multiple pathways to induce tolerance? J Cell Physiol 150:406–415

Kanei-Ishii C, Tankawa J, Nakai A, Morimoto RI, Ishii S (1997) Activation of heat shock transcription factor 3 by c-Myb in the absence of cellular stress. Science 277:246–248

Kim D, Ouyang H, Li GC (1995a) Heat shock protein hsp70 accelerates the recovery of heat-shocked mammalian cells through its modulation of heat shock transcription factor HSF1. Proc Natl Acad Sci USA 92:2126–2130

Kim D, Ouyang H, Yang S-H, Nussenzweig A, Burgman P, Li GC (1995b) A constitutive heat shock element-binding factor is immunologically identical to the Ku-autoantigen. J Biol Chem 270:15277–15284

Kim J, Nueda A, Meng Y-H, Dynan WS, Mivechi NF (1997) Analysis of the phosphorylation of human heat shock transcription factor-1 by MAP kinase family members. J Cell Biochem 67:43–54

Kline MP, Morimoto RI (1997) Repression of the heat shock factor 1 transcriptional activation domain is modulated by constitutive phosphorylation. Mol Cell Biol 17:2107–2115

Knauf U, Newton EM, Kyriakins J, Kingston RE (1996) Repression of human heat shock factor 1 activity at control temperature by phosphorylation. Genes Dev 10:2782–2793

Knuth MW, Gunderson SI, Thompson NE, Strasheim LA, Burgess RR (1990) Purification and characterization of proximal sequence element-binding protein 1, a transcription activating protein related to Ku and TREF that binds the proximal sequence element of the human U1 promoter. J Biol Chem 265:17911–17920

Konishi H, Matsuzaki H, Tanaka M, Ono Y, Tokunaga C, Kuroda S, Kikkawa U (1996) Activation of Rac-protein kinase by heat shock and hyperosmolarity stress through a pathway independent of phosphatidyl inositol 3-kinase. Proc Natl Acad Sci USA 93:7639–7643

Koong AC, Chen EY, Giaccia AJ (1994a) Hypoxia causes the activation of Nuclear Factor-kB through the phosphorylation of IKBa on tyrosine residues. Cancer Res 54:1425–1430

Koong AC, Chen EY, Mivechi NF, Denko NC, Stambrook P, Giaccia AJ (1994b) Hypoxic activation of Nuclear Factor-kB is mediated by a Ras and Raf-1 signaling pathway and does not involve MAP kinase (ERK1 or ERK2). Cancer Res 54:5273–5279

Kroeger PE, Morimoto RI (1994) Selection of new HSF1 and HSF2 DNA-binding sites reveals differences in trimer cooperativity. Mol Cell Biol 14:7592–7603

Kuhn A, Gottlieb TM, Jackson SP, Grummt I (1995) DNA-dependent protein kinase: a potent inhibitor of transcription by RNA polymerase I. Genes Dev 9:193–203

Kuhn A, Stefanovsky V, Grummt I (1993) The nucleolar transcription activator UBF relieves Ku-antigen-mediated repression of mouse ribosomal gene transcription. Nucleic Acids Res 21:2057–2063

Landry J, Bernier D, Chretien P, Nicole LM, Tanguay RM, Marceau N (1982) Synthesis and degradation of heat shock proteins during development and decay of thermotolerance. Cancer Res 42:2457–2461

Larson JS, Schuetz TJ, Kingston RE (1988) Activation in vitro of sequence-specific DNA binding by a human regulatory factor. Nature 335:372–375

Laszlo A (1988) The relationship of heat-shock proteins, thermotolerance, and protein synthesis. Exp Cell Res 178:401–414

Laszlo A (1992) The thermoresistant state: protection from initial damage or better repair? Exp Cell Res 202:519–531

Laszlo A, Li GC (1985) Heat-resistant variants of Chinese hamster fibroblasts altered in expression of heat shock protein. Proc Natl Acad Sci USA 82:8029–8033

Lee YJ, Dewey WC (1987) Induction of heat shock proteins in Chinese hamster ovary cells and development of thermotolerance by intermediate concentrations of puromycin. J Cell Physiol 132:1–11

Lees-Miller SP, Anderson CW (1991) The DNA-activated protein kinase, DNA-PK: a potential coordinator of nuclear events. Cancer Cells 3:341–346

Lees-Miller SP, Chen Y-R, Anderson CW (1990) Human cells contain a DNA-activated protein kinase that phosphorylates simian virus 40 T antigen, mouse p53, and the human Ku autoantigen. Mol Cell Biol 10:6472–6481

Leppa S, Pirkkala L, Saarento H, Sarge KD, Sistonen L (1997) Overexpression of HSF-beta inhibits hemin-induced heat shock gene expression and erythroid differentiation in K562 cells. J Biol Chem 272:15293–15298

Li G, Lee C-H (1992) Heat shock protein hsp70 protects cells from oxidative stress. ASTRO, San Diego, CA

Li GC (1983) Induction of thermotolerance and enhanced heat shock protein synthesis in Chinese hamster fibroblasts by sodium arsenite and by ethanol. J Cell Physiol 115:116–122

Li GC (1985) Elevated levels of 70000 dalton heat shock protein in transiently thermotolerant Chinese hamster fibroblasts and in their stable heat resistant variants. Int J Radiat Oncol Biol Phys 11:165–177

Li GC (1989) HSP 70 as an indicator of thermotolerance. 5th international symposium on hyperthermic oncology, Taylor and Francis, Kyoto, Japan, pp 256–260

Li GC, Hahn GM (1978) Ethanol-induced tolerance to heat and to adriamycin. Nature 274:699–701

Li GC, Li L, Liu RY, Rehman M, Lee WMF (1992) Heat shock protein hsp70 protects cells from thermal stress even after deletion of its ATP-binding domain. Proc Natl Acad Sci USA 89:2036–2040

Li GC, Liu RY, Li L, Shen G, Li X (1992) Hsp70: Role in thermotolerance. 6th international congress on hyperthermic oncology, Arizona Board of Regents, Tucson, AZ, pp 127–130

Li GC, Mivechi NF (1986) Thermotolerance in mammalian systems: a review. In: Anghileri LJ, Robert J (eds) Hyperthermia in cancer treatment, vol 1. CRC Press, Boca Raton, pp 59–77

Li GC, Petersen NS, Mitchell HK (1982) Induced thermal tolerance and heat shock protein synthesis in Chinese hamster ovary cells. Int J Radiat Oncol Biol Phys 8:63–67

Li GC, Werb Z (1982) Correlation between synthesis of heat shock proteins and development of thermotolerance in Chinese hamster fibroblasts. Proc Natl Acad Sci USA 79(10):3218–3222

Li GC, Yang S-H, Kim D, Nussenzweig A, Ouyang H, Wei J, Burgman P, Li L (1995) Suppression of heat-induced hsp70 expression by the 70-kDa subunit of the human Ku-autoantigen. Proc Natl Acad Sci USA 92:4512–4516

Li T, Hightower LE (1995) Effects of dexamethasone, heat shock, and serum responses on the inhibition of hsc70 synthesis by antisense RNA in NIH 3T3 cells. J Cell Physiol 164:344–355

Lindquist S (1986) The heat-shock response. Annu Rev Biochem 55:1151–1191

Lindquist S, Craig EA (1988) The heat shock proteins. Annu Rev Genet 22:631–677

Liu RY, Kim D, Yang S-H, Li GC (1993) Dual control of heat shock response: involvement of a constitutive heat shock element-binding factor. Proc Natl Acad Sci USA 90:3078–3082

Liu RY, Li X, Li L, Li GC (1992) Expression of human hsp70 in rat fibroblasts enhances cell survival and facilitates recovery from translational and transcriptional inhibition following heat shock. Cancer Res 52:3667–3673

Lowe SW, Ruley HE, Jacks T, Housman DE (1993a) p53-dependent apoptosis modulates the cytotoxicity of anticancer agents. Cell 74:957–967

Lowe SW, Schmitt EM, Smith SW, Osborne BA, Jacks T (1993b) p53 is required for radiation-induced apoptosis in mouse thymocytes. Nature 362:847–849

Martin J, Horwich AL, Hartl F-U (1992) Prevention of protein denaturation under heat stress by the chaperonin Hsp60. Science 258:995–998

Mathur SK, Sistonen L, Brown IR, Murphy SP, Sarge KD, Morimoto RI (1994) Deficient induction of human hsp70 heat shock gene transcription in Y79 retinoblastoma cells despite activation of heat shock factor 1. Proc Natl Acad Sci USA 91:8695–8699

McCarty JS, Buchberger A, Reinstein J, Bukau B (1995) The role of ATP in the functional cycle of the DnaK chaperone system. J Mol Biol 249:126–137

Michels AA, Nguyen VT, Konings AW, Kampinga HH, Bensaude O (1995) Thermostability of a nuclear-targeted luciferase expressed in mammalian cells. Destabilizing influence of the intranuclear microenvironment. Eur J Biochem 234:382–389

Milarski KL, Morimoto RI (1989) Mutational analysis of the human HSP70 protein: distinct domains for nucleolar localization and adenosine triphosphate binding. J Cell Biol 109:1947–1962

Mimori T, Akizuki M, Yamagata H, Inada S, Yoshida S, Homma M (1981) Characterization of a high molecular weight acidic nuclear protein recognized by autoantibodies in sera from patients with polymyositis-scleroderma overlap. J Clin Invest 68:611–620

Mimori T, Hardin JA (1986) Mechanism of interaction between Ku protein and DNA. J Biol Chem 261:10375–10379

Mivechi NF, Giaccia A (1995) Mitogen-activated protein kinase acts as a negative regulator of the heat shock response in NIH3T3 cell. Cancer 55:5512–5519

Mivechi NF, Murai T, Hahn GM (1994) Inhibitors of tyrosine and ser/thr phosphatases modulate the heat shock response. J Cell Biochem 54:186–197

Mivechi NF, Ouyang H, Hahn GM (1992) Lower heat shock factor activation and binding and faster rate of HSP-70 A messenger RNA turnover in heat sensitive human leukemias. Cancer Res 52:6815–6822

Mivechi NF, Park YM, Ouyang H, Shi XY, Hahn GM (1994) Selective expression of heat shock genes during differentiation of human myeloid leukemic cells. Leuk Res 18:597–608

Mivechi NF, Shi X-Y, Hahn GM (1995) Stable overexpression of human HSF-1 in murine cells suggests activation rather than expression of HSF-1 to be the key regulatory step in the heat shock gene expression. J Cell Biochem 59:266–280

Mizzen LA, Welch WJ (1988) Characterization of the thermotolerant cell. I. Effects on protein synthesis activity and the regulation of heat-shock protein 70 expression. J Cell Biol 106:1105–1116

Moodie SA, Willumsen BM, Weber MJ, Wolfman A (1993) Complexes of ras-GTP with raf-1 and mitogen-activated protein kinase kinase. Science 260:1658–1661

Morimoto RI, Tissieres A, Georgopoulos C (eds) (1990) Stress proteins in biology and medicine. Cold Spring Harbor Laboratory Press, Plainview

Mosser DD, Martin LH (1992) Induced thermotolerance to apoptosis in a human T lymphocyte cell line. J Cell Physiol 151:561–570

Munro S, Pelham HRB (1984) Use of peptide tagging to detect proteins expressed from cloned genes: deletion mapping functional domains of Drosophila hsp70. EMBO J 3:3087–3093

Nakai A, Morimoto RI (1993) Characterization of a novel chicken heat shock transcription factor, heat shock factor 3, suggests a new regulatory pathway. Mol Cell Biol 13:1983–1997

Nakai A, Tanabe M, Kawazoe Y, Inazawa J, Morimoto RI, Nagata K (1997) HSF4, a new member of the human heat shock factor family which lacks properties of a transcriptional activator. Mol Cell Biol 17:469–481

Nguyen VT, Morange M, Bensaude O (1989) Protein denaturation during heat shock and related stress. J Biol Chem 264:10487–10492

Nussenzweig A, Chen C, da Costa Soares V, Sanchez M, Sokol K, Nussenzweig MC, Li GC (1996) Requirement for Ku80 in growth and immunoglobulin V(D)J recombination. Nature 382:551–555

Nussenzweig A, Li GC (1993) Alteration of the heat shock response in hsc70 antisense transfected mammalian cells. 1st advanced workshop of the European Science Foundation, Paris, France, P-5

Rabindran SK, Giorgi G, Clos J, Wu C (1991) Molecular cloning and expression of a human heat shock factor, HSF1. Proc Natl Acad Sci USA 88:6906–6910

Rabindran SK, Haroun RI, Clos J, Wisniewski J, Wu C (1993) Regulation of heat shock factor trimer formation: role of a conserved leucine zipper. Science 259:230–234

Reeves WH (1992) Antibodies to p70/p80 (Ku) antigens in systemic lupus erythematosus. Rheum Dis Clin North Am 18:391–414

Reeves WH, Sthoeger ZM (1989) Molecular cloning of cDNA encoding the p70 (Ku) lupus autoantigen. J Biol Chem 264:5047–5052

Riabowol KT, Mizzen LA, Welch WJ (1988) Heat shock is lethal to fibroblasts microinjected with antibodies against hsp70. Science 242:433–436

Ritossa FM (1962) A new puffing pattern induced by heat shock and DNP in Drosophila. Experientia 18:571–573

Ritossa FM (1963) New puffs induced by temperature shock, DNP and salicylate in salivary chromosomes of Drosophila melanogaster. Drosophila Inf Service 37:122–123

Rouse J, Cohen P, Trigon S, Morange M, Alonso-Llamazares A, Zamanillo D, Hunt T, Nebreda AR (1994) A novel kinase cascade triggered by stress and heat shock that stimulates MAPKAP kinase-2 and phosphorylation of the small heat shock proteins. Cell 78:1027–1037

Rudiger S, Germeroth L, Schneider-Mergener J, Bukau B (1997) Substrate specificity of the DnaK chaperone determined by screening cellulose-bound peptide libraries. EMBO J 16:1501–1507

Sarge KD, Murphy SP, Morimoto RI (1993) Activation of heat shock gene transcription by heat shock factor 1 involves oligomerization, acquisition of DNA-binding activity, and nuclear localization and can occur in the absence of stress. Mol Cell Biol 13:1392–1407

Sarge KD, Park-Sarge OK, Kirby JD, Mayo KE, Morimoto RI (1994) Expression of heat shock factor 2 in mouse testis: potential role as a regulator of heat-shock protein gene expression during spermatogenesis. Biol Reprod 50:1334–1343

Sarge KD, Zimarino V, Holm K, Wu C, Morimoto RI (1991) Cloning and characterization of two mouse heat shock factors with distinct inducible and constitutive DNA-binding ability. Genes Dev 5:1902–1911

Schmid D, Baici A, Gehring H, Christien P (1994) Kinetics of molecular chaperone action. Science 263:971–973

Schuetz TJ, Gallo GJ, Sheldon L, Tempst P, Kingston RE (1991) Isolation of a cDNA for HSF2: evidence for two heat shock factor genes in humans. Proc Natl Acad Sci USA 88:6911–6915

Shi Y, Kroeger PE, Morimoto RI (1995) The carboxyl-terminal transactivation domain of heat shock factor 1 is negatively regulated and stress responsive. Mol Cell Biol 15:4309–4318

Shi Y, Mosser DD, Morimoto RI (1998) Molecular chaperones as HSF-1 specific transcriptional repressors. Genes Dev 12:654–666

Sistonen L, Sarge KD, Phillips B, Abravaya K, Morimoto RI (1992) Activation of heat shock factor 2 during hemin-induced differentiation of human erythroleukemia cells. Mol Cell Biol 12:4104–4111

Skowyra D, Georgopoulos C, Zylicz M (1990) The E. coli dnaK gene product, the hsp70 homolog, can reactivate heat-inactivated RNA polymerase in an ATP hydrolysis-dependent manner. Cell 62:939–944

Soncin F, Zhang X, Chu B, Zhong RD, Stevenson MA, Calderwood SK (1998) HSF-1 regulation by phosphorylation. Molecular Chaperones and Heat Shock Response, Cold Spring Harbor, p 247

Sorger PK, Lewis MJ, Pelham HRB (1987) Heat shock factor is regulated differently in yeast and HeLa cells. Nature 329:81–84

Sorger PK, Pelham HRB (1988) Yeast heat shock factor is an essential DNA-binding protein that exhibits temperature-dependent phosphorylation. Cell 54:855–864

Stevenson MA, Pollock SS, Coleman CN, Calderwood SK (1994) X-irradiation, phorbol esters, and H_2O_2 stimulate mitogen-activated protein kinase activity in NIH-3T3 cells through the formation of reactive oxygen intermediates. Cancer Res 54:12–15

Stokoe D, Macdonald SG, Cadwallader K, Symons M, Hancock JF (1994) Activation of Raf as a result of recruitment to the plasma membrane. Science 264:1463–1467

Strasser A, Anderson RL (1995) Bcl–2 and thermotolerance cooperate in cell survival. Cell Gr Diff 6:799–805

Subjeck JR, Sciandra JJ, Johnson RJ (1982) Heat shock proteins and thermotolerance: comparison of induction kinetics. Br J Radiol 55:579–584

Szabo M, Langer T, Schroder H, Flanagan J, Bukau B, Hartl FU (1994) The ATP hydrolysis-dependent reaction cycle of the Escherichia coli HSP-70 system DnaK, DnaJ and GrpE. Proc Natl Acad Sci USA 91:10345–10349

Westwick JK, Cox AD, Der CJ, Cobb MH, Hibi M, Karin M, Brenne DA (1994) Oncogenic Ras activates c-Jun via a separate pathway from the activation of extracellular signal-regulated kinases. Proc Natl Acad Sci USA 91:6030–6034

Westwood JT, Wu C (1993) Activation of Drosophila heat shock factor: conformational change associated with a monomer to trimer transition. Mol Cell Biol 13:3481–3486

Wiederrecht G, Seto D, Parker CS (1988) Isolation of the gene encoding the S. cerevisiae heat-shock transcription factor. Cell 54:841–853

Williams GT, McClanahan TK, Morimoto RI (1989) Ela transactivation of the human HSP70 promoter is mediated through the basal transcription complex. Mol Cell Biol 9:2574–2587

Wu B, Hunt C, Morimoto RI (1985) Structure and expression of the human gene encoding major heat shock protein hsp70. Mol Cell Biol 5:330–341

Wu B, Kingston RE, Morimoto RI (1986) Human HSP-70 promoter contains at least two distinct regulatory domain. Proc Natl Acad Sci USA 83:629–633

Wu C (1984) Activating protein factor binds in vitro to upstream control sequences in heat shock gene chromatin. Nature 311:81–84

Wu C (1995) Heat shock transcription factors: structure and regulation. Annu Rev Cell Dev Biol 11:441–469

Xia Z, Dickens M, Raingeaud J, Davis RJ, Greenberg ME (1995) Opposing effects of ERK and JNK-p38 MAP kinases on apoptosis. Science 270:1326–1331

Yaneva M, Busch H (1986) A 10 S particle released from deoxyribonuclease-sensitive regions of HeLa cell nuclei contains the 86-kilodalton-70-kilodalton protein complex. Biochemistry 25:5057–5063

Yang S-H, Nussenzweig A, Li L, Kim D, Ouyang H, Burgman P, Li GC (1996) Modulation of thermal induction of hsp70 expression by Ku autoantigen or its individual subunits. Mol Cell Biol 16:3799–3806

Yang SD, Lee S-C, Chang H-C (1997) Heat stress induces tyrosine phosphorylation/activation of kinase Fa/GSK-3a (a human carcinoma dedifferentiation modulator) in A431 cells. J Cell Biochem 65:16–26

Zimarino V, Wu C (1987) Induction of sequence-specific binding of Drosophila heat shock activator protein without protein synthesis. Nature 327:727–730

CHAPTER 4

Mitochondrial Molecular Chaperones hsp60 and mhsp70: Are Their Roles Restricted to Mitochondria?

B.J. Soltys and R.S. Gupta

A. Introduction

The 60-kDa heat shock protein (hsp60) and mitochondrial hsp70 (mhsp70) constitute two of the major molecular chaperones in eukaryotic organisms with essential functions in both stressed and non-stressed cells, which by binding and stabilizing unstable conformations of substrate proteins allow diverse proteins either to fold correctly, assemble into oligomeric complexes or be transported across membranes in a partially unfolded state (Ellis and van der Vies 1991; Craig et al. 1993; Ellis and Hartl 1996; Hartl 1996). They act very differently, however, and usually not in isolation from each other. Hsp60, also referred to as a chaperonin (Ellis and van der Vies 1991), has been regarded to be present and to function in protein folding only within organelles such as mitochondria and chloroplasts, which are of endosymbiotic origin, while the TCP-1 protein, which is only weakly sequence related to hsp60 (Gupta 1990a), functions as a chaperonin in the cytosolic compartment (Craig et al. 1993; Hartl 1996). In the case of hsp70, different hsp70 homologs encoded by separate nuclear genes are found in the cytosol, endoplasmic reticulum, mitochondria and chloroplasts. Of these, the mitochondrial hsp70 homolog is the most closely related to DnaK from gram negative bacteria, reflecting the endosymbiotic origin of this organelle (Gupta and Golding 1993).

Section B provides a brief overview of the structures of hsp60 and mhsp70, their substrate binding properties, mechanism of folding, established roles within mitochondria and in mitochondrial import, and functions under stress conditions. Consideration will also be given to cpn10, the co-factor or co-chaperonin for hsp60. Several excellent reviews are available on these first aspects for complementary or more in depth analysis of individual topics (Ellis and van der Vies 1991; Craig et al. 1993; Ellis and Hartl 1996; Hartl 1996). In Sect. C, we critically evaluate recent intriguing reports that have suggested that these chaperones, in addition to their expected localization and function within mitochondria of mammalian cells, may also localize, and by implication have ancillary roles, at certain extramitochondrial sites, including the cell surface. These include, but are not limited to, reports of surface expression in certain cases of stressed, apoptotic and tumor cells. Since this seemingly aberrant localization poses a dilemma in view of the established facts concerning the targeting of these nuclear encoded proteins to mitochondria, at issue is whether or not these are real phenomena. We present both sides of the case. The data compels us to explain the data by proposing novel trafficking mechanisms that are involved in the extramitochondrial distributions of these chaperone proteins. These putative trafficking pathways could prove to be new targets for pharmacological intervention in the pathological conditions in which these chaperones have been implicated. In Sect. D, we consider the reported involvement of these chaperones, particularly hsp60, in drug resistance and disease and Sect. E summarizes our view of prospects for future research.

B. Structure and Function

I. Studies with Purified Proteins

One of the main questions to date in research on the molecular chaperones hsp60 and mhsp70 has been the molecular mechanisms by which these proteins catalyse the proper folding and assembly of other proteins, primarily using in vitro systems and the bacterial homologs GroEL and DnaK. Both hsp60 and hsp70 have been highly conserved throughout evolution from bacteria to man (GUPTA and GOLDING 1993; GUPTA 1995). Hsp70 itself is the most highly conserved protein known at present and has been used extensively in phylogenetic analysis of prokaryotes and eukaryotes (GUPTA and GOLDING 1993). It is no surprise then that knowledge gained from GroEL and DnaK has proven to be highly relevant to hsp60 and hsp70 in higher eukaryotes.

1. Hsp70/DnaK

Hsp70 is an ATPase which functions as a 70 kDa monomer. It acts by binding hydrophobic residues on elongating nascent polypeptides and on unfolded proteins being imported into the mitochondrial matrix (CRAIG et al. 1993; HARTL 1996). This action can have three effects. First, since exposure of hydrophobic patches in proteins and hydrophobic interactions between proteins are the cause of protein aggregation, it prevents aggregation of the unfolded protein. Second, it prevents premature incorrect folding of the protein. Third, in mitochondrial import hsp70 acts as a molecular ratchet and/or molecular motor to bring unfolded preproteins into the mitochondrial matrix (see below). Inasmuch as mhsp70 is more closely related to DnaK than are the cytosolic and endoplasmic reticulum hsp70 homologs (GUPTA and GOLDING 1993), studies of DnaK structure and function are relevant in the first place to mhsp70. The crystal structure of the ATP binding domain of the constitutively expressed cytosolic form of hsp70 (hsc70) has been known for a number of years (FLAHERTY et al. 1990) while the crystal structure of the peptide-binding domain of DnaK has been more recently determined (ZHU et al. 1996). The substrate binds to hydrophobic sites which are located within a pocket and a flexible lid then sequesters the bound substrate, preventing the substrate from interacting with other proteins, which would result in aggregation. ATP hydrolysis correlates with closing of the lid; the lid is open when DnaK/hsp70 is in the ATP bound state, allowing substrate release, and is closed in the ADP bound state (ZHU et al. 1996). Substrate is usually released in a partially folded compact intermediate state, referred to as molten globules, which still run the risk of aggregating because of exposed surface hydrophobic patches. In such cases, further folding and prevention of aggregation require additional cycle(s) of binding and release or, more typically, transfer to the GroEL/hsp60 chaperonin for attainment of complete folding. In addition to protein folding, DnaK has been shown to solubilize protein aggregates in vitro (GETHING and SAMBROOK 1992), which may be relevant to reversal of heat-induced denaturation/aggregation of proteins.

2. Hsp60/GroEL

Both hsp60 in yeast and GroEL are large cylindrical oligomers comprised of 14 identical subunits and a total mass of ~800 kDa and it is in the large central cavity where proteins are normally folded sequestered from the environment. The subunits, each an ATPase, function in ATP dependent folding of proteins and are arranged into two 7-membered stacked rings, each with a large central cavity. Although hsp60 in yeast is composed of two stacked rings, apparently, however, mammalian hsp60 exists only as a single toroidal ring (PICKETTS et al. 1989; VIITANEN et al. 1992). The homologous protein in chloroplasts, Rubisco subunit binding protein, in contrast, is a hexadecamer composed of two distinct subunits, alpha and beta, which are highly related (ELLIS 1990). Cpn10/GroES, a co-factor or co-chaperonin for hsp60/GroEL, also forms a 7-membered ring of mass ~98 kDa and may bind to only one end of the hsp60/GroEL oligomer. In vivo genetic evidence has shown definitively that GroEL prevents protein aggregation in *Escherichia coli* (HORWICH et al. 1993). Approximately 30% of proteins in *Escherichia coli* require GroEL to attain their folded states (HARTL 1996). The crystal structure of GroEL has been available for several years (BRAIG et al. 1994) and the crystal structure of GroES has also been recently obtained (HUNT et al. 1996; MANDE et al. 1996). Electron microscopy has been very important in elucidating interactions with substrates and the effects of ATP hydrolysis (CHEN et al. 1994; ROSEMAN et al. 1996), the details of which are now also being addressed by crystallography (XU et al. 1997). Exposed hydrophobic regions on the protein to be folded first bind to one or more of the hydrophobic binding sites present in the 7 subunits in one of the GroEL rings. Then, cpn10 binding to one end of hsp60 is followed by substrate release into the central cavity of the hsp60 ring to which it is bound. Cpn10 also acts to cap that end of hsp60, preventing the release of substrate into the medium. An ATPase timer then causes release of the cpn10 cap (time ~10 s at 37°C) which halts folding and allows substrate exit. If hydrophobic patches are still exposed on the substrate's surface, however, substrate may rebind and another round of folding is begun (WEISSMAN et al. 1994).

A truncated GroEL recombinant protein, corresponding to residues 191–376, has been found to retain chaperonin activity – this "mini-chaperone" remains a monomer in solution and thus does not form the central cavity of the intact oligomer (BUCKLE et al. 1997). The crystal structure of the apical binding domain was determined (BUCKLE et al. 1997). The mini-chaperone was found to not form a "lid" to sequester substrate, as found in DnaK, and folding thus occurs free in solution. Such a mini-chaperone, however, should have exposed hydrophobic binding sites and may be prone to aggregation. In the cell, hsp60 thus far has only been found as an oligomer when isolated from mitochondria. The existence of a functional monomeric hsp60 in vivo has not been demonstrated.

Hsp60 may also be able to cause protein disassembly, or unfold proteins, as suggested from in vitro studies with GroEL (JACKSON et al. 1993). This would be achieved by binding to a hydrophobic patch and exerting a binding pressure to unfold. In vivo, this may allow, for example, incorrectly folded proteins to refold or be processed for degradation. A retrograde transfer of substrate from hsp60 to hsp70 may be involved in targeting for degradation (HAYES and DICE 1996).

It has recently been shown that GroEL is capable of direct membrane insertion in model lipid bilayers (TOROK et al. 1997). Insertion is dependent on the last 16 amino acids at the C-terminal end of GroEL. The insertion of GroEL alters the physical properties of the bilayer, increasing membrane order. More importantly, GroEL retains chaperonin activity in the membrane-associated state and functions as a "lipochaperonin" (TOROK et al. 1997). The membrane associated protein was also found to retain its oligomeric state. Hsp60 is highly conserved from prokaryotes to man (GUPTA 1995) and the C-terminal end motif thought to be involved in membrane insertion is found throughout evolution. Thus the mechanistic basis for membrane association of GroEL (TOROK et al. 1997) is probably retained in mammalian hsp60, although this will need to be directly shown.

In conclusion, an interesting feature of hsp60 is that its functional activity as a chaperone has been demonstrated in several structural forms. While both hsp60 and cpn10 are required for folding most proteins, some proteins can be folded by hsp60 without cpn10 (HÖHFIELD, HARTL 1994) and these two hsp60 forms must fold proteins by fundamentally different mechanisms. While hsp60 in yeast is also a double toroid (CRAIG et al. 1993), hsp60 is a single toroid in mammalian cells (PICKETTS et al. 1989; VIITANEN et al. 1992) and the differences in folding properties between these two is unknown. In vitro, a monomeric recombinant mini-chaperone retains chaperone activity (BUCKLE et al. 1997). Also, GroEL oligomer can insert into model membranes and retain activity (TOROK et al. 1997). GroEL can also fold, by unknown means, proteins which are too large to enter the central cavity (~5 nm wide) of the toroidal structure, an example being beta-galactosidase (AYLING and BANEYX

1996) which is a tetrameric protein with 116-kDa subunits. Substrates that fold within the central cavity are in the ~15–60 kDa range (HARTL 1996). Nothing is known about whether variations in hsp60 structural forms exist in the cell.

II. In Vivo and Mitochondrial Systems

In the first subcellular localization studies of hsp60 in mammalian cells (mammalian hsp60 was known at that time only as P1 protein), immunofluorescence studies localized hsp60 to mitochondria (GUPTA et al. 1985; GUPTA and DUDANI 1987) and biochemical fractionation revealed hsp60 is primarily present in the mitochondrial matrix compartment (GUPTA and AUSTIN 1987). Much of our knowledge concerning molecular chaperones, however, is derived primarily from studies on yeast, which afford both genetic and biochemical tools, and have been very productive systems for analysing the roles of mhsp70 and hsp60 in the import and folding of proteins in mitochondria (CRAIG et al. 1993). Indeed, it is through study of protein folding in mitochondria in yeast that the essential role of hsp60 as a chaperonin in eukaryotic cells was first determined (CHENG et al. 1989; OSTERMAN et al. 1989). Yeast temperature-sensitive mutants affected in hsp60 (CHENG et al. 1989, 1990; HALLBERG et al. 1993) showed that mutation affected the proper folding and assembly into complexes of several mitochondrial proteins. Yeast mutants have also identified the role of mhsp70 (KANG et al. 1990; GAMBILL et al. 1993; VOOS et al. 1994). These studies, together with in vitro studies of protein import and folding using mitochondrial systems (OSTERMAN et al. 1989; HARTL 1996; NEUPERT 1997) have contributed to the following understanding of these molecular chaperones.

Hsp60 and mhsp70 each only have a single functional gene, are synthesized with an N-terminal mitochondrial targeting sequence which requires that the gene product enter mitochondria and there is no evidence for alternate splicing of the mRNA for these gene products which might cause targetting to other cellular destinations. In the mitochondrial matrix, they normally act sequentially to fold imported proteins with imported preproteins being passed on from mhsp70 to hsp60 (MANNING-KRIEG et al. 1991).

Selected features of the mitochondrial import mechanism follow. This topic has recently been reviewed in detail (NEUPERT 1997). Studies in yeast and in vitro mitochondrial import reactions have shown that following synthesis of preproteins on cytosolic ribosomes, polypeptide chains are rapidly translocated to mitochondria while in association with cytosolic hsp70 (hsc70). In yeast mutants affected in SSA genes encoding different cytosolic hsp70 isoforms, preproteins accumulate in the cytosol (DESHAIES et al. 1988); transport of proteins to the endoplasmic reticulum and peroxisomes is also inhibited. The mitochondrial import channel is referred to as the translocase and is composed of outer membrane and inner membrane proteins, generally referred to as TOM and TIM proteins, respectively. The TOM and TIM proteins are now suggested to be dynamic molecular machines rather than simply passive components of a conventional transport channel (Pfanner and Meijer 1997). Studies in yeast first identified mhsp70 (SSC1 gene) as being involved in the mitochondrial import mechanism (Kang et al. 1990) in addition to its role in protein folding as a soluble protein in the matrix compartment. The entry of a preprotein into the matrix compartment is thought to have some similarities with the emergence of nascent chains from ribosomes. Approximately 10% of mhsp70 in the matrix compartment associates with TIM44 on the inner mito-chondrial membrane and functions to bring unfolded proteins into the matrix. The interaction of mhsp70 with Tim44 is dynamic and regulated by adenine nucleotides and the nucleotide exchange factor MGE (GrpE homolog, Mge1p in yeast). Mhsp70 with bound ATP associates with TIM44, while ADP-mhsp70 dissociates from it. Translocation of preprotein into the matrix de-pends on the mitochondrial membrane potential and upon the presence of mhsp70. In the absence of mhsp70, import is reversible. Matrix ATP hydroly-sis is required for mhsp70 to bind to and drive vectorial transport of the preprotein. ATP depletion in the mitochondrial compartment results in exit of the preprotein back into the cytoplasm. Two models have been proposed to explain mhsp70 driven vectorial transport including a (i) "molecular ratchet model" in which mhsp70 binding acts to prevent preprotein backsliding in the translocation channel and (ii) "translocation motor model" in which hsp70 acts as a molecular motor to pull the preprotein into the matrix (Pfanner and Meijer 1995). These models are not considered to be mutually exclusive.

After import, mhsp70 can mediate folding of certain proteins, particularly small ones, but the majority of proteins require further folding mediated by the hsp60 chaperonin system. For imported proteins destined for the inter-membrane compartment, hsp60 may function to keep the protein in an un-folded state (Koll et al. 1992), although certain proteins may translocate by an hsp60-independent pathway (Hallberg et al. 1993). In this context we also note the established function of GroEL in bacterial protein secretion (Bochkareva et al. 1988; Zeilstra-Ryalls et al. 1991) which, in view of the endosymbiotic origin of mitochondria (Margulis 1970; Gray 1992), may be a role in protein export that is retained by mitochondrial hsp60. Mhsp70 is also a chaperone for proteins synthesized on mitochondrial ribosomes, similarly to hsp70 in the cytoplasm, and also assists the assembly of inner membrane proteins (Neupert 1997). Mhsp70 also appears involved in protein degrada-tion and hsp60 can pass substrates over to mhsp70 for degradation (Hayes and Dice 1996).

Hsp70, in addition to roles in protein folding and translocation, also can induce the disassembly of protein complexes. Hsp70 is thought to interact with folded proteins which expose chaperone recognition motifs. In eukaryotic cells, the constitutive cytosolic hsp70 (hsc70) is one and the same as the clathrin-uncoating ATPase involved in dissassembly of the clathrin coats of endocytic vesicles. In *Escherichia coli*, DnaK is involved in initiation of DNA synthesis from various phage and plasmid origin of replication (Gething and Sambrook 1992; Georgopoulos et al. 1994).

Hsp70 functions under stress conditions are still inadequately understood. Conditions such as exposure to high temperatures induce protein aggregation. Although DnaK has been shown to solubilize protein aggregates in vitro (GETHING and SAMBROOK 1992), prevention of aggregation during unfolding may actually be the operating mechanism in vivo (SCHRÖDER et al. 1993).

C. Are hsp60 and mhsp70 Restricted to Mitochondria?

I. Subcellular Localization: The Unexplained Findings

While the evidence is quite conclusive that both hsp60 and mhsp70 are located and have functions primarily within mitochondria in both yeast and mammalian cells, a variety of evidence has, however, suggested that small amounts of these chaperones may also function elsewhere in the cell. Mhsp70 has been implicated to function in antigen presentation, having been identified as being one and the same as peptide-binding protein PBP72/74 (DOMANICO et al. 1993; DAHLSEID et al. 1994) and has also been implicated in cell senescence, having also been identified as "mortalin" (WADHWA et al. 1993, 1994), and both of these identities are likely inconsistent with an exclusive mitochondrial compartmentation. Mitochondrial cpn10, a co-factor for hsp60, has also been identified as being identical to early pregnancy factor (EPF), an important growth factor present in maternal serum (CAVANAGH and MORTON 1994), so there must be a physiological mechanism for secretion of this protein by maternal tissue and there must exist cellular receptors for cpn10 at target sites.There is also strong evidence for the involvement of hsp60 in peptide presentation (LUKACS et al. 1993, 1997; WELLS et al. 1997). In yeast temperature-sensitive mutants affected in hsp60, overexpression of the protein SCS1, which is an extramitochondrial protein, suppresses the mutant phenotype (SHU and HALLBERG 1995), a finding that is difficult to explain if hsp60 is present only within mitochondria [SCS1, also called Rts1p, is now known to be a yeast homolog of regulatory subunit B' of protein phosphatase 2A (SHU et al. 1997)]. The evidence supporting an association of hsp60 with the plasma membrane in mammalian cells, however, is the main data calling into question an exclusive mitochondrial localization, as follows.

One of the first demonstrations suggesting hsp60 was on the cell surface was that murine and human T cells which recognize mycobacterial hsp60 are specifically stimulated by a protein present on the surface of stressed macrophages (KOGA et al. 1989) and certain tumor cells (FISCH et al. 1990; SELIN et al. 1992; KAUR et al. 1993; FITZGERALD and KEAST 1994) and stimulation in the latter case has also been found to be blocked by polyclonal and monoclonal antibodies specific for hsp60 (KAUR et al. 1993). More definitive evidence was shown biochemically in Daudi lymphoma cells by immunoprecipitation of surface-iodinated proteins using polyclonal (FISCH et al. 1990) and monoclonal antibodies (KAUR et al. 1993) against hsp60. Another technique employing biotinylation of exterior cell surface proteins found evidence for cell surface hsp60 in cultured Chinese hamster ovary cells (CHO; Soltys and Gupta 1996) suggesting a cell surface presence is not restricted to certain types of tumor cells. Alternate means of identifying cell surface proteins have made further progress.

Using chemical crosslinking in living cells to analyse in vivo protein associations, hsp60 has been shown to interact in cells with P21[ras] (Ikawa and Weinberg 1992) – the identity of hsp60 was established in this study by microsequencing. Since P21[ras] is a plasma membrane protein involved in signal transduction, this finding implicates hsp60 in signal transduction events directly at the cell surface. It also appears that hsp60 molecules exert signal transduction effects if applied extracellularly: it has been shown that Hsp60 isolated from *Legionella pneumophila* has protein kinase C mediated signal transduction effects in macrophages (Retzlaff et al. 1996) and this finding also strongly suggests macrophages have cell surface receptors for hsp60.

Hsp60 has also been implicated in amino acid transport. In CHO-K1 cells exhibiting an increase in the A system of amino acid transport at the plasma membrane, a concomitant enhancement in the amount of hsp60 is observed (Jones et al. 1994). A hsp60 homolog has also been found associated with the L-system amino acid transporter in chronic lymphocytic leukemia B-lymphocytes (Woodlock et al. 1997). Recently, hsp60 has also been identified biochemically in the plasma membrane of a T lymphocyte cell line CEM-SS by Kammer and co-workers. These studies provide the first evidence that membrane associated hsp60 is phosphorylated in mammalian cells. Phosphorylation was shown to be due to type I protein kinase A activity and surface expression of hsp60 was found elevated following mitogen activation (Khan et al. 1998). Previously, phosphorylation of hsp60 has been demonstrated only in the case of GroEL in *Escherichia coli* following heat shock (Sherman and Goldberg 1992)- membrane-association was not evaluated.

In addition to the reported presence of hsp60 on tumor cells, cited above, several studies have indicated hsp60 becomes expressed on the cell surface in stressed cells and apoptic cells. Stressed aortic endothelial cells, exposed to cytokines or high heat, express hsp60 on their cell surface, as detected by fluorescence imaging, and are susceptible to complement-dependent lysis by hsp60-specific antibodies (Xu et al. 1993; Schett et al. 1995). In the case of apoptosis, it has recently been reported that both hsp60 and hsp70 are expressed on the cell surface of T cells undergoing apoptosis (Poccia et al. 1996). These results, and observations described above in tumor cells, have led to the concept that surface expression of these chaperones may arise in specific cases of stressed, apoptotic and in certain tumor cells. We and other laboratories have undertaken detailed electron microscopic (EM) studies of hsp localization. The results of immunogold EM localization studies and other findings, however, suggest surface expression may be more common.

In immunogold labeling of mammalian cells with anti-hsp60 antibody, the majority of hsp60 labeling is found primarily within mitochondria, as shown in Fig. 1a for CHO cells (Soltys and Gupta 1996). In addition to the intense labeling of mitochondria, hsp60 reactivity is also consistently detected as clusters of gold particles at discrete extramitochondrial sites. Several examples are indicated in Fig. 1a by arrowheads. Detailed investigation of extramito-chondrial reactivity were done in a variety of cultured cells, including primary human diploid fibroblasts, B-SC-1 monkey kidney cells, PC12 neuronal cells and Daudi Burkitt's lymphoma cells and all gave similar results (Soltys and Gupta 1996). Fig. 1b shows high magnification immunogold labeling in a CHO cell of a region near the cell surface where there are no mitochondria. Label-ing here is both on top of and below the cell surface, as indicated with open arrows. Additionally, there are two membraneous structures, indicated by closed arrows, that are gold-labeled and appear to be vesicles. Their identity is unknown. In the CHO cell in Fig. 1c, gold labeling is found at an electron-dense invagination of the plasma membrane (see arrow) that may represent a secretory or endocytic vesicle.

In related EM immunogold labeling studies, backscattered electron imag-ing of whole intact cells has suggested that the cell surface copy number of hsp60 is ~200–2000 molecules per cell, analysed in both CHO and human CEM-SS T lymphocytes, and it was estimated that this represents 1–10% of total cellular hsp60 (Soltys and Gupta 1997). Cell surface expression of hsp60 appears to be a general characteristic of both transformed and nontransformed mammalian cells, although there could be significant quanti-tative differences between cells of different origin.

In addition to hsp60 reactivity at the cell surface and in cytoplasmic vesicles, immunogold labeling in cultured cells has also consistently detected extramitochondrial hsp60 at discrete foci on endoplasmic reticulum and at sites on, or in close proximity, to the mitochondrial outer membrane (Soltys and Gupta 1996; see also below). The intermittent labeling of endoplasmic reticulum in cultured cells did not suggest hsp60 was in transit through the secretory pathway.

A number of mammalian (rodent) tissues have also been evaluated to determine hsp60 subcellular distribution. In pancreatic beta-cells, strong hsp60 reactivity has been observed in mature insulin secretory granules (Brudzynski et al. 1992a). Hsp60 antibodies specifically label the central core of mature insulin secretory granules, but not immature secretory granules, as shown in Fig. 2a. The fact that no labeling was found in immature secretory granules, or elsewhere along the secretory pathway (vis. Golgi and endoplas-mic reticulum) suggested hsp60 is imported only into mature insulin secretory vesicles. Figure 2b shows an example of a putative transport vesicle containing hsp60 reactivity that appears to be fusing with an insulin secretory vesicle. Hsp60 has also been localized in other types of secretory granules. Hsp60, cpn10 and hsp70 reactivity has been reported to be present all along the secretory pathway in pancreatic acinar cells, using polyclonal antibodies to

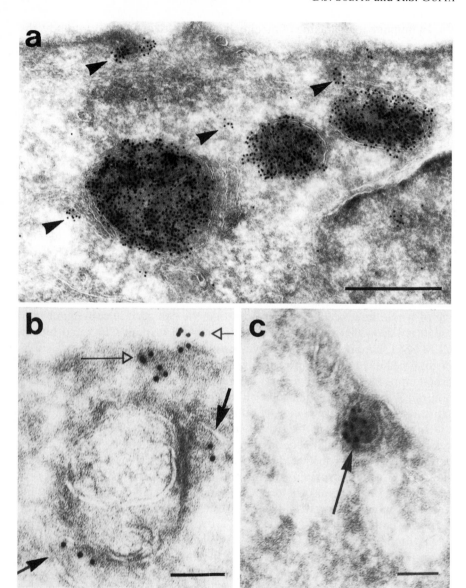

Fig. 1a–c. Electron microscopic visualization of hsp60 distribution in Chinese hamster ovary cells using immunogold labeling. Cryosections labeled with polyclonal antibody against CHO hsp60, followed by 10-nm colloidal gold markers. **a** Low magnification micrograph showing intense reactivity within the three mitochondria in the field of view. There is also reactivity at discrete extramitochondrial sites, with several examples *indicated by arrowheads*. *Bar*, 0.5 μm. **b,c** High magnification micrographs of cell surface regions. In **b** there is reactivity on and underneath the cell surface (*open arrows*) and in vesicular structures (*closed arrows*). In **c** reactivity is in a cell surface invagination, possibly a forming endocytic vesicle. *Bars in* **b** *and* **c** represent 0.1 μm. (From Soltys and Gupta 1996)

homologs from the photosynthetic bacterium *Chromatium vinosum* (VELEZ-GRANELL et al. 1994). Work in our own laboratory using highly specific polyclonal and monoclonal antibodies raised against mammalian hsp60, however, has shown that hsp60 reactivity in pancreatic acinar cells in restricted to zymogen granules (J. Chechetto and R.S. Gupta, unpublished). Hsp60 presence in secretory vesicles has also been found in growth hormone granules in the pituitary (J. Chechetto and R.S. Gupta, unpublished). However, hsp60 association with secretory vesicles is not ubiquitous in diverse tissues and has not been observed in other pituitary cell types.

In other tissues examined thus far, hsp60 was found to be present in rat liver in both mitochondria and peroxisomes (Fig. 2c; SOLTYS and GUPTA 1996), peroxisomes being an organelle involved in a variety of oxidative reactions. Hsp60 reactivity in the peroxisome in Fig. 2c is shown to be primarily associated with the crystalline inclusion or core material. The crystalline inclusion is known to be composed mainly of urate oxidase and is a distinguishing characteristic of rat liver peroxisomes (FAHIMI et al. 1993). Similar observations in rat liver were made with both polyclonal and monoclonal antibodies. Hsp60 has also been localized to mitochondria and peroxisomes in liver sections using antibody to GroEL (VELEZ-GRANELL et al. 1995). The imported Hsp60 was suggested to function in the assembly of the peroxisome core material (SOLTYS and GUPTA 1996). Studies of peroxisomes in typical cultured cells has been hampered thus far because they are typically few in number and as small as 100 nm in diameter.

Although the subcellular localization of hsp60 has been the most extensively studied, work on mHsp70 has also arrived at similar conclusions indicating an extramitochondrial presence. Mhsp70 has been localized at the plasma membrane and in cytoplasmic vesicles (VANBUSKIRK et al. 1991; SINGH et al. 1997), in unidentified cytoplasmic granules (SINGH et al. 1997) and localization at these sites may be relevant to mhsp70's role in peptide binding and antigen presentation (DOMANICO et al. 1993; DAHLSEID et al. 1994) and in cell senescence (WADHWA et al. 1993).

In conclusion, EM subcellular localization studies in various representative mammalian cultured cells have shown that 15–20% of hsp60 (SOLTYS and GUPTA 1996) and 10–20% of mhsp70 (SINGH et al. 1997) is located at extramitochondrial sites . These estimates are qualitatively supported by the results of EM localization studies in a wide variety of other cultured cells and tissues (GRIMM et al. 1991; BRUDZYNSKI et al. 1992a,b; BRUDZYNSKI 1993; ARIAS et al. 1994; VELEZ-GRANELL et al. 1994, 1995; ITOH et al. 1995; LE GALL and BENDAYAN 1996; SOLTYS and GUPTA 1997) even where the observed extramitochondrial labeling was not discussed (BOOG et al. 1992; LOHSE et al. 1993) or where, in the case of mhsp70, the identity of the antigen was not known (VANBUSKIRK et al. 1991; CARBAJAL et al. 1993).

Recent findings in our laboratory of novel uptake into live cells of antibodies to chaperones (Soltys, Gupta, unpublished data) have shown that antibodies against hsp60 and mhsp70 applied to living B-SC-1 monkey kidney cells,

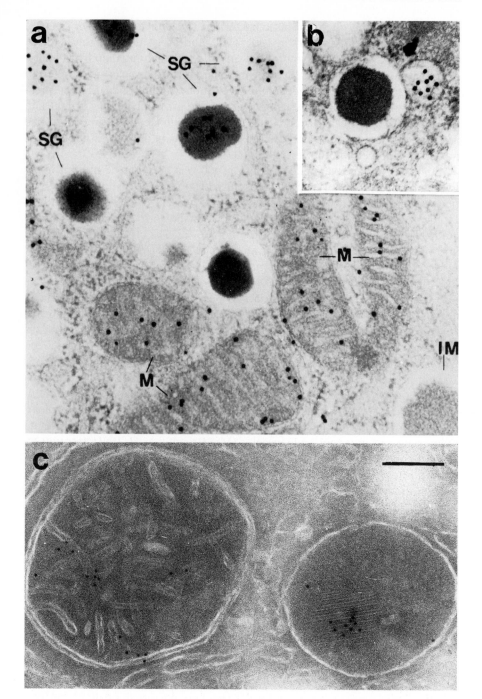

primary human diploid fibroblasts and CHO cells enter into cells and localize along mitochondria. The entry of antibodies was found to be temperature- and energy-independent, indicating entry is not by a classical endocytic mechanism nor via caveolae. The results have provided furthur evidence that hsp60 and mhsp70 are expressed on the cell surface. Furthermore, the surface molecules may be in equilibrium with the intracellular pools of these proteins and are internalized into the cell by a nonclassical mechanism. It is now recognized that proteins may translocate to and from the cell surface by pathways in-dependent of the ER-Golgi secretory pathway and classical endocytosis (SMALHEISER 1996) and this could be the case for these mitochondrial proteins.

II. Consideration of Possible Artifacts

Before we can consider the possible roles of hsp60 and mhsp70 at extramitochondrial sites, we must consider whether or not the localization of these proteins outside mitochondria may be an artifact resulting from adventitious cross-reaction of antibodies with unrelated proteins or from some other trivial possibility. We also need to address whether extramitochondrial labeling can be explained as being simply due to reaction with precursor protein that failed to enter mitochondria.

Hsp60 is nuclear encoded in eukaryotic organisms (JINDAL et al. 1989; PICKETTS et al. 1989; READING et al. 1989) and is synthesized in mammalian cells as a larger precursor form containing an N-terminal presequence (27 a.a.), which is necessary for its mitochondrial import and is cleaved during the maturation process in the mitochondrial matrix (JINDAL et al. 1989; SINGH et al. 1990). A formal possibility that has been considered is that the hsp60 reactivity at extramitochondrial sites represents a reaction with the precursor rather than with the mature protein (SOLTYS and GUPTA 1996). To evaluate this, CHO cells were labeled with (^{35}S) methionine either in the absence or presence of the potassium ionophore nonactin (Fig. 3a). Since treatment of cells with nonactin causes dissipation of the mitochondrial membrane potential, which is required for mitochondrial import and maturation (i.e. cleavage of the presequence) of precursor proteins (NEUPERT 1997), the precursor would accumulate outside of mitochondria. In control cells, immunoprecipitation with hsp60 antibody led to the precipitation of a single protein of 60kDa, which

---

Fig. 2a–c. Immunogold labeling of hsp60 in tissues. **a,b** Mouse pancreatic beta-cells. 20-nm gold markers. In **a** reactivity is within mitochondria (*M*) and the dense core of mature insulin secretory granules (*SG*) but not within immature insulin secretory granules (*IM*). In **b** a putative transport vesicle contains hsp60 reactivity and appears to be fusing with an insulin secretory vesicle (From BRUDZYNSKI et al. 1992a). **c** Rat liver. 10-nm gold markers. Reactivity is within the mitochondrion *on the left* and in the peroxisome *on the right*. In the peroxisome, identified by its single membrane and its striated electron-dense crystalline core, hsp60 reactivity is primarily in the core material. *Bar*, 0.2 μm. (From SOLTYS and GUPTA 1996)

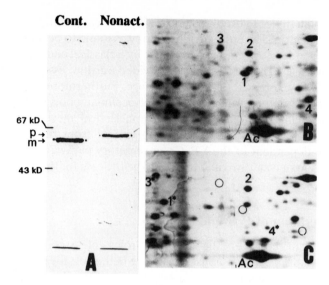

Fig. 3A–C. Mitochondrial import inhibition by nonactin and effect on hsp60 maturation. Control CHO cells or cells pretreated with 10 μg/ml nonactin for 2 h were labeled
with ³⁵S-methionine. **A** Hsp60 from labeled cells was immunoprecipitated using anti-
body-Sepharose beads and analyzed by SDS-PAGE. Fluorogram of the labeled gel is
shown. *Left lane,* control cells: only mature hsp60 is detected. *Right lane,* nonactin
treated cells: only precursor hsp60 is detected. **B,C** 2-D gel pattern of total ³⁵S-labeled
proteins in whole cell extracts. The protein spots marked *1, 2, 3* and *4* in control cells in
B identify the positions of hsp60, cytosolic hsc70, mitochondrial hsp70, and an
uncharacterized mitochondrial protein, respectively. **C** In cells treated with nonactin,
the spots *1, 3* and *4* disappear (their would-be position is *shown by open circles*) and
more basic precursor forms of these proteins (*indicated by asterisks*) are now observed.
Ac, actin spot. (From SOLTYS and GUPTA 1996)

corresponds to the mature hsp60 (identity confirmed by microsequencing). No
precursor form of hsp60 was detected under these conditions indicating
significant amounts are not present in cells under normal conditions. This
suggests the precursor is normally rapidly imported and converted into the
mature form. In contrast to the control cells which lack detectable precursor
hsp60, in cells treated with nonactin only the larger precursor was immunoprecipitated with the hsp60 antibodies. This indicated that the conversion of the precursor to the mature form was completely blocked. These
results also provided evidence that antibodies used in the study do not react
nonspecifically with any proteins, including any other 60 kDa protein that
possibly may have comigrated on the gel with mature hsp60. The total cellular
proteins from the control and nonactin treated cells were also analyzed by 2-
D gel electrophoresis (Fig. 3b). The identity of the numbered spots was earlier
determined by peptide mapping (GUPTA and AUSTIN 1987). Results of these
experiments again revealed that in nonactin treated cells there is no labeling of
the protein spot corresponding to mature hsp60 (or other mitochondrial proteins including mitochondrial hsp70), but that slightly larger, more basic forms

of these proteins, which are absent in control cells, accumulate under these conditions. These results provide evidence that mitochondrial targeting/import of the precursor hsp60 (or at least its presequence) is necessary for its conversion to the mature form and that precursor hsp60 is not present in any significant amounts at extramitochondrial sites.

What is the origin of the extramitochondrial hsp60? Presently, there is no evidence for the existence of more than one hsp60 gene or for alternate splicing of the mRNA for this gene product. These possibilities are also not supported by the results of the nonactin experiment, which clearly shows that upon abolishment of the mitochondrial membrane potential, only the precursor form of hsp60 accumulates in cells. The failure to see any other form of hsp60 in this experiment strongly suggests that the extramitochondrial Hsp60 is derived from the same precursor protein which is imported into mitochondria.

A number of possibilities were considered to explain the extramitochondrial labeling due to hsp60. First, the extramitochondrial labeling could be non-specific. However, this possibility is unlikely because several different monoclonal and polyclonal antibodies all gave similar results. Further, as indicated above, preadsorption of the antibodies with the purified recombinant human hsp60 abolished both mitochondrial as well as extramitochondrial labeling indicating its hsp60 specificity. Second, the extramitochondrial labeling could be due to cross-reactivity with some other antigen. This possibility is also considered unlikely because some of the antibodies employed have been shown previously to react specifically with hsp60 in 1- and 2-D gel blots (GUPTA et al. 1985; GUPTA and DUDANI 1987). They show no cross-reactivity with the cytosolic TCP-1 protein (unpublished results) which is distantly related to the hsp60 family of proteins (GUPTA 1995). In immunoprecipitation experiments these antibodies precipitate only hsp60. In experiments where maturation of hsp60 is inhibited by nonactin treatment, a larger precursor form of hsp60 is precipitated. These results strongly suggest that the only protein with which the antibodies react is mitochondrial hsp60. Third, the possibility that extramitochondrial labeling could be due to reactivity with the precursor form of hsp60 is also excluded by the fact that under normal growth conditions, precursor hsp60 is not detected in cells, indicating that a significant amount of precursor hsp60 is not present in cells. Fourth, the most likely possibility to explain the results of this investigation would be to suggest that although most of the hsp60 is localized in mitochondria, smaller amounts of this protein are indeed also present at other cellular sites.

The reality of extramitochondrial hsp60 has been further strengthened by microsequencing data showing that the identity of the cell surface protein associated with p21^{ras} (IKAWA and WEINBERG 1992), the A-system of amino acid transport (JONES et al. 1994) and the surface protein biotinylated in live Chinese hamster ovary cells (SOLTYS and GUPTA 1996), discussed in the preceding section, is in fact the bona fide mature form of hsp60. Also, microsequencing has established the identity of hsp60 that is phosphorylated in the plasma membrane of the T lymphocyte cell line CEM-SS (Khan et al. 1998).

In the case of mhsp70, one issue was whether or not our polyclonal antibody to mhsp70 also reacted with cytosolic hsp70 homologs. All studies with the mhsp70 antibody were also with affinity purified antibody and the specificity of this antibody was examined by two different experiments (Singh et al. 1997). In one experiment, the reactivity of mhsp70 antibody to purified recombinant mhsp70 and cytosolic hsc70 proteins was compared with a monoclonal antibody to cytosolic hsp70 (which recognizes both the cognate and heat induced forms of the proteins). Results of this experiment clearly showed that both these antibodies reacted only with the respective antigen and they showed no cross-reactivity to the other protein. The second approach involved examining the reactivity of these antibodies in 2-D gel blots of the total cellular proteins form CHO cells. Antibody to the mhsp70 showed no cross-reactivity towards the cytosolic hsp70 s and it reacted in a highly specific manner with only the protein spot corresponding to mhsp70. In immunocytochemistry, the labeling pattern observed with mhsp70 antibody (Singh et al. 1997) is totally distinct from that seen with antibody to cytosolic hsp70 (Ahmad et al. 1990).

Thus, the available data suggests that multicompartmentalization of hsp60 and mhsp70 is a real phenomenon and is not explainable by adventitious cross-reaction of antibodies with unrelated proteins. The results further imply, by consequence, that these molecular chaperones could have multifunctional roles in cell physiology. Mitochondrial molecular chaperones appear to be examples of cellular multicompartmentalization of proteins, an apparently physiological phenomenon of more general significance (Smalheiser 1996).

III. Possible Extramitochondrial Functions

Is hsp60 outside of mitochondria simply an aberrant protein that failed to enter mitochondria or is it involved in physiological functions? Although at present there is no direct evidence for hsp60 function outside of mitochondria, a number of observations suggest that the extramitochondrial hsp60 is likely involved in specific functions. The most suggestive of these observations is the presence of this protein in specialized compartments such as insulin secretory granules (Brudzynski et al. 1992a), zymogen granules (Velez-Granell et al. 1994; J. Chechetto and R.S. Gupta, unpublished) and growth hormone granules (J. Chechetto and R.S. Gupta, unpublished), and also in organelles such as peroxisomes (Velez-Granell et al. 1995; Soltys and Gupta 1996).

In insulin secretory granules, hsp60 was found to be specifically associated with the central insulin core of mature secretory granules, and not in immature granules. The fundamental difference between these two types of granules is that enzymatic conversion of proinsulin to insulin is followed by a poorly understood process of insulin condensation, giving rise to the highly compacted central core of the mature granule (HUTTON 1994). This ability of insulin to form a higher order structure is not shared with the proinsulin contained in immature granules. As such, the condensed insulin core within the insulin secretory granules represents a highly organized, supramolecular structure which serves to secrete functional insulin. Hsp60 could possibly have a chaperone role in core protein condensation – the established role of hsp60 in the formation of oligomeric protein complexes and in bacterial protein secretion is suggestive that the hsp60 within these granules is involved in similar functions.

Hsp60 has also been found in association with the urate oxidase crystalline cores of rat liver peroxisomes. The crystalloids within the peroxisomes, by analogy with mature insulin secretory granules, are also a higher order structure that may require a chaperone for assembly.

Numerous studies in bacteria of GroEL association with membranes have led to the suggestion that hsp60 homologs primitively function in the assembly of membrane proteins, and because of evolutionary conservation, the same may apply in eukaryotic cells. Membrane associations in bacteria has been shown by biochemical and immunochemical methods (GILLIS et al. 1985; KRISHNASAMY et al. 1988; VODKIN and WILLIAMS 1988; JAGER and BERGMAN 1990; JAKOB et al. 1993), flow cytometry (YANAGUCHI et al. 1996) and in a quantitative EM study in gram negative bacteria it was shown that approximately 30% of GroEL is associated with the cell envelope (SCORPIO et al. 1994). The same may apply to eukaryotic cells. In chloroplasts, stromal hsp60 interacts with the import intermediate-asssociated protein IAP100, an integral membrane protein (KESSLER and BLOBEL 1996). In mitochondria, chemical modification of cysteine groups in hsp60 with TFEC [S-(1, 1, 2, 2-tetrafluoroethyl)-L-cysteine] results in hsp60 association with the inner membrane (BRUSCHI et al. 1993; see below). A role for hsp60 in the assembly of membrane proteins, however, still awaits demonstration.

There is strong suggestive evidence that mhsp70 and cpn10 both have functions at extramitochondrial sites. Mitochondrial cpn10 has been identified as being identical to early pregnancy factor (EPF), an important growth factor present in maternal serum (CAVANAGH and MORTON 1994). Mhsp70 has been implicated to function in antigen presentation, having been identified as peptide-binding protein PBP72/74 (DOMANICO et al. 1993; DAHLSEID et al. 1994) and is also involved in cell senescence (WADHWA et al. 1993, 1994). A cell surface role for hsp70 has also been indicated in bacterial systems. For example, hsp70 may possibly function as an adhesion molecule on the cell surface of chlamydiae (RAULSTON et al. 1993), a pathogen which binds to and enters mammalian host cells.

IV. Proposed Transport Mechanisms

How hsp60 and mhsp70 might arrive at extamitochondrial locations is an important and unresolved problem. While there is precedence for primarily cytosolic proteins such as tubulin and actin being found at the cell surface

(SMALHEISER 1996) the challenge to explain hsp60 and mhsp70 presence at the cell surface is that these proteins must first exit or be exported from the mitochondrial matrix compartment, which is where these proteins are primarily present and first targetted to (SOLTYS and GUPTA 1996; SINGH et al. 1997). This is not without precedence. For example, the minor histocompatability antigen MTF is encoded by mitochondrial DNA and is also present on the cell surface (LOVELAND et al. 1990; FISCHER-LINDAHL et al. 1991). Table 1 provides a partial listing of mitochondrial proteins which have known functions outside of mitochondria and must also be exported. Although protein exit from mitochondria may now be coming to be recognized as a reality (POYTON et al. 1992; REED 1997), no transport mechanisms have been established.

There are at least four conceivable possiblities we can propose at present to explain exit of mitochondrially targetted molecular chaperones from mitochondria: (i) reverse operation of the mitochondrial import channel (UNGERMANN et al. 1994); (ii) existence of an unknown export pathway, by analogy with the bacterial-like Sec-type pathway found in chloroplasts (BAKER et al. 1996); (iii) hypothetical movement through lipids; (iv) vesicle-mediated export from mitochondria, by analogy with vesicular export from gram-negative bacteria (KADURUGAMUWA and BEVERIDGE 1995; BEVERIDGE and KADURUGAMUWA 1996). These possibilities are discussed, as follows.

Recent findings concerning mitochondrial import have shown that proteins going through the import channel may reverse their direction of movement once the mitochondrial targeting sequence is cleaved (UNGERMANN et al. 1994). This type of mechanism was first demonstrated in studies of fumarase in yeast and explains how 80–90% of fumarase molecules become destined for the cytosol (STEIN et al. 1994). Following the membrane potential driven transfer of the N-terminal segment of preproteins through the import channel, unless mhsp70 in the matrix compartment binds to the translocating preprotein, the protein can reverse direction as soon as the signal sequence is cleaved by the matrix resident peptidase and exit out of mitochondria (UNGERMANN et al. 1994; PFANNER and MEIJER 1995). Once in the cytosol, mature hsp60 may then rapidly partition into different membranes because of intrinsic membrane binding properties.

Mitochondria, however, are thought to be of endosymbiotic origin (MARGULIS, 1970; GRAY 1992) and another possibility is that mitochondria have retained a bacterial-like secretion system, or Sec-like pathway, from the original endosymbiont genome (BAKER et al. 1996). The original endosymbiont almost certainly had such a pathway and the question is whether it has been lost or is it still there but not yet identified? Chloroplasts are also of endosymbiotic origin and have indeed retained a Sec-pathway – proteins comprising this pathway in chloroplasts are also present in cyanobacteria (BAKER et al. 1996). However, no Sec-type proteins appear to be present in mitochondria of the yeast *Saccharomyces cerevisiae* (GLICK and VON HEIJNE 1996), where the complete genome has been sequenced. It appears imminent, nevertheless, that novel channel forming proteins, whether Sec-like or not, will

Table 1. Mitochondrial (mt) components that function at extramitochondrial locations

Component	Function	Localization	Reference
Early pregnancy factor or cpn10	Co-chaperonin, immunosuppressive factor, growth regulator	Serum, mt-matrix, extramitochondrial site(s) in platelets and regenerating liver	CAVANAGH and MORTON 1994
Fumarase	Enzyme	mt-matrix, cytosol	STEIN et al. 1994
MTF	Murine minor histocompatability antigen	Inner mt-membrane, cell surface, secretory pathway?	LOVELAND et al. 1990; FISCHER-LINDAHL et al. 1991
Aspartate aminotransferase	Fatty acid binding protein, mitochondrial enzyme	mt-matrix, cell surface	ISOLA et al. 1997
Mitochondrial large rRNA (*Drosophila*)	mt-ribosome structure/function, pole-cell forming factor during development	mt-matrix, polar granules in germ plasm	KOBAYASHI et al. 1997
Transforming growth factor-beta, (TGF-1)	Multiple effects: may regulate oncogenes, growth factor receptors and other cell surface	mt-matrix, cell surface	SPORN 1991
Cytochrome c	proteins apoptosis, electron transport protein	mt-intermembrane space, cytosol, nucleus?	YANG et al. 1997
Apoptosis-inducing factor	Apoptosis, protease	mt-intermembrane space, cytosol, nucleus?	SUSIN et al. 1996

soon be characterized in mammalian mitochondria. "Megachannel" formation is known to occur in the inner mitochondrial membrane during the phenomenon known as the mitochondrial permeability transition (REED 1997). The opening of this channel occurs almost universally during apoptosis, correlates with the dissipation of the mitochondrial membrane potential, and is associated with release of certain mitochondrial matrix proteins (IGBAVBOA et al. 1989) and of the intermembrane proteins cytochrome c (YANG et al. 1997) and the protease referred to as "apoptosis-inducing factor" (AIF) (SUSIN et al. 1996), the latter two of which induce apoptosis. Although such channels are not currently thought to form in non-apoptotic cells, transient megapore openings may be difficult to rule out. There is no evidence at present that work in this area is relevant to mitochondrial molecular chaperones.

A third possibility is that hsp60 and mhsp70 exit mitochondria by some form of direct movement through lipids. Unexplained crossing of the plasma membrane has been observed for certain cytosolic heat shock proteins, particularly hsp110, hsp71 and hsc73 (HIGHTOWER and GUIDON 1989) and externally applied antibodies to hsp60 and mhsp70 enter cells by an unconventional pathway (Soltys and Gupta, unpublished data). Homeoprotein entry into cells also follows an unconventional pathway that appears to be receptor-independent and not by endocytosis (PROCHIANTZ and THEODORE 1995; DEROSSI et al. 1996). It has been proposed that inverted micelle formation, involving the formation of a hydrophilic cavity at the cell surface that accomodates homeoproteins, is the most reasonable mechanism for explaining homeoprotein uptake (DEROSSI et al. 1996). This third possibility should be testable using model membranes. Since unfolded or partially folded precursor proteins enter mitochondria via putative import channels, the ability to cross a lipid bilayer directly may be restricted to the folded mature protein.

The facts that challenge the above three possibilities are that (i) hsp60 and mhsp70 outside mitochondria are found at very specific destinations (e.g. mature insulin secretory granules, cell surface, peroxisomes) and not everywhere as might be expected for a free cytosolic protein that partitions nonspecifically, (ii) the existence of mitochondrial molecular chaperones as free proteins in the cytosol is seriously questioned because hsp60, mhsp70 and cpn10 are not normally detected biochemically as free proteins in cytosolic fractions of mammalian cells (100000 × g supernatants; R.S. Gupta, unpublished) and (iii) the existence as free cytosolic proteins is also questioned by EM studies in which these proteins are always associated with membranous extramitochondrial components or electron-dense cytoplasmic granules. However, a very low concentration of hsp60 in the cytosolic compartment could in principle result from very rapid transit or cycling of hsp60.

Given the endosymbiotic origin of mitochondria from gram negative bacteria (GRAY 1989) and the fact that gram negative bacteria export certain proteins by means of membrane vesicle formation (KADURUGAMUWA and

BEVERIDGE 1995; BEVERIDGE and KADURUGAMUWA 1996), a fourth possibility for transport out of mitochondria is a vesicle-mediated process involving vesicle budding. This phenomenon in bacteria has been extensively studied by BEVERIDGE and coworkers (KADURUGAMUWA and BEVERIDGE 1995; BEVERIDGE and KADURUGAMUWA 1996), who have found it to be a general phenomenon in gram negative bacteria. Such shed vesicles qualify as transport vesicles because they also fuse with other bacteria. Both mitochondria and gram negative bacteria have two membranes, so a double-membraned structure is not nessarily a limitation to vesicle formation by mitochondria.

Figure 4a shows the submitochondrial distribution of Hsp60 in a cryosection of B-SC-1 cells. Both the outer and inner mitochondrial membranes are clearly seen and well preserved. Much of the immunogold labeling due to Hsp60 is observed to be in the matrix compartment. However, there is additional labeling on certain regions which lie on the cytoplasmic side of the mitochondrial outer membrane. Two examples of "granular-shaped" labeling on the cytoplasmic face of mitochondria are indicated by arrows. Despite the good preservation of membrane in these specimens, no membrane is seen around the cytoplasmic granules that are marked by arrows. However, preservation of endomembranes for EM studies is a common problem (SOLTYS et al. 1996) and the possibly that these granules are membrane limited cannot be ruled out. Figure 4b shows an example of budding of membrane vesicles from the gram negative bacterium *Pseudomonas aeruginosa* (KADURUGAMUWA and BEVERIDGE 1995). Such shed vesicles are able to fuse with other bacteria and

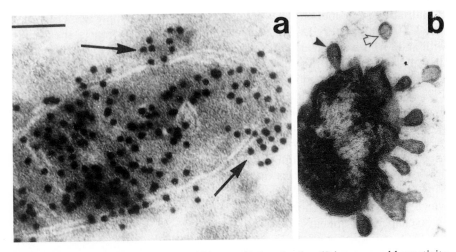

Fig. 4. a Mitochondrion in B-SC-1 kidney cell showing hsp60 immunogold reactivity both on the outer cytoplasmic face of the outer mitochondrial membrane (*arrows point to* granule-like reactivity) suggestive of hsp60 exit from mitochondria, and within the matrix compartment. 10-nm gold markers. *Bar,* 0.1 μm (From SOLTYS and GUPTA 1996). **b** Budding and release of membrane vesicles from the gram negative bacterium *Pseudomonas aeruginosa. Bar,* 0.1 μm. (From KADURUGAMUWA and BEVERIDGE 1995)

may also be referred to as transport veiscles. Apparent transport vesicles containing hsp60 have been observed in electron micrographs of pancreatic beta-cells in the process of fusion with insulin secretory vesicles, as shown above in Fig. 2b. This mechanism of transport could readily explain why hsp60 and mhsp70 are not found free in the cytosolic compartment and instead cosediment with membrane fractions, but remains challenged by the lack of direct evidence for budding of vesicles from mitochondria. The available data challenges but does not refute any of the four proposed mechanisms to explain chaperone export from mitochondria.

How might hsp60 and mhsp70, both highly soluble proteins, remain membrane-associated at the cell surface or elsewhere? Although much further work is needed on this topic, GroEL is capable of direct membrane insertion in model lipid bilayers (Torok et al. 1997). Insertion is dependent on the last 16 amino acids at the C-terminal end of GroEL. The insertion of GroEL alters the physical properties of the bilayer, increasing membrane order. The inserted oligomer also retains chaperonin activity in the membrane-associated state (Torok et al. 1997). The C-terminal end motif thought to be involved in membrane insertion is found throughout evolution. Thus the mechanistic basis for membrane association of GroEL (Torok et al. 1997) is probably retained in mammalian hsp60, although this will need to be directly shown. A second possible means of membrane insertion, as yet untested, could be a postranslational modification such as palmitoylation. Palmitoylation has been shown to allow alpha-tubulin, also a highly soluble protein, to become membrane-associated (Caron 1997).

Certain drug treatments cause relocalization of hsp60 to different mitochondrial compartments. The halogenated cysteine conjugate TFEC [S-(1,1,2,2-tetrafluoroethyl)-L-cysteine] administered to rats reacts specifically with and forms a conjugate with both hsp60 and mhsp70 in kidney mitochondria and the result of this modification is that 60% of hsp60 is found in the intermembrane space and 20% in the inner membrane fraction (Bruschi et al. 1993). Since hsp60 is considered to be a soluble protein in the matrix compartment, one explanation is that specific modifications can induce the association of hsp60 with mitochondrial membranes and exit from the matrix compartment. A second possibility is that this modification reaction actually traps hsp60 molecules in transit out of mitochondria. It will be of great interest to also examine what effect this drug treatment has upon expression of both hsp60 and mhsp70 at extramitochondrial sites. Further discovery of drugs which affect extramitochondrial distributions and the putative functions of these chaperones appears possible.

D. Hsp60 in Drug Resistance and Disease

Heat shock protein expression on the cell surface greatly simplifies the challenge in trying to understand the links between hsp60, drug resistance, cellular and humoral immunological function, and disease etiology.

Hsp60 was originally characterized in mammalian cells as a major protein, referred to as P1 in the first studies, which is modified in CHO cells resistant to the anti-tubulin drug podophyllotoxin (GUPTA et al. 1982, 1985). Subsequent cDNA cloning established its identity as the mammalian homolog of GroEL (JINDAL et al. 1989). How can a mutation in hsp60 lead to resistance to a drug that binds specifically to tubulin? The presence of both hsp60 and tubulin on the cell surface may hold the key to the answer.

The earlier work on these mutants, which represent the only available mammalian cell mutants with a specific mutation in hsp60, have been reviewed elsewhere (GUPTA 1990b).The resistance phenotype of these cells indcated they were not multidrug resistant mutants affected in P-glycoprotein (GUPTA 1990b). Furthermore, in biochemical experiments hsp60 was found to be co-released with tubulin from cellular fractions under certain conditions, suggesting that these two proteins interact with each other in the cell (GUPTA et al. 1982). However, the site(s) of this interaction was not determined. Since tubulin is not present within mitochondia, the results were an enigma (GUPTA et al. 1985; GUPTA 1990b). Since recent data reviewed here has provided strong evidence for the presence of a subpopulation of hsp60 on the cell surface and because tubulin is also known to be present on the cell surface (SMALHEISER 1996), we posit that the interaction between hsp60 and tubulin likely occurs at the cell surface where these two proteins may exist as a complex. This hypothesis has been strongly supported by our recent unpublished drug binding studies in live Chinese hamster ovary hsp60 mutant cells using ^{3}H-podophyllotoxin. These studies show that podophyllotoxin binding is greatly decreased in the mutant cells at drug concentrations at which the mutants are resistant and suggest that mutated hsp60 causes decreased drug-tubulin interaction. This reduced binding is specific for competitive anti-tubulin drugs only. Reduced podophyllotoxin binding is also found in cell extracts of these cells. Since the mutant cells do not take up the drug, both the receptor for these drugs, which is unpolymerized tubulin, and hsp60 are likely to be located at the cell surface and molecular changes in hsp60 can lead to alterations in drug binding to tubulin in the complex.

Hsp60 has been implicated in the development of autoimmune diseases (KAUFMANN 1992) including arthritis (VAN EDEN 1991), diabetes (COHEN 1991), multiple sclerosis (RAINE et al. 1996), Lyme disease (SIGAL 1997), Behçet disease (1997) and others. GroEL is also known as the bacterial common antigen. Current thought is that following the immune response to bacterial or parasitic infections, molecular mimicry due to the high sequence conservation of hsp60 homologs throughout evolution may be responsible for the development of an immune response directed against host hsp60. T cells reactive to self epitopes of hsp60 and antibodies to conserved regions of hsp60 have been detected in these diseases. However, ancilliary factors may also be involved in disease progression. Immunization using hsp60 produces the opposite of what one might expect – immunization may actually protect against these diseases rather than inducing them, as shown for adjuvant-induced arthritis (Billingham et al. 1990) and insulin-dependent diabetes (Elias et al. 1990, 1997).

Heat shock proteins have been implicated in antigen presentation (Suto and Srivastava 1995). The identification of the peptide binding protein PBP72/74 as mhsp70 (Domanico et al. 1993; Dahlseid et al. 1994), as discussed above, is very strong evidence for the physiological role of mhsp70 in this process. There is also strong evidence for the involvement of hsp60 in peptide presentation. In the case of MHC class 1 peptide presentation, transfection of B16 melanoma cells with mycobacterial hsp60 increases the amount of antigenic peptides at the cell surface and these cells are effectively lysed by cytotoxic T lymphocytes (Wells et al. 1997). In other studies, tumor cells transfected with mycobacterial hsp60 lose tumorigenicity and immunization with these cells enhances immunogenicity (Lukacs et al. 1993, 1997). Also, mice immunized with hsp60-transfected B16 melanoma cells are highly resistant to subsequent challenge with wild-type B16 cells (Wells et al. 1997). Although these studies used transfection with mycobacterial hsp60, similar responses might be expected using mammalian hsp60. The use of GroEL transfection of tumor cells as a means to treat cancer by unmasking tumor antigens for attack by the immune system (Lukacs et al. 1993, 1997) and the use of hsp60 protein therapies in autoimmune diseases (Elias et al. 1990; Elias and Cohen 1994; van Roon et al. 1997) are vivid examples of the potential roles these molecular chaperones may serve at extramitochondrial locations.

Hsp60 has also been implicated in other pathological states including cystic fibrosis (Polla et al. 1995), artherosclerosis (Xu and Wick 1996), prion diseases (encephalopathies; Kenward et al. 1996), aging and stroke (Mantle et al. 1995), myocardial infarction (Hoppichler et al. 1996), apoptosis (Poccia et al. 1996), while mhsp70 as mortalin has also been implicated in cell senescence (Wadhwa et al. 1994). Further insights into mitochondrial molecular chaperone trafficking and functions in the cell should prove very important to understanding these and other pathological conditions.

E. Future Prospects

A central implication of the subcellular localization studies, and the cumulative evidence pointing to an involvement of mitochondrial molecular chaperones in diverse cellular processes, is that specific mechanisms must exist for the export of these proteins from mitochondria and that these proteins also likely have important functions at specific extramitochondrial sites, particularly the cell surface. Mitochondrial molecular chaperones may prove to be one of the premier examples of cellular multicompartmentalization of proteins, the importance and extent of which as a more general phenomenon is now being recognized (Smalheiser 1996). There may also be trafficking of these mitochondrial proteins to and from the cell surface. Characterization of this trafficking pathway and the role of these molecular chaperones at extramitochondrial sites, in normal and pathological states, will be of great interest.

Acknowledgement. Work in the authors' laboratory is funded by a grant from the Medical Research Council of Canada.

References

Ahmad S, Ahuja R, Venner TJ, Gupta RS (1990) Identification of a protein altered in mutants resistant to microtubule inhibitors as a member of the major heat shock protein (hsp70) family. Mol Cell Biol 10:5160–5165

Arias AE, Velez-Granell CS, Torres-Ruíz JA, Bendayan M (1994) Involvement of molecular chaperones in the aberrant aggregation of secretory proteins in pancreatic acinar cells. Exp Cell Res 215:1–8

Ayling A, Baneyx F (1996) Influence of the GroE molecular chaperone machine on the in vitro refolding of Escherichia coli beta-galactosidase. Protein Sci 5:478–487

Baker A, Kaplan CP, Pool MR (1996) Protein targeting and translocation; a comparative survey. Biol Rev Camb Philos Soc 71:637–702

Beveridge TJ, Kadurugamuwa JL (1996) Periplasm. periplasmic spaces, and their relation to bacterial wall structure:novel secretion of selected periplasmic proteins from Pseudomonas aeruginosa. Microb Drug Resist 2:1–7

Billingham ME, Carney S, Butler R, Colston MJ (1990) A mycobacterial 65-kD heat shock protein induces antigen-specific suppression of adjuvant arthritis, but is not itself arthitogenic. J Exp Med 171:339–344

Bochkareva ES, Lissin NM, Girshovich AS (1988) Transient association of newly synthesized unfolded proteins with the heat-shock GroEL protein. Nature 336:254–257

Boog CJ, De Graeff-Meeder ER, Lucassen MA, van der Zee R, Voorhorst-Ogink MM, Kooten PJS, Geuze HJ, van Eden W (1992) Two monoclonal antibodies generated against human hsp60 show reactivity with synovial membranes of patients with juvenile chronic arthritis. J Exp Med 175:1805–1810

Braig K, Otwinowski Z, Hegde R, Boisvert DC, Joachimiak A, Horwich AL, Sigler PB (1994) The crystal structure of the bacterial chaperonin GroEL at 2.8 A. Nature 371:578–586

Brudzynski K, Martinez V, Gupta RS (1992a) Immunocytochemical localization of heat-shock protein 60-related protein in beta-cell secretory granules and its altered distribution in non-obese diabetic mice. Diabetologia 35:316–324

Brudzynski K, Martinez V, Gupta RS (1992b) Secretory granule autoantigen in insulin-dependent diabetes mellitus is related to 62 kDa heat-shock protein (hsp60). J Autoimmun 5:453–463

Brudzynski K (1993) Insulitis-caused redistribution of heat-shock protein HSP60 inside beta-cells correlates with induction of HSP60 autoantibodies. Diabetes 42:908–913

Bruschi SA, West KA, Crabb JW, Gupta RS, Stevens JL (1993) Mitochondrial HSP60 (P1 protein) and a HSP70-like protein (mortalin) are major targets for modification during S-(1,1,2,2-tetrafluoroethyl)-L-cysteine-induced nephrotoxicity. J Biol Chem 268:23157–23161

Buckle AM, Zahn R, Fersht AR (1997) A structural model for GroEL-polypeptide recognition. Proc Natl Acad Sci USA 94:3571–3575

Carbajal EM, Beaulieu J-F, Nicole LM, Tanguay RM (1993) Intramitochondrial localization of the major 70-kDa heat-shock cognate protein in Drosophila cells. Exp Cell Res 207:300–309

Caron JM (1997) Posttranslational modification of tubulin by palmitoylation: I. In vivo and cell-free studies. Mol Biol Cell 8:621–636

Cavanagh AC, Morton H (1994) The purification of early-pregnancy factor to homogeneity from human platelets and identification as chaperonin 10. Eur J Biochem 222:551–560

Chen S, Roseman AM, Hunter AS, Wood SP, Burston SG, Ranson NA, Clarke AR, Saibil HR (1994) Location of a folding protein and shape changes in GroEL-GroES complexes imaged by cryo-electron microscopy. Nature 371:261–264

Cheng MY, Hartl FU, Martin J, Pollock RA, Kalousek F, Neupert W, Hallberg EM, Hallberg RL, Horwich AL (1989) Mitochondrial heat-shock protein hsp60 is essential for assembly of proteins imported into yeast mitochondria. Nature 337:620–625

Cheng MY, Hartl FU, Horwich AL (1990) The mitochondrial chaperonin hsp60 is required for its own assembly. Nature 348:455–458

Cohen IR (1991) Autoimmunity to chaperonins in the pathogenesis of arthritis and diabetes. Annu Rev Immunol 9:567–589

Craig EA, Gambill BD, Nelson RJ (1993) Heat shock proteins: molecular chaperones of protein biogenesis. Microbiol Rev 57:402–414

Dahlseid JN, Lill R, Green JM, Xu X, Qiu Y, Pierce SK (1994) PBP74, a new member of the mammalian 70-kDa heat shock prtein family, is a mitochondrial protein. Mol Biol Cell 5:1265–1275

Derossi D, Calvet S, Trembleau A, Brunissen A, Chaassaing G, Prochiantz A (1996) Cell internalization of the third helix of the antennapedia homeodomain is receptor-independent. J Biol Chem 271:18188–18193

Deshaies RJ, Koch BD, Werner-Washburne M, Craig EA, Schekman R (1988) A subfamily of stress proteins facilitates translocation of secretory and mitochondrial precursor polypeptides. Nature 332:800–810

Domanico SZ, DeNagel DC, Dahlseid JN, Green JM, Pierce SK (1993) Cloning of the gene encoding peptide-binding protein 74 shows that it is a new member of the heat shock protein 70 family. Mol Cell Biol 13:3598–3610

Elias D, Markovits D, Reshef T, van der Zee R, Cohen IR (1990) Induction and therapy of autoimmune diabetes in the non-obese diabetic (NOD/Lt) mouse by a 65-kDa heat shock protein. Proc Natl Acad Sci USA 87:1576–1580

Elias D, Meilin A, Ablamunits V, Birk OS, Carmi P, Könen-Waisman S, Cohen IR (1997) Hsp60 peptide therapy of NOD mouse diabetes induces a Th2 cytokine burst and downregulates autoimmunity to various <b>-cell antigens. Diabetes 46:758–764

Elias D, Cohen IR (1994) Peptide therapy for diabetes in NOD mice. Lancet 343:704–706

Ellis RJ (1990) Chaperone function: the assembly of ribulose bisphosphate carboxylase-oxygenase. Annu Rev Cell Biol 6:125–149

Ellis RJ, Hartl F-U (1996) Protein folding in the cell:competing models of chaperonin function. FASEB J 10:20–26

Ellis RJ, van der Vies SM (1991) Molecular chaperones. Annu Rev Biochem 60:321–347

Fahimi HD, Baumgart E, Volkl A (1993) Ultrastructural spects of the biogenesis of peroxisomes in rat liver. Biochimie 75:201–208

Fisch P, Malkovsky M, Kovats S, Sturm E, Braakman E, Klein BS, Voss SD, Morrissey LW, DeMars R, Welch WJ, Bolhuis RLH, Sondel PM (1990) Recognition by Human V gamma 9/V delta 2 T cells of a GroEL Homolog on Daudi Burkitt's Lymphoma Cells. Science 250:1269–1273

Fischer-Lindahl KE, Hermel BE, Loveland BE, Wang C-R (1991) Maternally trabnsmitted antigen of mice: a model transplantation antigen. Annu Rev Immunol 9:351–372

Fitzgerald M, Keast D (1994) Fab fragments from the monoclonal antibody ML30 bind to treated human myeloid leukemia cells. FASEB J 8:259–261

Flaherty KM, DeLuca-Flaherty C, McKay DB (1990) Three-dimensional structure of the ATPase fragment of a 70K heat-shock cognate protein. Nature 346:623–628

Gambill BD, Voos W, Kang PJ, Miao B, Langer T, Craig EA, Pfanner N (1993) A dual role for mitochondrial heat shock protein 70 in membrane translocation of preproteins. J Cell Biol 123:109–117

Georgopoulos C, Liberek K, Zylicz M, Ang D (1994) Properties of the heat shock proteins of Escherichia coli and the autoregulation of the heat shock response. In: Morimoto RI, Tissieres A, Georgopoulos C (eds) The biology of heat shock proteins and molecular chaperones. Cold Spring Harbor Laboratory Press, Cold Spring Harbor, NY, pp 209–249

Gething MJ, Sambrook J (1992) Protein folding in the cell. Nature 355:33–45

Gillis TP, Miller RA, Young DB, Khanolkar SR, Buchanan TM (1985) Immunochemical characterization of a protein associated with mycrobacterium leprae cell wall. Infect Immun 49:371–377

Glick BS, von Heijne G (1996) Saccharomyces cerevisiae mitochondria lack a bacterial-type sec machinery. Protein Sci 5:2651–2652

Gray MW (1992) The endosymbiont hypothesis revisited. Int Rev Cytol 141:233–357

Grimm R, Speth V, Gatenby AA, Schafer E (1991) GroEL-related molecular chaperones are present in the cytosol of oat cells. FEBS 286:155–158

Gupta RS, Ho TK, Moffat MR, Gupta R (1982) Podophyllotoxin-resistant mutants of Chinese hamster ovary cells. Alteration in a microtubule-associated protein. J Biol Chem 257:1071–1078

Gupta RS, Venner TJ, Chopra A (1985) Genetic and biochemical studies with mutants of mammalian cells affected in microtubule-related proteins other than tubulin: mitochondrial localization of a microtubule-related protein. Can J Biochem Cell Biol 63:489–502

Gupta RS (1990a) Sequence and structural homology between a mouse t-complex protein TCP-and the 'chaperonin' family of bacterial (GroEL, 60-kDa heat shock antigen) and eukaryotic proteins. Biochem Intl 20:833–841

Gupta RS (1990b) Mitochondria, molecular chaperone proteins and the in vivo assembly of microtubules. Trends Biochem Sci 15:415–418

Gupta RS (1995) Evolution of the chaperonin families (hsp60, hsp10 and Tcp-) of proteins and the origin of eukaryotic cells. Mol Microbiol 15:1–11

Gupta RS, Austin RC (1987) Mitochondrial matrix localization of a protein altered in mutants resistant to the microtubule inhibitor podophyllotoxin. Eur J Cell Biol 45:170–176

Gupta RS, Dudani AK (1987) Mitochondrial binding of a protein affected in a mutant resistant to the microtubule inhibitor podophyllotoxin. Eur J Cell Biol 44:278–285

Gupta RS, Golding GB (1993) Evolution of HSP70 gene and its implications regarding relationships between archaebacteria, eubacteria, and eukaryotes. J Mol Evol 37:573–582

Hallberg EM, Shu Y, Hallberg RL (1993) Loss of mitochondrial hsp60 function: nonequivalent effects on matrix-targeted and intermembrane-targeted proteins. Mol Cell Biol 13:3050–3057

Hartl FU (1996) Molecular chaperones in cellular protein folding. Nature 381:571–580

Hayes SA, Dice JF (1996) Roles of molecular chaperones in protein degradation. J Cell Biol 132:255–258

Hightower LE, Guidon PT (1989) Selective release from cultures mammalian cells of heat-shock (stress) proteins that resemble glia-axon transfer proteins. J Cell Physiol 138:257–266

Höhfeld J, Hartl Fu (1994) Role of the chaperonin cofactor hsp10 in protein folding and sorting in yeast mitochondria. J Cell Biol 126:305–315

Hoppichler F, Lechleitner M, Traweger C, Schett G, Dzien A, Sturm W, Xu QB (1996) Changes of serum antibodies to heat-shock protein 65 in coronary heart disease and acute myocardial infarction. Atherosclerosis 126:333–338

Horwich AL, Low KB, Fenton WA, Hirshfield IN, Furtak K (1993) Folding in vivo of bacterial cytoplasmic proteins: role of GroEL. Cell 74:909–917

Hunt JF, Weaver AJ, Landry SJ, Gierasch L, Deisenhofer J (1996) The crystal structure of the GroES co-chaperonin at 2.8 A resolution. Nature 379:37–45
Hutton JC (1994) Insulin secretory granule biogenesis and the proinsulin-processing
  endopeptidases. Diabetologia 37:S48–S56
Igbavboa U, Zwizinski CW, Pfeiffer DR (1989) Release of mitochondrial matrix proteins through a Ca++-requiring, cyclosporin-sensitive pathway. Biochem Biophys
  Res Commun 161:619–625
Ikawa S, Weinberg RA (1992) An interaction between p21ras aand heat shock protein
  hsp60, a chaperonin. Proc Natl Acad Sci USA 89:2012–2016
Isola LM, Zhou S-L, Kiang C-L, Stump DD, Bradbury MW, Berk PD (1997) 3T3
  fibroblasts transfected with a cDNA for mitochondrial aspartate aminotransferase
  express plasma membrane fatty acid-binding protein and saturable fatty acid
  uptake. Proc Natl Acad Sci USA 92:9866–9870
Itoh H, Kobayashi R, Wakui H, Komatsuda A, Ohtani H, Miura AB, Otaka M,
  Masamune O, Andoh H, Koyama K, Sato Y, Tashima Y (1995) Mammalian 60-
  kDa stress protein (chaperonin homolog). Identification, biochemical properties
  and localization. J Biol Chem 270:13429–13435
Jackson GS, Staniforth RA, Halsall DJ, Atkinson T, Holbrook JJ, Clarke AR, Burston
  SG (1993) Binding and hydrolysis of nucleotides in the chaperonin catalytic cycle:
  implications for the mechanism of assisted protein folding. Biochem 32:2554–2563
Jager KM, Bergman B (1990) Localization of a multifunctional chaperonin (GroEL
  protein) in nitrogen-fixing Anabaena PCC 7120. Planta 183:120–125
Jakob U, Gaestel M, Engel K, Buchner J (1993) Small heat shock proteins are molecular chaperones. J Biol Chem 268:1517–1520
Jindal S, Dudani AK, Singh B, Harley CB, Gupta RS (1989) Primary structure of a
  human mitochondrial protein homologous to the bacterial and plant chaperonins
  and to the 65-kilodalton mycobacterial antigen. Mol Cell Biol 9:2279–2283
Jones M, Gupta RS, Englesberg E (1994) Enhancement in amount of P1 (hsp60) in
  mutants of Chinese hamster ovary (CHO-K1) cells exhibiting increases in the A
  system of amino acid transport. Proc Natl Acad Sci USA 91:858–862
Kadurugamuwa JL, Beveridge TJ (1995) Virulence factors are released from
  Pseudomonas aeruginosa in association with membrane vesicles during normal
  growth and exposure to gentamicin: a novel mechanism for enzyme secretion.
  J Bacteriol 177:3998–4008
Kang PJ, Ostermann J, Shilling J, Neupert W, Craig EA, Pfanner N (1990) Requirement for hsp70 in the mitochondrial matrix for translocation and folding of precur-
  sor proteins. Nature 348:137–143
Kaufmann SHE (1992) Heat shock proteins and the immune response. Immunol
  Today
Kaur I, Voss SD, Gupta RS, Schell K, Fisch P, Sondel PM (1993) Human peripheral
  gamma delta T cells recognize hsp60 molecules on Daudi Burkitt's lymphoma
  cells. J Immunol 150:2046–2055
Kenward N, Landon M, Laszlo L, Mayer RJ (1996) Heat shock proteins, molecular
  chaperones and the prion encephalopathies. Cell Stress Chaperones 1:18–22
Kessler F, Blobel G (1996) Interaction of the protein import and folding machineries of
  the chloroplast. Proc Natl Acad Sci USA 93:7684–7689
Khan IU, Wallin R, Gupta RS, Kammer GM (1998) Protein kinase A-catalyzed phosphorylation of heat shock protein 60 chaperone regulates its attachment
  to histore 2B in the T lymphocyte plasma membrane Proc Natl Acad Sci USA
  95:10425–10430
Kobayashi S, Amikura R, Okada M (1997) Presence of mitochondrial large ribosomal
  RNA outside mitochondria in germ plasm of Drsophila melanogaster. Science
  260:1521–1524
Koga T, Wand-Wurttenberger A, DeBruyn J, Munk ME, Schoel B, Kaufmann SHE
  (1989) T cells against a bacterial heat shock protein recognize stressed macrophages. Science 245:1112–1115

Koll H, Guiard B, Rassow J, Ostermann J, Horwich AL, Neupert W, Hartl F-U (1992) Antifolding activity of hsp60 couples protein import into the mitochondrial matrix with export to the intermembrane space. Cell 68:1163–1175

Krishnasamy S, Mannan RM, Krishnan M, Gnanam A (1988) Heat shock response of the chloroplast genome in Vigna sinensis. J Biol Chem 263:5104–5109

Le Gall IM, Bendayan M (1996) Possible association of chaperonin 60 with secretory proteins in pancreatic acinar cells. J Histochem Cytochem 44:743–749

Lehner T (1997) The role of heat shock protein, microbial and autoimmune agents in the aetiology of Behcet's disease. Int Rev Immunol 14:21–32

Lohse AW, Dienes HP, Herkel J, Hermann E, van Eden W, zum Buschenfelde K-HM (1993) Expression of the 60 kDa heat shock protein in normal and inflamed liver. J Hepatol 19:159–166

Loveland B, Wang C-R, Yonekawa H, Hermel E, Lindahl KF (1990) Maternally transmitted histocompatibility antigen of mice: a hydrophobic peptide of a mitochondrially encoded protein. Cell 60:971–980

Lukacs KV, Lowrie DB, Stokes RW, Colston MJ (1993) Tumor cells transfected with a bacterial heat-shock gene lose tumorigenicity and induce protection against tumors. J Exp Med 178:343–348

Lukacs KV, Nakakes A, Atkins CJ, Lowrie DB, Colston MJ (1997) In vivo gene therapy of malignant tumors with heat shock protein-gene. Gene Ther 4:346–350

Mande SC, Mehra V, Bloom BR, Hol WG (1996) Structure of the heat shock protein chaperonin-of Mycobacterium leprae. Science 271:203–207

Manning-Krieg UC, Scherer PE, Schatz G (1991) Sequential action of mitochondrial chaperones in protein import into the matrix. EMBO J 10:3273–3280

Mantle R, Singh B, Hachinski V (1995) Do serum antibodies to heat-shock protein 65 relate to age or stroke? Lancet 346:8991–8992

Margulis L (1970) Origin of eukaryotic cells. Yale Univ. Press, New Haven, CT.

Neupert W (1997) Protein import into mitochondria. Annu Rev Biochem 66:863–917

Osterman J, Horwich AL, Neupert W, Hartl F (1989) Protein folding in mitochondria requires complex formation with hsp60 and ATP hydolysis. Nature 341:125–130

Pfanner N, Meijer M (1995) Pulling in the proteins. Curr Biol 5:132–135

Pfanner N, Meijer M (1997) Mitochondrial biogenesis: the Tom and Tim machine. Curr Biol 7:R100–R103

Picketts DJ, Mayanil CS, Gupta RS (1989) Molecular cloning of a Chinese hamster mitochondrial protein related to the "chaperonin" family of bacterial and plant proteins. J Biol Chem 264:12001–12008

Poccia F, Piselli P, Vendetti S, Bach S, Amendola A, Placido R, Colizzi V (1996) Heat-shock protein expression on the membrane of T cells undergoing apoptosis. Immunol 88:6–12

Polla BS, Mariethoz E, Hubert D, Barazzone C (1995) Heat-shock proteins in host-pathogen interactions: implications for cystic fibrosis. Trends Microbiol 3:392–396

Poyton R, Duhl DMJ, Clarkson GHD (1992) Protein export from the mitochondrial matrix. Trends Cell Biol 2:369–375

Prochiantz A, Theodore L (1995) Nuclear/growth factors. Bioessays 17:39–44

Raine CS, Wu E, Ivanyi J, Katz D, Brosman CF (1996) Multiple sclerosis: a protective or a pathogenic role for heat shock protein in the central nervous system? Lab Invest 75:109–123

Raulston JE, Davis CH, Schmiel DH, Morgan MW, Wyrick PB (1993) Molecular characterization and outer membrane association of a Chlamydia trachomatis protein related to the hsp70 family of proteins. J Biol Chem 268:23139–23147

Reading DS, Hallberg RL, Myers AM (1989) Characterization of the yeast HSP60 gene coding for a mitochondrial assembly factor. Nature 337:655–659

Reed JC (1997) Double identity for proteins of the Bcl-family. Nature 387:773–776

Retzlaff C, Yamamoto Y, Okubo S, Hoffman PS, Friedman H, Klein TW (1996) Legionella pneumophila heat-shock protein-induced increase of interleukin-beta mRNA involves protein kinase C signalling in macrophages. Immunol 89:281–288

Roseman AM, Chen S, White H, Braig K, Saibil HR (1996) The chaperonin ATPase cycle:mechanism of allosteric switching and movements of substrate-binding domains in groEL. Cell 87:241–251

Schett G, Xu Q, Amberger A, van der Zee R, Recheis H, Wick G (1995) Autoantibodies against heat shock protein 60 mediate endothelial cytotoxicity. J Clin Invest 96:2569–2577

Schröder H, Langer T, Hartl F-U, Bukau B (1993) DnaK, DnaJ and GrpE form a cellular chaperone machinery capable of repairing heat-induced protein damage. EMBO J 12:4137–4144

Scorpio A, Johnson P, Laquerre A, Nelson DR (1994) Subcellular localization and chaperone activities of Borrelia burgdorferi Hsp60 and Hsp70. J Bacteriol 176:6449–6456

Selin LK, Stewart S, Shen C, Mao HQ, Wilkins JA (1992) Reactivity of gamma delta T cells induced by the tumor cell line RPMI 8226: functional heterogeneity of clonal populations and role of GroEL heat shock proteins. Scand J Immunol 36:107–117

Sherman MY, Goldberg AL (1992) Heat shock in Escherichia coli alters the protein-binding properties of the chaperonin groEL by inducing its phosphorylation. Nature 357:167–169

Shu Y, Yang H, Hallberg E, Hallberg R (1997) Molecular genetic analysis of Rts1p, a B' regulatory subunit of Saccharomyces cerevisiae protein phosphatase 2 A. Mol Cell Biol 17:3242–3253

Shu Y, Hallberg RL (1995) SCS1, a multicopy suppressor of hsp60-ts mutant alleles, does not encode a mitochondrially targeted protein. Mol Cell Biol 15:5618–56266

Sigal LH (1997) Lyme disease: a review of its immunology and immunopathogenesis. Annu Rev Immunol 15:63–92

Singh B, Patel HV, Ridley RG, Freeman KB, Gupta RS (1990) Mitochondrial import of the human chaperonin (HSP60) protein. Biochem Biophys Res Commun 169:391–396

Singh B, Soltys BJ, Wu Z-C, Patel HV, Freeman KB, Gupta RS (1997) Cloning and some novel characteristics of mitochondrial hsp70 from Chinese hamster cells. Exp Cell Res 234:205–216

Smalheiser NR (1996) Proteins in unexpected locations. Mol Biol Cell 7:1003–1014

Soltys BJ, Falah M, Gupta RS (1996) Identification of endoplasmic reticulum in the primitive eukaryote Giardia lamblia using cryoelectron microscopy and antibody to BiP. J Cell Sci 109:1909–1917

Soltys BJ, Gupta RS (1996) Immunoelectron microscopic localization of the 60-kDa heat shock chaperonin protein (hsp60) in mammalian cells. Exp Cell Res 222:16–27

Soltys BJ, Gupta RS (1997) Cell surface localization of the 60kDa heat shock chaperonin protein (hsp60) in mammalian cells. Cell Biol Int 21:315–320

Sporn MB (1991) Localization of transforming growth factor-beta1 in mitochondria of murine heart and liver. Cell Reg 2:467–477

Stein I, Peleg Y, Even-Ram S, Pines O (1994) The single translation product of the FUM1 gene (fumarase) is processed in mitochondria before being distributed between the cytosol and mitochondria in Saccharomyces cereevisiae. Mol Cell Biol 14:4770–4778

Susin SA, Zamzami N, Castedo M, Hirsch T, Marchetti P, Macho A, Daugas E, Geuskens M, Kroemer G (1996) Bcl-2 inhibits the mitochondrial release of an apoptogenic protease. J Exp Med 184:1331–1341

Suto S, Srivastava PK (1995) A mechanism for the specific immunogenicity of heat shock protein-chaperoned peptides. Science 269:1585–1588

Torok Z, Horvath I, Goloubinoff P, Kovacs E, Glatz A, Balogh G, Vigh L (1997) Evidence for a lipochaperonin:association of active protein-folding GroESL oligomers with lipids can stabilize membranes under heat shock conditions. Proc Natl Acad Sci USA 94:2192–2197

Ungermann C, Neupert W, Cyr D (1994) The role of hsp70 in conferring unidirectionality on protein translocation into mitochondria. Science 266:1250–1253

van Eden W (1991) Heat-shock proteins as immunogenic bacterial antigens with the potential to induce and regualte autoimmune arthritis. Immunol Rev 121:5–28

van Roon JAG, van Eden W, van Roy LAM, Lafeber JPG, Bijlsma WJ (1997) Stimulation of suppressive T cell responses by human but not bacterial 60-kD heat-shock protein in synovial fluid of patients with rheumatoid arthritis. J Clin Invest 100:459–463

VanBuskirk AM, DeNagel DC, Guagliardi LE, Brodsky FM, Pierce SK (1991) Cellular and subcellular distribution of PBP72/74, a peptide-binding protein that plays a role in antigen processing. J Immunol 146:500–506

Velez-Granell CS, Arias AE, Torres-Ruiz JA, Bendayan M (1994) Molecular chaperones in pancreatic tissue: the presence of cpn10, cpn60 and hsp70 in distinct compartments along the secretory pathway of the acinar cells. J Cell Sci 107:539–549

Velez-Granell CS, Arias AE, Torres-Ruiz JA, Bendayan M (1995) Presence of Chromatium vinosum chaperonins 10 and 60 in mitochondria and peroxisomes of rat hepatocytes. Biol Cell 85:67–75

Viitanen PV, Lorimer GH, Seetharam R, Gupta RS, Oppenheim J, Thomas JO, Cowan NJ (1992) Mammalian mitochondrial chaperonin 60 functions as a single toroidal ring. J Biol Chem 267:695–698

Vodkin MH, Williams JC (1988) A heat shock operon in Coxiella burnetii produces a major antigen homologous to a protein in both mycobacteria and Eschericha coli. J Bacteriol 170:1227–1234

Voos W, Gambill BD, Laloraya S, Ang D, Craig EA, Pfanner N (1994) Mitochondrial GrpE is present in a complex with hsp70 and preproteins in transit across membranes. Mol Cell Biol 14:6627–6634

Wadhwa R, Kaul SC, Ikawa Y, Sugimoto Y (1993) Identification of a novel member of mouse hsp70 family. Its association with cellular mortal phenotype. J Biol Chem 268:6615–6621

Wadhwa R, Kaul SC, Mitsui Y (1994) Cellular mortality to immortalization: mortalin. Cell Struct Funct 19:1–10

Weissman JS, Kashi Y, Fenton WA, Horwich AL (1994) GroEL-mediated protein folding proceeds by multiple rounds of binding and release of nonative forms. Cell 78:693–702

Wells AD, Rai SK, Salvato MS, Band H, Malkovsky M (1997) Restoration of MHC class I surface expression and endogenous antigen presentation by a molecular chaperone. Scand J Immunol 45:605–612

Woodlock TJ, Chen XX, Young DA, Bethlendy G, Lichtman MA, Segel GB (1997) Association of HSP60-like proteins with the L-system amino acid transporter. Arch Biochem Biophys 338:50–56

Xu Q, Kleindienst R, Waitz W, Dietrich H, Wick G (1993) Increased expression of heat shock protein 65 coincides with a population of infiltrating T lymphocytes in atherosclerotic lesions of rabbits specifically responding to heat shock protein 65. J Clin Invest 91:2693–2702

Xu Q, Wick G (1996) The role of heat shock proteins in protection and pathophysiology of the arterial wall. Mol Med Today 2:372–379

Xu Z, Horwich AL, Sigler PB (1997) The crystal structure of the asymmetric GroEL-GroES-(ADP)7 chaperonin complex. Nature 388:741–750

Yanaguchi H, Osaki T, Taguchi H, Hanawa T, Yamamoto T, Kamiya S (1996) Flow cytometric analysis of the heat shock protein 60 expressed on the cell surface of Heliobacter pylori. J Med Microbiol 45:270–277

Yang J, Liu X, Bhalla K, Kim CN, Ibrado AM, Cai J, Peng TI, Jones DP, Wang X
    (1997) Prevention of apoptosis by Bcl-:release of cytochrome c from mitochondria
    blocked. Science 275:1129–1132
Zeilstra-Ryalls J, Fayet O, Georgopoulos C (1991) The universally conserved GroE
    (Hsp60) chaperonins. Annu Rev Microbiol 454:301–325
Zhu X, Zhao X, Burkholder WF, Gragerov A, Ogata S, Gottesman ME, Hendrickson
    WA (1996) Structural analysis of substrate binding by the molecular chaperone
    DnaK. Science 272:1606–1614

CHAPTER 5

Role of Hsp27 and Related Proteins

A.-P. Arrigo and X. Préville

A. Introduction

Investigations of the cellular response to thermal and other types of stresses have allowed the identification of families of proteins (the heat shock or stress proteins, Hsp) whose expression is enhanced when environmental conditions become deleterious (reviewed in Georgopoulos and Welch 1993; Morimoto et al. 1994). Hsp are subdivided in two groups, based on their apparent molecular mass, i.e., the large and small heat shock proteins. Small heat shock proteins, now denoted small stress proteins (sHsp), are characterized by a domain of homology to αA,B-crystallin proteins from vertebrate eye (reviewed in Arrigo and Landry 1994). Despite this particular homology, sHsp are less conserved than the large Hsp (i.e., Hsp70) since, among species, they show greater variations in sequence, in number and in molecular mass. All sHsp analyzed so far share the extreme tendency to form oligomers. Like α-crystallin, sHsp are in the form of aggregates with heterodispersed native molecular masses (which can reach up to 800 kDa or more). This structural organization of sHsp depends on the physiology of the cell and probably also on the phosphorylation of these proteins (Siezen et al. 1978a; Arrigo 1987; Arrigo and Welch 1987; Arrigo et al. 1988; Kato et al. 1994; Lavoie et al. 1995; Mehlen and Arrigo 1994; Mehlen et al. 1995b,c).

sHsp expression enhances the cellular resistance to heat shock (Landry et al. 1989; Mehlen et al. 1993). This resistance is reminiscent of the in vitro chaperone activity associated with these proteins (Jakob et al. 1993; Jakob and Buchner 1994). However, in contrast to chaperones belonging to the Hsp70 family, sHsp act through an ATP-independent mechanism. In vivo, they are supposed to act in concert with other chaperones by creating a reservoir of folding intermediates (Ehrnsperger et al. 1997; Lee et al. 1997).

sHsp generate renewed interest because these proteins can confer protection against a variety of toxic chemicals, particularly those used in cancer chemotherapy, i.e., cisplatin and doxorubicin (Huot et al. 1991; Oesterreich et al. 1993; Richards et al. 1996; Garrido et al. 1996, 1997), and interfere with inflammatory mediators such as tumor necrosis factor (TNFα; Mehlen et al. 1995a, 1996a,b; Wang et al. 1996). Moreover, sHsp expression also appears correlated with the oncogenic status of the cell (Têtu et al. 1992, 1995). This phenomenon, which may be a result of the protection mediated by sHsp against TNFα, suggests that these stress proteins modulate the immuno-surveillance mediated by this cytokine. Another intriguing property of sHsp is their expression during early differentiation and at specific stages of development (reviewed in ARRIGO and TANGUAY 1991; DE JONG et al. 1993; ARRIGO and MEHLEN 1994; ARRIGO 1995). We recently showed that sHsp expression is essential to protect differentiating cells against apoptosis (MEHLEN et al. 1997b). Moreover, we and others have reported that these proteins modulate intracellular redox (MEHLEN et al. 1996a) and act as novel negative regulators of programmed cell death (MEHLEN et al. 1996a,b; SAMALI and COTTER 1996; ARRIGO 1998). Hence, sHsp are not only involved in the cellular defense mechanisms against aggression, but also participate in essential physiological processes in unstressed cells.

The present chapter summarizes recent observations concerning sHsp with particular emphasis on the discovery of the protective activity of these proteins against programmed cell death. We also discuss the physiological role of this important function of sHsp.

B. sHsp Genes and Control of Their Expression

I. The Family of sHsp and the Structure of the Genes Encoding These Proteins

The number of sHsp proteins varies between species: at least four major sHsp exist in *Drosophila*, mammalian cells contain three sHsp (αB-crystallin, Hsp20 and 27) as well as yeast (Hsp12, 26 and 42; WOTTOM et al. 1996), while plants contain more than 20 different sHsp. The molecular masses of these different proteins comprise between 15 and 30 kDa. All sHsp proteins analyzed so far are characterized by a conserved C-terminal domain (INGOLIA and CRAIG 1982; SOUTHGATE et al. 1983; WISTOW 1985), which is similar to the carboxy-terminal region of αA,B-crystallins from bovine lens (INGOLIA and CRAIG 1982; the "crystallin" domain). αA,B-crystallins (19–20 kDa) are major polypeptides which accumulate in the cytoplasm of lens cells; however, only αB-crystallin displays enhanced expression following heat shock and is therefore considered as a true sHsp (KLEMENZ et al. 1991). The four *Drosophila* sHsp (Dm-Hsp22,23,26 and 27) are encoded by distinct genes that are localized within a cluster of 12 kb of DNA at the locus 67B of chromosome 3L (AYME and TISSIÈRES 1985; PAULI and TISSÈRES 1990; PAULI et al. 1992; ARRIGO and TANGUAY 1991; ARRIGO and MEHLEN 1994). Human Hsp27 is encoded by a single gene located on chromosome 7; two additional pseudogenes are present on chromosome 3 and X (HICKEY et al. 1986; MCGUIRE et al. 1989). Murine Hsp25 is also encoded by only one gene (FROHLI et al. 1993; GAESTEL et al. 1993). Human αA- and αB-crystallins are encoded by single-copy genes located on chromosomes 21 and 11, respectively (QUAX-JEUKEN et al. 1985; NGO et al. 1989). In plants, sHsp are represented by at least two multigene families (RASCHKE et al. 1988).

II. Regulation of the Expression of sHsp Genes by Heat Shock

Synthesis of the different Hsp is coordinately induced by heat shock. This stress enhances the rate of sHsp synthesis by at least tenfold and sHsp accumulation can reach 1% of total cellular proteins according to the type of cells analyzed (ARRIGO and LANDRY 1994; WATERS et al. 1996). The heat-dependent induction of sHsp genes is under the control of specific regulatory sequences (HSE, heat shock element) localized upstream of the genes. In response to heat shock, a specific transcription factor, HSF-1, is activated through a trimerization process which triggers its binding to HSE elements. This leads to transcriptional activation of heat shock genes (MORIMOTO et al. 1994). HSE elements consist of inverted repeats of the pentameric nGAAn sequence. At least three HSE elements are necessary to activate heat shock genes by thermal stress. HSE elements have been found upstream of all sHsp genes including the four *Drosophila* sHsp, mammalian Hsp27 and αB-crystallin and plant sHsp (AYME et al. 1985; HICKEY et al. 1986; HOFFMAN and CORCES 1986; FROHLI et al. 1993; GAESTEL et al. 1993; KLEMENZ et al. 1991; WATERS et al. 1996). In contrast, αA-crystallin gene does not contain HSE element. This is in agreement with the fact that this gene is not induced by heat shock (KLEMENZ et al. 1991). Although less studied than in the case of Hsp70 gene, the binding of HSF to HSE elements appears essential to trigger the transcription of sHsp genes by heat shock. However, notorious differences have been observed when the kinetics of accumulation of the different stress proteins were analyzed. For example, during heat shock recovery, the synthesis of Hsp70 peaks and resumes more rapidly than that of Hsp27 (ARRIGO and WELCH 1987; LANDRY et al. 1991). Moreover, the temperature necessary to trigger sHsp synthesis is usually inferior to that necessary to induce the synthesis of Hsp70 (YOST et al. 1990). It is also interesting to note that in embryonic chicken cells, the strong induction of Hsp27 does not result in transcriptional activation but in increased translation of Hsp27 mRNA (EDINGTON and HIGHTOWER 1990). This phenomenon is reminiscent of that described for the induction of Hsp25 by cisplatin in Ehrlich tumoral cells (GOTTHARDT et al. 1996). Hence, mechanisms other than the binding of HSF to HSE may also play crucial role(s) in the control of sHsp expression in response to heat shock or other types of stress.

Several drugs or proteins can interfere or inhibit the inducibility of heat shock genes, including those encoding sHsp. Among others, one can cite flavonoids (i.e., quercitin; HANSEN et al. 1997), thiol compounds (HUANG et al. 1994) and PrPC prion protein (TATZELT et al. 1995). Oncogenic proteins that lead to cell transformation also alter sHsp expression. For example, the expression of Hsp27 is altered in adenovirus transformed rat cells (ZANTEMA et al. 1989) as well as in ras-transformed FR3T3 rat fibroblasts (ENGELBERG et al. 1994). Moreover, we recently reported that a common characteristic of murine NIH 3T3 fibroblasts transformed by either v-fos, v-src, T-antigen or v-Ha-ras 1 oncogenes is a delayed accumulation of Hsp27 (also denoted Hsp25 in mouse) after heat shock (FABRE-JONCA et al. 1995; GONIN et al. 1997). In contrast, prostaglandins and modulators of the arachidonic cascade enhance the stress-induced synthesis of Hsp27 and αB-crystallin (Ito et al. 1996, 1997a).

III. Regulation of the Constitutive and Hormone-Dependent Expression of sHsp Genes

Constitutive expression of sHsp has been described in different cells and tissues (Arrigo and Landry 1994). However, the level of expression of these proteins is highly variable and depends on the cell analyzed. For example, unstressed murine L929 and NIH 3T3 fibroblasts are devoid of constitutively expressed sHsp (Mehlen et al. 1995a). These proteins are only detected when these cells are exposed to heat shock. In contrast, murine embryonic stem cells (Mehlen et al. 1997b), monkey CV-1 or COS cells and human HeLa, MCF7, and T47D cells contain high levels of Hsp27 (Arrigo and Welch 1987; Arrigo et al. 1988; Mehlen et al. 1995c).

The molecular mechanisms regulating sHsp constitutive expression are not well understood. However, in some cells, sHsp expression depends on the presence of steroid hormones. For example, in embryonic *Drosophila* cells, sHsp are induced by the steroid molting hormone ecdysterone (Ireland and Berger 1982). In this regard, an ecdysterone receptor recognition sequence was discovered upstream of genes encoding these proteins (Riddihough and Pelham 1987). Similarly, murine Hsp25 gene contains an hormone regulatory element located close to the Pribnow box (Gaestel et al. 1993). Finally, human Hsp27 gene is regulated by both progesterone and estrogens (Ciocca et al. 1983b). Sequences responsible for this regulation are composed of an Sp1 element and a palindromic element that respond to estrogens (Porter et al. 1996).

Concerning αB-crystallin, Aoyama et al. (1993a) have discovered a glucocorticoid receptor sensitive sequence which lies upstream of the 5′ region of the gene. In addition, Scheier et al. (1996) have described a second promoter sequence that is essential for induction of αB-crystallin gene by glucocorticoids. Moreover, it has been shown that the sequence between positions –427 and –259 is necessary for the expression of αB-crystallin in muscular cells (Dubin et al. 1991), hence suggesting a muscle-specific control of αB-crystallin expression (Gopal-Srivastava and Piatigorsky 1993). Expression of this gene in lens cells was found to depend on a region situated close to position –115.

IV. Tissue-Specific sHsp Expression During Development and in Adult Organisms

Several studies have analyzed the developmental expression of the different *Drosophila* sHsp (Dm-Hsp27, -Hsp26, -Hsp23 and -Hsp22). It has been found that in contrast with their coordinate synthesis following heat shock, these sHsp display different patterns of expression during development. First, it was found that Dm-Hsp22 is not expressed during development while as mentioned above it is induced by heat shock (ARRIGO and TANGUAY 1991). Dm-Hsp27 is detected in neural embryo tubes and then in the central nervous system and the gonads of larvae (MARIN and TANGUAY 1996; PAULI et al. 1990). This protein is also expressed in differentiating imaginal disks of third instar larvae (PAULI et al. 1990). The pattern of expression of Dm-Hsp26 resembles that of Dm-Hsp27 (GLASER et al. 1986); however, some differences in the level expression of these proteins were detected (MARIN et al. 1993). Dm-Hsp23 is weakly expressed during embryogenesis and essentially in Midline Precursor Cells (ARRIGO and TANGUAY 1991). In young adult, different levels of Dm-Hsp23, -Hsp26 and -Hsp27 are present in the central nervous system and the gonads (MARIN et al. 1993).

In yeast, the expression of Hsp26 is detected during meiosis and ascospore development (KURTZ et al. 1986). sHsp are also expressed during plant development (HERNANDEZ and VIERLING 1993; WATERS et al. 1996). In unstressed *Xenopus* embryos, HEIKKILA et al. (1991) reported the presence of Hsp30 while 25 kDa Hsp-like proteins were observed in sensory ganglia from tadpoles but not from the adult bullfrog (HAMMERSCHLAG et al. 1989). In mammals, Hsp27/25 is also transiently expressed during development. For example, through days 13–20 of mouse development, Hsp25 accumulates in neurons of spinal cord and Purkinje cells GERNOLD et al. (1993). In young mice, this protein is no longer detectable in the brain but is still present in different muscles, including those of cardiac, stomach, colon, lung and bladder origins (TANGUAY et al. 1993; KLEMENZ et al. 1993). In rat, this polypeptide is observed at the level of neural tube closure (WALSH et al. 1991b). In adult humans, endometrial biopsies have revealed variations in the level of Hsp27 during the different phases of the menstrual cycle, a phenomenon probably regulated by steroid hormones (CIOCCA et al. 1983b).

During murine development, αB and αA-crystallins are maximally expressed in lens cells (WISTOW and PIATIGORSKY 1988). αB-crystallin expression occurs essentially in lens epithelial cells and was found to precede that of αA-crystallin, which is preferentially synthesized in lens fiber cells (ROBINSON and OVERBEEK 1996). Of interest, during mice development αB-crystallin is expressed in tissues other than the lens, including brain (glia cells; IWAKI et al. 1990), skin, skeletal and heart muscles (BENJAMIN et al. 1997; BHAT and NAGINENI 1989). In adult mice and rats, αB-crystallin is observed in several organs or tissues other than lens such as retina, heart, skeletal muscles, skin, kidneys, lungs, brain and bone marrow (BHAT and NAGINENI 1989; DUBIN et al. 1991; GERNOLD et al. 1993). αA-crystallin was also identified in spleen cells of adult mice (SRINIVASAN et al. 1992).

V. Specific sHsp Expression During Early Differentiation

sHsp share the intriguing property of being transiently expressed during early differentiation. This phenomenon, which was first reported in *Drosophila*, has been observed in all the organisms analyzed so far, hence revealing its ubiquitous nature (reviewed in ARRIGO and LANDRY 1994; ARRIGO 1995). For example, during *Drosophila* larval development, Dm-Hsp27 accumulates during the differentiation of imaginal disks; hence suggesting that this protein plays a role in this process (PAULI et al. 1990). Recent studies performed with mammalian cells have strengthened the hypothesis that sHsp play a role during early differentiation. Indeed, Hsp27 is transiently expressed during the early differentiation of numerous mammalian cells such as Ehrlich ascites cells (BENNDORF et al. 1988), embryonal carcinoma and stem cells (STAHL et al. 1992; MEHLEN et al. 1997b), normal B and B lymphoma cells (SPECTOR et al. 1992), osteoblasts, promyelocytic leukemia cells (SHAKOORI et al. 1992; SPECTOR et al. 1993, 1994; CHAUFOUR et al. 1996) and normal T cells (HANASH et al. 1993). Hsp27 accumulation usually occurs concomitantly with the differentiation-mediated decrease of cellular proliferation (SHAKOORI et al. 1992; SPECTOR et al. 1993, 1994; CHAUFOUR et al. 1996; MEHLEN et al. 1997b). It is not known whether other sHsp such as αB-crystallin are also transiently expressed during early differentiation. The molecular mechanism that controls Hsp27 expression was found to vary depending upon the differentiation signal. For example, during the phorbol ester mediated macrophagic differentiation of HL-60 cells, a transient increase in Hsp27 polypeptide is preceded by an increase in the corresponding mRNA (SPECTOR et al. 1993). In contrast, during granulocytic differentiation of the same cells, triggered by all-trans retinoic acid, the rise in Hsp27 results mostly in the change in half-life of this protein (SPECTOR et al. 1994). Hsp27 is a privileged substrate of the serine protease myeloblastin (SPECTOR et al. 1995), which is rapidly downregulated by retinoids and plays a role in growth arrest and differentiation of myeloid cells (BORIES et al. 1989). Hsp27 transient accumulation, phosphorylation and oligomerization, which occur during early granulocytic differentiation of human promyelocytic HL-60 cells, are summarized in Fig. 1.

VI. Pathological sHsp Expression and Associated Diseases

Particular expression profiles of sHsp are observed in association with various pathologies. For example, changes in the expression of mammalian Hsp25/27 have been observed in several cancer cells (CIOCCA et al. 1983a; NAVARRO et al. 1989; PUY et al. 1989). In the case of Ehrlich ascites tumoral cells, Hsp25 is overexpressed during the stationary phase of the tumor (GAESTEL et al. 1989). In this model, Hsp25 overexpression interferes with cell growth (KNAUF et al. 1992). Expression of this protein was also negatively correlated with oncogenicity of cells transformed by adenovirus (ZANTEMA et al. 1989). However, the reverse has been observed in breast cancer and leukemia (CIOCCA et al. 1993; KHALID et al. 1995). Abnormal Hsp27 expression in human breast cancer cells is usually estrogen-dependent (FUQUA et al. 1989). It has also recently been reported that sHsp expression modulates the growth of tumors that originated from L929 fibrosarcoma implanted in nude mice (BLACKBURN

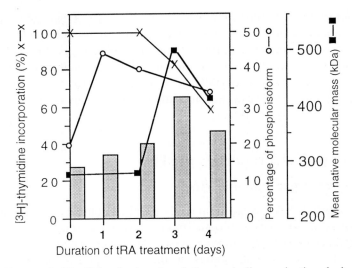

Fig. 1. Changes in Hsp27 level, phosphorylation and oligomerization during retinoic acid mediated granulocytic differentiation of human promyelocytic HL-60 cells. The level (*gray plots*) as well as the mean native molecular mass of Hsp27 was determined by western immunoblot of either total cell lysates or sizing chromatograph analysis of cytoplasmic extracts. Phosphorylation was determined by quantitative analysis of Hsp27 phosphoisoforms. [³H]-Thymidine incorporation is indicative of the decrease in DNA synthesis during early differentiation. (Reproduced from CHAUFOUR et al. 1996, by copyright permission of Churchill Livingstone)

et al. 1997). Moreover, genetically engineered rat colon carcinoma cells that constitutively express human Hsp27 can form more aggressive tumors than cells which do not contain this protein (C. Garrido, P. Mehlen, B. Chauffert and A.-P. Arrigo, unpublished result). Taken together, these observations suggest a link between sHsp expression and tumorigenicity. sHsp expression is therefore considered as a parameter to be taken into account in the prognosis of evolution of some human tumors (CIOCCA et al. 1992; TÊTU et al. 1992, 1995; THOR et al. 1991). For example, expression of Hsp27 is associated with a poor outcome in breast cancer (THOR et al. 1991) and osteosarcoma (UOZAKI et al. 1997) patients.

Concerning αB-crystallin, v-mos and Ha-ras oncogenes drastically increase the expression of this protein in NIH 3T3 cells (KLEMENZ et al. 1991). In vivo, abnormally high levels of αB-crystallin have been observed in glial tumors, including astrocytoma, glioblastoma and oligodendroglioma (AOYAMA et al. 1993b). Abnormally high levels of αB-crystallin are observed in some degenerative pathologies such as Alexander's (HEAD et al. 1993; IWAKI et al. 1989), Alzheimer's (RENKAWEK et al. 1993; SHINOHARA et al. 1993) or Creutzfeld-Jakob (KATO et al. 1992; RENKAWEK et al. 1992) diseases. This protein is also detected in Mallory's hepatic and Lewy bodies (LOWE et al. 1992). An increased αB-crystallin expression has been observed in benign tumors associated with tuberosis sclerosis (Iwaki and Tateishi 1991), and αB-crystallin autoantibodies have been detected in patients suffering from multiple sclerosis (Vannoort et al. 1995). Of interest, high levels of Hsp27 are also present in tissues from patients presenting these degenerative pathologies (Head et al. 1993; Iwaki et al. 1993; Kato et al. 1992; Renkawek et al. 1993; Shinohara et al. 1993). The mechanisms regulating this pathological sHsp overexpression are unknown.

C. Biochemical Properties of sHsp

I. Structural Organization of sHsp

Sequence analysis has revealed that the N-terminal domain of sHsp is not well conserved between species and only moderately conserved between sHsp of the same species (i.e., *Drosophila* sHsp). However, small sequences (i.e., MAPKAPK2 phosphorylation sites) in this domain are conserved and may therefore have important functions. In *Drosophila*, the first amino acids of the different sHsp are conserved and resemble peptide signal sequences (Southgate et al. 1983; Arrigo and Pauli 1988). sHsp carboxyl-terminal domain contains the crystallin box, the sequence of which is highly conserved between species (Ingolia and Craig 1982; De Jong et al. 1993). The remaining sequences in this domain are not well conserved except in *Drosophila* sHsp, where a significant homology was observed at the level of the 25 residues adjacent to the crystallin domain (Southgate et al. 1983).

Although the sequences of several sHsp are well known, little information is available concerning the structure of these proteins. Circular dichroism analysis has revealed that sHsp are mainly in a β-sheet with less than 5% of the protein in the form of α-helix (Surewicz and Olesen 1995). The presence of α-helix in amphiphilic conformation may confer properties to interact with membranes or other proteins (Plesofsky-Vig and Brambl 1990). Several models have been proposed to describe the sHsp tertiary structure. For example, Wistow (1985) has proposed that the organization of α-crystallin can be subdivided into two domains, the first corresponding to exon 1 of the gene, and the second to exons 2 and 3. Each of these areas represents motives which are folded according to axial symmetry. The conservation of structurally important amino acids between the different members of this family has allowed Wistow to extend his model to other members of the sHsp family. This model has been confirmed by Carver et al. (1993) using NMR (nuclear magnetic resonance) spectroscopy analysis of α-crystallin.

II. Quaternary Structure of sHsp

One of the main characteristics of sHsp (including α-crystallins) is their natural tendency to form oligomeric structures (with size ranging from 50 to 800 kDa) (Spector et al. 1971; Arrigo and Ahmad-Zadeh 1981; Arrigo et al.

1985, 1988; MEHLEN and ARRIGO 1994). In some cases, these aggregates can contain two different sHsp, for example, Hsp27 and αB-crystallin (ZANTEMA et al. 1992). The size of these aggregates varies according to cellular growth conditions, exposure to heat shock (ARRIGO et al. 1988) or tumor necrosis factor (MEHLEN and ARRIGO 1994; MEHLEN et al. 1995b; PREVILLE et al. 1998a). In addition, intracellular glutathione level modulates sHsp oligomerization (MEHLEN et al. 1997a). Different parameters such as pH, calcium (SIEZEN et al. 1980), or the aging process (HARDING et al. 1991) induce a hyperaggregation of these proteins. A similar phenomenon is observed in lens from patients suffering from cataract (SIEZEN et al. 1978b) or in cardiac muscles during ischemia (CHIESI et al. 1990). In the electron microscope sHsp oligomers have a globular and spherical appearance (ARRIGO et al. 1988; NOVER et al. 1989; LONGONI et al. 1990). However, the quaternary structure of these proteins is still unknown and will require X-ray analysis of crystallized oligomers. Current structural models essentially concern α-crystallins and are based on indirect methods of analysis (WALSH et al. 1991a; WISTOW 1993; MERCK et al. 1992; GROENEN et al. 1994). Recently, CARVER et al. (1994) have proposed that α-crystallin has a structure similar to the other chaperon proteins, which is a cylindrical form with a cavity in the middle. However, none of the models presented today explains the different structural organizations of sHsp observed in vitro (SIEZEN et al. 1980; SIEZEN and JGU 1982). Moreover, the dynamic changes in sHsp native size remain unexplained (VAN DEN OETELAAR et al. 1990; WISTOW 1993).

III. Phosphorylation of sHsp

sHsp are characterized by two prominent features: phosphorylation and oligomerization (KIM et al. 1984; WELCH 1985; ARRIGO et al. 1988; ARRIGO 1990a; OESTERREICH et al. 1993; ZHOU et al. 1993; MEHLEN and ARRIGO 1994; MEHLEN et al. 1995b). Human Hsp27 is phosphorylated at serine residues 15, 78 and 82 (LANDRY et al. 1992). In contrast, murine Hsp25 contains only two phosphorylated serine residues at positions 15 and 86 (GAESTEL et al. 1991). Hsp25/27 phosphorylation is induced by stimuli that include heat shock, TNFα and other forms of oxidative stress, interleukin-1, -3, and -6, okadaic acid, granulocyte-macrophage colony-stimulating factor, bradykinin, thrombin, histamine, ifosfamine, mitogens, tumor promoters, calcium ionophore, retinoic acid and cyclic nucleotide-dependent vasorelaxation (WELCH 1985; KAUR and SAKLATVALA 1988; ARRIGO 1990a; CRÊTE and LANDRY 1990; MENDELSOHN et al. 1991; LANDRY et al. 1992; SANTELL et al. 1992; ISSELS et al. 1993; KASAHARA et al. 1993; SPECTOR et al. 1993, 1994; AHLERS et al. 1994; FRESHNEY et al. 1994; BELKA et al. 1995; MEHLEN et al. 1995b; BEALL et al. 1997). Hsp25/27 phosphorylation is catalyzed by MAPKAP kinase 2 and 3pk (STOKOE et al. 1992; FRESHNEY et al. 1994; ROUSE et al. 1994; CUENDA et al. 1995; HUOT et al. 1995; LUDWIG et al. 1996). However, it is not excluded that in some cases sHsp phosphorylation may also result from the inactivation of specific phosphatases

(Gaestel et al. 1992; Guy et al. 1993). The oligomerization profile of unphosphorylatable mutants of human Hsp27 in which serines 15, 78 and 82 were replaced by either alanine, glycine or aspartic acid residues has been investigated. Depending on the amino acid which was used to substitute the serine sites, different patterns of Hsp27 oligomerization profile were observed. For example, alanine substitution generated large Hsp27 aggregates while glycine and aspartic acid did the reverse (Mehlen et al. 1997a). Similarly, Ser to Ala substitution of murine Hsp25 generated large oligomers (Preville et al. 1998a). Hence, phosphorylatable serine residues may regulate sHsp structural organization. Phosphorylation induced by either serum or TNFα essentially occurs at the level of small Hsp27 structures (<200 kDa) and is concomitant with increased native size (up to 700 kDa) of a fraction of this protein (Mehlen and Arrigo 1994; Mehlen et al. 1995b). Serum treatment also induces the association of a fraction of Hsp27, in the form of small and dephosphory-lated oligomers, with detergent-sensitive particulate fractions that may be of membranous origin (Mehlen and Arrigo 1994; Mehlen et al. 1995b). These interactions are no longer observed when growing cells are exposed to TNFα.

Concerning αB-crystallin, this protein also displays enhanced phosphory-lation when cells are exposed to various types of stimuli. Phosphorylation occurs at serines 19, 45 and 59 and is p38 MAP kinase dependent (Ito et al. 1997b).

IV. Cellular Localization of sHsp

In unstressed cells as well as in cells recovering from heat shock, mammalian Hsp27/25 is cytoplasmic (Lavoie et al. 1993a,b, 1995) and is sometimes con-centrated in a polarized perinuclear zone (Arrigo et al. 1988; Arrigo 1990b; Preville et al. 1996). A fraction of this protein is associated with detergent sensitive structures (Arrigo et al. 1988; Mehlen and Arrigo 1994) that do not appear to be of Golgi origin (Preville et al. 1996). As mentioned above, this fraction of Hsp27 is enriched in small unphosphorylated oligomers (Mehlen and Arrigo 1994). A similar locale has been observed for αB-crystallin while Hsp25 from chicken is more diffusely distributed in the cytoplasm (Collier and Schlesinger 1986).

Drosophila Dm-Hsp23 and tomato sHsp are concentrated in cytoplasmic granules (Duband et al. 1986; Nover et al. 1989). Drosophila Dm-Hsp 26 is cytoplasmic while Dm-Hsp27 is nuclear (Beaulieu et al. 1989; Pauli et al. 1990). Some şHsp are present in mitochondria (DmHsp22 and Neurospora crassa Hsp30) (Plesofsky-Vig and Brambl 1990). Moreover, some higher plants and algae sHsp are localized in chloroplasts (Vierling et al. 1988; Chen and Vierling 1991). During drastic heat shock or stress that induce Hsp synthesis, cytoplasmic sHsp share the tendency to accumulate into the nucleus where they form highly aggregated structures (Arrigo 1980, 1990b; Arrigo et al. 1980, 1988).

D. Functions of sHsp

I. sHsp Expression Induces Thermotolerance and Protects Cytoskeletal Architecture

Hsp expression by mild heat shock confers a transient state of resistance to thermal stress, termed thermotolerance (GERNER et al. 1975; LANDRY et al. 1982; LI and WERB 1982). Consequently, several studies have approached stress protein function by analyzing cells with deregulated Hsp expression. It was observed that overexpression of human and yeast Hsp70 enhanced the cellular resistance to heat shock (CRAIG and JACOBSEN 1984; LI et al. 1991), while inactivation of this protein induced thermosensitivity (RIABOVOL et al. 1988). A similar function was assigned to Hsp90 (BANSAL et al. 1991) and, in yeast, Hsp104 is a key protein which controls thermotolerance (SANCHEZ et al. 1990). Concerning sHsp, BERGER and WOODWARD (1983) were the first to report a putative role of these proteins in the cellular resistance to thermal stress. This hypothesis was confirmed by studies aimed at overexpressing these proteins in mammalian cells. It was then observed that either mammalian Hsp27 or *Drosophila* Dm-Hsp27 conferred heat shock resistance to different types of mammalian cells (LANDRY et al. 1989; ROLLET et al. 1992; MEHLEN et al. 1993). sHsp also confer protection against a variety of toxic chemicals used in cancer chemotherapy, i.e., cisplatin and doxorubicin (HUOT et al. 1991; OESTERREICH et al. 1993; RICHARDS et al. 1996; GARRIDO et al. 1996, 1997), a phenomenon which may be related to the ability of these proteins to protect against oxidative stress and programmed cell death (see Sect. D.IV below).

Other studies have suggested that sHsp could play a role in cell growth regulation. However, divergent results have been observed depending on the type of cell analyzed: in murine Ehrlich ascites tumor cells, overexpression of endogenous Hsp25 interfered with cell growth (KNAUF et al. 1992) while this protein had no effect when it was overexpressed in murine embryonic stem cells (MEHLEN et al. 1997b). In contrast, overexpression of human Hsp27 was reported to accelerate growth and senescence of bovine arterial endothelial cells (PIOTROWICZ et al. 1995).

sHsp also modulate cytoskeletal architecture. For example, α-crystallin expression modulates intermediate filament assembly (NICHOLS and QUINLAN 1994). Moreover, yeast Hsp26, gizzard Hsp25 and murine Hsp25 can inhibit actin polymerization in vitro (MIRON et al. 1991; BENNDORF et al. 1994; RAHMAN et al. 1995) and in vivo these proteins protect the integrity of actin microfilament architecture against different stresses (LAVOIE et al. 1993a,b, 1995; HUOT et al. 1996). Studies have also been performed to analyze whether phosphorylation regulates sHsp protective activity. In one case, a reduced modulation of actin microfilament dynamics and fluid phase pinocytosis was found to occur when the three phosphorylatable serines of human Hsp27 were replaced by glycine residues (LAVOIE et al. 1993b). In addition, overexpression of these triple mutants conferred a reduced resistance to heat shock and oxidative stress compared to wild type Hsp27 (Huot et al. 1996; Lavoie et al. 1995). In the other case, in vitro chaperone and in vivo thermoresistance-mediating activities of murine Hsp25 were not altered after the two phosphorylation sites, serines 15 and 86, were substituted by alanine (Knauf et al. 1994). It was then concluded that phosphorylation per se was not essential for these functions of Hsp25. However, as mentioned above, the substitution of sHsp phosphorylatable serine residues by different amino acids drastically altered the oligomerization property of these proteins. Moreover, only the small unphosphorylated oligomers of mammalian Hsp27/25 display actin polymerization-inhibiting activity (Benndorf et al. 1994).

II. sHsp Act as Protein Chaperones

The resistance mediated by sHsp expression against heat shock or other form of stress may result from the ability of these proteins to act as chaperones. Indeed, in vitro, these proteins can induce the renaturation of denatured polypeptides in an ATP-independent manner (Horwitz et al. 1992; Jakob et al. 1993). Moreover, the large oligomers of Hsp27 are those which display in vitro chaperone activity while, as mentioned above, actin capping-decapping activity appears associated with Hsp27 small oligomers (Benndorf et al. 1994). Recently, it has been proposed that non-native proteins interact with the large sHsp oligomers which accumulate during heat shock (Ehrnsperger et al. 1997; Lee et al. 1997). This phenomenon may create a reservoir of folding intermediates that could prevent further aggregation of non-native proteins and therefore enhance their refolding by Hsp70 and co-chaperones. Intracellular redox and/or glutathione could play an important role in maintaining the structural organization of these large structures and/or in allowing the renaturation of non-native proteins (see Sect. D.III, below). In heat shocked cells, sHsp chaperone activity may accelerate the recovery of heat-induced cellular damage, such as (i) nuclear protein aggregation (Kampinga et al. 1994), (ii) disruption of actin microfilament network (Lavoie et al. 1993a,b, 1995; Huot et al. 1996) and (iii) shut-off of RNA and protein synthesis (Carper et al. 1997).

III. sHsp Protection Against TNFα and Oxidative Stress Inducers

An interesting aspect of the relatively broad spectrum of protective activity mediated by sHsp concerns the effect of these proteins toward cells that undergo oxidative stress induced by different agents such as hydrogen peroxide, menadione, doxorubicin (Huot et al. 1991) or the pro-inflammatory cytokine tumor necrosis factor (TNFα; Mehlen et al. 1993, 1995a,b). TNFα is a 17-kDa multipotent cytokine produced primarily by activated macrophage which triggers the death of several transformed cells and in vivo induces the necrotic elimination of many tumor cells (Fiers et al. 1986, 1991; Sugarman et al. 1985; Beutler and Cerami 1989).

In mammalian cells, an early event that follows the binding of TNFα to its cell surface receptors is the phosphorylation of 27-kDa polypeptides (ROBAYE et al. 1989; KAUR and SAKLATVALA 1988; SCHUTZE et al. 1989), which were identified as Hsp27 phospho-isoforms (ARRIGO 1990a). The putative role of Hsp27 in response to TNFα was therefore investigated in sensitive cells that were genetically manipulated to express different levels of this protein. This was assessed in murine L929 or NIH 3T3-ras fibroblasts which are highly TNFα-sensitive and devoid of endogenous sHsp expression in absence of heat shock. We then demonstrated that human Hsp27 expression interfered with the cellular death induced by TNFα (MEHLEN et al. 1995a), a result confirmed by others (WANG et al. 1996). The protective function of sHsp against TNFα is conserved between species, hence suggesting that the active domain is localized in the conserved "crystallin domain" shared by these proteins. Moreover, this protection occurs independently of transcriptional events.

The binding of TNFα to its receptors induces a burst of reactive oxygen species (ROS), probably as a consequence of mitochondria dysfunction (YAMAUCHI et al. 1990; SCHULZE-OSTHOFF et al. 1992, 1993; GOOSSENS et al. 1995; MEHLEN et al. 1995a–c, 1996a). ROS can modulate gene expression, particularly those which are under the control of the transcription factor NF-κB (SCHRECK et al. 1991; SCHMIDT et al. 1995; KRETZ-REMY et al. 1996). At high concentration, as is the case in TNFα-treated murine L929 fibrosarcoma, ROS directly cause cytotoxic oxidative injuries (WONG et al. 1989; YAMAUCHI et al. 1990; MATSUDA et al. 1991; SCHUZE-OSTHOFF et al. 1992; HIROSE et al. 1993; MAYER and NOBLE 1994; BUTTKE and SANDSTROM 1994; GOOSSENS et al. 1995) that can be counteracted by the overexpression of detoxifiant enzymes or anti-oxidant drugs (WONG et al. 1989; MAYER and NOBLE 1994; BUTTKE and SANDSTROM 1994; SCHULZE-OSTHOFF et al. 1992, 1993; SCHMIDT et al. 1995; GOOSSENS et al. 1995). Since sHsp expression enhances the cellular resistance to hydrogen peroxide or menadione (HUOT et al. 1991, 1996; MEHLEN et al. 1993, 1996a; PREVILLE et al. 1998a,b), the activity of these proteins against TNFα is probably linked to a protective mechanism against ROS. However, no ROS detoxifiant activity has been found associated with sHsp. Hence, these proteins act probably indirectly as specialized chaperones toward detoxifiant enzymes or, on the other hand, enhance the repair of the damage caused by ROS.

An interesting feature of sHsp concerns the changes in their oligomerization and phosphorylation profile in response to oxidative stress. For example, a drastic increase in Hsp27 native molecular mass (up to 800 kDa) occurs during the 1st h of TNFα treatment. This phenomenon is transient, and by 4 h of treatment Hsp27 native size regresses drastically (<200 kDa). This transient shift in the size of Hsp27 oligomers is not observed in cells treated with anti-oxidants or overexpressing detoxifiant enzyme, suggesting that it is ROS dependent (MEHLEN et al. 1995b,c). Phosphorylation is also induced in response to TNFα and the phosphorylated Hsp27 isoforms are essentially recovered as small or medium-sized oligomers (<300 kDa). Analysis performed with non-phosphorylatable Hsp27 mutants allowed us to conclude that the large oligomers of Hsp27 represent the active form of the protein which mediates protection against TNFα. Moreover, the use of the P38 kinase inhibitor SB203580 led to the conclusion that phosphorylation, by inducing Hsp27 to concentrate in the form of small oligomers, triggers the inactivation of the protective activity of this protein against TNFα cytotoxicity (PREVILLE et al. 1998a).

In untreated murine L929 or NIH 3T3 cells, we recently observed that the expression of several sHsp, such as human Hsp27, Dm-Hsp27 or αB-crystallin, decreased the basal level of intracellular ROS; these proteins also inhibited the intracellular burst of ROS in response to TNFα. Consequently, ROS-dependent phenomena, such as lipid peroxidation, protein oxidation, and activation of the transcription factor NF-κB, were impaired in TNFα-treated L929 cells that express different kinds of sHsp (MEHLEN et al. 1996a). Oxidative stress also induces a rapid fragmentation of actin architecture, and murine Hsp25 expression protects against this effect (GUAY et al. 1997). In this respect, we have recently observed that the protection mediated by Hsp25 against actin network disruption by hydrogen peroxide is probably a consequence of the redox change mediated by this sHsp rather than a direct effect toward actin (PREVILLE et al. 1998b).

These protective properties of sHsp were found to depend on the tripeptide glutathione (MEHLEN et al. 1996a). Glutathione represents the major source of cellular thiol (MEISTER and ANDERSON 1983) and is therefore a potent detoxifiant that protects cells from oxidative stress (YAMAUCHI et al. 1990). As a consequence, sHsp do not protect against the oxidative stress mediated either by buthionine sulfoximine (BSO), a specific and essentially irreversible inhibitor of γ-glutamyl-cysteine synthetase, or diethyl maleate (DEM), which binds the free sulfhydryl groups of glutathione (MEHLEN et al. 1996a). Moreover, in cells that are devoid of constitutively expressed sHsp, for example, murine L929 or NIH 3T3 cells, sHsp expression induces a rise in glutathione but does not modify the reduced to oxidized ratio of this compound (MEHLEN et al. 1996a). It was also observed that sHsp induce glutathione to remain reduced during or immediately after oxidative stress; a phenomenon that stimulates the reducing power of the cell (X. Preville and A.-P. Arrigo, unpublished observation). How, in cells that are normally devoid of endogenous sHsp expression, exogenous sHsp can decrease the level of ROS and trigger glutathione to remain in its reduced form is not yet known. This particular property of sHsp may be related to their ability to act in vitro as protein chaperones. Indeed, sHsp may chaperone and, as such, could modulate the activity of enzymes involved in the ROS-glutathione pathway, such as glucose 6-phosphate dehydrogenase, Se-glutathione peroxidase, glutathione transferase or glutathione reductase. In this respect, preliminary results indicate that the activity of glucose 6-phosphate dehydrogenase is increased in sHsp-expressing cells and that this activity is protected during oxidative stress (X. Preville and A.-P. Arrigo, unpublished observation).

Moreover, GAENA and HARDING (1995, 1996) have reported that αB-crystallin can interfere with glucose-6-phosphate dehydrogenase inactivation by glycation or the inactivation of 6-phosphogluconate dehydrogenase by carbamylation. Hence, sHsp may act as specialized chaperones toward the NADPH(H)/glutathione pathway. Interestingly, the large oligomers of human or mouse Hsp27 which modulate ROS and glutathione levels also protect the cell against oxidative stress (MEHLEN et al. 1997; PREVILLE et al. 1998a) and display in vitro chaperone activity (JACOB and BUCHNER 1994). Moreover, a decrease in glutathione induces Hsp27 to concentrate as small oligomers while high levels of this redox modulator trigger Hsp27 to accumulate as large aggregates (MEHLEN et al. 1997). Hence, there exists a bidirectional relationship between Hsp27 and glutathione which may regulate the chaperone activity of this protein.

A model describing the putative role of Hsp27 against stress is presented in Fig. 2. It is postulated that glutathione depletion inhibits sHsp protective activity because it dissociates complexes formed between Hsp27 and non-native/oxidized proteins. In contrast, a rise in glutathione should increase the activity of these chaperone complexes by requiring more Hsp27 in the form of

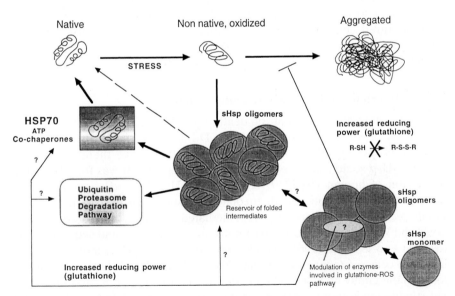

Fig. 2. Putative model of sHsp function during stress. Large sHsp oligomers modulate the activity of enzymes involved in glutathione-ROS pathway and modify their activity. This leads to a pro-reduced state that may interfere with protein aggregation and/or favors protein renaturation by chaperones. sHsp large oligomers bind denatured (or oxidized) proteins, hence creating a reservoir of polypeptides that will be either renatured by ATP-dependent chaperones or presented to the ubiquitin-dependent proteasome pathway. (Adapted in part from EHRNSPERGER et al. 1997 and LEE et al. 1997)

large oligomers. Hsp27 overexpression raises the level of glutathione and/or maintains glutathione in its reduced form probably because of an increased need of reductant to maintain the Hsp27 containing chaperone/non-native protein complexes and/or in stimulating the renaturation of non-native proteins by Hsp70 and co-chaperones.

IV. sHsp Expression Protects Against Apoptosis

Cells die by essentially two distinct processes, necrosis and apoptosis, which are characterized by distinct morphological characteristics. Necrosis is a passive degenerative event which is usually the consequence of toxic or physical insults. In this case, death leads to leakage of cell material through plasma membrane disruption, a phenomenon which in vivo induces inflammation. On the contrary, apoptosis is a more physiologic event which results in the elimination of cells through a definite program (reviewed in STELLER 1995) which requires ATP (TSUJIMOTO et al. 1997). Apoptosis, which is drastically controlled by molecular mechanisms that are conserved during ontogenesis, plays a fundamental role in the maintenance of the integrity of organisms by removing unnecessary cells (RAFF 1992; STELLER 1995; WHITE 1996; VAUX and STRASSER 1996). Moreover, various diseases are associated with dysregulated cell death (THOMPSON 1995). In vitro, apoptosis can be triggered in specific tissue cultured cells by agents or conditions which include, for example, growth factor deprivation, the Fas/APO-1/CD95 ligand, or the protein kinase inhibitor, staurosporine.

1. sHsp Interfere with In Vitro-Mediated Apoptosis

sHsp are potent antagonists of the death induced by different insults such as heat shock or oxidative stress (see Sects. D.I–III above). Usually, cell death induced by these insults is necrotic. We and others have therefore studied whether sHsp could also be effective against apoptosis. It was found that the expression of human Hsp27, *Drosophila* Dm-Hsp27 or αB-crystallin interfered with the apoptotic death mediated by staurosporine (MEHLEN et al. 1996b; see Fig. 3). A protective effect mediated by sHsp was also observed in U937 and Wehi-s cells exposed to camptothecin, etoposide and actinomycin D (SAMALI and COTTER 1996). We also investigated apoptosis induced by Fas ligand in cells overexpressing human Hsp27. Fas/APO-1/CD95 receptor belongs to the family of TNFα receptors; however, the type of death induced by Fas stimulation is apoptotic (SCHULZE-OSTHOFF et al. 1994). In vivo, TNFα and Fas have different functions: TNFα is very efficient in the necrotic eradication of tumoral cells, while Fas eliminates specific and non-pathological cells, for example, during the maturation and amplification of the immune system. It was then shown that human Hsp27 expression inhibited Fas apoptotic process in transfected L929 cell lines that constitutively express Fas/APO-1 receptor (MEHLEN et al. 1996b).

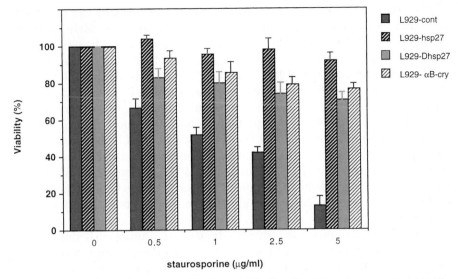

Fig. 3. Expression of human Hsp27, *Drosophila* Dm-Hsp27, or human αB-crystallin interferes with staurosporine-mediated apoptosis. Murine L929 fibrosarcoma, treated for 24 h with staurosporine (0.5–5 m*M*), was analyzed by crystal violet staining. *L929-cont.*, control cells; *L929-hsp27*, cells expressing human Hsp27; *L929-Dhsp27*, cells expressing *Drosophila* Dm-Hsp27; L929-αB-cry, cells expressing human αB-crystallin. Note the strong protection mediated by sHsp expression. (Reproduced from MEHLEN et al. 1996b, by copyright permission of the American Society for Biochemistry and Molecular Biology)

2. sHsp as Essential Anti-apoptotic Proteins During Early Cell Differentiation

We have analyzed the biological significance of the transient accumulation of Hsp27 during the early phase of several processes of differentiation. By using anti-sense technology, we observed that a 40% inhibition of Hsp27 accumulation interfered with the granulocytic differentiation of human promyelocytic HL-60 cells (CHAUFOUR et al. 1996). We also observed that an almost complete inhibition of the expression of Hsp25/27 aborted the differentiation of murine embryonic stem (ES) cells through an apoptotic process (MEHLEN et al. 1997b). Hsp25/27 can therefore be considered as a ubiquitously expressed protein during early differentiation which controls the division to differentiation transition by interfering with apoptosis.

3. Molecular Mechanisms Underlying the Anti-apoptotic Function of sHsp

The mechanism by which Hsp protect against apoptosis is not yet known. Preliminary results indicate that human Hsp27 act upstream of the caspase cascade. Another important parameter may be glutathione since the expression of sHsp as well as that of the anti-apoptotic protein Bcl-2 can modulate ROS and glutathione intracellular levels (Kane et al. 1993; Mehlen et al. 1996a; Arrigo 1998). However, the mode of action of these two proteins is different: (i) Bcl-2 is not a protein chaperone; (ii) Hsp27 is not mitochondria associated (Arrigo et al. 1988; Arrigo 1990b), as is the case for Bcl-2 (Hockenbery et al. 1990); and (iii), in contrast to Bcl-2 (Kane et al. 1993), sHsp do not protect against oxidative stress mediated by glutathione deprivation (Mehlen et al. 1996a). Apoptotic cells have been described to release reduced glutathione (Beaver and Waring 1995; van den Dobbelsteen et al. 1996) and/or to induce its oxidation independently of oxygen free radical production (Slater et al. 1995; Ghibelli et al. 1995). Consequently, increased glutathione levels can delay the apoptotic process (van den Dobbelsteen et al. 1996). Whether sHsp acts through glutathione modulation to interfere with apoptosis is an interesting possibility which should be investigated.

Concerning the anti-apoptotic activity of Hsp27 during early differentiation, it is possible that the transient increased oligomerization of this protein during early differentiation (Chaufour et al. 1996; Mehlen et al. 1997b, see also Fig. 1) reveals its interaction with specific proteins. This oligomerization occurs concomitantly with cell growth inhibition. It can therefore be speculated that Hsp25/27 prevents the aggregation of proteins that played a role during cell growth but which are no longer of use in differentiating cells (see Fig. 4). This process may ensure the proteolysis of these unwanted proteins. We have also observed that inhibition of sHsp synthesis during ES cell early differentiation diminished the transient increase in total cellular glutathione which is time correlated with Hsp27 expression (Mehlen et al. 1997b and Fig. 4A). Such a glutathione increase during early differentiation has already been observed in several cell systems and is essential for the differentiation process (Liang et al. 1991; Esposito et al. 1994; Atzori et al. 1994). Indeed, glutathione depriving drugs can block the differentiation program, probably by inhibiting transcription factors such as AP-1 and EGR-1 whose activation during differentiation is redox dependent (Esposito et al. 1994). The differentiation-mediated rise in glutathione may also regulate the metabolic changes that occur in differentiating cells. A hypothetical model describing Hsp27 function during early differentiation is presented in Fig. 4B.

E. Conclusions

sHsp are new negative regulators of programmed cell death with a broad spectrum of action. For example, sHsp interfere with TNFα-mediated cell death. Hence, in vivo, sHsp expression can (i) modulate inflammatory processes, (ii) be a parameter allowing healthy cells not to be destroyed by TNFα or (iii) allow some cancerous cells to escape the immunosurveillance mediated by this cytokine. In addition to being able to protect cells against stress-induced necrosis, sHsp share the conserved and fascinating property of inter-

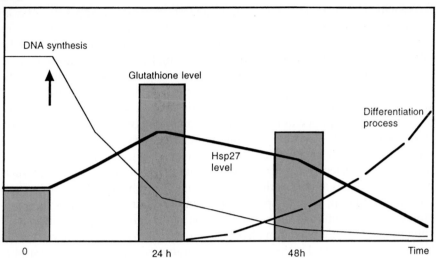

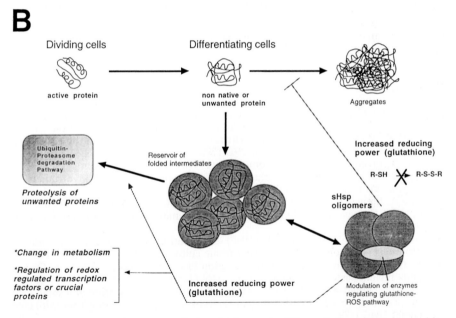

Fig. 4. A Hsp27 transient accumulation during early differentiation of murine embryonic stem cells correlates with increased level of total glutathione. This phenomenon is concomitant with increased oligomerization of Hsp27 (see also Fig. 1). *The arrow* specifies the beginning of the differentiation program (adapted from MEHLEN et al. 1997b). **B** Putative model of Hsp27 function during early differentiation. Hsp27 deprivation aborts the differentiation program and induces apoptosis. It is hypothesized that Hsp27 large oligomers bind unwanted (and probably denatured) proteins and present them to the ubiquitin-proteasome degradation pathway. The rise in glutathione (which is essential for the differentiation program) may play a role in this process and also interfere with protein aggregation. (Adapted in part from EHRNSPERGER et al. 1997 and LEE et al. 1997)

fering with apoptosis. This observation suggests that, in vivo, sHsp may play an important role in the development and integrity of organisms. An interesting example concerns the ubiquitous expression of sHsp during early cellular differentiation, a phenomenon which is essential for differentiating cells to escape death program. Hence, this particular sHsp expression represents the first example of a chaperone-mediated switch between differentiation and apoptotic programs. The massive apoptotic process which occurs after the maturation of lymphocytes also represents an interesting example where sHsp expression may be essential. This phenomenon allows the elimination of unwanted lymphocytes via mechanisms implying, among others, the Fas/APO-1 signal. The transient sHsp accumulation that takes place during the maturation of B or T cells as well as the protection mediated by these proteins against Fas suggest that, in vivo, sHsp participate in the control of lymphocyte populations. Studies aimed at understanding the molecular function of sHsp have enlightened the role played by these chaperones at the level of the intracellular redox potential and in the control of programmed cell death. Future studies will certainly give us a greater understanding of the molecular mechanism underlying the protective activity mediated by these fascinating proteins.

Acknowledgements. This work was supported by the Ligue contre le Cancer and the Région Rhône-Alpes.

References

Ahlers A, Engel K, Sott C, Gaestel M, Hermann F, Brach MA (1994) Interleukin 3 and granulocytic macrophage colony-stimulating factor induce activation of the MAPKAP kinase 2 resulting in in vitro serine phosphorylation of the small heat shock protein (Hsp27). Blood 83:1791–1798

Aoyama A, Frohli E, Schafer R, Klemenz R (1993a) Alpha B-crystallin expression in mouse NIH 3T3 fibroblasts: glucocorticoid responsiveness and involvement in thermal protection. Mol Cell Biol 13:1824–1832

Aoyama A, Steiger R, Frohli E, Schafer R, von Deimling A, Wiestler O, Klemenz R (1993b) Expression of alpha B-crystallin in human brain tumors. Int J Cancer 55:760–764

Arrigo A-P, Fakan S, Tissières A (1980) Localization of the heat shock induced proteins in Drosophila melanogaster tissue culture cells. Dev Biol 78:86–103

Arrigo A-P (1980) Investigation of the function of the heat shock proteins in Drosophila melanogaster tissue culture cells. Mol Gen Genet 178:517–524

Arrigo A-P, Ahmad-Zadeh C (1981) Immunofluorescence localization of a small heat shock protein (hsp23) in salivary gland cells of Drosophila melanogaster. Mol Gen Genet 184:73–79

Arrigo A-P, Darlix J-L, Simon M, Khandjian E, Spahr P-F (1985) Characterization of the prosome from Drosophila and its similarities with the cytoplasmic structures formed by the low molecular weight heat shock proteins. EMBO J 4:399–406

Arrigo A-P (1987) Cellular localization of hsp23 during Drosophila development and subsequent heat shock. Dev Biol 122:39–48

Arrigo A-P, Welch W (1987) Characterization and purification of the small 28,000-dalton mammalian heat shock protein. J Biol Chem 262:15359–15369

Arrigo A-P, Suhan J, Welch WJ (1988) Dynamic changes in the structure and locale of the mammalian low molecular weight heat shock protein. Mol Cell Biol 8:5059–5071

Arrigo A-P, Pauli D (1988) Characterization of HSP 27 and three immunologically related proteins during Drosophila development. Exp Cell Res 175:169–183

Arrigo A-P (1990a) Tumor necrosis factor induces the rapid phosphorylation of the mammalian heat-shock protein hsp28. Mol Cell Biol 10:1276–1280

Arrigo A-P (1990b) The monovalent ionophore monensin maintains the nuclear localization of the human stress protein hsp28 during heat shock recovery. J Cell Sci 96:419–427

Arrigo A-P, Tanguay RM (1991) Expression of heat shock proteins during development in Drosophila. In: Hightower L, Nover L (eds) Heat shock and development. Springer, Berlin Heidelberg New York, pp 106–119 (Results and problems in cell differentiation, vol 17)

Arrigo A-P, Mehlen P (1994) 1Expression, cellular location and function of low molecular weight heat shock proteins (hsp20 s) during development of the nervous system. In: Mayer J, Brown I (eds) Heat shock or cell stress proteins in the nervous system. Neuroscience Perspectives Series. Academic, New York, pp 145–167

Arrigo A-P, Landry J (1994) Expression and function of the low-molecular-weight heat shock proteins. In: Morimoto R, Tissières A, Georgopoulos C (eds) The biology of heat shock proteins and molecular chaperones. Cold Spring Harbor Laboratory Press, Cold Spring Harbor, NY, pp 335–373

Arrigo A-P (1995) Cell stress genes during development. Neuropathol Appl Neurobiol 21:488–491

Arrigo A-P (1998) Small stress proteins: chaperones that act as regulators of intracellular redox and programmed cell death. Biol Chem Hoppe Seyler 379:19–26

Atzori L, Dypbukt JM, Hybbinette SS, Moldeus P, Grafstrom RC (1994) Modification of cellular thiols during growth and squamous differentiation of cultured human bronchial epithelial cells. Exp Cell Res 211:115–120

Ayme A, Tissières A (1985) Locus 67B of Drosophila melanogaster contains seven, not four, closely related heat shock genes. EMBO J 4:2949–2954

Ayme A, Southgate R, Tissières A (1985) Nucleotide sequences responsible for the thermal inducibility of the Drosophila small heat-shock protein genes in monkey COS cells. J Mol Biol 182:469–475

Bansal GS, Norton PM, Latchman DS (1991) The 90Kd protein protects mammalian cells from the thermal stress but not from viral infection. Exp Cell Res 195:303–306

Beall AC, Kato K, Goldenring JR, Rasmussen H, Brophy CM (1997) Cyclic nucleotide-dependent vasorelaxation is associated with the phosphorylation of a small heat shock-related protein. J Biol Chem 272:11283–11287

Beaulieu JF, Arrigo A-P, Tanguay RM (1989) Interaction of Drosophila 27Kd heat shock protein with the nucleus of heat-shocked and ecdysterone-stimulated cultured cells. J Cell Sci 92:29–36

Beaver JP, Waring P (1995) A decrease in intracellular glutathione concentration precedes the onset of apoptosis in murine thymocyte. Eur J Cell Biol 68:47–54

Belka C, Ahlers A, Sott C, Gaestel M, Hermann F, Brach MA (1995) Interleukin (IL)-6 signaling leads to phosphorylation of the small heat shock protein in (Hsp)27 through the activation of the MAP kinase and MAPAKAP kinase 2 pathway in monocytes and monocytic leukemia cells. Leukemia 9:288–294

Benjamin IJ, Shelton J, Garry DJ, Richardson JA (1997) Temporospatial expression of the small HSP/alpha B-crystallin in cardiac and skeletal muscle during mouse development. Dev Dynam 208:75–84

Benndorf R, Kraft R, Otto A, Stahl J, Bohm H, Bielka H (1988) Purification of the growth-related protein p25 of Ehrlich ascites tumor and analysis of its isoforms. Biochem Int 17:225–234

Benndorf R, Hayess K, Ryazantsev S, Wieske M, Behlke J, Lutsch G (1994) Phosphorylation and supramolecular organization of murine small heat shock protein HSP25 abolish its actin polymerization-inhibiting activity. J Biol Chem 269:20780–20784

Berger EM, Woodward MP (1983) Small heat shock proteins in Drosophila may confer thermal tolerance. Exp Cell Res 147:437–442

Beutler B, Cerami A (1989) The biology of cachectin/TNF-α primary mediator of the host response. Annu Rev Immunol 7:625–655

Bhat SP, Nagineni CN (1989) αB subunit of lens-specific protein α-cristallin is present in other ocular and non-ocular tissues. Biochem Biophys Res Commun 158:319–325

Blackburn RV, Galoforo SS, Berns CM, Armour EP, McEachern D, Corry PM, Lee YJ (1997) Comparison of tumor growth between hsp25- and hsp27-transfected murine L929 cells in nude mice. Int J Cancer 72:871–877

Bories D, Raynal M-C, Solomon DH, Darzynkiewicz Z, Cayre YE (1989) Down-regulation of a serine protease, myeloblastin, causes growth arrest and differentiation of promyelocytic leukemia cells. Cell 59:959–968

Buttke TM, Sandstrom PA (1994) Oxidative stress as a mediator of apoptosis. Immunol Today 15:7–10

Carper SW, Rocheleau TA, Cimino D, Storm FK (1997) Heat shock protein 27 stimulates recovery of RNA and protein synthesis following a heat shock. J Cell Biochem 66:153–164

Carver JA, Aquilina JA, Truscott RJW (1993) An investigation into the stability of α-crystallin by NMR spectroscopy; evidence for a two-domain structure. Biochim Biophys Acta 1164:22–28

Carver JA, Aquilina JA, Truscott RJW (1994) A possible chaperone-like quaternary structure for alpha-crystallin. Exp Eye Res 59:231–234

Chaufour S, Mehlen P, Arrigo A-P (1996) Transient accumulation, phosphorylation and changes in the oligomerization of Hsp27 during retinoic acid-induced differentiation of HL-60 cells: possible role in the control of cellular growth and differentiation. Cell Stress Chap 1:225–235

Chen Q, Vierling E (1991) Analysis of conserved domains identifies a unique structural feature of a chloroplast heat shock protein. Mol Gen Genet 226:425–431

Chiesi M, Longoni S, Limbruno U (1990) Cardiac alpha-crystallin. Mol Cell Biochem 97:129–136

Ciocca DR, Adams DJ, Edwards DP, Bjercke RJ, McGuire WL (1983a) Distribution of an estrogen-induced protein with a molecular weight of 24,000 in normal and malignant human tissues and cells. Cancer Res 43:1204–1210

Ciocca DR, Asch RH, Adams DJ, McGuire WL (1983b) Evidence for modulation of a 24 K protein in human endometrium during the menstrual cycle. J Clin Endocrinol Metab 57:496–499

Ciocca DR, Fuqua SAW, Lock-Lim S, Toft DO, Welch WJ, McGuire WL (1992) Response of human breast cancer cells to heat shock and chemotherapeutic drugs. Cancer Res 52:3648–3654

Ciocca DR, Oesterreich S, Chamnes GC, McGuire WL, Fuqua SAW (1993) Biological and clinical implications of heat shock proteins 27000 (Hsp27): a review. J Natl Cancer Inst 85:1558–1570

Collier NC, Schlesinger M (1986) The dynamic state of heat shock proteins in chicken embryo fibroblasts. J Cell Biol 103:1495–1507

Craig EA, Jacobsen K (1984) Mutation of the heat inducible 70 kilodalton genes of yeast confer temperature sensitive growth. Cell 38:841–849

Crête P, Landry J (1990) Induction of HSP27 phosphorylation and thermoresistance in Chinese hamster cells by arsenite, cycloheximide, A23187 and EGTA. Radiat Res 121:320–327

Cuenda A, Rouse J, Doza Y, Meier R, Cohen P, Gallagher T, Young P, Lee J (1995) SB 203580 is a specific inhibitor of a MAP kinase homologue which is stimulated by cellular stresses and interleukin-1. FEBS Lett 364:229–233

de Jong W, Leunissen J, Voorter C (1993) Evolution of the alpha-crystallin/small heat-shock protein family. Mol Biol Evol 10:103–126

Duband J-L, Lettre F, Arrigo A-P, Tanguay RM (1986) Expression and cellular localization of HSP23 in unstressed and heat shocked Drosophila culture cells. Can J Genet Cytol 28:1088–1092

Dubin RA, Gopal-Srivastava R, Wawrousek EF, Piatigorsky J (1991) Expression of the murine αB-crystallin gene in lens and skeletal muscle: identification of a muscle-preferred enhancer. Mol Cell Biol 11:4340–4349

Edington B, Hightower L (1990) Induction of a chicken small heat shock (stress) protein: evidence of multilevel posttranscriptional regulation. Mol Cell Biol 10:4886–4898

Ehrnsperger M, Graber S, Gaestel M, Buchner J (1997) Binding of non-native protein to Hsp25 during heat shock creates a reservoir of folding intermediates for reactivation. EMBO J 16:221–229

Engelberg D, Zandi E, Parker CS, Karin M (1994) The yeast and mammalian ras pathways control transcription of heat shock genes independently of heat shock transcription factor. Mol Cell Biol 14:4929–4937

Esposito F, Agosti V, Morrone G, Morra F, Cuomo C, Russo T, Venuta S, Cimino F (1994) Inhibition of the differentiation of human myeloid cell lines by redox changes induced by glutathione depletion. Biochem J 301:649–653

Fabre-Jonca N, Gonin S, Diaz-Latoud C, Rouault JP, Arrigo AP (1995) Thermal sensitivity in NIH 3T3 fibroblasts transformed by the v-fos oncogene: correlation with reduced accumulation of 68-kDa and 25-kDa stress proteins after heat shock. Eur J Biochem 232:118–128

Fiers W, Brouckaert P, Devos R, Franzen L, Leroux-Roel G, Remaut E, Suffys P, Tavernier J, Van der Heyden J, Van Roy F (1986) Lymphokines and monokines in anti-cancer therapy. Cold Spring Harb Symp Quant Biol 51:587–595

Fiers W (1991) Tumor necrosis factor. Characterization at the molecular, cellular and in vivo level. FEBS Lett 285:199–212

Freshney NW, Rawlinson L, Guesdon F, Jones E, Cowley S, Hsuan J, Saklatvala J (1994) Interleukin-1 activates a novel protein kinase cascade that results in the phosphorylation of Hsp27. Cell 78:1039–1049

Frohli E, Aoyama A, Klemenz R (1993) Cloning of the mouse hsp25 gene and an extremely conserved hsp25 pseudogene. Gene 128:273–277

Fuqua SAW, Blum-Salingaros M, McGuire WL (1989) Induction of the estrogen-regulated "24K" protein by heat shock. Cancer Res 49:4126–4129

Gaena E, Harding JJ (1995) Molecular chaperones protect against glycation-induced inactivation of glucose-6-phosphate dehydrogenase. Eur J Biochem 231:181–185

Gaena E, Harding JJ (1996) Inhibition of 6-phosphogluconate by carbamylation and protection by α-crystallin, a chaperone-like protein. Biochem Biophys Res Commun 22:626–631

Gaestel M, Gross B, Benndorf R, Strauss M, Schunk W-H, Kraft R, Otto A, Bohm H, Stahl J, Drabsch H, Bielka H (1989) Molecular cloning, sequencing and expression in Escherichia coli of the 25-kDa growth-related protein of Ehrlich ascites tumor and its homology to mammalian stress proteins. Eur J Biochem 179:209–213

Gaestel M, Schröder W, Benndorf R, Lippman C, Buchner K, Huchot F, Ermann VA, Bielka H (1991) Identification of the phosphorylation sites of the murine small heat shock protein 25. J Biol Chem 266:14721–14724

Gaestel M, Gotthardt R, Muller T (1993) Structure and organization of a murine gene encoding small heat-shock protein Hsp25. Gene 128:279–283

Gaestel M, Benndorf R, Hayess K, Priemer E, Engel K (1992) Dephosphorylation of the small heat shock protein hsp25 by calcium/calmodulin-dependent (type 2B) protein phosphatase. J Biol Chem 267:21607–21611

Garrido C, Mehlen P, Fromentin A, Hammann A, Assem M, Arrigo A-P, Chauffert B (1996) Unconstant association between 27-kDa heat shock protein (Hsp27) content and doxorubicin resistance in human colon cancer cells. The doxorubicine-protecting effect of hsp27. Eur J Biochem 237:653–659

Garrido C, Ottavi P, Fromentin A, Hammann A, Hammann M, Arrigo A-P, Chauffert B, Mehlen P (1997) Hsp27 as a mediator of confluence-dependent resistance to cell death induced by anticancer drugs. Cancer Res 57:2661–2667

Georgopoulos C, Welch WJ (1993) Role of the major heat shock proteins as molecular chaperones. Annu Rev Cell Biol 9:601–634

Gerner EW, Schneider MJ (1975) Induced thermal resistance in HeLa cells. Nature 256:500–502

Gernold M, Knauf U, Gaestel M, Stahl J, Kloetzel P-M (1993) Development and tissue-specific distribution of mouse small heat shock protein hsp 25. Dev Genet 14:103–111

Ghibelli L, Coppola S, Rotilio G, Lafavia E, Maresca V, Ciriolo MR (1995) Non oxidative loss of glutathione in apoptosis via GSH extrusion. Biochem Biophys Res Commun 216:313–320

Glaser RI, Wolfner MF, Lis JT (1986) Spatial and temporal pattern of hsp26 expression during normal development. EMBO J 5:747–754

Gonin S, Fabre-Jonca N, Diaz-Latoud C, Rouault J-P, Arrigo A-P (1997) Transformation by T-antigen and other oncogenes delays Hsp25 accumulation in heat shock treated NIH 3T3 fibroblasts. Cell Stress Chap (in press)

Goossens V, Grooten J, De Vos K, Fiers W (1995) Direct evidence for tumor necrosis factor-induced mitochondrial reactive oxygen intermediates and their involvement in cytotoxicty. Proc Natl Acad Sci USA 92:8115–8119

Gopal-Srivastava R, Piatigorsky J (1993) The murine αB-crystallin/small heat shock protein enhancer: identification of αBE-2, αBE-3, and MRF control elements. Mol Cell Biol 13:7144–7152

Gotthardt R, Neininger A, Gaestel M (1996) The anti-cancer drug cisplatin induces Hsp25 in Ehrlich ascites tumor cells by a mechanism different from transcriptional stimulation influencing predominantly hsp25 translation. Int J Cancer 66:790–795

Groenen P, Merck K, de Jong W, Bloemendal H (1994) Structure and modifications of the junior chaperone alpha-crystallin. From lens transparency to molecular pathology. Eur J Biochem 225:1–19

Guay J, Lambert H, GingrasBreton G, Lavoie JN, Huot J, Landry J (1997) Regulation of actin filament dynamics by p38 map kinase-mediated phosphorylation of heat shock protein 27. J Cell Sci 110:357–368

Guy G, Cairns J, Ng S, Tan Y (1993) Inactivation of a redox-sensitive protein phosphatase during the early events of tumor necrosis factor/interleukin-1 signal transduction. J Biol Chem 268:2141–2148

Hammerschlag R, Maines S, Ando M (1989) Sensory glia from tadpoles but not adult bullfrogs synthesize heat shock-like proteins in vitro at non heat shock temperature. J Neurosci Res 23:416–424

Hanash SM, Strahler JR, Chan Y, Kuick R, Teichroew D, Neel JV, Hailat N, Keim DR, Gratiot-Deans J, Ungar D, Melhem R, Zhu XX, Andrews P, Loottspeich F, Eckerskorn C, Chu E, Ali I, Fox DA, Richardson BC (1993) Data base analysis of protein expression pattern during T-cell ontogeny and activation. Proc Natl Acad Sci USA 90:3314–3318

Hansen RK, Oesterreich S, Lemieux P, Sarge KD, Fuqua SAW (1997) Quercitin inhibits heat shock protein induction but not heat shock factor DNA binding in human breast carcinoma cells. Biochem Biophys Res Commun 239:851–856

Harding JJ (1991) Post-translational modification of lens protein in cataract. Exp Eye Res 52:205–212

Head M, Corbin E, Goldman J (1993) Overexpression and abnormal modification of the stress proteins alpha B- crystallin and HSP27 in Alexander disease. Am J Pathol 143:1743–1753

Heikkila JJ, Krone PH, Ovseneck N (1991) Regulation of heat shock gene expression during Xenopus development. In: Hightower L, Nover L (eds) Heat shock and development. Springer, Berlin Heidelberg New York, pp 120–137 (Results and problems in cell differentiation, vol 17)

Hernandez LD, Vierling E (1993) Expression of low molecular weight heat-shock proteins under field conditions. Plant Physiol 101:1209–1216

Hickey E, Brandon SE, Potter R, Stein G, Stein J, Weber LA (1986) Sequence and organization of genes encoding the human 27 kDa heat shock protein. Nucleic Acids Res 14:4127–4146

Hirose K, Longo DL, Oppenheim JJ, Matsushima K (1993) Overexpression of mitochondrial manganese superoxide dismutase promotes the survival of tumor cells exposed to interleukin-1, tumor necrosis factor, selected anticancers drugs, and ionizing radiation. FASEB J 7:361–368

Hockenbery D, Nunez G, Miliman C, Schreiber RD, Korsmeyer SJ (1990) Bcl-2 is an inner mitochondrial membrane protein that blocks programmed cell death. Nature 348:334–336

Hoffman E, Corces V (1986) Sequences involved in temperature and ecdysterone-induced ranscription are located in separate regions of a Drosophila melanogaster heat shock gene. Mol Cell Biol 6:663–673

Horwitz J, Huang Q-L, Ding L-L (1992) Alpha-crystallin can function as a molecular chaperone. Proc Natl Acad Sci USA 89:10449–10453

Huang LE, Zhang H, Bae SW, Liu AY (1994) Thiol reagents inhibit the heat shock response. Involvement of a redox mechanism in the heat shock signal transduction pathway. J Biol Chem 269:30718–30725

Huot J, Roy G, Lambert H, Chretien P, Landry J (1991) Increased survival after treatment with anti-cancer agents of Chinese hamster cells expressing the human 27,000 heat shock protein. Cancer Res 51:5245–5252

Huot J, Lambert H, Lavoie JN, Guimond A, Houle F, Landry J (1995) Characterization of 45-kDa/54-kDa HSP27 kinase, a stress sensitive kinase which may activate the phosphorylation-dependent protective function of mammalian 27-kDa heat shock protein HSP27. Eur J Biochem 227:416–427

Huot J, Houle F, Spitze DR, Landry J (1996) HSP27 phosphorylation-mediated resistance against actin fragmentation and cell death induced by oxidative stress. Cancer Res 56:273–279

Ingolia TD, Craig E (1982) Four small Drosophila heat shock proteins are related to each other and to mammalian alpha-crystallin. Proc Natl Acad Sci USA 79:2360–2364

Ireland RC, Berger EM (1982) Synthesis of low molecular weight heat shock peptides stimulated by ecdysterone in a cultured Drosophila cell line. Proc Natl Acad Sci USA 79:855–859

Issels RD, Meier F, Muller E, Multhoff G, Wilmanns W (1993) Ifosfamide induced stress response in human lymphocytes. Mol Aspects Med 14:281–286

Ito H, Hasegawa K, Inaguma Y, Kosawa O, Kato K (1996) Enhancement of stress-induced synthesis of hsp27 and alpha B crystallin by modulators of the arachidonic acid cascade. J Cell Physiol 166:332–339

Ito H, Okamoto K, Kato K (1997a) Prostaglandins stimulate the stress-induced synthesis of hsp27 and alpha B crystallin. J Cell Physiol 170:255–262

Ito H, Okamoto K, Nakayama H, Isobe T, Kato K (1997b) Phosphorylation of alpha-crystallin in response to various types of stress. J Biol Chem 272:29934–29941

Iwaki T, Kume-Iwaki A, Liem RKH, Goldman JE (1989) αB-crystallin is expressed in non-lenticular tissues and accumulates in alexander's disease brain. Cell 57:71–78

Iwaki T, Kume-Iwaki A, Goldman J (1990) Cellular distribution of alpha B-crystallin in non-lenticular tissues. J Histochem Cytochem 38:31–39

Iwaki T, Tateishi J (1991) Immunohistochemical demonstration of alphaB-crystallin in harmatomas of tuberous sclerosis. Am J Pathol 139:1303–1308

Iwaki T, Iwaki A, Tateishi J, Sakaki Y, Goldman J (1993) Alpha B-crystallin and 27-kd heat shock protein are regulated by stress conditions in the central nervous system and accumulate in Rosenthal fibers. Am J Pathol 143:487–495

Jakob U, Gaestel M, Engels K, Buchner J (1993) Small heat shock proteins are molecular chaperones. J Biol Chem 268:1517–1520

Jakob U, Buchner J (1994) Assisting spontaneity: the role of hsp90 and small hsp as molecular chaperones. Trends Biochem Sci 19:205–211

Kampinga HH, Brunsting JF, Stege GJ, Konings AW, Landry J (1994) Cells overexpressing Hsp27 show accelerated recovery from heat-induced nuclear protein aggregation. Biochem Biophys Res Commun 204:1170–1177

Kane DJ, Sarafian TA, Anton R, Hahn H, Gralla EB, Valentine JS, Ord T, Bredesen DE (1993) Bcl-2 inhibition of neural death: decreased generation of reactive oxygen species. Science 262:1274–1277

Kasahara K, Ikuta T, Chida K, Asakura R, Kuroki T (1993) Rapid phosphorylation of 28-kDa heat shock protein by treatment with okadaic acid and phorbol ester of BALBB/MK-2 mouse keratinocytes. Eur J Biochem 213:1101–1107

Kato S, Hirano A, Umahara T, Llena J, Herz F, Ohama E (1992) Ultrastructural and immunohistochemical studies on ballooned cortical neurons in Creutzfeldt-Jakob disease: expression of alpha B-crystallin, ubiquitin and stress-response protein 27. Acta Neuropathol (Berl) 84:443–448

Kato K, Hasegawa K, Goto S, Inaguma Y (1994) Dissociation as a result of phosphorylation of an aggregated form of the small stress protein, hsp27. J Biol Chem 269:11274–11278

Kaur P, Saklatvala J (1988) Interleukin 1 and tumor necrosis factor increase phosphorylation of fibroblast proteins. FEBS Lett 241:6–10

Khalid H, Tsutsumi K, Yamashita H, Kishikawa M, Yasunaga A, Shibata S (1995) Expression of the small heat shock protein (hsp) 27 in human astrocytomas correlates with histologic grades and tumor growth fractions. Cell Mol Neurobiol 15:257–268

Kim Y-J, Shuman J, Sette M, Przybyla A (1984) Nuclear localization and phosphorylation of three 25-kilodalton rat stress proteins. Mol Cell Biol 4:468–474

Klemenz R, Frohli E, Steiger RH, Schafer R, Aoyama A (1991) αB-Crystallin is a small heat shock protein. Proc Natl Acad Sci USA 88:3652–3656

Klemenz R, Andres A-C, Froehli E, Schaefer R, Aoyama A (1993) Expression of the murine small heat shock protein hsp25 and αB-crystallin in the absence of stress. J Cell Biol 120:639–645

Knauf U, Bielka H, Gaestel M (1992) Over-expression of the small heat-shock protein, hsp25, inhibits growth of Ehrlich ascites tumor cells. FEBS Lett 3:297–302

Knauf U, Bielka H, Gaestel M (1992) Over-expression of the small heat-shock protein, hsp25, inhibits growth of Ehrlich ascites tumor cells. FEBS Lett 3:297–302

Knauf U, Jakob U, Engel K, Buchner J, Gaestel M (1994) Stress- and mitogen-induced phosphorylation of the small heat shock protein Hsp25 by MAPKAP kinase 2 is not essential for chaperone properties and cellular thermoresistance. EMBO J 13:54–60

Kretz-Remy C, Mehlen P, Mirault M-E, Arrigo A-P (1996) Inhibition of IκB-α phosphorylation and degradation and subsequent NF-κB activation by glutathione peroxidase overexpression. J Cell Biol 133:1083–1093

Kurtz S, Rossi J, Petko L, Lindquist S (1986) An ancient developmental induction: heat-shock proteins induced in sporulation and oogenesis. Science 231:1154–1157

Landry JD, Bernier P, Chrétien P, Nicole L, Tanguay RM, Marceau N (1982) Synthesis and degradation of heat shock proteins during development and decay of thermotolerance. Cancer Res 42:2457–2461

Landry J, Chretien P, Lambert H, Hickey E, Weber LA (1989) Heat shock resistance conferred by expression of the human hsp27 gene in rodent cells. J Cell Biol 109:7–15

Landry J, Chretien P, Laszlo A, Lambert H (1991) Phosphorylation of HSP27 during development and decay of thermotolerance in Chinese hamster cells. J Cell Physiol 147:93–101

Landry J, Lambert H, Zhou M, Lavoie JN, Hickey E, Weber LA, Anderson CW (1992) Human HSP 27 is phosphorylated at serines 78 and 82 by heat shock and mitogen-activated kinases that recognize the same amino acid motif as S6 kinase II. J Biol Chem 267:794–803

Lavoie JN, Gingras-Breton G, Tanguay RM, Landry J (1993a) Induction of Chinese hamster HSP27 gene expression in mouse cells confers resistance to heat shock. HSP27 stabilization of the microfilament organization. J Biol Chem 268: 3420–3429

Lavoie JN, Hickey E, Weber LA, Landry J (1993b) Modulation of actin microfilament dynamics and fluid phase pinocytosis by phosphorylation of Heat Shock Protein 27. J Biol Chem 268:24210–24214

Lavoie JN, Lambert H, Hickey E, Weber LA, Landry J (1995) Modulation of cellular thermoresistance and actin filament stability accompanies phosphorylation-induced changes in the oligomeric structure of heat shock protein 27. Mol Cell Biol 15:505–516

Lee GJ, Roseman AM, Saibil HR, Vierling E (1997) A small heat shock protein stably binds heat-denatured model substrates and can maintain a substrate in a folding-competent state. EMBO J 16:659–671

Li GC, Werb Z (1982) Correlation between synthesis of heat shock proteins and development of thermotolerance in chinese hamster fibroblasts. Proc Natl Acad Sci USA 79:3918–3922

Li GC, Li L, Liu YU, Mak JY, Chen L, Lee WMF (1991) Thermal response of rat fibroblasts stably transfected with the human hsp70-kDa heat shock protein en-coding gene. Proc Natl Acad Sci USA 88:1681–1685

Liang SM, Liang CM, Hargrove ME, Ting CC (1991) Regulation by glutathione of the effects of lymphokines on differentiation of primary activated lymphocytes. Influence of glutathione on cytotoxic activity of CD3-AK cells. J Immunol 146:1909–1913

Longoni S, Lattonen S, Bullock G, Chiesi M (1990) Cardiac alpha-crystallin. II. Intra-cellular localization. Mol Cell Biochem 97:121–128

Lowe J, McDermott H, Pike I, Spendlove I, Landon M, Mayer RJ (1992) αB crystallin expression in non-lenticular tissues and selective presence in ubiquinated inclusion bodies in human disease. J Pathol 166:61–68

Ludwig S, Engel K, Hoffmeyer A, Sithanandam G, Neufeld B, Palm D, Gaestel M, Rapp UR (1996) 3pK, a novel MAP kinase activated protein kinase is targeted by three MAP kinase pathways. Mol Cell Biol 16:6687–6697

Marin R, Valet J, Tanguay RM (1993) hsp23 and hsp26 exhibit distinct spatial and temporal patterns of constitutive expression in Drosophila adults. Dev Genet 14:69–77

Marin R, Tanguay RM (1996) Stage-specific localization of the small heat shock protein Hsp27 during oogenesis in Drosophila melanogaster. Chromosoma 105: 142–149

Martins LM, Earnshaw WC (1997) Apoptosis: alive and kicking in 1997. Trends Cell Biol 7:111–114

Matsuda M, Masutani H, Nakamura H, Miyajima S, Yamauchi A, Yonehara S, Uchida A, Irimajiri K, Horiuchi A, Yodoi J (1991) Protective activity of adult T cell leukemia-derived factor (ADF) against tumor necrosis factor-dependant cytotox-icity on U937 cells. J Immunol 147:3837–3841

Mayer M, Noble M (1994) N-Acetyl-L-cysteine is a pluripotent protector against cell death and enhancer of trophic factor-mediated cell survival in vitro. Proc Natl Acad Sci USA 91:7496–7500

McGuire SE, Fuqua SAW, Naylor SL, Helin-Davis DA, McGuire WL (1989) Chromo-somal assignement of the human 27-kDa heat shock protein gene family. Somat Cell Mol Genet 15:167–171

Mehlen P, Briolay J, Smith L, Diaz-Latoud C, Fabre N, Pauli D, Arrigo A-P (1993) Analysis of the resistance to heat and hydrogen peroxide stresses in COS cells transiently expressing wild type or deletion mutants of the Drosophila 27-kDa heat shock protein. Eur J Biochem 215:277–284

Mehlen P, Arrigo A (1994) The serum-induced phosphorylation of mammalian hsp27 correlates with changes in its intracellular localization and levels of oligomeriza-tion. Eur J Biochem 221:327–334

Mehlen P, Preville X, Chareyron P, Briolay J, Klementz R, Arrigo AP (1995a) Constitutive expression of Human hsp27, Drosophila hsp27, or Human αB-

Crystallin confers resistance to TNF- and oxidative stress-induced cytotoxicity in stably transfected murine L929 fibroblasts. J Immunol 154:363–374

Mehlen P, Mehlen A, Guillet D, Preville X, Arrigo A-P (1995b) Tumor necrosis factor-α induces changes in the phosphorylation, cellular localization, and oligomerization of human hsp27, a stress protein that confers cellular resistance to this cytokine. J Cell Biochem 58:248–259

Mehlen P, Kretz-Remy C, Briolay J, Fostan P, Mirault M-E, Arrigo A-P (1995c) Intracellular reactive oxygen species as apparent modulators of hsp27 structural organization and phosphorylation in basal and tumor necrosis factor alpha-treated T47D human carcinoma cells. Biochem J 312:367–375

Mehlen P, Kretz-Remy C, Preville X, Arrigo AP (1996a) Human hsp27, Drosophila hsp27 and αB-crystallin expression-mediated increase in glutathione is essential for the protective activity of these proteins against TNFα-induced cell death. EMBO J 15:2695–2706

Mehlen P, Schulze-Osthoff K, Arrigo AP (1996b) Small stress proteins as novel regulators of apoptosis. J Biol Chem 271:16510–16514

Mehlen P, Weber L, Hickey E, Arrigo A-P (1997a) Large unphosphorylated aggregates as the active form of hsp27 which controls intracellular reactive oxygen species and glutathione levels and generates a protection against TNFα in NIH-3T3-ras cells. Biochem Biophys Res Commun (in press)

Mehlen P, Mehlen A, Godet J, Arrigo A-P (1997b) Hsp27 as a shift between differentiation and apoptosis in embryonic stem cells. J Biol Chem 31657–31665

Mendelsohn ME, Zhu Y, O'Neill S (1991) The 29 kDa proteins phosphorylated in thrombin activated human platelets are forms of the estrogen receptor-related 27 kDa heat shock protein. Proc Natl Acad Sci USA 88:11212–11216

Meister A, Anderson ME (1983) Glutathione. Annu Rev Biochem 52:711–760

Merck K, De Haard-Hoekman W, Oude Essink B, Bloemendal H, De Jong W (1992) Expression and aggregation of recombinant alpha A-crystallin and its two domains. Biochim Biophys Acta 1130:267–276

Miron T, Vancompernolle K, Vandekerckhove J, Wilchek M, Geiger B (1991) A 25-kD inhibitor of actin polymerization is a low molecular mass heat shock protein. J Cell Biol 114:255–261

Morimoto RI, Tissières A, Georgopoulos C (1994) Heat shock proteins and molecular chaperones. Cold Spring Harbor Laboratory Press, Cold Spring Harbor

Navarro D, Cabrera JJ, Falcon O, Jimenez P, Ruiz A, Chirino R, Lopez A, Rivero JF, Diaz-Chico JC, Diaz-Chico BN (1989) Monoclonal antiboby characterization of progesterone receptors, estrogen receptors and the stress-responsive protein of 27 kDa (SRP27) in human uterine leiomyoma. J Steroid Biochem 34:491–498

Ngo JT, Klisak I, Dubin RA, Piatigorsky J, Mohandas T, Sparkes RS, Bateman JB (1989) Assignment of the alpha B-crystallin gene to human chromosome 11. Genomics 5:665–669

Nicholl ID, Quinlan RA (1994) Chaperone activity of α-crystallins modulates intermediate filament assembly. EMBO J 13:954–953

Nover L, Scharf K-D, Neumann D (1989) Cytoplasmic heat shock granules are formed from precursor particles and are associated with a specific set of mRNAs. Mol Cell Biol 9:1298–1308

Oesterreich S, Weng C-N, Qiu M, Hilsenbeck SG, Osborne CK, Fuqua SAW (1993) The small protein hsp27 is correlated with growth and drug resistance in human breast cancer cell lines. Cancer Res 53:4443–4448

Pauli D, Tissières A (1990) Developmental expression of the heat shock genes in Drosophila melanogaster. In: Morimoto R, Tissières A, Georgopoulos C (eds) Stress proteins in biology and medicine. Cold Spring Harbor Press, New York, pp 361–378

Pauli D, Tonka C-H, Tissières A, Arrigo A-P (1990) Tisue-specific expression of the heat shock protein HSP 27 during Drosophila melanogaster development. J Cell Biol 111:817–828

Pauli D, Arrigo A-P, Tissières A (1992) Heat shock response in Drosophila. Experientia 48:623–627

Piotrowicz RS, Weber LA, Hickey E, Levin EG (1995) Accelerated growth and senescence of arterial endothelial cells expressing the small molecular weight heat shock protein Hsp27. FASEB J 9:1079–1084

Plesofsky-Vig N, Brambl R (1990) Gene sequence and analysis of hsp30, a small heat shock protein of Neurospora crassa which associates with mitochondria. J Biol Chem 265:15432–15440

Porter W, Wang F, Wang W, Duan R, Safe S (1996) Role of estrogen receptor/Sp1 complexes in estrogen-induced heat shock protein 27 gene expression. Mol Endocrinol 10:1371–1378

Preville X, Mehlen P, Fabre-Jonca N, Chaufour S, Kretz-Remy C, Michel MR, Arrigo A-P (1996) Biochemical and immunofluorescence analysis of the constitutively expressed hsp27 stress protein in growing monkey CV-1 cells. J Biosci 21: 1–14

Preville X, Schultz H, Knauf U, Gaestel M, Arrigo A-P (1998a) Analysis of the role of Hsp25 phosphorylation reveals the importance of the oligomerization state of this small heat shock protein in its protective function against TNFα- and hydrogen peroxide induced cell death. J Cell Biochem (in press)

Preville X, Gaestel M, Arrigo A-P (1998b) Phosphorylation is not essential for Hsp25 protective activity against H_2O_2-mediated disruption of L929 cells actin cytoskeleton; a protection which appears related to the redox change mediated by Hsp25. Cell Stress Chap (in press)

Puy LA, Castro GL, Olcese JE, Lotfi HO, Brandi HR, Ciocca DR (1989) Analysis of a 24-Kilodalton protein in the human uterine cervix during abnormal growth. Cancer 64:1067–1073

Quax-Jeuken Y, Quax W, van Rens GLN, Meera Khan P, Bloemendal H (1985) Assignement of the human αA-crystallin gene (CRYA1) to chromosome 21. Cytogenet Cell Biol 40:727–728

Raff MC (1992) Social control on cell survival and cell death. Nature 356:397–400

Rahman DRJ, Bentley NJ, Tuite MF (1995) The Saccharomyces cerevisiae small heat shock protein Hsp26 inhibits actin polymerisation. Biochem Soc Trans 23:S77

Raschke EG, Baumann G, Schöffl B (1988) Nucleotide sequence analysis of soybean small heat shock protein genes belonging to two different multigenic famillies. J Mol Biol 199:549–557

Renkawek K, de Jong W, Merck K, Frenken C, van Workum F, Bosman G (1992) alpha B-crystallin is present in reactive glia in Creutzfeldt-Jakob disease. Acta Neuropathol (Berl) 83:324–327

Renkawek K, Bosman G, Gaestel M (1993) Increased expression of heat-shock protein 27 kDa in Alzheimer disease: a preliminary study. Neuroreport 5:14–16

Riabovol KT, Mizzen LA, Welch WJ (1988) Heat shock is lethal to fibroblasts microinjected with antibody against Hsp70. Science 242:433–436

Richards E, Hickey E, Weber L, Master J (1996) Effect of overexpression of the small heat shock protein HSP27 on the heat and drug sensitivities of human testis tumor cells. Cancer Res 56:2446–2451

Riddihough G, Pelham HRB (1987) An ecdysone response element in the Drosophila hsp27 promoter. EMBO J 6:3729–3734

Robaye B, Hepburn A, Lecocq R, Fiers W, Boeynaems JM, Dumont JE (1989) Tumor necrosis factor-α induces the phosphorylation of 28 kDa stress proteins in endothelial cells: possible role in protection against cytotoxicity? Biochem Biophys Res Commun 163:301–308

Robinson ML, Overbeek PA (1996) Differential expression of alpha A- and alpha B-crystallin during murine ocular development. Invest Ophthalmol Visual Sci 37:2276–2284

Rollet E, Lavoie JN, Landry J, Tanguay RM (1992) Expression of Drosophila's 27 kDa heat shock protein into rodent cells confers thermal resistance. Biochem Biophys Res Commun 185:116–120

Rouse J, Cohen, Trigon S, Morange M, Alonsollamazares A, Zamanillo D, Hunt T, Nebreda AR (1994) A novel kinase cascade triggered by stress and heat shock that stimulates MAPKAP kinase-2 and phosphorylation of the small heat shock proteins. Cell 78:1027–1037

Samali A, Cotter TG (1996) Heat shock proteins increase resistance to apoptosis. Exp Cell Res 223:163–170

Sanchez Y, Lindquist S (1990) Hsp104 is required for induced thermotolerance. Science 248:1112–1115

Santell L, Bartfield NS, Levin EG (1992) Identification of a protein transiently phosphorylated by activators of endothelial cell function as the heat shock protein HSP27. A possible role for protein kinase C. Biochem J 284:705–710

Scheier B, Foletti A, Stark G, Aoyama A, Dobbeling U, Rusconi S, Klemenz R (1996) Glucocorticoids regulate the expression of the stress protein alpha B-crystallin. Mol Cell Endocrinol 123:187–198

Schmidt KN, Amstad P, Cerutti P, Baeuerle PA (1995) The roles of hydrogen peroxide and superoxide as messengers in the activation of transcription factor NF-κB. Chem Biol 2:13–22

Schreck R, Rieber P, Baeuerle PA (1991) Reactive oxygen intermediates as apparently widely used messengers in the activation of the NF-κB transcription factor and HIV-1. EMBO J 10:2247–2258

Schulze-Osthoff K, Bakker AC, Vandevoorde V, Baeyaert R, Haegeman G, Fiers W (1992) Cytotoxic activity of tumor necrosis factor is mediated by early damage of mitochondrial functions. Evidence for the involvement of mitochondrial radical generation. J Biol Chem 267:5317–5323

Schulze-Osthoff K, Baeyaert R, Vandevoorde V, Haegerman G, Fiers W (1993) Depletion of the mitochondrial electron transport abrogates the cytotoxic and gene-inducible effects of TNF. EMBO J 12:3095–3104

Schulze-Osthoff K, Krammer PH, Droge W (1994) Divergent signalling via APO/Fas and the TNF receptor, two homologous molecules involved in physiological cell death. EMBO J 13:4587-4596

Schutze S, Scheurich P, Pfizenmaier K, Kronke M (1989) Tumor necrosis factor signal transduction. Tissue-specific serine phosphorylation of a 26-kDa cytosolic protein. J Biol Chem 264:3562–3567

Shakoori AR, Oberdorf AM, Owen TA, Weber LA, Hickey E, Stein JL, Lian JB, Stein GS (1992) Expression of heat shock genes during differentiation of mammalian osteoblasts and promyelocytic leukemia cells. J Cell Biochem 48:277–287

Shinohara H, Inaguma Y, Goto S, Inagaki T, Kato K (1993) Alpha B crystallin and HSP28 are enhanced in the cerebral cortex of patients with Alzheimer's disease. J Neurol Sci 119:203–208

Siezen R, Bindels J, Hoenders H (1978a) The quaternary structure of bovine alpha-crystallin. Size and charge microheterogeneity: more than 1000 different hybrids? Eur J Biochem 91:387–396

Siezen RJ, Bindel JG, Hoenders HJ (1978b) The quaternary structure of bovine alpha-crystallin from old human eye lenses. Eur J Biochem 91:387–396

Siezen R, Bindels J, Hoenders H (1980) The quaternary structure of bovine alpha-crystallin. Effects of variation in alkaline pH, ionic strength, temperature and calcium ion concentration. Eur J Biochem 111:435–444

Siezen R, Jgu B (1982) Stepwise dissociation/denaturation and reassociation/renaturation of bovine alpha-crystallin in urea and guanidine hydrochloride: sedimentation, fluorescence, near-ultraviolet and far-ultraviolet circular dichroism studies. Exp Eye Res 34:969–983

Slater AFG, Nobel CSI, Maellaro E, Bustamante J, Kimland M, Orrenius S (1995) Nitrone spin traps and a nitroxide antioxidant inhibit a common pathway of thymocyte apoptosis. Biochem J 306:771–778

Southgate R, Ayme A, Voellmy R (1983) Nucleotide sequence analysis of the drosophila small heat shock gene cluster at locus 67B. J Mol Biol 165:35–57

Spector A, Li LK, Augusteyn RC, Schneider A, Freund T (1971) α-cristallin: the isolation and characterization of distinct macromolecular fractions. Biochem J 124:337–343

Spector NL, Samson W, Ryan C, Gribben J, Urba W, Welch WJ, Nadler LM (1992) Growth arrest of human B lymphocytes is accompanied by the induction of the low molecular weight mammalian heat shock protein (Hsp28). J Immunol 148:1668–1673

Spector N, Ryan C, Samson W, Nadler LM, Arrigo A-P (1993) Heat shock protein is a unique marker of growth arrest during macrophage differentiation of HL-60 cells. J Cell Physiol 156:619–625

Spector N, Mehlen P, Ryan C, Hardy L, Samson W, Nadler LM, Fabre N, Arrigo A-P (1994) Regulation of the mammalian 28 kDa heat shock protein by retinoic acid during differentiation of human leukemic HL-60 cells. FEBS Lett 337:184–188

Spector N, Hardy L, Ryan C, Miller WH, Humes JL, Nadler LM, Luedke E (1995) 28-kDa mammalian heat shock protein, a novel substrate of a growth regulatory protease involved in differentiation of human leukemia cells. J Biol Chem 270:1003–1006

Srinivasan A, Nagineni C, Bhat S (1992) alpha A-crystallin is expressed in non-ocular tissues. J Biol Chem 267:23337–23341

Stahl J, Wobus AM, Ihrig S, Lutsch G, Bielka H (1992) The small heat shock protein hsp25 is accumulated in P19 embryonal carcinoma cells and embryonic stem cells of line BLC6 during differentiation. Differentiation 51:33–37

Steller H (1996) Mechanisms and gene of cell suicide. Science 267:1445–1449

Stokoe D, Engel K, Campbell D, Cohen P, Gaestel M (1992) Identification of MAPKAP kinase 2 as a major enzyme responsible for the phosphorylation of the small mammalian heat shock proteins. FEBS Lett 313:307–313

Sugarman BJBB, Aggarwal PE, Hass IS, Figari MA, Palladino H, Shepard HM (1985) Recombinant human tumor necrosis factor-α: effects on proliferation of normal and transformed cells in vitro. Science 230:943–945

Surewicz W, Olesen P (1995) On the thermal stability of alpha-crystallin: a new insight from infrared spectroscopy. Biochemistry 34:9655–9660

Tanguay RM, Wu Y, Khandjian EW (1993) Tissue-specific expression of heat shock proteins of the mouse in the absence of stress. Dev Genet 14:112–118

Tatzelt J, Zuo J, Voellmy R, Scott M, Hartl U, Prusiner SB, Welch WJ (1995) Scrapie prions selectively modify the stress response in neuroblastoma cells. Proc Natl Acad Sci USA 92:2944–2948

Têtu B, Lacasse B, Bouchard H-L, Lagacé R, Huot J, Landry J (1992) Prognostic influence of HSP-27 expression in malignant fibrous histiocytoma: a clinico-pathological and immunohistochemical study. Cancer Res 52:2325–2328

Têtu B, Brisson J, Landry J, Huot J (1995) Prognostic significance of heat-shock protein-27 in node-positive breast carcinoma: an immunohistochemical study. Breast Cancer Res Treat 36:93–97

Thompson CB (1995) Apoptosis in the pathogenesis and treatment of deseases. Science 267:1456–1462

Thor A, Benz C, Moore D, Goldman E, Edgerton S, Landry J, Schwartz L, Mayall B, Hickey E, Weber LA (1991) Stress response protein (srp-27) determination in primary human breast carcinomas: clinical, histologic, and prognostic correlations. J Natl Cancer Inst 83:170–178

Tsujimoto Y (1997) Apoptosis and necrosis: intracellular ATP level as a determinant for cell death modes. Cell Death Differ 4:427–434

Uozaki H, Oriuchi H, Ishida T, Iijima T, Imamura T, Machinami R (1997) Overexpression of resistance-related proteins (Metallothioneins, Glutathione-S-transferase π, Heat shock protein 27 and lung resistance-related protein) in osteosarcoma. Cancer 79:2336–2344 van den Dobbelsteen DJ, Nobel CSI, Schlegel J, Cotgreave IA, Orrenius S, Slater AFG (1996) Rapid and specific efflux of reduced glutathione during apoptosis induced by anti-Fas/APO-1 antibody. J Biol Chem 271:15420–15427

van den Oetelaar P, van Someren P, Thomson J, Siezen R, Hoenders H (1990) A dynamic quaternary structure of bovine alpha-crystallin as indicated from inter-molecular exchange of subunits. Biochemistry 29:3488–3493

Vannoort JM, Vansechel AC, Bajramovic JJ, Elouagmiri M, Polman CH, Lassmann H, Ravid R (1995) The small heat-shock protein alpha B-crystallin as candidate autoantigen in multiple sclerosis. Nature 375:798–801

Vaux DL, Strasser A (1996) The molecular biology of apoptosis. J Biol Chem 93:2239–2244

Vierling E, Nagao RT, Derocher E, Harris LM (1988) A heat shock protein localized to chloroplasts is a member of an eukaryotic superfamilly of heat shock proteins. EMBO J 7:575–581

Walsh MT, Sen AC, Chakrabarti B (1991a) Micellar subunit assembly in a three-layer model of oligomeric α-crystallin. J Biol Chem 266:20079–20084

Walsh D, Li K, Crowther C, Marsh D, Edwards M (1991b) Thermotolerance and heat shock response during early development of the mammalian embryo. In: Hightower L, Nover L (eds) Heat shock and development. Springer, Berlin Heidelberg New York, pp 58–70 (Results and problems in cell differentiation, vol 17)

Wang G, Klostergard J, Khodadadian M, Wu J, Wu TW, Fung KP, Carper SW, Tomasovic SP (1996) Murine cells transfected with human Hsp27 cDNA resist TNF-induced cytotoxicity. J Immunother Emphasis Tumor Immunol 19:9–20

Waters ER, Lee GJ, Vierling E (1996) Evolution, structure and function of the small heat shock proteins in plants. J Exp Bot 47:325–338

Welch WJ (1985) Phorbol ester, calcium ionophore, or serum added to quiescent rat embryo fibroblasts cells all result in the elevated phosphorylation of two 28,000-Dalton mammalian stress proteins. J Biol Chem 260:3058–3062

White E (1996) Life, death, and the pursuit of apoptosis. Genes Dev 10:1–15

Wistow G (1985) Domain structure and evolution in α-crystallins and small heat-shock proteins. FEBS Lett 181:1–6

Wistow G (1993) Possible tetramer-based quaternary structure for alpha-crystallins and small heat shock proteins. Exp Eye Res 56:729–732

Wistow GJ, Piatigorsky J (1988) Lens crystallins: the evolution and expression of proteins for a highly specialized tissue. Annu Rev Biochem 57:479–504

Wong GHW, Elwell JE, Oberby LW, Goeddel D (1989) Manganous superoxide dismutase is essential for cellular resistance to tumor necrosis factor. Cell 58:923–931

Wotton D, Freeman K, Shore D (1996) Multimerization of Hsp42p, a novel heat shock protein of Saccharomyces cerevisae, is dependent on a conserved carboxyl-terminal sequence. J Biol Chem 271:2717–2723

Yamauchi N, Kuriyama YH, Watanabe N, Neda H, Maeda M, Himeno T, Tsuji Y (1990) Suppressive effects of intracellular glutathione on hydroxyl radical produc-tion induced by tumor necrosis factor. Int J Cancer 46:884–888

Yost H, Petersen R, Lindquist S (1990) RNA metabolism: strategies for regulation in the heat shock response. Trends Genet 6:223–227

Zantema A, Jong ED, Lardenoije R, Eb AJVD (1989) The expression of heat shock protein hsp27 and a complexed 22-kilodalton protein is inversely correlated with oncogenicity of adenovirus transformed cells. J Virol 63:3368–3375

Zantema A, Vries MV-D, Maasdam D, Bol S, Eb AVD (1992) Heat shock protein 27 and αB-crystallin can form a complex, which dissociates by heat shock. J Biol Chem 267:12936–12941

Zhou M, Lambert H, Landry J (1993) Transient activation of a distinct serine protein kinase is responsible for 27-kDa heat shock protein phosphorylation in mitogen-stimulated and heat-shocked cells. J Biol Chem 268:35–43

CHAPTER 6

Ubiquitin and the Stress Response

C.M. PICKART

A. Introduction

Cells respond to stresses such as elevated temperature, heavy metals, and amino acid analogs by inducing the transcription of a set of genes whose products, known as stress proteins, enhance survival under stress conditions. The major shared property of conditions and agents which induce the stress response is the ability to damage cellular proteins. Protein damage is a key event in the induction of the stress response, as indicated by the finding that forced production of denatured proteins triggers the synthesis of heat shock proteins at sub-heat shock temperatures (GOFF and GOLDBERG 1985; ANANTHAN et al. 1986). Moreover, reducing intracellular levels of damaged proteins is a principal objective of the stress response, because protein damage can be highly toxic: the denaturation of a protein not only causes the loss of function of that specific molecule, but may also, through the improper exposure of hydrophobic amino acid side chains, lead to the aggregation of other proteins. Levels of damaged proteins can be reduced in two ways (PARSELL and LINDQUIST 1993; SHERMAN and GOLDBERG 1996; GOTTESMAN et al. 1997). On the one hand, specific molecular chaperones can prevent the aggregation of damaged proteins, and catalyze their refolding. On the other hand, specific proteases can degrade damaged proteins. These two strategies for coping with stress-denatured proteins, namely salvage and elimination, may not be fully independent. Nor are the essential functions of chaperones and proteases restricted to stress conditions.

In bacteria, the ATP-dependent Lon protease is a heat shock protein and plays a prominent role in the elimination of damaged proteins in stress conditions (GOFF and GOLDBERG 1985). The ATP-dependent Clp protease also plays a role in the stress response (PARSELL and LINDQUIST 1993; SHERMAN and GOLDBERG 1996; GOTTESMAN et al. 1997). Several lines of evidence suggest that the ubiquitin-proteasome pathway, which is absent in prokaryotes, serves the corresponding function in eukaryotic cells. The ubiquitin pathway is essential, and it represents the principal mechanism for degrading short-lived proteins in eukaryotic cells (CIECHANOVER et al. 1984). Although recent investigations have highlighted the pathway's role in controlling the levels of key regulators of normal cellular function, such as cyclins (KING et al. 1996) and transcription factors (HOCHSTRASSER 1996), the pathway was first described as a mechanism for the selective degradation of abnormal proteins containing amino acid analogs (Hershko et al. 1982; Ciechanover et al. 1984). These abnormal proteins model the damaged species produced in stressed cells. The inference that the ubiquitin pathway plays an important role in the eukaryotic stress response was subsequently confirmed: certain components of the pathway are stress proteins, and the covalent tagging of proteins by ubiquitin, which is the molecular signal for substrate recognition by the proteasome, is augmented in heat shock and other stress conditions. Accumulating evidence suggests that the pathway is responsible for the clearance of damaged proteins from multiple subcellular compartments in stressed cells. This chapter will focus on the role of the ubiquitin-proteasome pathway in the eukaryotic stress response. Although parallels in well-characterized prokaryotic systems will be mentioned, the reader is directed to excellent earlier reviews for a detailed discussion of the relationship between proteolysis and the stress response in prokaryotes (Parsell and Lindquist 1993; Hayes and Dice 1996; Sherman and Goldberg 1996; Gottesman et al. 1997).

B. The Ubiquitin-Proteasome Pathway

In order to discuss how the ubiquitin-proteasome pathway functions in the stress response, a brief review of the pathway's salient features is necessary. Substrate degradation in the pathway may be divided into two phases, both of which require the hydrolysis of ATP. In the first step, ubiquitin is covalently conjugated to the substrate; in the second step, the ubiquitinated substrate is degraded by the 26 S proteasome and functional ubiquitin is regenerated (Fig. 1). The outcome of the initial conjugation phase is the formation of a covalent bond between the ε-amino group of a substrate lysine residue, and ubiquitin's C-terminal carboxylate (Hershko et al. 1980; see Hochstrasser 1996). Usually the substrate is conjugated to multiple molecules of ubiquitin, in the form of a polymeric chain linked through Gly76-Lys48 isopeptide bonds between successive ubiquitin molecules (Chau et al. 1989; Fig. 1). Chains of this structure constitute the principal signal for targeting substrates to the proteasome (Chau et al. 1989; Finley et al. 1994; see Pickart 1997).

Conjugation usually requires the sequential actions of three enzymes (Fig. 2). First, ubiquitin activating enzyme (E1) activates the C-terminus of ubiquitin through the formation of a ubiquitin adenylate intermediate, which gives rise to an E1-ubiquitin thiol ester. Second, ubiquitin is transferred to a cysteine residue at the active site of a ubiquitin conjugating enzyme (E2). Finally, a ubiquitin-protein ligase (E3) transfers ubiquitin from the E2 to the substrate lysine residue. In some cases, this final step occurs through the formation of an E3-ubiquitin thiol ester intermediate (see Hochstrasser 1996). Repetition of the third step leads to the assembly of the branched polyubiquitin chain.

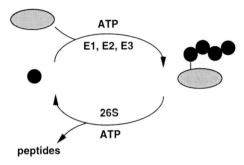

Fig. 1. The ubiquitin-proteasome pathway. *The gray oval* represents a substrate protein molecule; *the black circle* represents a ubiquitin molecule. *E1, E2,* and *E3* are ubiquitin activating, conjugating, and ligase enzymes, respectively; *26 S* is the 26 S proteasome. The substrate molecule *on the right* is linked to a polyubiquitin chain

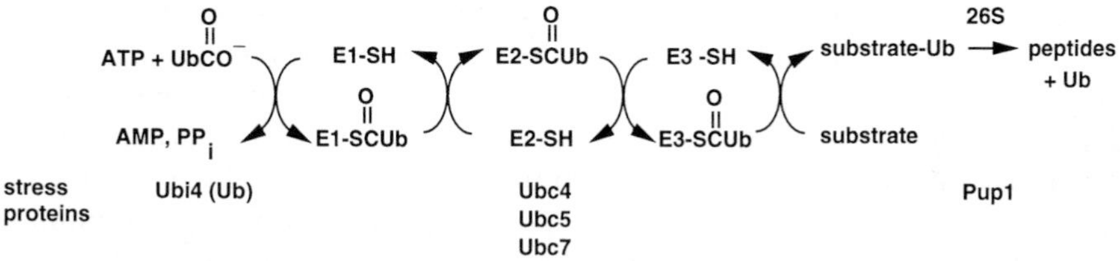

$$\text{ATP} + \text{Ub}\overset{\overset{O}{\|}}{\text{C}}\text{O}^- \quad \rightleftharpoons \quad \text{E1-SH} \quad \rightleftharpoons \quad \text{E2-S}\overset{\overset{O}{\|}}{\text{C}}\text{Ub} \quad \rightleftharpoons \quad \text{E3 -SH} \quad \rightleftharpoons \quad \text{substrate-Ub} \xrightarrow{26S} \text{peptides} + \text{Ub}$$

AMP, PP$_i$ — E1-SCUb — E2-SH — E3-SCUb — substrate

stress proteins   Ubi4 (Ub)   Ubc4 Ubc5 Ubc7   Pup1

Fig. 2. Mechanism of ubiquitin-protein conjugation and identities of known stress proteins in the ubiquitin-proteasome pathway. The enzymatic pathway of ubiquitin conjugation is shown in detail. As noted in the text, not all ligase (E3) enzymes form thiol ester intermediates with ubiquitin (see text). *At the bottom* are shown the names of components of the yeast ubiquitin pathway, which are known stress proteins. *Ubi4 (Ub)* is ubiquitin itself, expressed via the polyubiquitin gene; *Ubc4, Ubc5,* and *Ubc7* are ubiquitin conjugating enzymes (E2s); *Pup1* is an active site-containing beta subunit of the catalytic (20 S) module of the proteasome

The hallmark of the ubiquitin-proteasome pathway is high specificity: most short-lived proteins are substrates of the pathway, but the turnover of individual proteins can be independently regulated. Selectivity arises primarily at the stage of ubiquitin conjugation, through the existence of substrate-based ubiquitination signals, and conjugating enzymes which recognize these signals. The budding yeast *S. cerevisiae* has a single E1 gene, at least ten E2 genes, and an undetermined number of E3 genes (HOCHSTRASSER 1996). (It is likely that the number of E3s exceeds the number of E2s, but the E3s have been harder to identify, because they belong to multiple protein families.) In the hierarchy of conjugating enzymes, the E3 appears to be the primary specificity factor, by virtue of its ability to bind the substrate directly; the E2 also makes a contribution to substrate specificity. The current model for selective ubiquitin conjugation postulates that a given E3 enzyme functions in cooperation with one (or a few) E2 enzymes to ubiquitinate substrates which are recognized through the presence of a common primary sequence element, or ubiquitination signal. This model nicely explains the ubiquitination of cyclins and Cdk inhibitors (KING et al. 1996; HOCHSTRASSER 1996). As expected based on the organization just outlined, inactivation of E1 is globally inhibitory for ubiquitin-dependent proteolysis (CIECHANOVER et al. 1984; FINLEY et al. 1984; KULKA et al. 1988), while substrate-specific inhibition is usually observed following inactivation of individual E2 or E3 enzymes (HOCHSTRASSER 1996).

Once the substrate is ubiquitinated, it is recognized and degraded by the 26 S proteasome. This complex enzyme is made up of the 20 S proteasome together with the 19 S regulatory complex (see RUBIN and FINLEY 1995; COUX et al. 1996). The 20 S proteasome, consisting of 28 subunits arranged as a stack of four heptameric rings, contributes the proteolytic machinery (LÖWE et al. 1995; GROLL et al. 1996). Each of its two outer rings is composed of seven alpha-type subunits, while each inner ring is composed of seven beta-type subunits. In eukaryotes, several of the subunits in each beta ring have active sites. These sites line a chamber that is accessible only through a narrow pore at either end of the barrel. Substrates must therefore be unfolded before they can be degraded. Except in special cases, the 26 S proteasome does not degrade folded proteins unless they are first conjugated to a polyubiquitin chain, indicating that unfolding and translocation of the substrate into the active site channel are coupled to the recognition of the polyubiquitin chain signal. This coupling ensures that only substrates recognized by the conjugation cascade are susceptible to the action of the proteasome. The processes of recognition, unfolding, and translocation are carried out by the 19 S complex, but the molecular mechanisms of these processes are poorly understood. Among the 15–20 subunits of the 19 S complex are six members of the AAA-ATPase protein family. This protein family also includes several bona fide chaperones (see GOTTESMAN et al. 1997). The ATPase subunits of the 19 S complex probably mediate substrate unfolding (RUBIN and FINLEY 1995). The necessity for unfolding is presumably the basis of the ATP requirement in the degradative phase of the pathway (Fig. 1), because the proteasome degrades unfolded proteins in an ATP-independent fashion (e.g., DICK et al. 1991).

C. The Ubiquitin Pathway and the Stress Response

A role for the ubiquitin-proteasome pathway in the eukaryotic stress response first became evident about a decade ago as a result of several key findings. Taking advantage of the availability of ts85 cells, which harbor a thermolabile E1 enzyme, Varshavsky and co-workers showed that the ubiquitin pathway is responsible for the turnover of abnormal proteins formed through the incorporation of amino acid analogs (CIECHANOVER et al. 1984; FINLEY et al. 1984). Such a role had already been suspected based on the finding that these abnormal proteins are preferentially conjugated to ubiquitin in intact reticulocytes and are degraded in an ATP-dependent manner (HERSHKO et al. 1982). The subsequent identification of several components of the pathway as stress proteins confirmed that the pathway plays a major role in the cellular response to stress.

I. Stress Proteins in the Ubiquitin Pathway

1. Ubiquitin

All cells have multiple ubiquitin genes, each of which encodes a fusion between ubiquitin and another protein (ÖZKAYNAK et al. 1987). In some cases the other protein is ubiquitin itself: most organisms have at least one "polyubiquitin" gene which specifies multiple copies of the 76-residue ubiquitin coding sequence in a spacerless head-to-tail array (DWORKIN-RASTL et al. 1984; ÖZKAYNAK et al. 1984; BOND and Schlesinger 1985). The polyubiquitin gene typically carries an upstream heat shock element, and is subject to transcriptional induction by diverse stresses, including heat shock (BOND and SCHLESINGER 1985; FINLEY et al. 1987), heavy metals (arsenic and cadmium; BOND et al. 1988; MÜLLER-TAUBENBERGER et al. 1988; JUNGMANN et al. 1993), amino acid analogs (FINLEY et al. 1987; MÜLLER-TAUBENBERGER et al. 1988), DNA-damaging agents (TREGER et al. 1988), and oxidants (CHENG et al. 1994). (Space limitations preclude a more complete listing of the many reports of stress induction of polyubiquitin gene expression.) Expression of the *S. cerevisiae* polyubiquitin gene, known as *UBI4*, can also be induced by ubiquitin depletion or by inhibition of the ubiquitin-proteasome pathway. This latter induction is probably due to the accumulation of denatured proteins (Sect. C.IV).

Ubiquitin gene expression has been most thoroughly characterized in budding yeast, where there is one polyubiquitin gene (*UBI4*) and three other ubiquitin genes. Ubiquitin expression is required for viability (FINLEY et al. 1994): the *UBI1*, *UBI2*, and *UBI3* genes provide ubiquitin for vegetative growth, while the *UBI4* gene is specifically required for survival in stress conditions (below). The *UBI1-UBI3* genes encode fusions between ubiquitin (at its C-terminus) and small basic proteins (ÖZKAYNAK et al. 1987). The resulting fusion proteins are co-translationally processed by specific proteases (see HOCHSTRASSER 1996) to generate the 76-residue ubiquitin polypeptide and the carboxyl-extension peptides, which become subunits of the ribosome (FINLEY et al. 1989; REDMAN and RECHSTEINER 1989). The extension peptides are also essential proteins. Expressing them separately from the fused ubiquitin moiety causes a defect in ribosome biogenesis, apparently because the fused ubiquitin acts to augment the levels of the extension peptides (FINLEY et al. 1989). The molecular basis of this "chaperone" function of ubiquitin remains to be elucidated, but one possibility is that the rapid folding of the emerging ubiquitin module prevents co-translational degradation of the fusion partner. Consistent with this model, an N-terminal ubiquitin moiety can facilitate the expression of diverse heterologous proteins in prokaryotic and eukaryotic cells (reviewed in BAKER et al. 1994). In eukaryotes the transient lifetime of the fusion protein is important for this chaperone function, because the ubiquitin moiety of a metabolically stable fusion protein can act as a degradation signal by initiating the synthesis of a branched polyubiquitin chain (JOHNSON et al. 1992). It is somewhat ironic that a single alpha-linked ubiquitin can promote the folding of a fusion partner, while a branched polyubiquitin chain constitutes a degradation signal.

The yeast polyubiquitin gene encodes five tandem repeats of the ubiquitin coding sequence, followed by one extra amino acid (ÖZKAYNAK et al. 1984; FINLEY et al. 1987). The function of this C-terminal residue, which is universally encoded by polyubiquitin genes, is unclear. Possibly it serves to inhibit conjugation of linear polyubiquitin to cellular proteins in the event of a failure in ubiquitin-processing mechanisms. The *UBI4* gene is dispensable for vegetative growth, but is essential for survival at high temperature and in other stress conditions, including growth in the presence of the arginine analog canavanine (FINLEY et al. 1987). The *UBI4* gene is also required for sporulation (FINLEY et al. 1987), and is induced in meiosis (TREGER et al. 1988) and in stationary phase (FINLEY et al. 1987). As discussed below (Sect. C.II), the imposition of stress is usually accompanied by a dramatic redistribution of ubiquitin, such that most of the ubiquitin in the cell becomes conjugated to substrate proteins. The requirement for *UBI4* gene expression in stressed cells reflects the need to replenish the pool of free ubiquitin; in the absence of such replenishment, the ubiquitination of denatured proteins is inhibited and these species accumulate to toxic levels (Sect. C.III). The unique structure of the polyubiquitin gene serves the purpose of rapidly generating a high level of ubiquitin protein, as suggested by the finding that stressed *ubi4* null cells are viable when transformed with a high-copy plasmid carrying a "mono-ubiquitin" gene under the control of the *UBI4* promoter (FINLEY et al. 1987).

2. Ubiquitin-Conjugating Enzymes

Among the multiple E2 enzymes in yeast, the closely-related and functionally redundant enzymes encoded by the *UBC4* and *UBC5* genes play a predominant role in the ubiquitination and turnover of normal short-lived proteins (SEUFERT and JENTSCH 1990). However, these two enzymes also play a vital role in the cellular response to stress. Transcription of both genes is induced by heat shock (SEUFERT and JENTSCH 1990) and by cadmium (JUNGMANN et al. 1993). Double disruptants are profoundly defective in the turnover of canavanyl proteins, and do not survive growth on canavanine (SEUFERT and JENTSCH 1990). The working model is that augmented expression of Ubc4p and Ubc5p allows for the ubiquitination (and thus elimination) of denatured proteins in stressed cells. Ubc4/5-like E2s appear to function with multiple E3 enzymes (HOCHSTRASSER 1996). The E3 enzymes presumed to mediate the selective recognition of denatured proteins in stressed cells have not been identified (Sects. C.II and C.V).

A different E2, encoded by the *UBC7* gene, is required for the survival of yeast cells at elevated concentrations of cadmium (JUNGMANN et al. 1993). Cadmium (but not other heavy metals) stimulates the transcription of the *UBC7* gene, as well as the *UBC5* and *UBI4* genes. Cadmium stress thus holds certain features in common with heat shock (above), but cadmium stress must also invoke unique deleterious effects which can be selectively ameliorated by Ubc7p-mediated ubiquitination, because the *UBC7* gene is not subject to thermal induction (JUNGMANN et al. 1993). One possibility is that the unique function of Ubc7p is related to cadmium-induced damage of resident proteins of the endoplasmic reticulum (ER). The ubiquitination of damaged ER proteins frequently depends upon the activity of Ubc7p, as discussed below (Sect. C.III). However, Ubc7p has also been implicated in the turnover of cytosolic proteins in unstressed cells (CHEN et al. 1993; SADIS et al. 1995), and it cannot be excluded that this substate pool is relevant to Ubc7p function in stressed cells.

3. Other Pathway Components

The expression of ubiquitin and specific E2 enzymes is up-regulated in stressed cells (above). Levels of abnormal and denatured proteins also increase in stressed cells. The outcome of these effects is an increase in the level of ubiquitin conjugates (Sect. C.II). It is not known whether the increase in the level of conjugates is sufficient to saturate the proteasome, but if it is, the activity of the proteasome might be expected to increase in stressed cells. That partial inactivation of the proteasome can induce the stress response (Sect. C.IV) suggests that proteasomal capacity is not in gross excess even in unstressed cells, and there are some indications that proteasome activity increases in stress. The mechanism of this increase is of some interest, because augmenting the concentration of 26 S proteasomes could in principle require the coordinated induction of about 40 genes (Sect. B).

A modest increase in the concentration of yeast proteasomes was observed in response to mutational inhibition of the 20 S beta subunit encoded by the *PUP1* gene, and may account for the cadmium resistance associated with this mutation (ARENDT and HOCHSTRASSER 1997). It was suggested that denatured proteins, which accumulated due to the decreased activity of the proteasome, led to the induction of the *PUP1* gene via a consensus heat shock element in its promoter. However, the promoters of the other six beta subunit genes lack such an element (M. Hochstrasser, personal communication). One possible explanation for the observed increase in the concentrations of other proteasome subunits is that many subunits are normally synthesized in excess of Pup1p, but are degraded because they do not assemble into functional 26 S proteasomes. In this case, increasing the concentration of the limiting subunit could increase the level of proteasomes by titrating the excess subunits into functional complexes.

It is also possible that the activity of the proteasome could increase (or be otherwise altered) in the absence of marked changes in the overall concentration of the enzyme. Specific activity could be modulated by replacing specific subunits, or by covalently modifying pre-existing subunits. Phosphorylated forms of certain subunits have been reported, but the impact of these modifications on activity is unknown. That subunit reshuffling can occur in mammals is known from studies of antigen presentation, where replacement of certain 20 S subunits serves to modulate proteolytic cleavage site specificity (see Coux et al. 1996). The replacement of 20 S subunits is also possible in yeast, because one of the yeast alpha subunit genes is nonessential (Emori et al. 1991), and dominant 20 S proteasome mutations are known (e.g., Gerlinger et al. 1997); whether subunit replacement represents a natural regulatory mechanism in yeast is unknown. There are indications that the composition of the 19 S regulatory complex may be flexible in higher eukaryotes; the levels of a subset of ATPase subunits, and the overall activity of the 26 S proteasome, increase dramatically in insect cells undergoing programmed cell death (Dawson et al. 1995; Jones et al. 1995). The substrates of ubiquitin conjugation in stressed cells are probably partially unfolded, so a change in the unfolding components of the 19 S complex (presumably the ATPase subunits) in stressed cells would not be too surprising.

II. Ubiquitin Conjugation in Stressed Cells

Cellular ubiquitin distribution changes in stressed cells: the fraction of ubiquitin that is unconjugated decreases, while the fraction of ubiquitin in conjugates increases (Carlson et al. 1987; Finley et al. 1987; Parag et al. 1987; Bond et al. 1988; Kulka et al. 1988; Lee et al. 1996; Shang et al. 1997). This effect, which is transient unless expression of the polyubiquitin gene is blocked, can be extreme: in unstressed cells, about 50% of the ubiquitin is unconjugated (Haas and Bright 1985; Haas 1988), whereas less than 10% of the ubiquitin is unconjugated in heat-shocked HeLa cells (Carlson et al. 1987). Cycloheximide does not interdict stress-induced ubiquitin redistribution, suggesting that this event is mainly due to increased susceptibility of proteins to ubiquitination, as opposed to increased levels of ubiquitin or conjugating enzymes. However, post-translational regulation of the activities of rate-limiting conjugating enzymes could also contribute to increased conjugate levels (Shang et al. 1997).

Damaged proteins are the presumptive substrates of ubiquitination in stressed cells. How might such proteins be selectively recognized by the ubiquitin pathway? We do not yet know. As discussed above (Sect. B), the selection of substrates for ubiquitin conjugation usually depends upon the recognition of a substrate-based ubiquitination signal by a cognate E3 enzyme. Known ubiquitination signals include short stretches of about ten to thirty amino acids, such as the "destruction box" of the mitotic cyclins or the "delta domain" of c-Jun (GLOTZER et al. 1991; TREIER et al. 1994). In other instances, specific phosphorylation can serve as a ubiquitination signal (e.g., SKOWYRA et al. 1997; FELDMAN et al. 1997; see HOCHSTRASSER 1996). Signals which have been identified to date bring about the ubiquitination of one protein or a small set of proteins. A signal of this type is unlikely to be useful in mediating the recognition of damaged proteins, since nearly all proteins in the cell have the potential to be damaged by stress. Whether "stress-dependent" ubiquitination signals exist, what their properties are, and whether there are specific E3s which recognizes them, are key unanswered questions.

It is possible that clusters of hydrophobic amino acids represent one such signal. Hydrophobic side chains are typically sequestered in the cores of globular proteins, and their exposure to solvent is a marker of unfolding, as indicated by the finding that molecular chaperones bind specifically to hydrophobic side chains (e.g., BLOND-ELGUINDI et al. 1993). This kind of ubiquitination signal would specifically couple degradation to unfolding, rather than coupling degradation to the recognition of specific covalent modifications that might be associated with protein damage (e.g., thiol oxidation). That hydrophobic side chains may indeed serve as ubiquitination signals is suggested by the results of a search for short peptide sequences capable of targeting a reporter construct to the yeast 26 S proteasome (SADIS et al. 1995). Of three "synthetic signals" thus identified, two included clusters of bulky hydrophobic amino acids. The turnover of reporter molecules bearing such signals depended on Ubc4p, Ubc5p, and Ubc7p. All of these E2 enzymes have been implicated in the stress response (Sect. C.I.2).

No E3 enzymes have (yet) been identified as stress proteins, nor has any E3 been shown to play an important role in the stress response. Nonetheless, the induction of specific E2s in stressed cells makes it very likely that one or more "stress E3s" exist, because E2s rarely (if ever) function autonomously *in vivo*. As discussed above (Sect. B), a given E2 usually functions with one E3 or a small set of E3s. However, yeast Ubc4p and Ubc5p, and higher Ubc4/5 homologs, have been implicated in conjugation mediated by a significant number of E3s (HOCHSTRASSER 1996). Thus, the identification of these E2s as stress proteins has not provided much insight into the identities or properties of the presumptive stress E3s. So far no E3 has been identified which functions in Ubc7p-dependent ubiquitination.

Although stress-induced ubiquitination probably serves mainly to target damaged proteins to the proteasome (Sect. C.III), the results of recent studies on polyubiquitin chains raise the possibility that some of the ubiquitination in stressed cells serves a different purpose. As discussed above (Sect. B), the principal signal for targeting substrates to the proteasome is a polyubiquitin chain linked through Gly76-Lys48 isopeptide bonds. However, at least four other lysine residues of ubiquitin can serve as sites of chain initiation (see PICKART 1997), and one kind of alternative polyubiquitin chain has been implicated in the stress response. The relevant observation is that overexpression of K63R-ubiquitin does not complement the heat and canavanine sensitivities of *ubi4* null cells, whereas ubiquitins carrying single arginine substitutions at the six other lysine residues complement these defects (ARNASON and ELLISON 1994). Because the incorporation of K63R-ubiquitin will specifically terminate the elongation of a polyubiquitin chain linked through Lys63, these results suggest that Lys63-linked chains play a specific role in the stress response. Although *ubi4* null cells overexpressing K63R-ubiquitin are canavanine-sensitive, they are nonetheless competent in the turnover of canavanyl proteins. These observations suggest that the signaling function of K63-linked chains is either non-proteolytic, or restricted to a low-abundance pool of proteolytic substrates. The existence of a novel function for Lys63-linked chains is not restricted to stress conditions: in yeast, Lys63-linked chains have been implicated in DNA repair at normal temperatures (SPENCE et al. 1995).

The E2 enzymes implicated in the assembly of Lys63-linked chains are the stress-induced enzymes Ubc4p and Ubc5p (ARNASON and ELLISON 1994; SPENCE et al. 1995). These enzymes apparently serve a dual function in stress, because they are also needed for polyubiquitination, presumably involving Lys48, that targets a large pool of substrates for degradation in stressed cells (SEUFERT and JENTSCH 1990). The ability of Ubc4p/Ubc5p to serve more than one function in stressed cells probably reflects the ability of these E2s to cooperate with more than one E3 enzyme (above).

What function might Lys63-linked chains serve in stressed cells? In the absence of direct evidence, only speculation is possible. If they were selectively conjugated to specific substrates, and had a higher affinity for the proteasome than Lys48-linked chains, Lys63-linked chains could prioritize these substrates for turnover. The scope of any such proteolytic signaling function must be limited (above). Conversely, if they bound in a proteolytically-nonproductive mode to the proteasome, Lys63-linked chains could spare unfolded proteins from turnover and allow an increased time for substrate refolding by (unidentified) proteasome-associated chaperones. Lys63-linked chains may also direct a completely novel interaction(s) that is beneficial in stressed cells. The potential for such interactions is suggested by recent findings which implicate Lys63-linked chains in the targeting of a plasma membrane protein to the yeast vacuole (GALAN and HAGUENAUER-TSAPIS 1997; HICKE 1997).

III. Ubiquitin-Mediated Degradation in Stressed Cells

As discussed above (Sect. A), proteolysis represents one part of a two-pronged response to protein damage in stressed cells. The model is that damaged proteins which are not refolded by chaperones will be degraded by proteases; in eukaryotes, the ubiquitin-proteasome pathway is responsible for such degradation. That damaged proteins are more susceptible to ubiquitination (Sect. C.II) supports this model, as does the presence of ubiquitin, conjugating enzymes, and a proteasome subunit in the arsenal of heat shock proteins (Fig. 2). The concentration of ubiquitin conjugates rises in stressed cells (Sect. C.II). The expectation is that flux through the proteasome will similarly increase. However, the potential occurrence of non-proteolytic ubiquitination (involving Lys63 chains, above) indicates a need for caution in equating increased ubiquitination with increased proteasomal turnover.

The strongest evidence in support of the hypothesis that stress-induced ubiquitination reflects stress-induced degradation derives from studies with amino acid analogs. Such abnormal proteins are preferential substrates for ubiquitination (HERSHKO et al. 1982; CIECHANOVER et al. 1984; SEUFERT and JENTSCH 1990). The resulting conjugates are rapidly degraded in a manner that is dependent upon the presence of a functional proteasome: partial inactivation of the yeast 26 S proteasome renders cells unable to grow on the arginine analog canavanine, and this defect is associated with strong inhibition of the turnover of pulse-labeled canavanyl proteins (HEINEMEYER et al. 1991, 1993; HILT et al. 1993).

In the case of other stresses, the degradation data are less compelling. Although an increase in the level of ubiquitin conjugates occurs in virtually all types of stress (Sect. C.II), protein turnover has been observed to exhibit responses ranging from inhibition (MUNRO and PELHAM 1984; CARLSON et al. 1987) to modest stimulation (PARAG et al. 1987; BOND et al. 1988; SHANG et al. 1997). It is likely that some apparent contradictions reflect the use of differing experimental regimens, especially in regard to heat shock (see BOND et al. 1988). Pulse-chase methods are generally used to monitor turnover. If pre-existing (unlabeled) proteins make up a large fraction of the substrates for ubiquitination in stressed cells (Sect. C.II), and if the proteasome becomes saturated with conjugates, then the turnover of pulse-labeled proteins could decrease, through competition effects, even though overall flux through the proteasome increases. It is also possible that heat-denatured proteins aggregate, either before or after conjugation to ubiquitin (CARLSON et al. 1987), causing proteasomal turnover to become dependent upon slow disaggregation mediated by chaperones. Despite these uncertainties, the most attractive model is that the heat-damaged proteins which undergo ubiquitination are subsequently degraded by the proteasome. In support of this model, partial inactivation of the proteasome can cause hypersensitivity to heat stress (HEINEMEYER et al. 1991, 1993; HILT et al. 1993; OHBA 1994; GERLINGER et al. 1997).

The discussion above has been framed around the canonical substrates of the ubiquitin pathway, which are soluble proteins of the cytosol and nucleus. A major development of the past several years has been the recognition that the ubiquitin-proteasome pathway is intimately involved in protein quality control in the ER (see KOPITO 1997; SOMMER and WOLF 1997). Abnormal, unfolded proteins in the ER lumen and membrane can undergo retrograde transport via the ER protein channel, and then be degraded by the proteasome (BIEDERER et al. 1996; HILLER et al. 1996; WERNER et al. 1996; WIERTZ et al. 1996). In general the ubiquitination of ER-resident proteins requires the E2 encoded by the *UBC7* gene. Ubc7p is localized to the cytosolic face of the ER membrane through association with an integral membrane protein known as Cue1p (BIEDERER et al. 1997). In some cases the *UBC6*-encoded E2 enzyme, which is an integral ER membrane protein, is also required for the ubiquitination of ER proteins (BIEDERER et al. 1996). Interestingly, retrograde transport through the ER protein channel appears to be coupled to ubiquitination of the transported substrate (BIEDERER et al. 1997). It is possible that the proteasome, through its ability to recognize the substrate's polyubiquitin tag and initiate unfolding and translocation, provides part of the driving force for retrograde transport (KOPITO 1997; SOMMER and WOLF 1997).

Besides its probable involvement in stress-induced protein turnover in the ER, ubiquitination also targets plasma membrane proteins to the yeast vacuole (equivalent to the mammalian lysosome). The degradation of these substrates involves vacuolar (lysosomal) hydrolases rather than the proteasome (see HICKE 1997). Studies in mammalian ts85 cells had previously implicated ubiquitin conjugation in the heat stress-induced turnover of proteins in autophagic (lysosomal) vacuoles (GROPPER et al. 1991). Thus, ubiquitination is a key event which invokes stress-induced proteolysis in multiple subcellular compartments.

IV. The Ubiquitin Pathway and Induction of the Stress Response

A variety of conditions which inhibit or overload the ubiquitin-proteasome pathway cause cells to mount a stress response. For example, thermal inactivation of E1 in ts85 cells, which globally interdicts ubiquitination, causes heat shock protein synthesis at an abnormally low temperature (FINLEY et al. 1984). Deletion of the yeast *UBC4* and *UBC5* genes, which presumably interdicts ubiquitination in a more selective manner, stimulates the expression of the *UBI4* polyubiquitin gene and the Hsp70 gene at sub-heat shock temperatures (SEUFERT and JENTSCH 1990). Artificial depletion of ubiquitin, through the expression of a ubiquitin derivative that irreversibly traps ubiquitin in branched polyubiquitin chains (AMERIK et al. 1997), leads to transcriptional induction of the yeast *UBI4* gene (ARNASON and ELLISON 1994). And chemical inhibition of proteasomes stimulates transcription of the polyubiquitin, Bip, and Hsp70 genes (BUSH et al. 1997; LEE and GOLDBERG 1998).

These and other data implicate defective functioning of the ubiquitin-proteasome pathway in the induction of the stress response. An early model postulated that the ubiquitin pathway specifically controlled the level of a positive regulator of the heat shock response (MUNRO and PELHAM 1984; FINLEY et al. 1984). Such a model pertains in bacteria, where turnover of the heat shock sigma factor is an important feature in the regulated expression of heat shock genes (see PARSELL and LINDQUIST 1993; SHERMAN and GOLDBERG 1996 see also ZHOU et al. 1996). However, the transcription factors controlling the expression of eukaryotic heat shock genes are stable proteins whose activities are regulated post-translationally. Mammalian heat-shock factors are normally sequestered in transcriptionally-inactive complexes with Hsp70 (MORIMOTO et al. 1996). That interfering with the ubiquitin-proteasome pathway can induce the stress response probably most often reflects the general role of the pathway in the clearance of damaged proteins. Specifically, when levels of damaged proteins rise, as will occur if the ubiquitin pathway is inhibited or overloaded, an increased fraction of cellular Hsp70 will partition into complexes with damaged proteins, with the consequent release of (active) heat shock transcription factors (MORIMOTO et al. 1996). Thus, the level of denatured proteins, as opposed to the specific activity of the ubiquitin pathway, is the critical variable which determines whether or not the stress response will be invoked. This model can best accommodate the finding that proteins which cannot be ubiquitinated, due to chemical blockade of their lysine residues, can efficiently induce the stress response when injected into *Xenopus* oocytes; the ability of such microinjected proteins to induce stress protein synthesis correlates positively with ability to form aggregates (and, thus, complexes with Hsp70), rather than with ubiquitination potential (MIFFLIN and COHEN 1994a,b).

V. Involvement of Molecular Chaperones in Ubiquitin-Dependent Degradation

In so far as substrate unfolding must precede substrate degradation by the 26 S proteasome, proteasome activity may be considered to depend absolutely on the activities of intrinsic, stoichiometric molecular chaperones, presumably the six ATPase subunits of the 19 S regulatory module. The Clp protease of *E. coli* represents a similar case, in which the proteolytic P subunits must assemble with ATPase subunits in order to achieve the degradation of proteins. In the case of Clp, the P subunit can associate with several distinct ATPase subunits, with consequent modulation of proteolytic substrate specificity; certain Clp ATPase subunits can also function as free-standing chaperones (see GOTTESMAN et al. 1997). In the case of the proteasome, substrate specificity is determined primarily at the stage of ubiquitin attachment (Sect. B), while the identities and properties of the ATPase subunits change only under special conditions, if at all (Sect. C.I.3). There is no evidence that the proteasome's ATPase subunits ever function as free-standing chaperones.

In some cases, however, conventional molecular chaperones can facilitate degradation in the ubiquitin-proteasome pathway. Chaperones can also facilitate proteolysis by energy-dependent proteases in bacteria (SHERMAN and GOLDBERG 1996) and mitochondria (WAGNER et al. 1994). The majority of reports in eukaryotes involve the constitutively expressed "cognates" of heat shock proteins (known as hscs). Hsc70 is required for the ubiquitination of certain denatured protein substrates in reticulocyte lysate (BERCOVICH et al. 1997). Hsc70 also facilitates the proteasome-dependent turnover of improperly lipidated apolipoprotein B in cultured hepatic cells (FISHER et al. 1997).

Overexpression of an Hsc70 homolog restores the ubiquitin-dependent turn-over of an artificial substrate in yeast cells carrying a mutation in a protea-some subunit (OHBA 1994), and the DnaJ homolog Ydj1p is required for the ubiquitination of several natural and artificial substrates of the ubiquitin path-way in yeast cells (LEE et al. 1996; YAGLOM et al. 1996). A different yeast DnaJ homolog, Sis1p, is apparently required for the turnover of ubiquitinated forms of one artificial substrate in yeast cells (cited in SHERMAN and GOLDBERG 1996). The latter substrate, a ubiquitin-fused derivative of beta-galactosidase (JOHNSON et al. 1992) is a correctly-folded, enzymatically active tetramer even after conjugation to ubiquitin (J. Piotrowski and C. Pickart, unpublished ex-periments), so it is perhaps not surprising that extra assistance is necessary to achieve its unfolding and translocation into the proteasome. In most of the above-described cases, the chaperone was shown to form a complex with the substrate. Substrate complexation is presumably important for degradation (below).

These results suggest that engagement with stress-induced chaperones and degradation by proteases are not mutually exclusive fates for stress-damaged proteins. Specifically, the fate of a stress-denatured protein may depend on kinetic partitioning between refolding and degradation: if the chap-erone fails to achieve folding of the substrate, the chaperone could facilitate either ubiquitination of the substrate or (if it is already ubiquitinated) its turnover by the proteasome (KOPITO 1997). Consistent with a role for chaper-ones in facilitating ubiquitin-dependent degradation of stress-damaged pro-teins, mutations in the *YDJ1* gene inhibit the turnover of amino acid analog-containing proteins in yeast cells; the data indicate that Ydj1p plays a role in the selective ubiquitination of these well-characterized substrates (LEE et al. 1996). The results of another recent study implicate Hsp90 in the ubiquitin-mediated turnover of heat-denatured luciferase in reticulocyte lysate (SCHNEIDER et al. 1996).

Although these observations are suggestive, the scope of the proteolytic assistance provided by chaperones in stressed cells remains to be determined. Moreover, the term "assistance" can accommodate a number of mechanisms. The existing data are most consistent with a model in which the chaperone functions to keep the substrate in a disaggregated state in which the presump-tive ubiquitination signal remains available for recognition (Fig. 3). Whether ubiquitin conjugating enzymes act on the free substrate, versus chaperone-associated forms of the substrate (i.e., whether turnover proceeds along path A or B in Fig. 3) may depend upon the relative levels of substrates, chaper-ones, and proteolytic components, and on the properties of individual com-plexes. For example, the degradation of misfolded mitochondrial proteins by the yeast Pim1 protease (a homolog of the *E. coli* Lon protease) requires that Hsp70 and DnaJ homologs be present in the mitochondrion, but degradation is inhibited if the dissociation of the Hsp70-substrate complex is blocked (WAGNER et al. 1994). Evidently the chaperone-substrate complex cannot be acted on by the protease (i.e., the route analogous to path B in Fig. 3 is

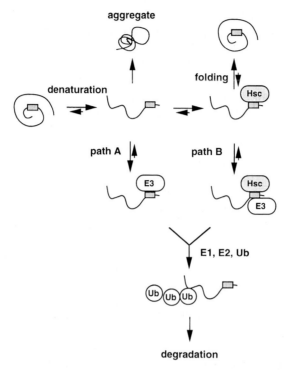

Fig. 3. Kinetic partitioning model for involvement of chaperones and proteases in the salvage and degradation of stress-damaged proteins. *The shaded oval* is a chaperone molecule; *the white oval* is a ubiquitin-protein ligase (E3) molecule; *the white circles* are ubiquitin molecules. *The shaded box* corresponds to a common substrate-based determinant of the binding of a chaperone and an E3 enzyme

inoperative). This may be explained if the bound chaperone blocks a substrate-based signal for Pim1p, or if the chaperone blocks movement of the substrate into the protease active site. A different situation pertains with Hsp90. Here inhibiting the dissociation of the Hsp90-luciferase complex (and the folding of luciferase) stimulates the proteasome-dependent turnover of this substrate (SCHNEIDER et al. 1996). Presumably in this case chaperone binding leaves accessible a determinant for ubiquitin conjugation. (It remains to be shown that Hsp90 facilitates the ubiquitination of luciferase, as opposed to the degradation of luciferase conjugates; however, the chaperone could equally well act to maintain ubiquitin conjugates in a soluble, proteolysis-competent form). As discussed above (Sect. C.II), certain chaperones and conjugating enzymes may recognize the same "signals," specifically hydrophobic side chains, in denatured substrates. In this case the binding of conjugating enzymes and chaperones would be mutually exclusive and degradation would occur via path A in Fig. 3. (In a limiting case of path A, the rate-limiting step in degradation would be the release of the substrate from the chaperone;

Wagner et al. 1994.) Mutually exclusive binding could be beneficial in tending to ensure that refolding would be attempted whenever chaperone levels are high.

In this model, the fate of the substrate following chaperone binding depends upon the relative rates of refolding and degradation: the longer the chaperone maintains the substrate in a soluble form without refolding it, the greater is the chance that the substrate will be recognized by the degradation apparatus. Although speculative, the model of Fig. 3 is generally consistent with experimental data indicating that altering the relative rates of refolding and degradation alters the partitioning between refolding and turnover in the predicted fashion (Wagner et al. 1994; Schneider et al. 1996; Fisher et al. 1997). Currently available data do not exclude a different model, in which chaperone-substrate complexes bind specifically to components of the ubiquitin pathway (e.g., E3 enzymes). However, there is no positive evidence to support the latter model.

D. Outstanding Questions

A functional ubiquitin-proteasome pathway is an essential feature of the stress response in eukaryotes. The role of the pathway in eliminating damaged proteins in the cytosol, nucleus, and endoplasmic reticulum forms the basis of this requirement. However, a number of critical questions about the role of the pathway in the stress response remain unanswered. How damaged proteins are selectively recognized for ubiquitin conjugation is poorly understood; the E3 enzymes presumed to mediate this recognition remain unidentified. It is unclear whether the activity of the 26 S proteasome is strongly modified in stressed cells, and the function of alternatively-linked polyubiquitin chains in stressed cells remains to be determined. The degree to which chaperones facilitate the ubiquitin-dependent degradation of stress-damaged proteins, and the specific mechanisms by which they may achieve this effect, are also open questions. The ubiquitin-proteasome pathway, like certain molecular chaperones, is essential in unstressed cells. With answers to these questions should come a better understanding of how the unique attributes of the ubiquitin pathway have been exploited to achieve the selective recognition of damaged proteins in stressed cells.

Acknowledgements. Our work on the ubiquitin pathway is supported by a grant from the National Institutes of Health (DK46984). I am grateful to Robert Cohen for helpful comments on the manuscript.

References

Amerik AY, Swaminathan S, Krantz BA, Wilkinson KD, Hochstrasser M (1997) In vivo disassembly of free polyubiquitin chains by yeast Ubp14 modulates rates of protein degradation by the proteasome. EMBO J 16:4826–4838

Ananthan J, Goldberg AL, Voellmy R (1986) Abnormal proteins serve as eukaryotic stress signals and trigger the activation of heat shock genes. Science 232:522–524

Arendt CS, Hochstrasser MH (1997) Identification of the yeast 20 S proteasome catalytic centers and subunit interactions required for active-site formation. Proc Natl Acad Sci USA 94:7156–7161

Arnason T, Ellison MJ (1994) Stress resistance in Saccharomyces cerevisiae is strongly correlated with assembly of a novel type of multiubiquitin chain. Mol Cell Biol 14:7876–7883

Baker RT, Smith SA, Marano R, McKee J, Board PG (1994) Protein expression using cotranslational fusion and cleavage of ubiquitin. J Biol Chem 269:25381–25386

Bercovich B, Stancovski I, Mayer A, Blumenfeld N, Laszlo A, Schwartz AL, Ciechanover A (1997) Ubiquitin-dependent degradation of certain protein substrates in vitro requires the molecular chaperone Hsc70. J Biol Chem 272:9002–9010

Biederer T, Volkwein C, Sommer T (1996) Degradation of subunits of the Sec61p complex, an integral component of the ER membrane, by the ubiquitin-proteasome pathway. EMBO J 15:2069–2076

Biederer T, Volkwein C, Sommer T (1997) Role of Cue1p in ubiquitination and degradation at the ER surface. Science 278:1806–1809

Blond-Elguindi S, Cwiria SE, Dower WJ, Lipshutz RJ, Sprang SR, Sambrook JF, Gething M-J (1993) Affinity panning of a library of peptides displayed on bacteriophages reveals the binding specificity of BiP. Cell 75:717–728

Bond U, Schlesinger MJ (1985) Ubiquitin is a heat shock protein in chicken embryo fibroblasts. Mol Cell Biol 5:949–956

Bond U, Agell N, Haas AL, Redman K, Schlesinger MJ (1988) Ubiquitin in stressed chicken embryo fibroblasts. J Biol Chem 263:2384–2388

Bush KT, Goldberg AL, Nigam SK (1997) Proteasome inhibition leads to a heat-shock response, induction of endoplasmic reticulum chaperones, and thermotolerance. J Biol Chem 272:9086–9092

Carlson N, Rogers S, Rechsteiner M (1987) Microinjection of ubiquitin: changes in protein degradation in HeLa cells subjected to heat shock. J Cell Biol 104:547–555

Chau V, Tobias JW, Bachmair A, Marriott D, Ecker DJ, Gonda DK, Varshavsky A (1989) A multiubiquitin chain is confined to specific lysine in a targeted short-lived protein. Science 243:1576–1583

Chen P, Johnson P, Sommer T, Jentsch S, Hochstrasser M (1993) Multiple ubiquitin-conjugating enzymes participate in the in vivo degradation of the yeast MATα2 repressor. Cell 74:357–369

Cheng L, Watt R, Piper PW (1994) Polyubiquitin gene expression contributes to oxidative stress resistance in respiratory yeast (Saccharomyces cerevisiae). Mol Gen Genet 243:358–362

Ciechanover A, Finley D, Varshavsky A (1984) Ubiquitin dependence of selective protein degradation demonstrated in the mammalian cell cycle mutant ts85. Cell 37:57–66

Coux O, Tanaka K, Goldberg AL (1996) Structure and functions of the 20 S and 26 S proteasomes. Annu Rev Biochem 65:801–847

Dawson SP, Arnold JE, Mayer NJ, Reynolds SE, Billett MA, Gordon C, Colleaux L, Kloetzel PM, Tanaka K, Mayer RJ (1995) Developmental changes of the 26 S proteasome in abdominal intersegmental muscles of Manduca sexta during programmed cell death. J Biol Chem 270:1850–1858

Dick LR, Moomaw CR, DeMartino GN, Slaughter CA (1991) Degradation of oxidized insulin B chain by the multiproteinase complex macropain (proteasome). Biochemistry 30:2725–2734

Dworkin-Rastl E, Shrutkowski A, Dworkin MB (1984) Multiple ubiquitin mRNAs during Xenopus laevis development contain tandem repeats of the 76 amino acid coding sequence. Cell 39:321–325

Emori Y, Tsukahara T, Kawasaki H, Ishiura S, Sugita H, Suzuki K (1991) Molecular cloning and functional analysis of three subunits of yeast proteasome. Mol Cell Biol 11:344–353

Feldman RMR, Correll CC, Kaplan KB, Deshaies RJ (1997) A complex of Cdc4p, Skp1p, and Cdc53p/cullin catalyzes ubiquitination of the phosphorylated CDK inhibitor Sic1p. Cell 91:221–230

Finley D, Ciechanover A, Varshavsky A (1984) Thermolability of ubiquitin-activating enzyme from the mammalian cell cycle mutant ts85. Cell 37:57–66

Finley D, Özkaynak E, Varshavsky A (1987) The yeast polyubiquitin gene is essential for resistance to high temperatures, starvation, and other stresses. Cell 48:1035–1046

Finley D, Bartel B, Varshavsky A (1989) The tails of ubiquitin precursors are ribosomal proteins whose fusion to ubiquitin facilitates ribosome biogenesis. Nature 338:394–401

Finley D, Sadis S, Monia BP, Boucher P, Ecker DJ, Crooke ST, Chau V (1994) Inhibition of proteolysis and cell cycle progression in a multiubiquitination-deficient yeast mutant. Mol Cell Biol 14:5501–5509

Fisher EA, Zhou M, Mitchell DM, Wu X, Omura S, Wang H, Goldberg AL, Ginsberg HN (1997) The degradtaion of apolipoprotein B100 is mediated by the ubiquitin-proteasome pathway and involves heat shock protein 70. J Biol Chem 272:20427–20434

Galan J-M, Haguenauer-Tsapis R (1997) Ubiquitin Lys63 is involved in ubiquitination of a yeast plasma membrane protein. EMBO J 16:5847–5854

Gerlinger U-M, Gückel R, Hoffmann M, Wolf DH, Hilt W (1997) Yeast cycloheximide-resistant crl mutants are proteasome mutants defective in protein degradation. Mol Biol Cell 8:2847–2499

Glotzer M, Murray AW, Kirschner MW (1991) Cyclin is degraded by the ubiquitin pathway. Nature 349:132–138

Goff SA, Goldberg AL (1985) Production of abnormal proteins in E. coli stimulates transcription of lon and other heat shock genes. Cell 41:587–595

Gottesman S, Wickner S, Maurizi MR (1997) Protein quality control: triage by chaperones and proteases. Genes Dev 11:815–823

Groll M, Ditzel L, Lowe J, Stock D, Bochtler M, Bartunik HD, Huber R (1997) Structure of 20 S proteasome from yeast at 2.4 Å resolution. Nature 386:463–471

Gropper R, Brandt RA, Elias S, Bearer CF, Mayer A, Schwartz AL, Ciechanover A (1991) The ubiquitin-activating enzyme, E1, is required for stress-induced lysosomal degradation of cellular proteins. J Biol Chem 266:3602–3610

Haas AL, Bright PM (1985) The immunochemical detection and quantitation of intracellular ubiquitin-protein conjugates. J Biol Chem 260:12464–12473

Haas AL (1988) Immunochemical probes of ubiquitin pool dynamics. In: Rechsteiner M (ed) Ubiquitin. Plenum, New York, pp 173–206

Hayes SA, Dice JF (1996) Roles of molecular chaperones in protein degradation. J Cell Biol 132:255–258

Heinemeyer W, Kleinschmidt JA, Saidowsky J, Escher C, Wolf DH (1991) Proteinase yscE, the yeast proteasome/multicatalytic-multifunctional proteinase: mutants unravel its function in stress induced proteolysis and uncover its necessity for cell survival. EMBO J 10:555–562

Heinemeyer W, Gruhler A, Möhrle V, Mahé Y, Wolf DH (1993) PRE2, highly homologous to the human major histocompatibility complex-linked RING10 gene, codes for a yeast proteasome subunit necessary for chymotryptic activity and degradation of ubiquitinated proteins. J Biol Chem 268:5115–5120

Hershko A, Eytan E, Ciechanover A, Haas AL (1982) Immunochemical analysis of the turnover of ubiquitin-protein conjugates in intact cells: relationship to the breakdown of abnormal proteins. J Biol Chem 257:13964–13970

Hershko A, Ciechanover A, Heller H, Haas AL, Rose IA (1980) Proposed role of ATP in protein breakdown: conjugation of proteins with multiple chains of the polypeptide of ATP-dependent proteolysis. Proc Natl Acad Sci USA 77:1783–1786

Hicke L (1997) Ubiquitin-dependent internalization and down-regulation of plasma membrane proteins. FASEB J 11:1215–1226

Hiller MM, Finger A, Schweiger M, Wolf DH (1996) ER degradation of a misfolded luminal protein by the cytosolic ubiquitin-proteasome pathway. Science 273:1725–1728

Hilt W, Enenkel C, Gruhler A, Singer T, Wolf DH (1993) The PRE4 gene codes for a subunit of the yeast proteasome necessary for peptidylglutamyl-peptide-hydrolyzing activity. J Biol Chem 268:3479–3486

Hochstrasser M (1996) Ubiquitin-dependent protein degradation. Annu Rev Genet 30:405–439

Johnson ES, Bartel B, Seufert W, Varshavsky A (1992) Ubiquitin as a degradation signal. EMBO J 11:497–505

Jones MEE, Haire MF, Kloetzel P-M, Mykles DL, Schwartz LM (1995) Changes in the structure and function of the multicatalytic proteinase (MCP) during programmed cell death in the intersegmental muscles of the hawkmoth, Manduca sexta. Dev Biol 169:436–447

Jungmann J, Reins H-A, Schobert C, Jentsch S (1993) Resistance to cadmium mediated by ubiquitin-dependent proteolysis. Nature 361:369–371

King RW, Deshaies RF, Peters JM, Kirschner MW (1996) How proteolysis drives the cell cycle. Science 274:1652–1659

Koptio RR (1997) ER quality control: the cytoplasmic connection. Cell 88:427–430

Kulka RG, Raboy B, Schuster R, Parag HA, Diamond G, Ciechanover A, Marcus M (1988) A Chinese hamster cell cycle mutant arrested at G$_2$ phase has a temperature-sensitive ubiquitin-activating enzyme, E1. J Biol Chem 263:15726–15731

Lee DH, Sherman MY, Goldberg AL (1996) Involvement of molecular chaperone Ydj1 in the ubiquitin-dependent degradation of short-lived and abnormal proteins in Saccharomyces cerevisiae. Mol Cell Biol 16:4773–4781

Lee DH, Goldberg AL (1998) Proteasome inhibitors cause induction of heat shock proteins and trehalose, which together confer thermotolerance in Saccharomyces cerevisiae, Mol Cell Biol 18:30–38

Löwe J, Stock D, Jap B, Zwickl P, Baumeister W, Huber R (1995) Crystal structure of the 20 S proteasome from the archaeon T. acidophilum at 3.4 Å resolution. Science 268:533–539

Mifflin LC, Cohen RE (1994a) Characterization of denatured protein inducers of the heat shock (stress) response in Xenopus laevis oocytes. J Biol Chem 269:15710–15717

Mifflin LC, Cohen RE (1994b) hsc70 moderates the heat shock (stress) response in Xenopus laevis oocytes and binds to denatured protein inducers. J Biol Chem 269:15718–15723

Morimoto RI, Kroeger PE, Cotto JJ (1996) The transcriptional regulation of heat shock genes: a plethora of heat shock factors and regulatory conditions. EXS 77:139–163

Müller-Taubenberger A, Hagmann J, Noegel A, Gerisch G (1988) Ubiquitin gene expression in Dictyostelium is induced by heat and cold shock, cadmium, and inhibitors of protein synthesis. J Cell Sci 90:51–58

Munro S, Pelham HRB (1984) Use of peptide tagging to detect proteins expressed from cloned genes: deletion mapping functional domains of Drosophila hsp70. EMBO J 2:3087–3093

Ohba M (1994) A 70-kDa heat shock cognate protein suppresses the defects caused by a proteasome mutation in Saccharomyces cerevisiae. FEBS Lett 251:263–266

Özkaynak E, Finley D, Varshavsky A (1984) The yeast ubiquitin gene: head-to-tail repeats encoding a polyubiquitin precursor. Nature 312:663–666

Özkaynak E, Finley D, Solomon MJ, Varshavsky A (1987) The yeast ubiquitin genes: a family of natural gene fusions. EMBO J 6:1429–1439

Parag HA, Raboy B, Kulka, RG (1987) Effect of heat shock on protein degradation in mammalian cells: involvement of the ubiquitin system. EMBO J 6:55–61

Parsell DA, Lindquist S (1993) The function of heat-shock proteins in stress tolerance: degradation and reactivation of damaged proteins. Annu Rev Genet 27:437–496

Pickart CM (1997) Targeting of substrates to the 26 S proteasome. FASEB J 11:1055–1066

Redman KL, Rechsteiner M (1989) Identification of the long ubiquitin extension as ribosomal protein S27a. Nature 338:438–440

Rubin DM, Finley D (1995) The proteasome: a protein-degrading organelle? Curr Biol 5:854–858

Sadis S, Atenza C, Finley D (1995) Synthetic signals for ubiquitin-dependent proteolysis. Mol Cell Biol 15:4086–4094

Schneider C, Sepp-Lorenzino L, Nimmesgern E, Ouerfelli O, Danishefsky S, Rosen N, Hartl FU (1996) Pharmacologic shifting of a balance between protein refolding and degradation mediated by Hsp90. Proc Natl Acad Sci USA 93:14536–14541

Seufert W, Jentsch S (1990) Ubiquitin-conjugating enzymes UBC4 and UBC5 mediate selective degradation of short-lived and abnormal proteins. EMBO J 9:543–550

Shang F, Gong X, Taylor A (1997) Activity of ubiquitin-dependent pathway in response to oxidative stress: ubiquitin-activating enzyme is transiently up-regulated. J Biol Chem 272:23086–23093

Sherman MY, Goldberg AL (1996) Involvement of molecular chaperones in intracellular protein breakdown. EXS 77:57–78

Skowyra D, Craig KL, Tyers M, Elledge SJ, Harper JW (1997) F-box proteins are receptors that recruit phosphorylated substrates to the SCF ubiquitin-ligase complex. Cell 91:209–219

Sommer T, Wolf DH (1997) Endoplasmic reticulum degradation: reverse protein flow of no return. FASEB J 11:1227–1233

Spence J, Sadis S, Haas AL, Finley D (1995) A ubiquitin mutant with specific defects in DNA repair and multiubiquitination. Mol Cell Biol 15:1265–1273

Treger JM, Heichman KA, McEntee K (1988) Expression of the yeast UBI4 gene increases in response to DNA-damaging agents and in meiosis. Mol Cell Biol 8:1132–1136

Treier M, Staszewski LM, Bohmann D (1994) Ubiquitin-dependent c-Jun degradation in vivo is mediated by the delta domain. Cell 78:787–798

Wagner I, Arlt H, van Dyck L, Langer T, Neupert W (1994) Molecular chaperones cooperate with PIM1 protease in the degradation of misfolded proteins in mitochondria. EMBO J 13:5135–5145

Werner EL, Brodsky JL, McCracken AA (1996) Proteasome-dependent endoplasmic reticulum-associated protein degradation: an unconventional route to a familiar fate. Proc Natl Acad Sci USA 93:13797–13801

Wiertz EJHJ, Jones TR, Sun L, Bogyo M, Geuze HJ, Ploegh HL (1996) The human cytomegalovirus US11 gene product dislocates the MHC class I heavy chains from the endoplasmic reticulum to the cytosol. Cell 84:769–779

Yaglom JA, Goldberg AL, Finley D, Sherman MY (1996) The molecular chaperone Ydj1 is required for the p34 CDC23-dependent phosphorylation of the cyclin Cln3 that signals its degradation. Mol Cell Biol 16:3679–3684

Zhou M, Wu X, Ginsberg, HN (1996) Evidence that a rapidly turning over protein, normally degraded by proteasomes, regulates hsp72 gene transcription in HepG2 cells, J Biol Chem 40:24769–24775

CHAPTER 7
Regulation of Heat Shock Genes by Cytokines

A. Stephanou and D.S. Latchman

A. Introduction

The genes encoding heat shock or stress proteins (Hsps) are among the most highly conserved proteins in the animal kingdom, so homologous that prokaryotic and eukaryotic species have at least 50% identity at the genomic level (Lindquist 1988). It was originally reported that when the chromosomes of the giant salivary gland of *Drosophila* were exposed to elevated temperature, this resulted in areas of swelling on the chromosome known as puffs (Ritossa 1962). These puffs indicated transcriptionally active genes belonging to the Hsps family. After the observations with *Drosophila*, Hsps were discovered to be a universal phenomenon. Exposure of prokaryotic, eukaryotic and even plant cells to potentially damaging environmental stresses, selected toxins, oxygen depravation, inhibitors of energy metabolism, cytokines, viral and microbial infections caused enhanced synthesis of Hsps (Lindquist 1986). In most cell types 1–2% of total proteins consist of Hsps, even prior to stress suggesting an important role for these proteins in the biology and physiology of the unstressed cell. The identification of different genes and forms of Hsps has led to the classification of Hsp gene families according to the molecular weight of the encoded protein. It is widely accepted that most members of the Hsps bind to unfolded proteins, preventing their aggregation and misfolding during cell stress and are therefore also termed "molecular chaperones" (Gething and Sambrook 1992). Other Hsps (Hsp90) are known to maintain the steroid receptors in the inactive form and also to bind transcription factors, suggesting a role for Hsps in signal transduction.

The induction of Hsps in response to various stresses or non-stressful conditions is dependent on the activation of a group of specific transcription factors, the heat shock factors (HSFs), which bind to the heat shock element (HSE) in the promoters of Hsps (Morimoto 1993; Lis and Wu 1993). Four HSFs (HSF-1 to 4) have been cloned from a number of organisms and their roles have been characterized. Only HSF-1 has been shown to be involved in regulating Hsps in response to thermal stress, whereas HSF-2, HSF-3 and HSF-4 may be involved in Hsps regulation in unstressed cells and their levels are regulated in response to a wide variety of biological processes such as immune activation and cellular differentiation (Morimoto et al. 1992; Morimoto 1993). In general, however, the stimuli which induce such alterations in Hsps gene expression under non-stress conditions have been poorly characterized and the mechanisms by which they act are unclear. In this chapter, we discuss recent studies indicating that Hsps are not only regulated by the HSFs, but also by transcription factors that are induced by specific cytokines.

B. Cytokines

Under normal circumstances, maintenance of homeostasis in mammals, including man, is achieved by a number of mechanisms. One such mechanism includes the release of cytokines in response to stressful challenges, such as tissue injury and infection, and a prominent feature of these responses is the induction of a group of proteins called the acute-phase proteins, which are involved in restoration of homeostasis (Baumann and Gauldie 1994). Cytokines are important mediators of the acute-phase response and their signal is transmitted into the cell via different cytokine-receptors and intracellular signalling pathways. Responses to infection also lead to activation of inflammatory and immune-derived cytokine genes that also act on various organs in the body to maintain homeostasis (Dinarello 1996). Accumulating evidence has demonstrated that signalling via cytokines leads to the activation of a relatively small number of similar transcription factors. One of the characteristic features of cytokines is their functional pleiotropy and redundancy, i.e., one cytokine shows a wide variety of biologic functions on various tissues and cells, and several different cytokines exert similar and overlapping functions on a certain cell type.

Interleukin-6 (IL-6) is a typical example of such a multifunctional cytokine. It was originally identified as a B-cell differentiation factor (BSF-2) that induced the final maturation of B cells into antibody-producing cells (Miyaura et al. 1988). Further studies have shown that IL-6 can also act on other cell types such as T cells, macrophages, hepatocytes and neuronal cells (Taga and Kishimoto 1992; Wagner 1996). Several studies have shown an autocrine or paracrine action of IL-6 in human myeloma growth and IL-6 overexpressing transgenic mice develop B cell plasmacytomas (Kawano et al. 1988; Suematsu et al. 1989). In vitro or in vivo studies have shown that anti-IL-6 antibodies could inhibit the growth of myeloma cells. In contrast, IL-6 itself inhibits the growth of a murine myeloid leukemia cell line (M1) and induces their differentiation into macrophages (Miyaura et al. 1988). The pleiotropic functions of IL-6 are summarized in Table 1.

The IL-6 receptor (IL-6R) belongs to a group of receptors that share a common feature among other cytokine receptors, in that the intra-cytoplasmic regions lack any known signal transduction motif such as a tyrosine or serine/threonine kinase domain. Further studies demonstrated the presence of a molecule associated with IL-6R responsible for signal transduction. Binding of IL-6 to IL-6R triggers the association of a non-ligand-binding 130kDa signal transducing molecule, gp130 (Hibi et al. 1990). The expression pattern of

Table 1. Pleiotropic function of IL-6

Cell type	Effect
B cells	Immunoglobulin production; proliferation of myeloma cells
T cells	Proliferation of cytotoxic T cells; proliferation and differentiation of T cells
Hepatocytes	Acute-phase protein synthesis
Bone	Stimulation of osteoclast formation and induction of bone resorption
Adult heart cells	Negative iontrophic effect on heart
Neuronal cells	Neural differentiation of PC12 cells; activates the hypothalamic-pituitary axis

gp130 was shown to be ubiquitous in many different organs and did not necessarily parallel the expression of IL-6R (SAITO et al. 1992). This suggested that gp130 may be used as a signal transducer for other cytokine receptors. For example, the heart has been shown to express high levels of gp130 but not IL-6R, suggesting the presence of an unknown natural ligand for gp130 in heart tissues. Recently, cardiotropin-1 (CT-1) has been identified as a cytokine in the heart that uses gp130 to induce intracellular signalling (PENNICA et al. 1995a,b). Moreover, gp130 knock-out mice develop haematological and myocardial abnormalities and die shortly after birth, suggesting a very important role for gp130 (YORSHIDA et al. 1996). In contrast, IL-6 knock-out mice show no overt developmental abnormalities (POLI et al. 1994).

As mentioned above, the pleiotropic and redundant properties of IL-6 and other cytokines could be explained by the use of a common signal transducer gp130. It has recently been demonstrated that other cytokine receptors are also associated with gp130 and these include leukaemia inhibitory factor (LIF), oncostatin M (OM), interleukin-11 (IL-11), ciliary neurotropic factor (CNTF) and CT-1 receptors (GEARING et al. 1992; PENNICA et al. 1995b). LIF was also discovered as an inhibitor of M1 cell growth and to also act on the liver to induce the release of acute-phase proteins (HILTON and GOUGH 1991). Similar effects were also observed for OM (ROSE and BRUCE 1991). CNTF is a cytokine that maintains survival of ciliary neurons as well as motor neurons and that induces differentiation of oligodendrocytes into astrocytes (STOKLI et al. 1989). IL-11 was originally identified as a plasmacytoma growth factor that displayed very similar biologic activities to IL-6. CT-1 was described as a cytokine that was induced in cardiac myocyte hypertropy (PENNICA et al. 1995a).

C. Transcription Factors Activated by the IL-6 Receptor Family

I. C/EBPs

The IL-6 intracellular signalling pathway has been extensively studied and two groups of transcription factors NF-IL6 (C/EBPβ), NF-IL6β (C/EBPδ) and

STAT3 are the nuclear targets of IL-6 response (Akira and Kishimoto 1992; Akira et al. 1994). NF-IL6 was originally identified as a nuclear factor that binds to a 14-bp palindromic sequence within the promoter region of the human IL-6 gene (Isshiki et al. 1990). Following the cloning of NF-IL6, several other C/EBP family members have been cloned. At present there are five members including C/EBPα, NF-IL6 (C/EBPβ), NF-IL6 (C/EBPδ), C/EBPγ (Gadd153) and CHOP-10.

C/EBPα is expressed in adipose, liver and placental tissues that play vital roles in energy metabolism. C/EBPα expression is important in adipocyte differentiation and also plays an important role in controlling the expression of genes critical to liver function (McKnight et al. 1989). Moreover, mice deficient in C/EBPα become profoundly hypoglycaemic and die several hours after birth due to deficient glycogen stores in the liver (Wang et al. 1995). In addition, the hepatocytes and adipocytes of the mutant mice fail to accumulate lipid.

C/EBPβ (NF-IL6) and C/EBPδ (NF-IL6β) are expressed at low levels in normal tissues but are both inducible by the stimulation by IL-6, but also by tumour necrosis factor (TNF), interleukin-1 (IL-1) and lipopolysaccharide (LPS; Kinoshita et al. 1992). They both also play a role in energy metabolism and adipocyte differentiation, but also in modulating the immune system (Spiegelman and Flier 1996; Samuelsson et al. 1991). C/EBPβ deficient mice are viable but highly susceptible to facultative intracellular organisms (Tanaka et al. 1995; Screpanti et al. 1995). C/EBPβ is a 35 kDa protein that is also alternatively spliced to produce a 21 kDa protein that is able to bind to DNA but has no transcriptional activity. C/EBPβ activity is regulated by phosphorylation of the thr-235 residue by the ras-dependent MAP kinase cascade (Nakajima et al. 1993).

C/EBPγ was originally cloned as a nuclear factor that binds to functionally important C/EBP binding sites. Like the short 21 kDa C/EBPβ, C/EBPγ also has no transcriptional activity. C/EBPγ is ubiquitously expressed in normal adult tissues but is most abundant in B lymphocytes (Roman et al. 1990). CHOP-10 was cloned as a protein interacting with the bZIP domain of C/EBPβ. Although CHOP can form heterodimers with other members of the C/EBP family, the heterodimers cannot bind to known C/EBP binding sites due to having several amino acid substitutions in the highly conserved basic regions present in all bZIP proteins (Ron and Habener 1992).

II. STATs

As mentioned above, cytokines are released in response to inflammatory stimuli, but are also important together with other growth factors in regulating multiple aspects of cell growth and differentiation. Studies of the transcriptional response to interferons have identified the Janus kinase-signal *transduc*ers and *a*ctivators of *t*ranscription (STATs). Characterization of the interferon (INF-α and INF-γ) signalling pathway have identified a group of receptor-associated tyrosine kinases belonging to the JAK family (JAK1, JAK2 and JAK3; DARNELL et al. 1994). Upon ligand binding the JAKs become activated and this leads to the recruitment of latent cytoplasmic transcription factors or STATs that become phosphorylated and form homo- or heterodimers which are then translocated to the nucleus and bind to STAT-responsive genes. In the case of INF-α or INF-γ, it was shown that both STAT1 and STAT2 are activated (DARNELL et al. 1994). Later it was demonstrated that IL-6-related cytokines activated STAT-3 (also called APRF) (AKIRA et al. 1994). STAT4 has been shown to be activated by IL-12 (YAMAMOTO and QUELLE 1994; KAPLAN et al. 1996), and STAT5 activation was demonstrated by IL-3, IL-5 and prolactin (MUI et al. 1996). STAT6 has been reported to be activated by IL-4 (TAKEDA et al. 1996).

There are six STATs known at present and they share several conserved structural and functional domains including a DNA binding domain and a potential phosphotyrosine-binding motif. DNA binding by STATs is totally dependent on phosphorylation and they recognize a similar DNA binding sequence. The consensus sequence has been deduced and shown to be TTCC(G/C)GGAA, although a number of recent STAT binding sites have been characterized and shown to vary slightly (IHLE 1996). The functions of STATs may also be influenced by serine phosphorylation via the MAPK pathway, suggesting that both the JAK and MAPK activity are able to modulate STAT activation (WEN et al. 1995; ZANG et al. 1995).

D. Role of Interleukin-6 Family of Cytokines in Regulating Hsps

Cytokines play a pivotal, but a paradoxical role both in immunity and inflammation and are also implicated as potent mediators of the pathology of diseases. Interleukin-1 (IL-1) was the first cytokine discovered as a protein capable of inducing pyrexia (endogenous pyrogen; DINARELLO 1996). IL-1 was later shown to have many more actions. Tumour necrosis factor (TNF) was also demonstrated to have a similar pattern of activity (MOSHAGE 1997), whereas IL-6 shares only some of these proinflammatory actions. These differences in activity were shown to be due to alternative intracellular signalling pathways activated by these cytokines (AKIRA and KISHIMOTO 1992). All three cytokines stimulate the liver to induce the synthesis of acute-phase proteins, with IL-6 being the most potent.

It is generally accepted that NF-IL6 and STAT-3 signalling pathways allow IL-6 to activate two distinct sets of genes each of which is responsive to one of these pathways. Thus, class I acute-phase proteins (such as α-acetic glycoprotein, haptaglobin, C-reactive protein and serum amyloid) contain responsive elements to NF-IL6 and these factors have been shown to be involved in the activation of these genes following IL-6 treatment (AKIRA and KISHIMOTO 1992). In agreement with this idea, these genes are stimulated by exposure of cells to IL-1, TNFα and LPS, which also stimulates NF-IL6 activity without affecting STAT-3 (GANTER et al. 1989). In contrast, class II acutephase genes such as fibrinogen, thiostatin and α-microglobin are not inducible by IL-1 and lack binding sites for NF-IL-6. Instead these genes contain STAT-3 responsive elements and allow the binding of STAT-3, which is responsible for activation of these genes in response to IL-6 (OLIVIERO and CORTESE 1989).

In the following section, we discuss and provide evidence to show that cytokines are also capable of inducing stress proteins or the Hsps. Our initial studies began by investigating the elevated levels of Hsp90 in systemic lupus erythematosis (SLE), which will be discussed further in Sect. G of this chapter. Interestingly, elevated levels of circulating IL-6 have been reported in a number of different autoimmune diseases such as rheumatoid arthritis (EASTGATE et al. 1988), juvenile chronic arthritis (DE BENEDETTI et al. 1991) and SLE (LINKER-ISRAELI et al. 1991), and the levels have been shown to be correlated with disease activity, being highest in patients with active disease. These findings therefore suggested that IL-6 might play a role in the pathogenesis of autoimmune diseases. Furthermore, infusion of an antibody to IL-6 can relieve disease symptoms in lupus-prone NZB/NZW F1 mice (FINCK et al. 1994). Therefore a role for IL-6 in disease pathogenesis is likely to involve the induction of the expression of specific genes within its target cells. We and others have shown that elevated levels of Hsp90 in peripheral blood mononuclear cells (PBMCs) from a specific subset of SLE patients correlated with disease activity in some organs or systems (DEGUCHI et al. 1987; NORTON et al. 1994; TWOMEY et al. 1993; LATCHMAN and ISENBERG 1994). We then started to investigate the role of IL-6 in the activation of Hsp90 by studying the effect of IL-6 on Hsp90 protein levels and on the Hsp90 gene promoter. IL-6 was shown to induce the accumulation of Hsp90 in both liver cells and in PBMCs (STEPHANOU et al. 1997). Interestingly, IL-6 also induced the expression of Hsp70 in liver cells but not in PBMCs. At least in liver cells this effect is mediated by IL-6 activation of the Hsp90β gene promoter, which can also be produced in different cell types by the overexpression of the IL-6-induced transcription factors NF-IL6 and NF-IL6β. Moreover, several NF-IL6 binding sites are present in the region of the promoter from −1044 to −300, which mediates the response to IL-6 itself, and also mediates its activation by NF-IL6. A short region of the Hsp90 promoter (−643 to −623) containing a binding site for these factors can confer responsiveness to IL-6 on a heterologous promoter (STEPHANOU et al. 1998).

Additional studies have demonstrated that LIF can also enhance Hsp70 and Hsp90 protein levels in liver cells (Stephanou et al., unpublished data). CT-1 and LIF have also recently been shown to enhance the expression of Hsp70 and Hsp90 protein levels in neonatal cardiomyocytes and hence protect against subsequent exposure to severe thermal or ischaemic stress. These results therefore suggest that CT-1 and LIF may have therapeutic potential in the protection of the heart from stress, particularly if the protective effects of

CT-1 can be dissected away from the potential damaging induction of cardiac hypertrophy (STEPHANOU et al. 1998b). Recently, CT-1 has been reported to reduce programmed cell death or apopotosis in neonatal cardiomyocytes via the MAPK pathway and not the STAT-3 pathway (SHENG et al. 1997). Taken together these observations indicate that the activation of the gp130 pathway by IL-6, CT-1 and LIF can result in elevated Hsp expression.

Further studies demonstrated that the Hsp90β promoter is also activated by the STAT-3 signalling pathway (STEPHANOU et al. 1998). Furthermore, additional analysis of the Hsp90 promoter revealed that the short region between −643 and −623 which was previously shown to confer responsiveness to NF-IL6 also contained two STAT-3-like binding sites and was also activated by STAT-3 overexpression in transfection experiments (Fig. 1). Moreover, NF-IL6 and STAT-3 synergized strongly in activating the Hsp90β promoter (STEPHANOU et al. 1998). In addition, experiments with dominant negative mutants of these factors showed that the effect of IL-6 itself on the Hsp90β promoter is strongly dependent on the synergistic interaction of NF-IL6 and STAT-3. Hence the Hsp90β promoter appears to have a novel pattern of inducibility which is dependent upon both the IL-6 activated pathways involving the threonine phosphorylation of NF-IL6 by MAP kinases and the tyrosine phosphorylation of STAT-3 by JAK family kinases.

Further studies have demonstrated that both NF-IL6 and STAT-3 are able to interact differently with HSF-1 or a heat shock (STEPHANOU et al. 1998).

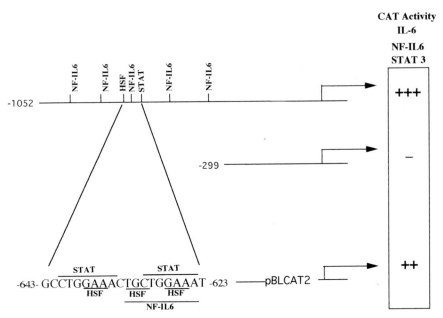

Fig. 1. Hsp90β promoter reporter constructs showing responsiveness to IL-6, NF-IL-6 and STAT 3. pBLCAT2 is a heterologous promoter in which Hsp90β HSE/NF-IL6/STAT binding sites were ligated

Thus overexpressed NF-IL6 and HSF-1/ or heat shock stimuli cooperate and enhance the activity of the Hsp90β promoter, while overexpressed STAT-3 and HSF-1/or heat shock stimuli antagonize each other. Further studies were performed to determine which of these opposite interactions of IL-6-stimulated transcription factors with HSF-1 predominated when cells were exposed to both heat shock and IL-6 in the absence of any transcription factors. It was shown that both heat shock and IL-6 individually activated the Hsp90β promoter; however, when both stimuli were applied together a much weaker increase in promoter activity was observed compared to that seen with either stimulus alone (Stephanou et al. 1998). Hence the synergistic interaction between HSF-1 and NF-IL6 which we observed in transfection experiments appears to be overcome by the antagonistic interaction of HSF-1 and STAT-3 when the transcription factors are activated by the appropriate stimuli rather than by overexpression. Moreover, IL-1, which activates only the NF-IL6 and not the STAT-3 pathway, was able to synergize with heat shock and produce a strong activation of the Hsp90β promoter (Stephanou et al. 1998). It is clear therefore that the activity of Hsp90β gene is also influenced by other pathways other than the heat shock-activated pathway and that these results suggest that Hsp90 gene regulation is more complex than has been supposed. However, these results render this promoter distinct from those of the liver acute-phase protein genes, which appear to fall in two separate classes which are predominantly regulated either by the NF-IL6 pathway or the STAT-3 pathway (Fig. 2).

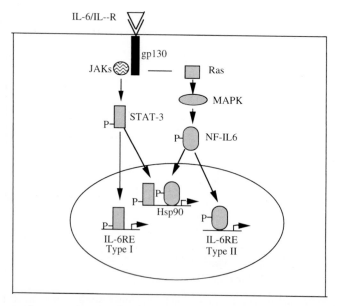

Fig. 2. Two pathways activated by IL-6 stimulate either type I or type II genes. Hsp90β gene is activated by both pathways

It is possible that this difference may reflect the tissue specific expression of Hsp90 in all cell types. Indeed the great majority of studies of IL-6 inducible genes have focused on genes encoding proteins which are expressed in only a limited range of cell types such as the acute-phase proteins (GANTER et al. 1989; AKIRA and KISHIMOTO 1992), or the immunoglobins (RAYNAL et al. 1989). It is unlikely, however, that this difference is responsible for the unique response of the Hsp90β gene promoter since we observed identical responses of the promoter in both liver cell lines, which express acute-phase protein genes and neuronal cells which do not (STEPHANOU et al. 1997).

Other groups have also shown the induction of Hsps by cytokines in different cell types. For example, IL-4, which is known to specifically activate STAT 6 (TAKEDA et al. 1996), has been shown to enhance the expression of Hsp90α in human lymphocytes (METZ et al. 1996). IL-1 beta also increases the levels of Hsp70 and Hsp90 in rat islet pancreatic cells (HELQUEST et al. 1991). Both TNFα and INFγ have also been reported to enhance the levels of Hsp70 in granulosa-luteal cells (KIM et al. 1996). In addition, IL-4 and IFNγ increased the levels of Hsp27 in human renal carcinoma cells (SULLIVAN et al. 1997).

E. Role of IFN-γ in Regulating Hsp Expression

IFN-γ is also a multifunctional cytokine that is known to have anti-viral and anti-tumour properties by inducing specific IFN-γ responsive genes (PESTFA et al. 1987; SEN and RANSHFF 1993; DARNELL et al. 1994). In contrast to IL-6, IFN-γ specifically activates the STAT-1 signalling pathway via the JAKs (SCHINDLER et al. 1995). Like STAT-3, STAT-1 is also alternatively spliced to produce STAT-1β that lacks the carboxyl 38 amino acids. This variant is also recruited to the receptor complex, becomes phosphorylated and binds DNA, but does not activate gene transcription. Therefore it was suggested that STAT-1β may act as a naturally occurring dominant negative form of STAT-1 (IHLE and KERR 1995). It is also important to note at this stage that although all the STATs that have been identified all bind very similar DNA sequences, the divergent nature of the carboxyl region is likely to be critical in affecting individual gene expression and provide specificity for each STAT (IHLE 1996). Recently the role of STAT proteins in cytokine signalling has been assessed by STAT gene knock-out mice. STAT-1 knock-out mice show no overt developmental abnormalities and are indistinguishable from their normal counterparts on the basis of size, activity, or ability to reproduce. However, these animals display a complete lack of responsiveness to either INF-β or INF-γ and are highly sensitive to infection by microbial pathogens and viruses (MERAZ et al. 1996; DURBIN et al. 1996). This is in contrast to STAT-3 knock-out, which resulted in embryonic lethality prior to gastrulation, suggesting that STAT-3 has an important role in development (TAKEDA et al. 1997). In vitro studies using dominant negative STAT proteins have recently been assessed. Dominant negative STAT-3 introduced in M1 cells was shown to abolish the

IL-6-induced growth arrest and macrophage differentiation, indicating that STAT-3 is a critical molecule which determines the cellular decision from cell proliferation to cell growth and differentiation in M1 cells (NAKAJIMA et al. 1996). In contrast, a STAT-1 dominant negative mutant had no effect on cellular differentiation in this system.

Recently we investigated whether the STAT-1 pathway may also play a role in activating the expression of Hsps. IFN-γ treatment was shown to induce the expression of Hsp70 and Hsp90 in the IFN-γ responsive HepG2 cell line. In addition, overexpression of STAT-1 enhanced the activities of the Hsp70 and Hsp90β promoter. Furthermore, in studies with a STAT-1 deficient cell line U3A, INF-γ was unable to activate either the Hsp70 or Hsp90 promoter. However, in studies with the U3A-STAT1 cell line in which STAT-1 was re-introduced, activation of both the Hsp70 and Hsp90 promoter was established in response to INF-γ. In addition, INF-γ increased the levels of both Hsp70 and Hsp90 protein. Interestingly, STAT-1 and HSF-1 were shown to co-operate in activating the Hsp70 and Hsp90β promoter. In vivo protein-binding studies demonstrated protein-protein interaction between STAT-1 and HSF-1 but not HSF-1 and STAT-3 (Stephanou et al., unpublished data). These studies therefore indicate differential interactions between STAT-1 or STAT-3 with HSF-1 and identify HSF-1 as an interaction partner with STAT-1 (Fig. 3). This is the first report showing that HSF-1 is able to interact directly with another transcription factor and may explain the mechanism whereby the induction of Hsps is observed during cytokine stimulation. Interestingly, HSF-3 has been shown to interact and co-operate with c-Myb and it was suggested that this was the mechanism linking c-Myb in cellular proliferation and Hsp induction during the cell cycle (KANEI-ISHII et al. 1997).

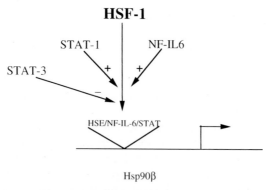

Fig. 3. Interaction of HSF-1 with NF-IL6, STAT 1 and STAT 3, showing either a stimulatory (+) response or an inhibitory (−) response in modulating the Hsp90β promoter

F. Elevation of C/EBPs and STATs and Hsps Expression During Inflammatory Pathological States

Some members of the Hsp70, Hsp27 and Hsp90 families have been suggested to play a defined role in cancer and it has been reported that Hsps are overexpressed in patients with malignant tumours compared to healthy controls and this overexpression does show some correlation with disease features. The relationship between Hsps and cancer is discussed in more detail in Chap. 19. In vitro, breast cancer cells require exogenous serum-derived factors for optimal growth. Protein factors responsible for this activity resulted in the isolation of growth factors that were important for the growth of these cells. These factors included epidermal growth factor (EGF) and its receptor (EGFR), which unlike the the IL-6 receptor family, contains an intracellular domain with tyrosine kinase activity and is also able to activate the STAT-3 pathway in a JAK independent manner (DARNELL et al. 1994).

However, recently members of the IL-6 receptor family including IL-6, LIF, OM, CNTF and IL-11 have been shown to be expressed in breast cancer cells (MIRANDA et al. 1996; DOUGLAS et al. 1997). In addition, these cytokines were shown to increase the proliferation of breast cancer cells. Moreover, the gp130 subunit was also demonstrated in these cells (DOUGLAS et al. 1997). The finding that breast cancer cells are also able to secrete IL-6, LIF and OM suggests that these cytokines may be important in regulating the growth of breast cells. To investigate the relationship between elevated Hsps and IL-6 cytokine family in breast cancer, we have studied the levels of Hsps and the transcription factors C/EBPβ, C/EBPδ, STAT-1 and STAT-3. As shown in Fig. 4, the increased levels of Hsp70 and Hsp90 paralleled the enhanced expression of C/EBPβ, C/EBPδ, STAT-1 and STAT-3. These data therefore suggest that expression of Hsps may be regulated by transcription factors involved in the IL-6 signalling pathway in breast cancer. In addition, transfection studies using an Hsp90 promoter construct demonstrated an increase in promoter activity in breast cancer cell lines treated with IL-6 and LIF (Stephanou et al., unpublished data).

It has been reported that immune activation by phorbol esters or by other immunomodulators can lead to increased expression of Hsps. However, the mechanism resulting in this enhanced expression of Hsps is unclear. We have started to investigate the role of C/EBPs and STATs in regulating Hsps in lymphocytes after stimulation with PHA, which is a non-specific T-lymphocyte mitogen that mimics a T cell inflammatory response. As shown in Fig. 5, PHA resulted in a dramatic induction in the expression of C/EBPβ, C/EBPδ, STAT-1 and STAT-3. The increased levels of these transcription factors also paralleled the increased expression of Hsp70 and Hsp90. These results differ with PBMCs cells treated with IL-6 where only Hsp90 was shown to be induced but not Hsp70 or Hsp27. Therefore these results suggest that the induction of Hsps may depend upon the levels and expression of transcription factors in response to a mitogen or cytokine stimulation. These studies also

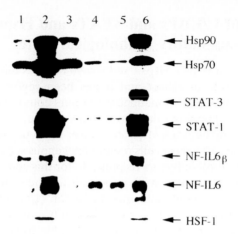

Fig. 4. Expression of Hsps, NF-IL6, NF-IL6β, STAT 1, STAT 3 and HSF-1 in six different breast cancer tissues

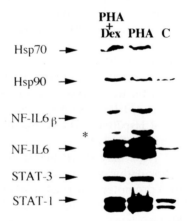

Fig. 5. Induction of Hsps, NF-IL6, NF-IL6β, STAT 1 and STAT 3 in phytohemagglutinin (*PHA*) stimulated lymphocytes. Dexamethasone (*Dex*) was also examined together with PHA. *The asterisk* depicts the alternative splice forms of NF-IL6

indicate that the altered levels of Hsps observed in disease or inflammatory states may be modulated by cytokine-induced transcription factors.

G. Role of IL-6 and Hsps in SLE

A number of studies have suggested a role of Hsps in disease states and conservation of structure and function of many Hsps may provide a link between immunity and autoimmune diseases. The role of Hsps in autoimmune diseases will be discussed in more detail in Chaps. 15, 16, and 17. Hsps have been shown to be immunodominant antigens during bacterial and mycobacterial infections. Peripheral T cells carrying the γ/δ T cell receptor (TCR) recognize the mycobacterial 65 kDa Hsp (HOLOSHITZ et al. 1989; O'BRIEN et al. 1989). There are major structural similarities between mycobacterial and human Hsps. Infection may lead to increased expression of both mycobacterial and self Hsps. Thus it has been hypothesized that these Hsps could be involved in the aetiopathogenesis of autoimmunity. Abnormalities in the expression or localization of the endogenous human Hsps may be required to trigger autoimmune reactivity, from a response that was initially directed against bacterial Hsps (KINDAS-MÜGGE et al.; REES et al. 1988). As has been widely discussed, the development of an autoimmune condition will depend upon a number of additional coexistent factors, including genetic susceptibility, hormonal milieu and diet, for example (SHOENFELD and ISENBERG 1989). However if, in some individuals, these factors are "in place" following initial exposure to exogenous Hsps, a second event such as a viral infection (LA THANGUE and LATCHMAN 1988) could lead to the upregulation of human Hsps and/or their surface localization, which in turn induces antibodies and T cells primed against the bacterial or even protozoan proteins reacting against human proteins which might be an important triggering event for autoimmunity (LATCHMAN and ISENBERG 1994). Although much attention has focused on Hsp60/65 in animal models of rheumatoid arthritis observed in certain rat strains, there is now evidence suggesting a role for Hsp90 in patients with systemic lupus erythematosus (SLE). Thus we and others have demonstrated increased levels of Hsp90 in peripheral blood mononuclear cells (PBMCs) of patients with SLE and MRL/lpr mice (DEGUCHI et al. 1987; NORTON et al. 1994; TWOMEY et al. 1993; STEPHANOU et al. 1997).

The elevation of Hsp90 which is observed in SLE patients and MRL/lpr mice may be involved in the production of autoantibodies to this protein which has been observed both in SLE patients and in MRL/lpr mice, where it occurs after the initial elevation of Hsp90 levels (CONROY et al. 1994, 1996). This possibility is supported by similar results in human breast cancer patients, in which elevated levels of Hsp90 in tumour tissue were similarly paralleled by the development of autoantibodies to Hsp90 (CONROY et al. 1995). Such data suggest elevated IL-6 levels can induce the specific elevation of Hsp90 protein levels, which in turn results in autoantibody production. The complex series of events which occur in both human SLE and MRL/lpr mice renders this hypothesis difficult to test in this situation. However, our previous studies strongly suggest that IL-6 plays a critical role in the induction of the Hsp90 promoter in vitro (STEPHANOU et al. 1997, 1998).

In order therefore to test directly the role of IL-6 in regulating Hsp90 expression in vivo, we have recently used mice which have been artificially engineered to express elevated levels of IL-6 either by being made transgenic for extra copies of the IL-6 gene (SUEMATSU et al. 1989) or by inactivation of the gene encoding the transcription factor C/EBPβ, which also results in the elevation of IL-6 levels in these mice (Screpanti et al. 1995). In these experiments elevated levels of Hsp90 were observed in both the IL-6 transgenic and the C/EBPβ knock-out mice. Hence the elevated IL-6 levels induced in these animals are indeed paralleled by increased levels of Hsp90 compared to normal control mice (Stephanou et al. 1998c). In addition it was also observed that both IL-6 transgenic and C/EBPβ knock-out animals produced autoantibodies to Hsp90 (Stephanou et al. 1998c). It is also of interest that inactivation of the IL-6 gene in the C/EBPβ knock-out mice resulted in the suppression and

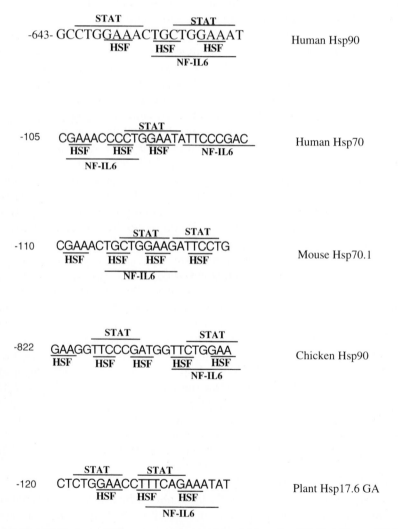

Fig. 6. The HSEs from different species contain neighbouring NF-IL6-like and STAT-like binding sites

severity of Castleman-like disease observed in these animals and a reduction in the production of autoantibodies (SCREPANTI et al. 1996). Therefore these studies strongly suggest that IL-6 is likely to play a critical role in the regulation of Hsp90 levels and autoantibody production, both in autoimmune disease states and potentially in normal cells in vivo. It is also important to note that the induction of Hsp90 by IL-6 in PBMCs and also in SLE and MLR/lpr mice and in IL-6 transgenic mice seem to be specific, since no changes in the levels of other Hsps were observed (FAULDS et al. 1994; STEPHANOU et al. 1997, 1998c).

H. Conclusion

In this chapter we have reviewed recent work demonstrating that cytokines belonging to the IL-6 receptor family or IFN-γ are able to stimulate the expression of Hsps by activating specific transcription factors which bind and transactivate the Hsp genes. Thus, these studies indicate that transcription factors other than the HSFs may also play a role in modulating Hsp expression. In addition our studies reveal an unexpected complexity in the regulation of Hsps by non-stressful stimuli. The ability of the Hsp genes to respond to the cytokines is likely to be responsible for its elevated expression in disease states such as SLE and breast cancer. Interestingly, analysis of the promoters of Hsps from different species reveals similar but not identical DNA binding sequences between HSFs and STATs or NF-IL6 (Fig. 6). Thus, this would suggest a composite response element for these transcription factors which have evolved together and may be an explaination at the molecular level of how HSF-1 is able to interact and cooperate with the STATs and NF-IL6. We hope that further studies will provide a clear understanding of the molecular mechanism and our hypothesis of the mechanisms of regulation of Hsps by cytokines.

References

Akira S, Kishimoto T (1992) IL-6 and NF-IL6 in acute-phase response and viral infection. Immunol Rev 127:25–45

Akira S, Nishio Y, Inoue M, Wang X-J, Wei S, Matsusaka S, Yoshida K, Sudo T, Naruto M, Kishimoto T (1994) Molecular cloning of APRF, a novel IFN-stimulating gene factor 3 p91 related transcription factor involved in the gp 130 mediated signalling pathway. Cell 77:63–71

Baumann H, Gauldie J (1994) The acute phase response. Immunol Today 15:74–80

Conroy SE, Faulds GB, Williams W, Latchman DS, Isenberg DA (1994) Detection of autoantibodies to the 90kD heat shock protein in SLE and other autoimmune diseases. Br J Rheum 33:923–926

Conroy SE, Gibson SL, Brunstrom G, Isenberg DA, Luqmani Y, Latchman DS (1995) Autoantibodies to the 90kD heat shock protein in sera of breast cancer patients. Lancet 345:126

Conroy SE, Tucker L, Latchman DS, Isenberg DA (1996) Incidence of anti-hsp90 and 70 antibodies in children with SLE, dermatomyositis, and juvenile chronic arthritis. Clin Exp Rheum 35:99–104

Darnell JE Jr, Kerr IM, Stark GR (1994) Jak-STAT pathways and transcriptional activation in response to IFNs and other extracellular signalling proteins. Science 264:1414–1417

De Benedetti F, Massa M, Robbion R, Ravelli A, Burgio GR, Martini A (1991) Correlation of serum IL-6 levels with joint involvement and thrombocytosis in systemic juvenile rheumatoid arthritis. Arthritis Rheum 34:1158–1163

Deguchi Y, Negoro S, Kishimoto S (1987) Heat shock protein synthesis by human peripheral mononuclear cells from SLE patients. Biochem Biophys Res Commun 148:1063–1068

Dinarello CA (1996) Biologic basis for interleukin-1 in disease. Blood 87:2095–2147

Douglas AM, Goss GA, Sutherland RL, Hilton DJ, Berndt MC, Nicola NA Begley CG (1997) Expression and function of members of the cytokine receptor superfamily on breast cancer cells. Ongogene 14:661–669

Durbin JE, Hackenmiller E, Benjamin C, Simon MC, Levy DE (1996) Targeted disruption of the mouse Stat1 gene results in compromised innate immunity to viral disease. Cell 84:443–448

Eastgate JA, Symons JA, Wood NC, Grinlinton FMO, Giovine FS, Duff GW (1988) Correlation of plasma IL-1 levels with disease activity in rheumatoid arthritis. Lancet ii:706–709

Faulds GB, Isenberg DA, Latchman DS (1994) The tissue-specific elevation in synthesis of the 90kD heat shock protein precedes the onset of disease in lupus prone MRL/lpr mice. J Rheum 21:234–238

Ferrarini M, Heltai S, Zocchi MR, Rugarli C (1992) Unusual expression and localization of heat shock proteins in human tumor cells. Int J Cancer 51:613–619

Finck BK, Chan B, Wofsy D (1994) Interleukin 6 promotes murine lupus in NZB/NZW F1 mice. J Clin Invest 945:585–591

Ganter U, Arcone R, Toniatti C, Morrone G, Ciliberto G (1989) Dual control of C-reactive protein gene expression by IL-1 and IL-6. EMBO J 8:3773–3779

Gearing DP, Comeau MR, Friend DJ, Gimpel SD, Thut CJ, McGourtry J, Brasher KK, King JA, Gillis S, Mosley B, Zeigler SF, Cosman D (1992) The IL-6 signal transducer, gp130: an oncostatin M receptor and affinity converter for the LIF receptor. Science 255:1434–1437

Gething MJ. Sambrook J (1992) Protein folding in the cell. Nature 335:33–45

Helquist S, Polla BS, Johannesen J, Nerup J (1991) Heat shock protein induction in rat pancreatic islets by IL-1 beta. Diabetologia 34:150–156

Hibi M, Murakami M, Saito M, Hirano T, Taga T, Kishimoto T (1990) Molecular cloning and expression of an IL-6 signal transducer, gp130. Cell 63:1149–1157

Hilton DJ, Gough NM (1991) Leukemia inhibitory factor; a biological prespective. J Cell Biochem 46:21–26

Holoshitz J, Koning F, Colligan JE, De Bruyn J, Strober S (1989) Isolation of CD4⁻ CD8⁻ mycobacteria-reactive T lymphocyte clones from rheumatoid arthritis synovial fluid. Nature 339:226–229

Ihle J (1996) STATs: signal transducers and activators of transcription. Cell 84:331–334

Isenberg DA, Ehrenstein M, Longhurst C, Kalsi J (1994) The origin, sequence, structure and consequences of developing anti-DNA antibodies – a human perspective. Arthritis Rheum 27:169–80

Isshiki H, Akira S, Tanabe O, Nakajima T, Shimamoto T, Hirano T, Kishimoto T (1990) Constitutive and IL-1 inducible factors interact with the IL-1 responsive elements in the IL-6 gene. Mol Cell Biol 10:2757–2764

Kanie-ishii C, Tanikawa J, Nakai A, Morimoto RI, Ishii S (1997) Activation of HSF-3 by c-Myb in the absence of cellular stress. Science 277:246–248

Kaplan MH, Sun Y-L, Hoey T, Grubsy SJ (1996) Impaired IL-12 responses and enhanced development of Th2 cells in STAT4-deficient mice. Nature 382:174–177

Kawano M, Hirano T, Matsuda T, Taga T, Horii Y, Iwato K, Asaoka H, Tang B, Tanabe O, Tanaka H, Kishimoto T (1988) Autocrine generation and essential requirement of IL-6 for human multiple myeloma. Nature 332:83–86

Kim AH, Khanna A, Aten RF, Olive DL, Behrman HR (1996) Cytokine induction of heat shock proteins in human granulosa-luteal cells. Mol Hum Reprod 2:549–554

Kinoshita S, Akira S, Kishimoto T (1992) A member of the C/EBP family, NF-IL6β, forms heterodimer and transcriptionally synergizes with IL-6. Proc Natl Acad Sci USA 87:1473–1476

Kishimoto T, Akira S, Narazaki M, Taga T (1995) Interleukin-6 family of cytokines and gp130. Blood 86:1243–1254

Latchman DS, Isenberg DA (1994) The role of HSP90 in SLE. Autoimmunity 19:211–218

La Thangue NB, Latchman DS (1988) A cellular protein related to heat shock protein 90 accumulates during herpes simplex virus infection and is overexpressed in transformed cells. Exp Cell Res 178:169–179

Lindquist S (1986) The heat shock response. Annu Rev Biochem 55:1151–1191

Lindquist S (1988) The heat shock proteins. Annu Rev Genet 22:631–677

Linker-Israeli M, Deans RJ, Wallace DJ, Prehn J, Ozeri-Chen T, Kinenberg JR (1991) Elevated levels of endogenous IL-6 in SLE. A putative role in pathogenesis. J Immunol 147:117–123

Lis J, Wu C (1993) Protein traffic on the heat shock promoter: parking, stalling and trucking along. Cell 74:1–4

McKnight SL, Lane MD, Gluecksohn-waelsch S (1989) Is CCAAT/enhancer-binding protein a central regulator of energy metabolism ? Genes Dev 3:2021–2024

Meraz MA, White JM, Sheehan K C-F, Bach EA, Rodig SJ, Dighe AS, Kaplan DH, Riley JK, Greenlund AC, Campbell D, Carver-Moore K, DuBois RN, Clark R, Aguet M, Schreiber RD (1996) Targeted disruptiuon of the Stat1 gene in mice reveals unexpected physiological specificity in the JAK-STAT signalling pathway. Cell 84:431–436

Metz K, Ezernieks J, Sebald W, Duschl A (1996) Interlerkin-4 upregulates the heat shock protein Hsp90α and enhances transcription of a reporter gene coupled to a single heat shock element. FEBS Lett 385:25–28

Mirander B, Crichton JE, Zhao Y, Bulun SE, Simpson ER (1996) Expression of transcripts of IL-6 related cytokines by human breast tumors, breast cancer cells and adipose stromal cells. Mol Cell Endocrinol 118:215–220

Minota S, Koyasu S, Yahara I, Weinfeld J (1988) Autoantibodies to the heat-shock protein hsp90 in systemic lupus erythematosus. J Clin Invest 81:106–109

Miyaura C, Onozaki K, Akiyama Y, Taniyama T, Hirano T, Kishimoto T, Suda T. (1988) Recombinant human IL-6 is a potent inducer of differentiation of mouse myeloid leukemic cells (M1). FEBS Lett 234:17–22

Morimoto RI, Sarge KD, Abravaya K (1992) Transcriptional regulation of the heat shock genes. J Biol Chem 267:21087–21990

Morimoto RI (1993) Cells in stress: transcriptional activation of the heat shock genes. Science 259:1409–1410

Moshage H (1997) Cytokines and the hepatic acute phase response. J Pathol 181:257–266

Mui AL, Wakao H, O'farrell A, Harada N, Miyajima H (1996) IL-3 GM-CSF and IL-5 signals through two STAT5 homologs. EMBO J 14:1166–1175

Nakajima T, Kinoshita S, Sasagawa T, Sasaki K, Naruto M, Kishimoto T, Akira S (1993) Phosphorylation at threonine-235 by ras-dependent mitogen-activated protein kinase cascade is essential for transcription factor NF-IL6. Proc Natl Acad Sci USA 90:2207–2211

Nakajima K, Yamanaka Y, Nakae K, Kojima H, Ichiba M, Kiuchi N, Kitaoka T, Fukada T, Hibi M, Hirano T (1996) A central role for Stat3 in IL-6- induced regulation of growth and differentiation in M1 leukemic cells. EMBO J 15:3651–3658

Norton PM, Isenberg DA, Latchman DS (1988) Elevated levels of the 90kD heat shock protein in a proportion of SLE patients with active disease. J Autoimmun 2:187–195

O'Brien RL, Happ MP, Dallas A, Palmer E, Kubo R, Born WK (1989) Stimulation of a major subset of lymphocytes expressing T cell receptor $\gamma\delta$ by an antigen derived from Mycobacterium tuberculosis. Cell 57:667–674

Oliviero S, Cortese R (1989) The human hepatoglobin gene promoter: IL-6 -responsive elements interact with a DNA binding protein induced by IL-6. EMBO J 8:1145–1151

Paul SR, Bennett F, Calvetti JA, Kelleher K, Wood CR, O'Hara RM, Leary AC, Sibley B, Clark SC, Williams DA, Yang Y-C (1990) Molecular cloning of a cDNA encoding IL-11, a stromal cell-derived lymphopoeitic and hematopoietic cytokine. Proc Natl Acad Sci USA 87:7512–7516

Poli V, Balena R, Fattori E, Markatos A, Yamamoto M, Tanaka H, Cilliberto G, Rodan GA, Costantini F (1994) IL-6 deficient mice are portected from bone loss caused by estrogen depletion. EMBO J 13:1189–1196

Pennica D, Shaw KJ, Swanson TA, Moore MW, Shelton DL, Zioncheck KA, Rosenthal A, Taga T, Paoni NF, WoodW (1995a) CT-1: biological activities and binding to LIF receptor/gp130 signaling receptor complex J Biol Chem 270:10915–10919

Pennica D, King K, Shaw KJ, Luis E, Rullamas J, Luoh S-M Darbonne WC, Knutzon DS, Yen R, Chien KR, baker JB, Wood W (1995b) Expression cloning of cardiotrophin 1, a cytokine that induces cardiac myocyte hypertrophy. Proc Natl Acad Sci USA 92:1142–1146

Raynal MC, Liu Z, Hirano T, Mayer L, Kishimoto T, Chen-Kiang S (1989) IL-6 induces secretion of IgG1 by coordinated transcriptional activation and differential mRNA accumulation. Proc Natl Acad Sci USA 86:8024–8028

Ritossa R (1962) A new puffing pattern induced by temperature shock and DNP in Drosophila. Experientia 18:571–573

Roman C, Platero JS, Shuman J, Calame K (1990) IgEBP-1: a ubiquitous expressed immunoglobulin enhancer binding protein that is similar to C/EBP and hetero-dimerizes with C/EBP. Genes Dev 4:1404–1415

Ron D, Habener JF (1992) CHOP, a novel developmentally regulated nuclear protein that dimierises with C/EBP and LAP and functions as a dominant-negative inhibitor of gene transcription. Genes Dev 6:439–453

Rose TM, Bruce AG (1991) Oncostatin M is a member of a cytokine family that includes LIF, GCSF and IL-6. Proc Natl Acad Sci USA 88:8641–8645

Saito M, Yoshida K, Hibi M, Taga T, Kishimoti T (1992) Molecular cloning of a murine IL-6 receptor-associated signal transducer, gp130, and its regulated expression in vivo. J Immunol 148:4066–4071

Samuelsson L, Stomgberg K, Vikman K, Bjursell G, Enerback S (1991) The CCAAT/ enhancing binding protein and its role in adipocyte differentiation: evidence for direct involvement in terminal adipocyte development. EMBO J 10:3787–3793

Screpanti I, Romani L, Musiani P, Modesti A, Fattoro E, Lazzaro D, Sellitto C, Scarpa S, Bellavia D, Lattanzio G, Bistoni F, Frati L, Cortese R, Gulino A, Ciliberto G, Costani F, Poli V (1995) Lymphoproliferative disorder and imbalanced T-helper response in C/EBP β-deficient mice. EMBO J 14:1932–1941

Screpanti I, Musiani P, Musiani P, Bellavia D, Capeletti M, Aiello FB, Maroder M, Frati L, Modesti A, Gulino A, Poli V (1996) Inactivation of the IL-6 gene prevents development of the multicentre Castleman's disease in C/EBP β-deficient mice. J Exp Med 184:1561–1566

Sheng Z, Knowlton K, Chen T, Hoshijima M, Brown JH, Chien KR (1997) CT-1 inhibition of cardiac myocyte apoptosis via a MAPK- dependent pathway Divergent from downstream CT-1 signals for myocardial cell hypertrophy. J Biol Chem 272:5788–5791

Shoenfeld Y, Isenberg DA (1989) The mosaic of autoimmunity. Elsevier, Amsterdam

Spiegelman BM, Flier JS (1996) Adipogenesis and obesity: rounding out the big picture. Cell 87:377–389

Stephanou A, Amin V, Isenberg DA, Akira S, Kishimoto T, Latchman DS (1997) IL-6 activates heat shock protein 90β gene expression. Biochem J 321:103–106

Stephanou A, Isenberg DA, Akira S, Kishimoto T, Latchman DS (1998a) NF-IL6 and STAT-3 signalling pathways co-operate to mediate the activation of the Hsp90β gene by IL-6 but have opposite effects on its inducibility by heat shock. Biochem J 330:189–195

Stephanou A, Brar B, Heads R, Marber MS, Pennica D, Latchman DS (1998b) Cardiotrophin-1 induces heat shock protein synthesis in cultured cardiac cells and protects them from stressful stimuli. J Mol Cell Cardiol 30:849–855

Stephanou A, Conroy S, Isenberg DA, Poli V, Ciliberto G, Latchman DS (1998c) Elevation of IL-6 in transgenic mice results in increased levels of the 90 KD heat shock protein and production of the anti-Hsp90 antibodies. J Autoimmun 11:249–253

Stokli KA, Lottspeich F, Sendtner M, Masaiakowski P, Carrol P, Gotz R, Lindholm D, Thoesen H (1989) Molecular cloning, expression and regional distribution of rat CNTF. Nature 342:20–23

Suematsu S, Matsuda T, Aozasa K, Akira S, Nakano N, Ohno S, Miyazaki J-I, Yamamura K-I, Hirano T, Kishimoto T (1989) IgG1 plamacytosis in IL-6 transgenic mice. Proc Natl Acad Sci USA 86:7547–7551

Sullivan CM, Smith DM, Matsui NM, Andrews LE, Clauser KR, Chapeaurouge A, Burlingame AL, Epstein LB (1997) Identification of constitutive and gamma-interferon-and IL-4 regulated proteins in human renal carcinoma cell line ACHN. Cancer Res 57:1137–1143

Taga T, Kishimoto T (1992) Cytokine receptors and signal transduction. FASEB J 6:3387–3396

Takeda K, Tanaka T, Shi W, Matsumoto M, Minami M, Kashiwamura S-I, Nakanishi K, Yoshida N, Kishimoto T, Akira S (1996) Essential role of STAT6 in IL-4 signalling. Nature 380:627–630

Takeda K, Noguchi K, Shi W, Tanaka T, Matsumoto M, Yoshida N, Kishimoto T, Akira S (1997) Targeted disruption of the mouse Stat3 gene leads to embryonic lethality. Proc Nat Acad Sci USA 94:3801–3804

Tanaka K, Akira S, Yoshida K, Umemoto M, Yoneda Y, Shirafuji N, Fugiwara H, Suematsu S, Yoshida N, Kishimoto T (1995) Target disruption of the NF-IL6 gene discloses its essential role in bacterial killing and tumour cytotoxicity by macrophages. Cell 80:353–361

Twomey BM, Dhillon VB, McCallum S, Isenberg DA, Latchman DS (1993) Elevated levels of the 90kD heat shock protein in patients with systemic lupus erythematosus are dependent upon enhanced transcription of the hsp90b gene. J Autoimmun 6:495–506

Wagner JA (1996) Is IL-6 both a cytokine and a neurotrophic factor? J Exp Med 183:2417–2419

Wang N-D Finegold MJ, Bradley A, Ou CN, Abdelesayed SV, Wilde MD, Taylor LR, Wilson DR, Darlington GJ (1995) Impaired energy homeostasis in C/EBPα knockout mice. Science 269:1108–1112

Wen Z, Zong Z, Darnell JE Jr (1995) Maximal activation of transcription by STAT1 and STAT3 requires both tyrosine and serine phosphorylation. Cell 82:241–245

Yamamoto K, Quelle FW (1994) STAT4, a novel interferon activation site binding protein expressed in early myeloid differentiation. Mol Cell Biol 14:4342–4347

Yoshida K, Taga T, Saito M, Suematsu S, Kumanagoh A, Tanaka T, Fujiwara H, Hirata M, Yamagami T, Nakahata T, Hirabayashi Y, Yomoda Y, Wang WZ, Mori C, Shiota K, Yoshida N, Kishimoto T (1996) Targeted disruption of gp130, a common signal transducer for IL-6 family of cytokines, leads to myocardial and haematological disorders. Proc Natl Acad Sci USA 93:407–411

Zang X, Blenis J, Li H, Schindler C, Chen-Kiang S (1995) Requirement of serine phosphorylation for formation of STAT-promoter complexes. Science 267:1990–1894

CHAPTER 8
Regulation of Heat Shock Genes by Ischemia

T.S. Nowak, Jr., Q. Zhou, W.J. Valentine, J.B. Harrub, and H. Abe

A. Introduction

The evolutionarily conserved and ubiquitous nature of the heat shock response is well established. However, perhaps nowhere is the potential heterogeneity of this response more evident than in the brain. Several recent reviews provide detailed evaluations of the heat shock response following ischemia and other brain insults (Abe and Nowak 1996b; Massa et al. 1996; Planas et al. 1997). As documented in these, as well as below and elsewhere in this volume (see Chaps. 9, 11, and 12), different insults induce a given heat shock gene with differing cell type specificities, and an ischemic insult can result in divergent patterns of expression of individual heat shock genes, implying distinct regulatory mechanisms. Furthermore, impaired translation is a prominent feature of ischemic injury in brain, and this imposes a significant limitation on the expression of proteins encoded by ischemia-induced mRNAs.

This presentation will first provide a brief overview of the postischemic heat shock response in brain with an emphasis on those issues relevant to heat shock regulation. Attention will focus on expression of the 70 kDa heat shock protein, hsp72, in global ischemia models, which has been most extensively studied, but particular emphasis will be placed on comparisons with expression patterns of other ischemia-inducible genes, including other heat shock proteins, and with results obtained following focal ischemic insults. These results will provide evidence that hsp72 induction occurs as cells approach a threshold for lethal injury. Discrepancies between hsp72 mRNA and protein expression will be discussed in the context of well-known postischemic protein synthesis deficits, and in relation to recent findings that hsp72 protein is refractory to detection in acutely injured neurons when examined by conventional immunocytochemical methods.

Data regarding the mechanism of postischemic hsp72 induction will be presented, including in vivo studies demonstrating heat shock factor (HSF) localization in brain, as well as HSF activation following ischemia. Finally, an in vitro model of anoxia/aglycemia in hippocampal slices will be shown to reproduce patterns of stress response activation after global ischemia in vivo. Preliminary pharmacological studies in hippocampal slices, consistent with results in other heat shock models, demonstrate an attenuation of hsp72 induction by calcium chelators and sulfhydryl reducing agents. While by no means providing a completely satisfying understanding of mechanisms controlling the stress response, these results establish the suitability of this in vitro approach for further studies of postischemic heat shock regulation in brain.

B. Patterns of Heat Shock Gene Expression After Global and Focal Ischemia

I. Gene Expression and Neuronal Vulnerability After Global Ischemia

In view of the selective neuronal vulnerability that is the most prominent pathological feature of transient ischemic insults, one of the main interests in the study of ischemia-inducible genes has been to elucidate the relationship between changes in gene expression and the underlying neuronal pathophysiology. Studies of protein expression after ischemia are complicated by the prolonged and severe translational deficits that occur after both global and focal insults (COOPER et al. 1977; NOWAK et al. 1985; MIES et al. 1991), particularly in those cells destined to be lost (DIENEL et al. 1980; THILMANN et al. 1986; MIES et al. 1991; WIDMANN et al. 1991). There can be additional problems with the detection of specific proteins, notably including the stress protein hsp72 (HARRUB and NOWAK 1998), as detailed below. For such reasons, induced gene expression after ischemia is most reliably evaluated at the level of mRNA expression, although immunocytochemical results can be interpreted with appropriate caution. Initial studies established that mRNAs expressed in postischemic brain yielded a translation product with characteristics of the 70kDa stress protein, hsp72, employing insults sufficient to result in virually complete loss of CA1 neurons in dorsal hippocampus (NOWAK 1985; DIENEL et al. 1986), and in vivo labeling studies were consistent with the induction of hsp72 and other stress proteins (KIESSLING et al. 1986). Probes selective for stress-inducible members of the hsp70 family detected pronounced increases in mRNA levels in postischemic brain (NOWAK et al. 1990). There was also evidence for increased expression of mRNAs encoding constitutively expressed hsc70/hsc73 (ABE et al. 1991). Increases in expression of other heat shock genes have been observed in a series of studies in related models, including modest increases in mRNAs encoding ubiquitin (NOWAK et al. 1990; NOGA and HAYASHI 1996), hsp90 (KAWAGOE et al. 1993) and the mitochondrial stress protein hsp60 (ABE et al. 1993), and striking induction of hsp32 (heme oxygenase 1, HO-1; PASCHEN et al. 1994; TAKEDA et al. 1994) and the small heat shock protein, hsp25/27 (KATO et al. 1994).

As shown in Fig. 1, in situ hybridization demonstrates prominent hsp72 mRNA expression in a wide range of neuron populations at early recirculation intervals after global ischemia, consistently observed in studies employing both gerbils and rats (NOWAK 1991; KAWAGOE et al. 1992c; ABE et al. 1993; SAITO et al. 1995). While the initial pattern is not predictive of eventual

Global   Focal

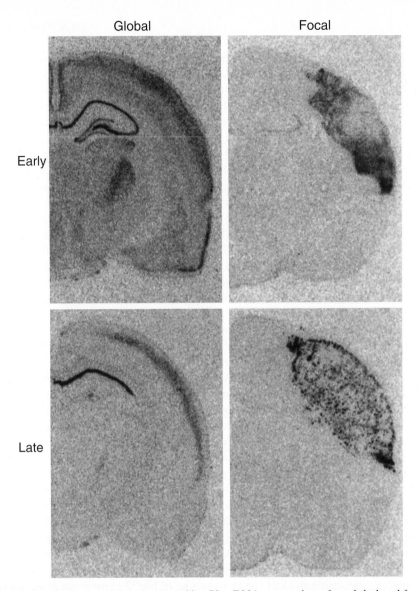

Early

Late

Fig. 1. Distribution and time course of hsp72 mRNA expression after global and focal ischemia. Rats were subjected to transient global ischemia produced by 10 min cardiac arrest or focal ischemia produced by permanent middle cerebral artery occlusion, and hsp72 mRNA expression evaluated by in situ hybridization with ^{35}S-labeled probes at the level of dorsal hippocampus. The hybridization signal was at the limits of detection in control sections (not shown). Global ischemia resulted in prominent hsp72 induction in major neuron populations of hippocampus, with signal in cortex and thalamus also evident at the level of the illustrated section obtained during early (6h) recirculation. At a later interval (24h), signal persisted only in vulnerable hippocampal CA1 and cortical neurons. Focal ischemia resulted in prominent hsp72 mRNA expression in the margin of the affected territory, with notable absence of expression within the severely ischemic core. Prominent intense foci of expression were evident throughout the infarct 24h following the insult, reflecting expression in surviving vascular elements

pathology, there is a progressive loss of hsp72 mRNA from most regions during 24–48 h, with persistent expression confined to the vulnerable hippocampal CA1 neurons and certain other populations that subsequently will be lost (Nowak 1991; Kawagoe et al. 1992c). The diffuse initial pattern of neuronal hsp72 expression is consistent with the general vulnerability of neurons vs. other cell types to transient ischemic insults, and lasting expression in CA1 neurons may indicate that severely affected cell populations experience a more sustained inducing stress. Alternatively, mechanisms that would normally terminate the stress response may be impaired in injured neurons. It should be noted that a similar pattern of prolonged expression in CA1 neurons has been observed for several other ischemia-inducible mRNAs, including, for example, hsc70 and hsp60 (Abe et al. 1993) as well as coppper-zinc superoxide dismutase (Matsuyama et al. 1993), reminiscent of the superinduction frequently observed when translational inhibition disrupts feedback mechanisms regulating transcription (Greenberg et al. 1986; Ginty et al. 1990). These issues will be considered further below in the context of heat shock factor regulation after ischemia.

Immunocytochemical studies of hsp72 expression are broadly consistent with the in situ hybridization studies, with the exception that limited accumulation of the protein occurs in vulnerable CA1 neurons in which deficits in translation are most pronounced. While preferential expression of heat shock proteins has been suggested in some models it is clear that severe stresses preclude the translation of heat shock mRNAs as well, and this appears to be the case in postischemic brain (Nowak 1985). Such deficits are typically more complete in the gerbil model (Vass et al. 1988; Kirino et al. 1991), while in the rat significant hsp72 immunoreactivity is detected in CA1 neurons even after insults severe enough to produce eventual neuron loss (Chopp et al. 1991; Simon et al. 1991; Deshpande et al. 1992; Tomioka et al. 1993). In both species, short insults produce preferential hsp72 expression in CA1 neurons (Kirino et al. 1991; Simon et al. 1991) and a shift toward this distribution is also observed after neuroprotective interventions that effectively lessen the impact of the ischemic insult (Chopp et al. 1992; Aoki et al. 1993; Bergstedt 1993; Liu et al. 1993). Methodological complications unique to hsp72 detection complicate the precise determination of the time course of hsp72 expression from immunocytochemical studies, as will be discussed in detail below. In addition, cells that are rapidly lost following global ischemia may be unable to mount even a transcriptional response (Saito et al. 1995). Nevertheless, the above observations support the general correlation between hsp72 induction and ischemic stress experienced by a given cell population. The absence of hsp72 expression in ischemia-resistant interneurons is likewise consistent with this interpretation (Ferrer et al. 1995). It should be noted that small regions of acute injury resembling microinfarcts are occasionally seen after global ischemia and involve hsp72 expression in multiple cell types (Chopp et al. 1991; Gonzalez et al. 1991; Nowak and Abe 1994), as will be considered below in the context of focal ischemia.

Ubiquitin expression has provided an index of cell stress after ischemia that exhibits a pattern of responses largely consistent with that of hsp72. While overall ubiquitin mRNA is only modestly increased above constitutive levels in postischemic brain (Nowak et al. 1990; Noga and Hayashi 1996), in situ hybridization has indicated significant increases in signal intensity, particularly in major hippocampal neuron populations, during 2–6 h after transient ischemia (Noga and Hayashi 1996). Expression of the UbC polyubiquitin transcript was particularly responsive. At the protein level, an early study demonstrated a striking initial loss of ubiquitin immunoreactivity throughout hippocampus, followed by a progressive recovery in dentate granule cells and CA3 neurons during 24–72 h, with minimal recovery in CA1 neurons destined to be lost (Magnusson and Wieloch 1989). CA1 neurons protected by hypothermia or made tolerant by a priming insult showed eventual restoration of ubiquitin immunoreactivity (Yamashita et al. 1991; Kato et al. 1993). Other studies in a rat hypoxia/ischemia model that produced less severe insults indicated an early increase in ubiquitin immunoreactivity in vulnerable neurons (Gubellini et al. 1997). Biochemical studies have shown that there is a dramatic increase in ubiquitin conjugate levels after ischemia, distributed in a detergent-insoluble particulate subcellular fraction, accompanied by a depletion of free ubiquitin (Hayashi et al. 1991, 1992a,b, 1993). Recent immunocytochemical and immunoblot results with several antibodies having different selectivities for free and conjugated ubiquitin have clearly localized a persistent depletion of free ubiquitin within vulnerable CA1 neurons, with sustained accumulation of ubiquitin conjugates (Morimoto et al. 1996). Factors determining this state presumably include increased rates of protein ubiquitination, perhaps outstripping proteolytic capacity, accompanied by reduced ubiquitin synthesis in neurons with compromised protein synthetic machinery, although the relative contributions of such effects remain to be determined.

In contrast to the prominent neuronal stress response indicated by the changes in hsp72 and ubiquitin expression, recent immunocytochemical studies of the low molecular weight heat shock protein, hsp27, indicate a combined neuronal and glial localization (Kato et al. 1994). Mild insults induced a predominantly glial expression after 1–3 days, which was even more pronounced after severe insults resulting in neuron loss and reactive gliosis, while neuronal expression was restricted to vulnerable populations in hippocampus and cortex. This observation suggests that, in addition to an association with neuronal stress, hsp27 expression is a component of astrocyte activation, accompanying the increase in expression of glial fibrillary acidic protein (GFAP) that is a classical indicator of this response (Kindy et al. 1992). Since this occurs in cells that do not exhibit other components of the postischemic stress response, hsp27 induction in astroglia must involve distinct regulatory mechanisms.

Blot hybridization studies of HO-1/hsp32 have demonstrated striking induction in rat brain during 24 h recirculation after severe (20–30 min) global ischemia (Paschen et al. 1994; Takeda et al. 1994). The time course of expression paralleled that of hsp72 expression, and in situ hybridization studies demonstrated a neuronal component of HO-1 expression, but there was also a significant glial response (Takeda et al. 1994). This may reflect a microglial distribution, since recent studies have demonstrated induction of HO-1 in association with microglial activation (Turner et al. 1998), and a microglial component of HO-1 expression has been observed after focal ischemia (see below).

Some constitutively expressed heat shock genes show modest responsiveness to ischemia, and a generalized increase in expression that is not well-correlated with neuronal vulnerability or reactive changes in other cell types. Hsc70 mRNA is expressed with a prominent neuronal distribution in control brain, and shows moderately increased expression without a notable change in distribution after transient ischemia (Kawagoe et al. 1992c). There is evidence that hsc70 is induced by mild stimuli such as seizure activity that do not induce hsp72 (Wong et al. 1992), and hsc70 may be preferentially induced by shorter intervals of ischemia (Kawagoe et al. 1992a). Postischemic changes in hsc70 expression may therefore be regulated by mechanisms independent of heat shock activation. In situ hybridization and immunocytochemistry studies of hsp90 have likewise demonstrated prominent constitutive neuronal expression in rat brain (Izumoto and Herbert 1993; Gass et al. 1994), and this signal is relatively uniformly increased in hippocampal neurons after ischemia (Kawagoe et al. 1993).

II. Gene Expression After Focal Ischemia

The correlation between cellular vulnerability and gene expression observed in global ischemia models holds as well after focal ischemia, with the added complexity that multiple cell types are involved. As shown in Fig. 1, hsp72 mRNA expression is restricted to the region of ischemic injury, characterized initially by a rim of expression surrounding the ischemic core, and later displaying intense foci of hybridization scattered throughout the established infarct (Welsh et al. 1992; Nowak and Jacewicz 1994). The attenuated expression within severely ischemic territory reflects the absence of transcription in areas in which blood flow is reduced below the threshold for energy failure, the prominence of which can be expected to vary with the effectiveness of occlusion and degree of collateral flow within a given model. The distribution of subsequent hsp72 protein expression is further limited by the extent of protein synthesis inhibition, which exhibits a lower CBF threshold than overt energy failure, and provides a better predictor of the extent of the eventual infarct (Mies et al. 1991). The net result of such effects is a narrow rim of hsp72 immunoreactivity in that territory in which blood flow is below the threshold for mRNA induction, but above the threshold for protein synthesis failure (Kamiya et al. 1995). The cellular localization of hsp72 expression within this "penumbra" can be further distinguished to consist of an inner margin of glial involvement and an outer margin of selective neuronal expression, while within the ischemic core hsp72 immunoreactivity is limited to surviving microvessels (GONZALEZ et al. 1989; SHARP et al. 1991).

Consistent with such observations after permanent occlusion, transient intervals of focal ischemia result in duration-dependent variations in the cellular distribution of hsp72 expression. Blot hybridization studies established that hsp72 mRNA expression in an ischemic hemisphere increased with ischemia duration between 15 min and 2 h when evaluated at 8 h recirculation (WANG et al. 1993), and quantitative estimates based on the polymerase chain reaction have demonstrated progressive accumulation of both hsp72 and hsc73 mRNAs between 3 and 8 h after 30 min insults (ABE et al. 1995). In situ hybridization demonstrated sustained hsp72 mRNA expression in the territory made ischemic by a transient 1 h insult, declining gradually after 24 h, but with more rapid loss in regions exhibiting infarction (KINOUCHI et al. 1993). In another study, brief insults below the threshold for infarction resulted in progressive accumulation of hsp72 mRNA between 1 h and 4 h recirculation (SORIANO et al. 1995). After short periods of occlusion, immunoreactive protein was found in neurons within the ischemic territory, while longer insults resulted in a pattern of combined neuronal and glial expression surrounding a core of vascular staining as infarction progressed (KINOUCHI et al. 1993). The glial response in such models predominantly involves microglia rather than astrocytes (LI et al. 1992; SORIANO et al. 1994). Finally, it should be noted that hsp72 expression in models of combined hypoxia/ischemia in developing brain may exhibit features of either global or focal ischemia in the adult, depending on the age of the animal and the severity of the insult (FERRIERO et al. 1990; BLUMENFELD et al. 1992; MUNELL et al. 1994; GILBY et al. 1997). As after global ischemia, it may be considered that hsp72 immunoreactivity provides a reliable indicator of cells at risk which have been able to survive for some time following the insult.

A striking dissociation is evident between the distribution of hsp72 induction and the responses of many other genes inducible after focal ischemia. Immediate-early genes are induced after focal ischemia, with distributions throughout the ipsilateral cortex as well as in certain ipsilateral and contralateral subcortical structures (WELSH et al. 1992; KINOUCHI et al. 1994; NOWAK and JACEWICZ 1994). While the cortical expression is readily understood in the context of the brief depolarization of spreading depression that occurs after focal insults, the signals responsible for remote expression in structures such as hippocampus remain to be identified. Pharmacological studies indicate that c-fos expression in the region immediately surrounding the ischemic focus is not readily attenuated by N-methyl-D-aspartate (NMDA) antagonists that blunt expression elsewhere in the ipsilateral hemisphere (UEMURA et al. 1991; GASS et al. 1992), suggesting that there is a mechanistically distinct component of expression in this region of overlap with hsp72 induction, that is also characterized by repeated depolarizations (IIJIMA et al. 1992). Modest hsp72 induction has also been observed to overlap a component of remote c-fos induction in hippocampus after focal ischemia (Kawagoe et al. 1992b; Welsh et al. 1992; Kinouchi et al. 1993; Nowak and Jacewicz 1994).

As noted after global ischemia, diverse signals appear to regulate heat shock genes following focal ischemia. Two studies have compared the induction of several heat shock genes after middle cerebral artery occlusion (Higashi et al. 1994; Wagstaff et al. 1996). The first employed samples encompassing both the ischemic core and adjacent cortex, and demonstrated two distinct patterns of expression, with the hsp70 family (grp78, hsc70, and two hsp72 transcripts) reaching maximal expression within a few hours, while transcripts designated hsp27 and hsp47 progressively increased during 24–48 h (Higashi et al. 1994). Changes in hsp90 expression were not detected. Relative increases in grp78 mRNA expression were moderate, as might be expected given its significant constitutive expression, consistent with findings in another blot hybridization study (Wang et al. 1993). The second study sampled tissue from severely ischemic territory during a time course of 24 h, with the main findings being robust, progressive induction of hsp72 and hsp27 (Wagstaff et al. 1996). Hsp60 mRNA was constitutively expressed, and was also elevated relative to tubulin mRNA expression in the region of injury, while hsp90 and hsp56 did not show notable changes. Immunoblot analysis indicated increased levels of the corresponding hsp72, hsp27 and hsp60 proteins, as well as an increase in hsp56. In situ hybridization has demonstrated that, in contrast to the peri-infarct distribution of hsp72, hsp27 was induced throughout ipsilateral cortex after focal ischemia (Higashi et al. 1994). Consistent with findings noted above in the context of global ischemia, an astroglial localization of hsp27 has been demonstrated after focal insults (Kato et al. 1995), and recent results have satisfyingly demonstrated that such expression can be induced by spreading depression (Plumier et al. 1997). In contrast, a pronounced microglial expression has been noted for HO-1/hsp32, with a component at the rim of striatal infarcts after brief focal ischemia, and a more diffuse distribution extending outside the ischemic territory after longer occlusions (Nimura et al. 1996). While hsp32 expression was not detected in cortical neurons that expressed hsp72 following short focal occlusions it was coexpressed with hsp72 in neurons after longer insults, as well as in endothelial cells within infarcts. Hsc70 mRNA, constitutively expressed in neurons, is induced with a distribution overlapping that of hsp72 after transient focal ischemia (Kawagoe et al. 1992b). The above results demonstrate a heterogeneous distribution of heat shock protein induction after focal ischemia, with expression of hsp72 remaining the response most closely associated with the potential for ischemic injury.

III. Cryptic hsp72 Expression After Ischemia

In addition to the evident impact of postischemic protein synthesis deficits on accumulation of translation products of ischemia-inducible mRNAs, a more subtle discrepancy between transcriptional induction and apparent protein expression has emerged in the case of hsp72. It was observed in early studies that hsp72 was a prominent translation product of active brain polyribosomes within a few hours recirculation after transient ischemia (NOWAK 1985). However, initial immunocytochemical results in the gerbil failed to detect appreciable signal in neurons until at least 16 h recirculation after 10 min ischemia (VASS et al. 1988), although very rarely a weak signal has been detected at earlier intervals (HARRUB and NOWAK 1998), and 24–48 h has been the typical interval chosen to evaluate hsp72 expression in most studies of ischemia or other insults (SIMON et al. 1991; TAKEMOTO et al. 1995; GILBY et al. 1997). Fos, Jun and other products of immediate-early genes are detected at much earlier intervals in dentate granule cells and CA3 neurons (KIESSLING et al. 1993; NOWAK et al. 1993; TAKEMOTO et al. 1995), closely reflecting the time course of protein synthesis recovery in these cell populations. Based on such observations it was hypothesized that cryptic hsp72 expression might be associated with an interval of active stress, with detection allowed during subsequent "relaxation" in surviving cells (NOWAK 1991). This transition was suggested to reflect changes in accessibility of the antigenic site in the course of protein-protein interactions intrinsic to the chaperone function of hsp70 proteins, as previously suggested to explain variations in detectability of hsp70 proteins during the cell cycle (MILARSKI et al. 1989).

Recent immunoblot results showed that hsp72 rapidly accumulated in gerbil hippocampus after transient ischemia, with a broad plateau of maximal expression during 6–48 h, and increased hsp72 levels were evident within as little as 2 h recirculation (HARRUB and NOWAK 1998). Parallel immunocytochemistry in paraformaldehyde-fixed vibratome sections confirmed the absence of detectable hsp72 expression in neurons at 6 h recirculation. While rates of hsp72 synthesis and degradation undoubtedly vary among the several neuron populations that express hsp72 mRNA, and while minimal translation would be expected in CA1, this observation strongly supports the conclusion that hsp72 is expressed, but undetectable by conventional immunocytochemistry, in dentate granule cells and CA3 neurons at early postischemic intervals.

Further studies have succeeded in unmasking cryptic hsp72 immunoreactivity in such material (HARRUB and NOWAK, manuscript in preparation). Antigen retrieval techniques employing heating in citrate buffer, often with the use of microwave irradiation, have been used for some time to improve detection of various antigens in paraffin-embedded tissue (CATTORETTI et al. 1993). As shown in Fig. 2, application of such a treatment to vibratome sections of postischemic gerbil brain allows the demonstration of robust hsp72 expression in hippocampal CA3 neurons 6 h after an ischemic insult, that was absent from sections processed by conventional methods. Identical results have been obtained with commercially available monoclonal and polyclonal antibodies specific for inducible hsp70 proteins (product numbers SPA-810 and SPA-812, StressGen Biotechnologies, Victoria, BC).

The nature of the interactions that restrict antigen accessibility in brain sections remain to be identified. A characteristic feature of hsp72 protein

Untreated   Citrate/Heat

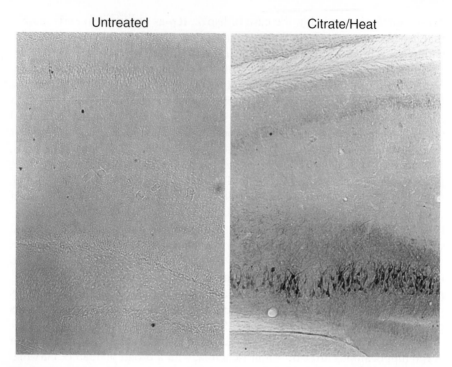

Fig. 2. Cryptic hsp72 expression after transient global ischemia. Vibratome sections were prepared from a gerbil brain perfusion-fixed with phosphate-buffered paraformaldehyde at 6 h recirculation following 10 min ischemia. Hippocampal hsp72 immunoreactivity was not detected in sections processed by conventional methods, but prominent expression in CA3 neurons was evident when the same tissue was heated in 10 mM citrate-buffered saline at 80°C for 30 min, followed by routine processing

expression in other models has been its nuclear localization during early intervals following a stress, followed by subsequent cytoplasmic distribution (Velazquez and Lindquist1984; Welch and Feramisco 1984). However, it is only the cytoplasmic pool of hsp72 that has been reliably detected by immunocytochemistry in postischemic neurons (Vass et al. 1988). Interestingly, one study in which immunoblot detection of hsp72 showed a delayed time course in postischemic brain employed a preliminary centrifugation step so that only the supernatant fraction was analyzed (Nishi et al. 1993), suggesting that cryptic hsp72 missed in the analysis was localized to a particulate fraction in brain homogenates, and this has been demonstrated at early intervals after hyperthermic or hypoxic insults to astrocyte cultures (Bergeron et al. 1996). On the other hand, the above antigen retrieval data indicate a relatively diffuse distribution of unmasked hsp72 in soma and processes of neurons. The subcellular distribution of hsp72 during early postischemic recirculation is currently under investigation. As considered above, hsp70s associate with a wide range of cellular proteins (BECKMANN et al. 1990; SRIVASTAVA 1993; TAMURA et al. 1997). While there are differences in peptide binding specificities of different members of this family (FOURIE et al. 1994), functional consequences of such specificities have yet to be demonstrated (JAMES et al. 1997). In addition, there is evidence for interactions between hsp70s and HSFs that could contribute to heat shock regulation (ABRAVAYA et al. 1992; BALER et al. 1992; MOSSER et al. 1993; NUNES and CALDERWOOD 1995; BALER et al. 1996), and association of hsp72 with hsc73 is considered to contribute to nuclear localization (BROWN et al. 1993). Fixation of any such complexes involving hsp72 could conceivably stabilize a conformation in which recognized epitopes remain inaccessible to antibody during routine immunocytochemistry.

C. Regulation of the Postischemic Heat Shock Response

I. Injury Thresholds and the Stress Response

1. Thresholds for Expression of hsp72 and Other Ischemia-Inducible Genes

From the above results it is clear that induction of hsp72 represents a response to ischemic insults that is strongly correlated with the evolving pattern of injury. Prolonged expression of hsp72 mRNA is evident in vulnerable cell populations after severe ischemia (NOWAK 1991), while immunocytochemical studies after short periods of ischemia, or after insults attenuated by neuroprotective approaches, demonstrate selective hsp72 accumulation in more vulnerable neuron populations (KIRINO et al. 1991; SIMON et al. 1991). Together with selective localization of hsp72 expression within the ischemic territory after focal ischemia, such results suggest that hsp72 induction occurs as cells approach a critical injury threshold, which may or may not result in death. A similar conclusion has been drawn from comparisons of electrical stimulation thresholds for neuronal injury and hsp72 expression (SLOVITER and LOWENSTEIN 1992), and from studies of seizure thresholds for hsp72 induction (VASS et al. 1989). However, it has been difficult to precisely compare induction thresholds for individual transcriptional responses in global ischemia models, due to the inherent variability in injury produced when occlusion duration is the only index of insult severity. For example, while c-fos mRNA induction has been observed following as little as 1 min occlusion in one gerbil study (KINDY et al. 1991), comparison of c-fos and hsp72 expression in another report indicated no induction of either mRNA after 1 min ischemia, while both were expressed after 2 min or 5 min insults (IKEDA et al. 1994). The clear dissociation between expression of hsp72 and immediate-early genes after focal ischemia demonstrates that these transcriptional responses differ considerably in their induction thresholds (WELSH et al. 1992). In recent studies we have monitored the duration of ischemic depolarization during varied intervals of global ischemia as a precise index of the insult severity experienced by

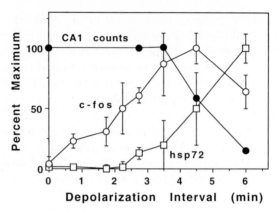

Fig. 3. Thresholds for neuronal injury and transcriptional responses after global ischemia. Gerbils were subjected to carotid artery occlusions of varied duration, during which ischemic depolarizations were monitored by glass microelectrodes in each hippocampus as an index of insult severity. Brains were either perfusion fixed for quantitative histological evaluation of CA1 neuron survival at 1 week, or were fresh frozen and processed for in situ hybridization at 1 h recirculation. Induction of c-fos and other immediate-early genes in CA1 neurons was evident after mild insults, well below the threshold for injury, while appreciable hsp72 expression was only detected as the duration of ischemic depolarization approached the threshold for neuron loss

gerbil hippocampus (ABE and NOWAK 1996a,b). As shown in Fig. 3, the depolarization threshold for hsp72 induction in hippocampal CA1 neurons is only slightly lower than the threshold for neuron loss, while c-fos and other immediate-early genes are induced after much shorter intervals. These observations are also relevant to mechanisms of induced ischemic tolerance reported in a number of rodent models following mild insults below the threshold for injury, in which an involvement of the postischemic heat shock response has been hypothesized (KITAGAWA et al. 1990; KIRINO et al. 1991). The depolarization threshold for induction of ischemic tolerance in the gerbil corresponds precisely to that for induction on immediate-early genes, arguing strongly that prior hsp72 induction is not required for expression of tolerance in such models (ABE and NOWAK 1996a,b; ABE et al., manuscript in preparation). This is also consistent with evidence of cortical protection after intervals of spreading depression that would not induce hsp72 expression (KOBAYASHI et al. 1995).

2. Temperature Effects on hsp72 Expression After Global Ischemia

Among the variables that influence brain injury following ischemia and other insults, temperature appears to be the most critical. Decreased temperature during an interval of occlusion profoundly attenuates neuronal loss (BUSTO et al. 1987; CHURN et al. 1990; WELSH et al. 1990; ANDOU et al. 1992; NURSE and CORBETT 1994), and has been a significant artifact in pharmacological studies

(BUCHAN and PULSINELLI 1990). Conversely, moderate hyperthermia during ischemia worsens damage (BUSTO et al. 1987; CHURN et al. 1990; DIETRICH et al. 1990b) and expands injury to involve the microvasculature (DIETRICH et al. 1990a). Particularly important are observations that delayed temperature modulation restricted to the postischemic recirculation period can influence pathological outcome, as recently reviewed (COLBOURNE et al. 1997). In the gerbil model, spontaneous hyperthermia during early recirculation contributes to neuron loss (KUROIWA et al. 1990), although temperature elevation is not required to observe damage after severe insults (KATO et al. 1991; COLBOURNE et al. 1993). Brief postischemic hypothermia may not be overtly protective (WELSH and HARRIS 1991; DIETRICH et al. 1993), but modest, prolonged cooling during recirculation reduces injury (COLBOURNE and CORBETT 1995), and can account for neuroprotection observed following some pharmacological treatments (NURSE and CORBETT 1996).

It is now clear that temperature variations during recirculation can strongly modulate the postischemic stress response. Mild hyperthermia of a magnitude comparable to that which can occur spontaneously during recirculation in the gerbil (39–40°C) dramatically increases hsp72 mRNA expression after brief ischemia in this model, affecting both the early, generalized neuronal expression and the later component restricted to CA1 (SUGA and NOWAK 1998). This observation provides direct evidence that temperature-sensitive signals responsible for hsp72 induction are generated during postischemic recirculation. It should be noted that the temperature threshold for hsp72 induction in response to hyperthermic stress in rodent brain is higher than that reached in the above studies (NOWAK et al. 1990), and that the distribution of heat-induced hsp72 expression in brain is predominantly glial and vascular (MARINI et al. 1990; McCABE and SIMON 1993), with a more modest neuronal component (BLAKE et al. 1990; PARDUE et al. 1992), suggesting that temperature elevation is not directly responsible for the postischemic heat shock response. Possible mechanisms by which temperature could impact heat shock signaling after ischemia will be considered in more detail below.

II. Heat Shock Factor Activation After Global Ischemia

An initial report in a focal ischemia model demonstrated heat shock factor 1 (HSF1) activation following a permanent occlusion (HIGASHI et al. 1995), consistent with the generalization that HSF1 accounts for increases in heat shock element (HSE) binding activity observed following most stresses, including hypoxic/ischemic insults to various cell types in vitro (BENJAMIN et al. 1990; BERGERON et al. 1996). Since infarction after focal ischemia produces injury and stress responses involving multiple cell types, it was not clear whether the HSF1 activation observed in this model necessarily occurred in neurons, or whether it was contributed by glia or the microvasculature. As noted above, hyperthermic insults in brain result in preferential glial and vascular hsp72 induction (Marini et al. 1990; McCabe and Simon 1993). Studies in vitro demonstrate a robust glial response to hyperthermia, which has proved difficult to observe in cultured neurons (Nishimura et al. 1988, 1991; Marini et al. 1990). There is evidence that this can be explained by an absence of appreciable HSF1 expression in neurons under these conditions (Marcuccilli et al. 1996), although recent results indicate that a delayed hsp72 expression accompanied by HSF1 activation can be observed in some neuron populations (Nishimura and Dwyer 1996). It was therefore of interest to examine the role of heat shock factors in the neuronal response to global ischemia in brain.

HSF1 and HSF2 immunoreactivities are both expressed in gerbil hippocampus (Fig. 4), with obvious expression in major neuron populations, and with an apparent nuclear localization. This demonstrates that mature neurons of the adult rodent brain express both HSF1 and HSF2, and suggests that the lack of HSF1in cultured neurons from neonatal animals may reflect developmental differences in HSF1 expression, or may arise as a result of culture conditions. This result indicates that neurons also account for a considerable proportion of HSF2 expression in brain, which is expressed at levels comparable to those seen in other rodent tissues (Goodson et al. 1995). Nuclear localization of the

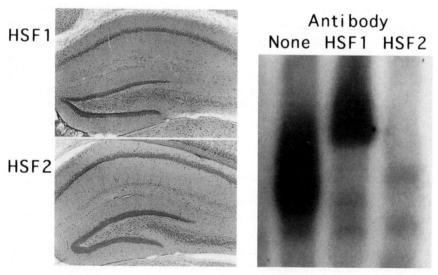

Fig. 4. Heat shock factor expression and activation in gerbil brain. HSF1 and HSF2 were readily detected by immunocytochemistry in control gerbil hippocampus, with a prominent distribution in nuclei of major neuron populations as well as other cell types. Heat shock element binding activity induced after transient ischemia was quantitatively shifted in the presence of HSF1 antibody, while binding was abolished by antibody directed against HSF2

native, inactive factors is contrary to results of many biochemical studies, but is consistent with immunocytochemical results in *Drosophila* cells in which it was also shown that HSF1 is rapidly lost from the nuclear compartment of unstressed cells during fractionation procedures (WESTWOOD et al. 1991). As shown in Fig. 4, transient ischemia resulted in increased HSE binding activity that was quantitatively shifted by incubation with antibody recognizing HSF1, while such studies have consistently demonstrated a loss of protein-HSE complexes upon addition of HSF2-specific antibodies. These results support the conclusion that HSF1 is involved in the regulation of neuronal heat shock responses after global ischemia, and leave open the possibility that HSF2 may also contribute to transcriptional regulation in response to this challenge. Interestingly, while HSF2 has generally been considered to have a role in developmental regulation of the stress response, synergistic effects of HSF1 and HSF2 activation on human hsp70 induction have been described (SISTONEN et al. 1994). In other work, transient transfection studies of constructs derived from an inducible rat hsp70 gene indicated that while either of two heat shock elements in its promoter was sufficient for hyperthermic inducibility, both were required for responsiveness to ischemia (MESTRIL et al. 1994). Together, these results are consistent with the suggestion that postischemic hsp72 induction may involve both HSF1 and HSF2, although further studies are necessary to confirm this hypothesis. Finally, it should be noted that quantitative estimates of HSF activity in gerbil hippocampus suggest a biphasic time course, with an early peak at approximately 1 h and an interval of sustained activation through 24 h (NOWAK 1998; W. VALENTINE, H. ABE, T. SORIMACHI and T. S. NOWAK, Jr., submitted manuscript), suggesting that ongoing transcriptional activation may contribute to the prolonged upregulation of hsp72 expression in vulnerable neurons.

III. Heat Shock Regulation After Anoxia/Aglycemia in Hippocampal Slices

1. Hsp72 Induction After In Vitro Anoxia/Aglycemia

In vivo models present considerable difficulties with respect to elucidating signaling mechanisms involved in regulation of gene expression. Hippocampal slices have been used in a number of studies investigating neuronal injury, employing varying degrees of oxygen and/or glucose deprivation to model ischemic insults in vitro, as recently reviewed (SICK and SOMJEN 1998), including changes in gene expression (CHARRIAULT-MARLANGUE et al. 1992; HASEGAWA et al. 1997), and we have examined the possibility that such preparations could be used in mechanistic studies of the heat shock response. Initial studies indicated that induction of hsp72, as well as immediate-early genes, could routinely occur as a consequence of the ischemia and trauma associated with slice preparation (NOWAK et al. 1994; ZHOU et al. 1995). In view of the striking protective effects of hypothermia during in vivo ischemia noted above, and prior reports of beneficial effects of cooling in slices (Newman et al. 1992), we reduced body temperature to 32°C prior to slice preparation. This, together with other improvements, yielded an optimized slice preparation from adult rat hippocampus in which hsp72 expression was at the limits of detection. In such slices microtubule associated protein 2 (MAP2) immunocytochemistry and histological integrity were preserved, while there were modest increases in expression of immediate-early genes (Zhou et al. 1995), consistent with those that might be expected after brief ischemia, below the threshold for neuron injury (Nowak et al. 1993; Sommer et al. 1995).

Since the evolution of hippocampal neuron loss after ischemia occurs too slowly to be easily studied within the practical survival time of acutely prepared slices from adult animals, we chose to define an appropriate model of transient global ischemia in the slice preparation based on the dual criteria of sustained histological integrity and characteristic changes in hsp72 expression. In preliminary studies it was found that even 2 min of transient anoxia/aglycemia resulted in rapid and severe neuron death following restoration of oxygen and glucose. Calcium loading would be expected to be much greater under in vitro conditions, given the large volume bathing the slices in comparison with the limited extracellular space in brain, and neurons were found to tolerate an insult of at least 6 min duration when calcium levels were reduced from 1.5 to 0.2 mM during anoxia/aglycemia. Under such conditions insults of 2–6 min duration resulted in prominent hsp72 mRNA induction in hippocampal neurons, which increased progressively between 30 min and 3 h (Zhou and Nowak 1996). Figure 5 illustrates a typical comparison of hsp72 expression in slices evaluated before and after anoxia/aglycemia. The hybridization signal is reproducibly weaker in CA1 than in dentate granule cells and CA3 neurons, as also observed at early recirculation intervals after global ischemia in vivo (Ikeda et al. 1994). Such results establish the hippocampal slice as a tool in which the postischemic stress response can be reliably modeled in vitro.

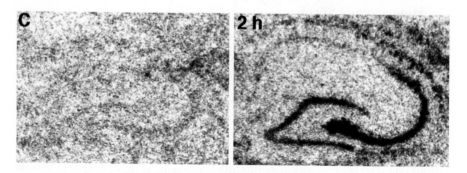

Fig. 5. Hsp72 mRNA induction in hippocampal slices after in vitro anoxia/aglycemia. Rat hippocampal slices were incubated under control conditions or subjected to 4 min anoxia/aglycemia followed by 2 h recovery, after which frozen sections were prepared and hybridized with ^{35}S-labeled probe. Hsp72 mRNA was at the limits of detection in control slices but was prominently induced after the in vitro insult

2. Pharmacological Manipulation of hsp72 Expression

The signaling pathways by which HSF activation can occur remain under investigation, and a complete consideration of the manipulations that can induce or modulate the stress response is beyond the scope of this presentation. However, studies of heat shock regulation after hyperthermic exposure of cell cultures have identified several treatments that can attenuate the stress response, implicating specific signals in the regulatory pathway. A number of results suggest a calcium dependence of heat shock activation (MOSSER et al. 1990; PRICE and CALDERWOOD 1991; KIANG et al. 1994), as well as an inhibitory effect of the tyrosine kinase inhibitor, genistein (PRICE and CALDERWOOD 1991). Inhibitory effects of the sulfhydryl reducing agent, dithiothreitol (DTT), have also been reported (HUANG et al. 1994). We have substantially reproduced these observations in hippocampal slices. In view of the calcium dependence of anoxic/aglycemic injury in the slices, it was not surprising that slices incubated under calcium-free conditions during and after such insults showed undetectable hsp72 mRNA expression. Preliminary results with genistein have been equivocal, with a trend toward reduced expression in CA3, but no significant effect on expression in dentate granule cells at concentrations up to 0.3 mM (not shown). The dose-response relationship for DTT inhibition of hsp72 induction after anoxia/aglycemia is illustrated in Fig. 6, demonstrating half-maximal inhibition at 0.3 mM. While such results must be interpreted with caution, they demonstrate the general comparability of pharmacological effects on the postischemic stress response in hippocampal slices and on hyperthermia-induced responses in other models.

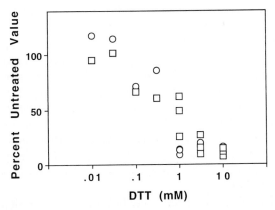

Fig. 6. Attenuation of anoxic/aglycemic hsp72 induction in hippocampal slices by dithiothreitol. Relative hsp72 mRNA expression was evaluated in rat hippocampal slices subjected to 4 min anoxia/aglycemia and 2 h recovery in the presence of varied concentrations of dithiothreitol (*DTT*). A comparable effect was evident in CA3 neurons (*circles*) and dentate granule cells (*squares*) with 50% reduction at approximately 0.3 mM DTT

It should be noted that, in addition to effects on tyrosine kinases, genistein has antioxidant properties (WEI et al. 1995), and such an action could also contribute to its reported effect to attenuate the heat shock response in some models. The potential contribution of reactive oxygen species to stress response signaling is indicated by a range of studies demonstrating at least some components of heat shock activation following exposure to hydrogen peroxide and other oxidants (COURGEON et al. 1988; BRUCE et al. 1993; FREEMAN et al. 1995). Recent results suggest that oxidative stress can affect ubiquitination pathways (JAHNGEN-HODGE et al. 1997), and that proteasome inhibition can upregulate hsp72 expression (ZHOU et al. 1996). Intriguing recent observations suggest that there may be more widespread hsp72 expression in transgenic mice overexpressing copper-zinc superoxide dismutase following various insults, including focal ischemia (KAMII et al. 1994), global ischemia (KONDO et al. 1996) or excitotoxic treatment (KONDO et al. 1997), even though smaller lesions are eventually produced (YANG et al. 1994). These observations are consistent with a scenario in which increasing flux through superoxide dismutase may contribute to heat shock signaling mechanisms. It should also be noted that there is a strong temperature dependence of hydroxyl radical production after transient ischemia (KIL et al. 1996) which may be correlated with the enhanced induction of hsp72 mRNA in association with postischemic hyperthermia noted above (SUGA and NOWAK 1998). The hippocampal slice should provide a useful model in which to further test such speculative hypotheses regarding regulation of the postischemic stress response.

D. Summary and Conclusions

Studies summarized above document complex patterns of heat shock gene induction after global and focal ischemia, of which it may be considered that hsp72 expression is most strongly associated with cellular vulnerability following such insults. Depletion of free ubiquitin and accumulation of ubiquitin conjugates, together with ubiquitin mRNA induction, are closely associated both temporally and anatomically with the pattern of hsp72 expression after global ischemia, and restoration of free ubiquitin is indicative of cellular recovery and survival. Changes in the expression of other heat shock proteins, although undoubtedly of functional importance, either represent quantitatively moderate increases above contitutive levels (hsc70, hsp90, and hsp60), or occur predominantly in association with reactive responses in astrocytes (hsp27) or microglia (hsp32).

The most severely injured cells may be lost so acutely that neither transcriptional or translational responses are possible, but significant hsp72 mRNA expression is generally observed in vulnerable populations after ischemic insults typically studied. Impaired translation is a significant factor affecting the synthesis of all proteins encoded by ischemia-induced mRNAs. While successful expression of hsp72 immunoreactivity is observed in surviving neurons and other cell types affected in a given model, it is clear that accumulation of hsp72 immunoreactivity does not guarantee survival. This may become even more apparent as antigen retrieval techniques are applied to unmask cryptic hsp72 protein expression in various models. Evaluations of depolarization thresholds for cell death, hsp72 induction and ischemic tolerance in a gerbil model further indicate that hsp72 expression occurs in association with injury and is not required for production of tolerance after mild insults.

The heterogeneity of heat shock protein induction after ischemic insults implies that several independent mechanisms must contribute to the patterns of expression observed. Both HSF1 and HSF2 appear to be ubiquitously expressed in brain, and both may contribute to HSE binding activity induced after global ischemia. While the signaling mechanisms responsible for heat shock regulation remain to be fully elucidated, hippocampal slices subjected to anoxic/aglycemic insults in vitro provide a model system that faithfully recapitulates the characteristics of hsp72 induction observed after transient ischemia in brain, that should prove useful in further mechanistic studies of the postischemic stress response.

References

Abe H, Nowak TS Jr (1996a) Gene expression and induced ischemic tolerance following brief insults. Acta Neurobiol Exp 56:3–8

Abe H, Nowak TS Jr (1996b) The stress response and its role in cellular defense mechanisms after ischemia. In: Siesjö BK, Wieloch T (eds) Cellular and molecular mechanisms of ischemic brain damage. Lippincott-Raven, Philadelphia, p 451 (Advances in neurology, vol 71)

Abe K, Kawagoe J, Aoki M, Kogure K (1993) Changes of mitochondrial DNA and heat shock protein gene expressions in gerbil hippocampus after transient forebrain ischemia. J Cereb Blood Flow Metab 13:773–780

Abe K, Kogure K, Itoyama Y (1995) Rapid and semiquantitative analysis of HSP72 and HSC73 heat shock mRNAs by mimic RT-PCR. Brain Res 683:251–253

Abe K, Tanzi RE, Kogure K (1991) Induction of HSP70 mRNA after transient ischemia in gerbil brain. Neurosci Lett 125:166–168

Abravaya K, Myers MP, Murphy SP, Morimoto RI (1992) The human heat shock protein hsp70 interacts with HSF, the transcription factor that regulates the heat shock response. Genes Dev 6:1153–1164

Andou Y, Mitani A, Masuda S, Arai T, Kataoka K (1992) Re-evaluation of ischemia-induced neuronal damage in hippocampal regions in the normothermic gerbil. Acta Neuropathol (Berl) 85:10–14

Aoki M, Abe K, Liu X-H, Lee T-H, Kato H, Kogure K (1993) Reduction of HSP70 and HSC70 mRNA inductions by bifemaline hydrochloride after transient ischemia in gerbil brain. Neurosci Lett 154:69–72

Baler R, Welch WJ, Voellmy R (1992) Heat shock gene regulation by nascent polypeptides and denatured proteins: hsp70 as a potential autoregulatory factor. J Cell Biol 117:1151–1159

Baler R, Zou J, Voellmy R (1996) Evidence for a role of Hsp70 in the regulation of the heat shock response in mammalian cells. Cell Stress Chaperones 1:33–39

Beckmann RP, Mizzen LA, Welch WJ (1990) Interaction of hsp70 with newly synthesized proteins: implications for protein folding and assembly. Science 248:850–854

Benjamin IJ, Kröger B, Williams RS (1990) Activation of the heat shock transcription factor by hypoxia in mammalian cells. Proc Natl Acad Sci USA 87:6263–6267

Bergeron M, Mivechi NF, Giaccia AJ, Giffard RG (1996) Mechanism of heat shock protein 72 induction in primary cultured astrocytes after oxygen-glucose deprivation. Neurol Res 18:64–72

Bergstedt K (1993) Ischemic and hypoglycemic brain damage. Studies on protein synthesis and heat-shock protein expression in the rat brain, thesis, Lund University, Sweden

Blake MJ, Nowak TS Jr, Holbrook NJ (1990) In vivo hyperthermia induces expression of HSP70 mRNA in brain regions controlling the neuroendocrine response to stress. Mol Brain Res 8:89–92

Blumenfeld KS, Welsh FA, Harris VA, Pesenson MA (1992) Regional expression of c-fos and heat shock protein-70 mRNA following hypoxia-ischemia in immature rat brain. J Cereb Blood Flow Metab 12:987–995

Brown CR, Martin RL, Hansen WJ, Beckmann RP, Welch WJ (1993) The constitutive and stress inducible forms of hsp 70 exhibit functional similarities and interact with one another in an ATP-dependent fashion. J Cell Biol 120:1101–1112

Bruce JL, Price BD, Coleman N, Calderwood SK (1993) Oxidative injury rapidly activates the heat shock transcription factor but fails to increase levels of heat shock proteins. Cancer Res 53:12–15

Buchan A, Pulsinelli WA (1990) Hypothermia but not the N-methyl-D-aspartate antagonist, MK-801, attenuates neuronal damage in gerbils subjected to transient global ischemia. J Neurosci 10:311–316

Busto R, Dietrich WD, Globus MY-T, Valses I, Scheinberg P, Ginsberg MD (1987) Small differences in intraischemic brain temperature critically determine the extent of ischemic neuronal injury. J Cereb Blood Flow Metab 7:729–738

Cattoretti G, Pileri S, Parravicini C, Becker MHG, Poggi S, Bifulco C, Key G, D'Amato L, Sabattini E, Feudale E, Reynolds F, Gerdes J, Rilke F (1993) Antigen unmasking on formalin-fixed, paraffin-embedded tissue sections. J Pathol 171:83–98

Charriaut-Marlangue C, Pollard H, Kadri-Hassani N, Khrestchatisky M, Moreau J, Dessi F, Kang KI, Ben-Ari Y (1992) Increase in specific proteins and mRNAs following transient anoxia-aglycaemia in rat CA1 hippocampal slices. Eur J Neurosci 4:766–776

Chopp M, Li Y, Dereski MO, Levine SR, Yoshida Y, Garcia JH (1991) Neuronal injury and expression of 72-kDa heat-shock protein after forebrain ischemia in the rat. Acta Neuropathol (Berl) 83:66–71

Chopp M, Li Y, Dereski MO, Levine SR, Yoshida Y, Garcia JH (1992) Hypothermia reduces 72-kDa heat-shock protein induction in rat brain after transient forebrain ischemia. Stroke 23:104–107

Churn SB, Taft WC, Billingsley MS, Blair RE, DeLorenzo RJ (1990) Temperature modulation of ischemic neuronal death and inhibition of calcium/calmodulin-dependent protein kinase II in gerbils. Stroke 21:1715–1721

Colbourne F, Corbett D (1995) Delayed postischemic hypothermia: a six month survival study using behavioral and histological assessments of neuroprotection. J Neurosci 15:7250–7260

Colbourne F, Nurse SM, Corbett D (1993) Spontaneous postischemic hyperthermia is not required for severe CA1 ischemic damage in gerbils. Brain Res 623:1–5

Colbourne F, Sutherland G, Corbett D (1997) Postischemic hypothermia. A critical appraisal with implications for clinical treatment. Mol Neurobiol 14:171–201

Cooper HK, Zalewska T, Hossmann K-A, Kleihues P (1977) The effect of ischemia and recirculation on protein synthesis in the rat brain. J Neurochem 28:929–934

Courgeon A-M, Rollet E, Becker J, Maisonhaute C, Best-Belpomme M (1988) Hydrogen peroxide (H_2O_2) induces actin and some heat-shock proteins in Drosophila cells. Eur J Biochem 171:163–170

Deshpande J, Bergstedt K, Lindén T, Kalimo H, Wieloch T (1992) Ultrastructural changes in the hippocampal CA1 region following transient cerebral ischemia: evidence against programmed cell death. Exp Brain Res 88:91–105

Dienel GA, Kiessling M, Jacewicz M, Pulsinelli WA (1986) Synthesis of heat shock proteins in rat brain cortex after transient ischemia. J Cereb Blood Flow Metab 6:505–510

Dienel GA, Pulsinelli WA, Duffy TE (1980) Regional protein synthesis in rat brain following acute hemispheric ischemia. J Neurochem 35:1216–1226

Dietrich WD, Busto R, Alonso O, Globus MY-T, Ginsberg MD (1993) Intraischemic but not postischemic brain hypothermia protects chronically following global forebrain ischemia in rats. J Cereb Blood Flow Metab 13:541–549

Dietrich WD, Busto R, Halley M, Valdes I (1990a) The importance of brain temperature in alterations of the blood-brain barrier following cerebral ischemia. J Neuropathol Exp Neurol 49:486–497

Dietrich WD, Busto R, Valdes I, Loor Y (1990b) Effects of normothermic versus mild hyperthermic forebrain ischemia in rats. Stroke 21:1318–1325

Ferrer I, Soriano MA, Vidal A, Planas AM (1995) Survival of parvalbumin-immunoreactive neurons in the gerbil hippocampus following transient forebrain ischemia does not depend on HSP-70 protein induction. Brain Res 692:41–46

Ferriero DM, Soberano HQ, Simon RP, Sharp FR (1990) Hypoxia-ischemia induces heat shock protein-like (hsp72) immunoreactivity in neonatal rat brain. Dev Brain Res 53:145–150

Fourie AM, Sambrook JF, Gething M-JH (1994) Common and divergent peptide binding specificities of hsp70 molecular chaperones. J Biol Chem 269:30470–30478

Freeman ML, Borrelli MJ, Syed K, Senisterra G, Stafford DM, Lepock JR (1995) Characterization of a signal generated by oxidation of protein thiols that activates the heat shock transcription factor. J Cell Physiol 164:356–366

Gass P, Schröder H, Prior P, Kiessling M (1994) Constitutive expression of heat shock protein 90 (HSP90) in neurons of the rat brain. Neurosci Lett 182:188–192

Gass P, Spranger M, Herdegen T, Bravo R, Köck P, Hacke W, Kiessling M (1992) Induction of FOS and JUN proteins following focal ischemia in the rat cortex: differential effect of MK-801. Acta Neuropathol (Berl) 84:545–553

Gilby KL, Armstrong JN, Currie WR, Robertson HA (1997) The effects of hypoxia-ischemia on expression of c-Fos, c-Jun and Hsp70 in the young rat hippocampus. Mol Brain Res 48:87–96

Ginty DD, Marlowe M, Pekala PH, Seidel ER (1990) Multiple pathways for the regulation of ornithine decarboxylase in intestinal epithelial cells. Am J Physiol 258:G454–G460

Gonzalez MF, Lowenstein D, Fernyak S, Hisanaga K, Simon R, Sharp FR (1991) Induction of heat shock protein 72-like immunoreactivity in the hippocampal formation following transient global ischemia. Brain Res Bull 26:241–250

Gonzalez MF, Shiraishi K, Hisanaga K, Sagar SM, Mandabach M, Sharp FR (1989) Heat shock proteins as markers of neural injury. Mol Brain Res 6:93–100

Goodson ML, Park-Sarge O-K, Sarge KD (1995) Tissue-dependent expression of heat shock factor 2 isoforms with distinct transcriptional activities. Mol Cell Biol 15:5288–5293

Greenberg ME, Hermanowski AL, Ziff EB (1986) Effect of protein synthesis inhibitors on growth factor activation of c-fos, c-myc, and actin gene transcription. Mol Cell Biol 6:1050–1057

Gubellini P, Bisso GM, Ciofi-Luzzatto A, Fortuna S, Lorenzini P, Michalek H, Scarsella G (1997) Ubiquitin-mediated stress response in a rat model of brain transient ischemia/hypoxia. Neurochem Res 22:93–100

Harrub JB, Nowak TS Jr (1998) Cryptic expression of the 70 kDa heat shock protein, hsp72, in gerbil hippocampus after transient ischemia. Neurochem Res 23:703–708

Hasegawa K, Litt L, Espanol MT, Gregory GA, Sharp FR, Chan PH (1997) Effects of neuroprotective dose of fructose-1,6-bisphosphate on hypoxia-induced expression of c-fos and hsp70 mRNA in neonatal rat cerebrocortical slices. Brain Res 750:1–10

Hayashi T, Takada K, Matsuda M (1991) Changes in ubiquitin and ubiquitin-protein conjugates in the CA1 neurons after transient sublethal ischemia. Mol Chem Neuropathol 15:75–82

Hayashi T, Takada K, Matsuda M (1992a) Post-transient ischemia increase in ubiquitin conjugates in the early reperfusion. Neuroreport 3:519–520

Hayashi T, Takada K, Matsuda M (1992b) Subcellular distribution of ubiquitin-protein conjugates in the hippocampus following transient ischemia. J Neurosci Res 31:561–564

Hayashi T, Tanaka J, Kamikubo T, Takada K, Matsuda M (1993) Increase in ubiquitin conjugates dependent on ischemic damage. Brain Res 620:171–173

Higashi T, Nakai A, Uemura Y, Kikuchi H, Nagata K (1995) Activation of heat shock factor 1 in rat brain during cerebral ischemia or after heat shock. Mol Brain Res 34:262–270

Higashi T, Takechi H, Uemura Y, Kikuchi H, Nagata K (1994) Differential induction of mRNA species encoding several classes of stress proteins following focal cerebral ischemia in rats. Brain Res 650:239–248

Huang LE, Zhang H, Bae SW, Liu AY-C (1994) Thiol reducing reagents inhibit the heat shock response. Involvement of a redox mechanism in the heat shock signal transduction pathway. J Biol Chem 269:30718–30725

Iijima T, Mies G, Hossmann K-A (1992) Repeated negative DC deflections in rat cortex following middle cerebral artery occlusion are abolished by MK-801: effect on volume of ischemic injury. J Cereb Blood Flow Metab 12:727–733

Ikeda J, Nakajima T, Osborne OC, Mies G, Nowak TS Jr (1994) Coexpression of c-fos and hsp70 mRNAs in gerbil brain after ischemia: induction threshold, distribution and time course evaluated by in situ hybridization. Mol Brain Res 26:249–258

Izumoto S, Herbert J (1993) Widespread constitutive expression of HSP90 messenger RNA in rat brain. J Neurosci Res 35:20–28

Jahngen-Hodge J, Obin MS, Gong X, Shang F, Nowell TR Jr, Gong J, Abasi H, Blumberg J, Taylor A (1997) Regulation of ubiquitin-conjugating enzymes by glutathione following oxidative stress. J Biol Chem 272:28218–28226

James P, Pfund C, Craig EA (1997) Functional specificity among molecular chaperones. Science 275:387–389

Kamii H, Kinouchi H, Sharp FR, Koistinaho J, Epstein CJ, Chan PH (1994) Prolonged expression of hsp70 mRNA following transient focal cerebral ischemia in transgenic mice overexpressing CuZn-superoxide dismutase. J Cereb Blood Flow Metab 14:478–486

Kamiya T, Jacewicz M, Pulsinelli WA, Nowak TS Jr (1995) CBF thresholds for RNA and protein synthesis after focal ischemia and the effect of MK-801. J Cereb Blood Flow Metab 15:S1

Kato H, Araki T, Kogure K (1991) Postischemic spontaneous hyperthermia is not a major aggravating factor for neuronal damage following repeated brief cerebral ischemia in the gerbil. Neurosci Lett 126:21–24

Kato H, Chen T, Liu X-H, Nakata N, Kogure K (1993) Immunohistochemical localization of ubiquitin in gerbil hippocampus with induced tolerance to ischemia. Brain Res 619:339–343

Kato H, Kogure K, Liu X-H, Araki T, Kato K, Itoyama Y (1995) Immunohistochemical localization of the low molecular weight stress protein HSP27 following focal cerebral ischemia in the rat. Brain Res 679:1–7

Kato H, Liu Y, Kogure K, Kato K (1994) Induction of 27-kDa heat shock protein following cerebral ischemia in a rat model of ischemic tolerance. Brain Res 634:235–244

Kawagoe J, Abe K, Aoki M, Kogure K (1993) Induction of HSP90α heat shock mRNA after transient global ischemia in gerbil hippocampus. Brain Res 621:121–125

Kawagoe J, Abe K, Kogure K (1992a) Different thresholds of HSP70 and HSC70 heat shock mRNA induction in post-ischemic gerbil brain. Brain Res 599:197–203

Kawagoe J, Abe K, Sato S, Nagano I, Nakamura S, Kogure K (1992b) Distributions of heat shock protein (HSP) 70 and heat shock cognate protein (HSC) 70 mRNAs after transient focal ischemia in rat brain. Brain Res 587:195–202

Kawagoe J, Abe K, Sato S, Nagano I, Nakamura S, Kogure K (1992c) Distributions of heat shock protein-70 mRNAs and heat shock cognate protein-70 mRNAs after transient global ischemia in gerbil brain. J Cereb Blood Flow Metab 12:794–801

Kiang JG, Carr FE, Burns MB, McClain DE (1994) HSP-72 synthesis is promoted by increase in $[Ca^{2+}]_i$ or activation of G proteins but not pH_i or cAMP. Am J Physiol 267:C104–C114

Kiessling M, Dienel GA, Jacewicz M, Pulsinelli WA (1986) Protein synthesis in postischemic rat brain: a two-dimensional electrophoretic analysis. J Cereb Blood Flow Metab 6:642–649

Kiessling M, Stumm G, Xie Y, Herdegen T, Aguzzi A, Bravo R, Gass P (1993) Differential transcription and translation of immediate early genes in the gerbil hippocampus after transient global ischemia. J Cereb Blood Flow Metab 13:914–924

Kil HY, Zhang J, Piantadosi CA (1996) Brain temperature alters hydroxyl radical production during cerebral ischemia/reperfusion in rats. J Cereb Blood Flow Metab 16:100–106

Kindy M, Bhat AN, Bhat NR (1992) Transient ischemia stimulates glial fibrillary acid protein and vimentin gene expression in the gerbil neocortex, striatum and hippocampus. Mol Brain Res 13:199–206

Kindy MS, Carney JP, Dempsey RJ, Carney JM (1991) Ischemic induction of protooncogene expression in gerbil brain. J Mol Neurosci 2:217–228

Kinouchi H, Sharp FR, Chan PH, Koistinaho J, Sagar SM, Yoshimoto T (1994) Induction of c-fos, junB, c-jun, and hsp70 mRNA in cortex, thalamus, basal ganglia, and hippocampus following middle cerebral artery occlusion. J Cereb Blood Flow Metab 14:808–817

Kinouchi H, Sharp FR, Hill MP, Koistinaho J, Sagar SM, Chan PH (1993) Induction of 70-kDa heat shock protein and hsp70 mRNA following transient focal cerebral ischemia in the rat. J Cereb Blood Flow Metab 13:105–115

Kirino T, Tsujita Y, Tamura A (1991) Induced tolerance to ischemia in gerbil hippocampal neurons. J Cereb Blood Flow Metab 11:299–307

Kitagawa K, Matsumoto M, Tagaya M, Hata R, Ueda H, Niinobe M, Handa N, Fukunaga R, Kimura K, Mikoshiba K, Kamada T (1990) "Ischemic tolerance" phenomenon found in brain. Brain Res 528:21–24

Kobayashi S, Harris VA, Welsh FA (1995) Spreading depression induces tolerance of cortical neurons to ischemia in rat brain. J Cereb Blood Flow Metab 15:721–727

Kondo T, Murakami K, Honkaniemi J, Sharp FR, Epstein CJ, Chan PH (1996) Expression of hsp70 mRNA is induced in the brain of transgenic mice overexpressing human CuZn-superoxide dismutase following transient global cerebral ischemia. Brain Res 737:321–326

Kondo T, Sharp FR, Honkaniemi J, Mikawa S, Epstein CJ, Chan PH (1997) DNA fragmentation and prolonged expression of c-fos, c-jun, and hsp70 in kainic acid-induced neuronal cell death in transgenic mice overexpressing human CuZn-superoxide dismutase. J Cereb Blood Flow Metab 17:241–256

Kuroiwa T, Bonnekoh P, Hossmann K-A (1990) Prevention of postischemic hyperthermia prevents ischemic injury of CA_1 neurons in gerbils. J Cereb Blood Flow Metab 10:550–556

Li Y, Chopp M, Garcia JH, Yoshida Y, Zhang ZG, Levine SR (1992) Distribution of the 72-kd heat-shock protein as a function of transient focal cerebral ischemia in rats. Stroke 23:1292–1298

Liu Y, Kato H, Nakata N, Kogure K (1993) Temporal profile of heat shock protein 70 synthesis in ischemic tolerance induced by preconditioning ischemia in rat hippocampus. Neuroscience 56:921–927

Magnusson K, Wieloch T (1989) Impairment of protein ubiquitination may cause delayed neuronal death. Neurosci Lett 96:264–270

Marcuccilli CJ, Mathur SK, Morimoto RI, Miller RJ (1996) Regulatory differences in the stress response of hippocampal neurons and glial cells after heat shock. J Neurosci 16:478–485

Marini AM, Kozuka M, Lipsky RL, Nowak TS Jr (1990) 70-Kilodalton heat shock protein induction in cerebellar astrocytes and cerebellar granule cells in vitro: comparison with immunocytochemical localization after hyperthermia in vivo. J Neurochem 54:1509–1516

Massa SM, Swanson RA, Sharp FR (1996) The stress gene response in brain. Cereb Brain Metab Rev 8:95–158

Matsuyama T, Michishita H, Nakamura H, Tsuchiyama M, Shimizu S, Watanabe K, Sugita M (1993) Induction of copper-zinc superoxide dismutase in gerbil hippocampus after ischemia. J Cereb Blood Flow Metab 13:135–144

McCabe T, Simon RP (1993) Hyperthermia induces 72 kDa heat shock protein expression in rat brain in non-neuronal cells. Neurosci Lett 159:163–165

Mestril R, Chi S-H, Sayen R, Dillmann WH (1994) Isolation of a novel inducible rat heat-shock protein (HSP70) gene and its expression during ischaemia/hypoxia and heat shock. Biochem J 298:561–569

Mies G, Ishimaru S, Xie Y, Seo K, Hossmann K-A (1991) Ischemic thresholds of cerebral protein synthesis and energy state following middle cerebral artery occlusion in rat. J Cereb Blood Flow Metab 11:753–761

Milarski KL, Welch WJ, Morimoto RI (1989) Cell cycle-dependent association of HSP70 with specific cellular proteins. J Cell Biol 108:413–423

Morimoto T, Ide T, Ihara Y, Tamura A, Kirino T (1996) Transient ischemia depletes free ubiquitin in the gerbil hippocampal CA1 neurons. Am J Pathol 148:249–257

Mosser DD, Duchaine J, Massie B (1993) The DNA binding activity of the human heat shock transcription factor is regulated in vivo by hsp70. Mol Cell Biol 13:5427–5438

Mosser DD, Kotzbauer PT, Sarge KD, Morimoto RI (1990) In vitro activation of heat shock transcription factor DNA-binding by calcium and biochemical conditions that affect protein conformation. Proc Natl Acad Sci USA 87:3748–3752

Munell F, Burke RE, Bandele A, Gubits RM (1994) Localization of c-fos, c-jun, and hsp70 mRNA expression in brain after neonatal hypoxia-ischemia. Dev Brain Res 77:111–121

Newman GC, Qi H, Hospod FE, Grundmann K (1992) Preservation of hippocampal brain slices with in vivo or in vitro hypothermia. Brain Res 575:159–163

Nimura T, Weinstein PR, Massa SM, Panter S, Sharp FR (1996) Heme oxygenase-1 (HO-1) protein induction in rat brain following focal ischemia. Mol Brain Res 37:201–208

Nishi S, Taki W, Uemura Y, Higashi T, Kikuchi H, Kudoh H, Satoh M, Nagata K (1993) Ischemic tolerance due to the induction of HSP70 in a rat ischemic recirculation model. Brain Res 615:281–288

Nishimura RN, Dwyer BE (1996) Evidence for different mechanisms of induction of HSP70i: a comparison of cultured rat cortical neurons with astrocytes. Mol Brain Res 36:227–239

Nishimura RN, Dwyer BE, Clegg K, Cole R, de Vellis J (1991) Comparison of the heat shock response in cultured cortical neurons and astrocytes. Mol Brain Res 9:39–45

Nishimura RN, Dwyer BE, Welch W, Cole R, de Vellis J, Liotta K (1988) The induction of the major heat-stress protein in purified rat glial cells. J Neurosci Res 20:12–18

Noga M, Hayashi T (1996) Ubiquitin gene expression following transient forebrain ischemia. Mol Brain Res 36:261–267

Nowak TS Jr (1985) Synthesis of a stress protein following transient ischemia in the gerbil. J Neurochem 45:1635–1641

Nowak TS Jr (1991) Localization of 70kDa stress protein mRNA induction in gerbil brain after ischemia. J Cereb Blood Flow Metab 11:432–439

Nowak TS Jr (1998) Heat shock responses in global ischemia. In: Ginsberg MD, Bogousslavsky J (eds) Cerebrovascular disease: pathophysiology, diagnosis and management. Blackwell Science, Malden, p 565

Nowak TS Jr, Abe H (1994) Postischemic stress response in brain. In: Morimoto RI, Tissières A, Georgopoulos C (eds) The biology of heat shock proteins and molecular chaperones. Cold Spring Harbor Laboratory, Plainview, NY, p 553

Nowak TS Jr, Bond U, Schlesinger MJ (1990) Heat shock RNA levels in brain and other tissues after hyperthermia and transient ischemia. J Neurochem 54:451–458

Nowak TS Jr, Fried RL, Lust WD, Passonneau JV (1985) Changes in brain energy metabolism and protein synthesis following transient bilateral ischemia in the gerbil. J Neurochem 44:487–494

Nowak TS Jr, Jacewicz M (1994) The heat shock/stress response in focal cerebral ischemia. Brain Pathol 4:67–76

Nowak TS Jr, Osborne OC, Suga S (1993) Stress protein and proto-oncogene expression as indicators of neuronal pathophysiology after ischemia. In: Kogure K, Hossmann K-A, Siesjö BK (eds) Neurobiology of ischemic brain damage. Elsevier Science, Amsterdam, p 195 (Progress in brain research, vol 96)

Nowak TS Jr, Zhou Q, Voulalas PJ, Sarvey J (1994) Gene expression as an index of pathophysiology associated with slice preparation. In: Schurr A, Rigor BM (eds) Brain slices in basic and clinical research. CRC Press, Boca Raton, p 257

Nunes SL, Calderwood SK (1995) Heat shock factor – 1 and the heat shock cognate 70 protein associate in high molecular weight complexes in the cytoplasm of NIH-3T3 cells. Biochem Biophys Res Commun 213:1–6

Nurse S, Corbett D (1994) Direct measurement of brain temperature during and after intraischemia hypothermia: correlation with behavioral, physiological, and histological endpoints. J Neurosci 14:7726–7734

Nurse S, Corbett D (1996) Neuroprotection after several days of mild, drug-induced hypothermia. J Cereb Blood Flow Metab 16:474–480

Pardue S, Groshan K, Raese JD, Morrison-Bogorad M (1992) Hsp70 mRNA induction is reduced in neurons of aged rat hippocampus after thermal stress. Neurobiol Aging 13:661–672

Paschen W, Uto A, Djuricic B, Schmitt J (1994) Hemeoxygenase expression after reversible ischemia of rat brain. Neurosci Lett 180:5–8

Planas AM, Soriano MA, Estrada A, Sanz O, Martin F, Ferrer I (1997) The heat shock response after brain lesions: induction of 72kDa heat shock protein (cell types involved, axonal transport, transcriptional regulation) and protein synthesis inhibition. Prog Neurobiol 51:607–636

Plumier J-CL, David J-C, Robertson HA, Currie WR (1997) Cortical application of potassium chloride induces the low-molecular weight heat shock protein (hsp27) in astrocytes. J Cereb Blood Flow Metab 17:781–790

Price BD, Calderwood SK (1991) Ca^{2+} is essential for multistep activation of the heat shock factor in permeabilized cells. Mol Cell Biol 11:3365–3368

Saito N, Kawai K, Nowak TS Jr (1995) Reexpression of developmentally regulated MAP2c mRNA after ischemia: colocalization with hsp72 mRNA in vulnerable neurons. J Cereb Blood Flow Metab 15:205–215

Sharp FR, Lowenstein D, Simon R, Hisanaga K (1991) Heat shock protein hsp72 induction in cortical and striatal astrocytes and neurons following infarction. J Cereb Blood Flow Metab 11:621–627

Sick TJ, Somjen GG (1998) Tissue slice: application to study of cerebral ischemia. In: Ginsberg MD, Bogousslavsky J (eds) Cerebrovascular disease: pathophysiology, diagnosis and management. Blackwell Science, Malden, p 137

Simon RP, Cho H, Gwinn R, Lowenstein DH (1991) The temporal profile of 72-kDa heat-shock protein expression following global ischemia. J Neurosci 11:881–889

Sistonen L, Sarge KD, Morimoto RI (1994) Human heat shock factors 1 and 2 are differentially activated and can synergistically induce hsp70 gene transcription. Mol Cell Biol 14:2087–2099

Sloviter RS, Lowenstein DH (1992) Heat shock protein expression in vulnerable cells of the rat hippocampus as an indicator of excitation induced neuronal stress. J Neurosci 12:3004–3009

Sommer C, Gass P, Kiessling M (1995) Selective c-JUN expression in CA1 neurons of the gerbil hippocampus during and after acquisition of an ischemia-tolerant state. Brain Pathol 5:135–144

Soriano MA, Ferrer I, Rodriguez FE, Planas AM (1995) Expression of c-fos and inducible hsp-70 mRNA following a transient episode of focal ischemia that had non-lethal effects on the rat brain. Brain Res 670:317–320

Soriano MA, Planas AM, Rodríguez-Farré E, Ferrer I (1994) Early 72-kDa heat shock protein induction in microglial cells following focal ischemia in the rat brain. Neurosci Lett 182:205–207

Srivastava PK (1993) Peptide-binding heat shock proteins in the endoplasmic reticulum: role in immune response to cancer and in antigen presentation. Adv Cancer Res 62:153–177

Suga S, Nowak TS Jr (1998) Postischemic hyperthermia increases expression of hsp72 mRNA after brief ischemia in the gerbil. Neurosci Lett 243:57–60

Takeda A, Onodera H, Sugimoto A, Itoyama Y, Kogure K, Shibahara S (1994) Increased expression of heme oxygenase mRNA in rat brain following transient forebrain ischemia. Brain Res 666:120–124

Takemoto O, Tomimoto H, Yanagihara T (1995) Induction of c-fos and c-jun gene products and heat shock protein after brief and prolonged cerebral ischemia in gerbils. Stroke 26:1639–1648

Tamura Y, Peng P, Liu K, Daou M, Srivastava PK (1997) Immunotherapy of tumors with autologous tumor-derived heat shock protein preparations. Science 278:117–120

Thilmann R, Xie Y, Kleihues P, Kiessling M (1986) Persistent inhibition of protein synthesis precedes delayed neuronal death in postischemic gerbil hippocampus. Acta Neuropathol (Berl) 71:88–93

Tomioka C, Nishioka K, Kogure K (1993) A comparison of induced heat-shock protein in neurons destined to survive and those destined to die after transient ischemia in rats. Brain Res 612:216–220

Turner CP, Bergeron M, Matz P, Zegna A, Noble LJ, Panter SC, Sharp FR (1998) Heme oxygenase-1 is induced in glia throughout brain by subarachnoid hemoglobin. J Cereb Blood Flow Metab 18:257–273

Uemura Y, Kowall NW, Moskowitz MA (1991) Focal ischemia in rats causes time-dependent expression of c-fos protein immunoreactivity in widespread regions of ipsilateral cortex. Brain Res 552:99–105

Vass K, Berger ML, Nowak TS Jr, Welch WJ, Lassmann H (1989) Induction of stress protein HSP70 in nerve cells after status epilepticus in the rat. Neurosci Lett 100:259–264

Vass K, Welch WJ, Nowak TS Jr (1988) Localization of 70 kDa stress protein induction in gerbil brain after ischemia. Acta Neuropathol (Berl) 77:128–135

Velazquez JM, Lindquist S (1984) hsp70: nuclear concentration during environmental stress and cytoplasmic storage during recovery. Cell 36:655–662

Wagstaff MJD, Collaco-Moraes Y, Aspey BS, Coffin RS, Harrison MJG, Latchman DS, de Belleroche JS (1996) Focal cerebral ischaemia increases the levels of several classes of heat shock proteins and their corresponding mRNAs. Mol Brain Res 42:236–244

Wang S, Longo FM, Chen J, Butman M, Graham SH, Haglid KG, Sharp FR (1993) Induction of glucose regulated protein (grp78) and inducible heat shock protein (hsp70) mRNAs in rat brain after kainic acid seizures and focal ischemia. Neurochem Int 23:575–582

Wei H, Bowen R, Cai Q, Barnes S, Wang Y (1995) Antioxidant and antipromotional effects of the soybean isoflavone genistein. Proc Soc Exp Biol Med 208:124–130

Welch WJ, Feramisco JR (1984) Nuclear and nucleolar localization of the 72 000-dalton heat shock protein in heat-shocked mammalian cells. J Biol Chem 259:4501–4513

Welsh FA, Harris VA (1991) Postischemic hypothermia fails to reduce ischemic injury in gerbil hippocampus. J Cereb Blood Flow Metab 11:617–620

Welsh FA, Moyer DJ, Harris VA (1992) Regional expression of heat shock protein-70 mRNA and c-fos mRNA following focal ischemia in rat brain. J Cereb Blood Flow Metab 12:204–212

Welsh FA, Sims RE, Harris VA (1990) Mild hypothermia prevents ischemic injury in gerbil hippocampus. J Cereb Blood Flow Metab 10:557–563

Westwood JT, Clos J, Wu C (1991) Stress-induced oligomerization and chromosomal relocalization of heat-shock factor. Nature 353:822–827

Widmann R, Kuroiwa T, Bonnekoh P, Hossmann K-A (1991) [^{14}C]Leucine incorporation into brain proteins in gerbils after transient ischemia: relationship to selective vulnerability of hippocampus. J Neurochem 56:789–796

Wong M-L, Weiss SRB, Gold PW, Doi SQ, Banerjee S, Licinio J, Lad R, Post RM, Smith MA (1992) Induction of constitutive heat shock protein 73 mRNA in the dentate gyrus by seizures. Mol Brain Res 13:19–25

Yamashita K, Eguchi Y, Kajiwara K, Ito H (1991) Mild hypothermia ameliorates ubiquitin synthesis and prevents delayed neuronal death in the gerbil hippocampus. Stroke 22:1574–1581

Yang G, Chan PH, Chen J, Carlson E, Chen SF, Weinstein P, Epstein CJ, Kamii H (1994) Human copper-zinc superoxide dismutase transgenic mice are highly resistant to reperfusion injury after focal cerebral ischemia. Stroke 25:165–170

Zhou M, Wu X, Ginsberg HN (1996) Evidence that a rapidly turning over protein, normally degraded by proteasomes, regulates hsp72 gene transcription in HepG2 cells. J Biol Chem 271:24769–24775

Zhou Q, Abe H, Nowak TS Jr (1995) Immunocytochemical and in situ hybridization approaches to the optimization of brain slice preparations. J Neurosci Meth 59:85–92

Zhou Q, Nowak TS Jr (1996) Induction of hsp72 mRNA following in vitro anoxia/aglycemia in rat hippocampal slices. In: Krieglstein J, Oberpichler-Schwenk H (eds) Pharmacology of cerebral ischemia 1996. Wissenschaftliche Verlagsgesellschaft, Stuttgart, p 131

CHAPTER 9

Regulation of Heat Shock Transcription Factors by Hypoxia or Ischemia/Reperfusion in the Heart and Brain

J. Nishizawa and K. Nagata

A. Introduction

Despite advances in modern medicine, the pathologic processes induced by a lack of blood supply or ischemia remain the most prevalent cause of death in developed countries, including myocardial infarction, cerebral ischemia and embolic vascular occlusions of other tissues. Recent advances in diagnosis and treatment have allowed, at an early stage, the rapid return of blood flow by surgical, interventional or pharmacological means, to prevent infarction or cell necrosis in many of these ischemic tissues, and to reduce mortality or morbidity. However, if the treatment is administered too late, the beneficial effects in mortality and morbidity diminish (ISIS-2 1988; GUSTO 1993; YELLON and MARBER 1994), because prolonged artery occlusion or ischemia results in such severe tissue infarction or necrosis that the return of flow or reperfusion produces few beneficial effects (REIMER and JENNINGS 1979). In addition, reperfusion of ischemic areas contributes to further tissue damage called reperfusion injury (for review see McCORD 1985; KARMAZYN 1991; MAXWELL and LIP 1997). For instance, severe arrhythmia or myocardial stunning occurs in the heart.

Thus, protection of ischemic tissues has been the subject of experimental and clinical research for a long time especially in the heart and brain, and numerous investigators have attempted to reduce injury due to ischemia/ reperfusion. However, few pharmacological means have yet been established for clinical use that are effective in protection against ischemia/reperfusion. Therefore, it is of considerable importance to understand the mechanisms by which cells or tissues are damaged during ischemia/reperfusion, to identify compensatory responses that may augment cell survival, and to exploit the method of clinical application for the protection of ischemic tissues. Likewise, there is room for improvement in strategies to protect the ischemic tissues during high-risk cardiovascular surgery or interventional procedures, or to preserve the organs prior to transplantation.

All cells and organisms, from bacteria to higher eukaryotes, share a common response to hyperthermia, ischemia/reperfusion, or other physiological stresses that are unfavorable to their survival through the induction of a group of protective proteins termed heat shock or stress proteins (Hsps). These proteins are highly conserved throughout evolution and play essential roles as molecular chaperones under normal conditions as well as under stressed conditions by facilitating the folding, intracellular transport, assembly, and disassembly of other cellular proteins (for review see Morimoto et al. 1994). The induction of Hsps has been demonstrated to be directly involved in the acquisition of resistance to ischemia (Marber et al. 1995; Plumier et al. 1995; Hutter et al. 1996; Radford et al. 1996).

The heat shock/stress response is mainly regulated at the transcriptional level by the activation of a pre-existing transcription factor, the heat shock factor (HSF). The HSF binds to the heat shock element (HSE), a cis-acting promoter sequence present upstream of all heat shock genes. The HSE is composed of at least three pentanucleotide modules (nGAAn) arranged as contiguous inverted repeats (Perisic et al. 1989; for review see Chap. 10, this volume, and Morimoto et al. 1994).

In this chapter, we will summarize the mechanisms by which cells or tissues are damaged during hypoxia or ischemia/reperfusion, review the induction of the DNA-binding activity of HSF under these conditions mainly in the heart and the brain, and discuss the mechanisms by which this activity is induced by these insults.

B. Regulation of Heat Shock Gene Transcription

I. Family of Heat Shock Factors

Although there is only a single HSF-encoding gene in yeast and *Drosophila* (Wiederrecht et al. 1988; Clos et al. 1990), in vertebrate cells, a family of HSFs has been identified and it was suggested that distinct HSF family members may differentially regulate the transcription of heat shock genes (Rabindran et al. 1991; Sarge et al. 1991; Schuetz et al. 1991; Nakai and Morimoto 1993; Nakai et al. 1997). The HSF family includes HSF1, HSF2, HSF3, and HSF4. HSF1 acquires DNA-binding activity in response to elevated temperatures, ischemia/reperfusion, oxidative stress, and exposure to heavy metals and amino acid analogues (Baler et al. 1993; Sarge et al. 1993; Fawcett et al. 1994). Thus, HSF1 seems to be the general stress-response factor (Morimoto et al. 1994). HSF2 appears to be a developmental HSF and is activated during erythroid differentiation of human K562 erythroleukemia cells with hemin treatment (Sistonen et al. 1992, 1994) and is constitutively activated in mouse embryonic stem cells and in spermatogenic cells of mouse testes (Murphy et al. 1994; Sarge et al. 1994). HSF3 was cloned only in chicken cells (Nakai and Morimoto 1993). Although HSF1 and HSF3 are activated by various stresses, such as heat shock and sodium arsenite, the kinetics and the thresholds of the activation of HSF1 and HSF3 are different (Nakai et al. 1995; Tanabe et al. 1997). HSF1 responds to stress rapidly, whereas HSF3 responds slowly. In addition, HSF3 was reported to be activated by heat shock at higher temperatures or with higher concentrations of sodium arsenite than HSF1. HSF3 was demonstrated to be involved in persistent and burst activations of stress genes during severe stress. Recently, human HSF4 that lacks activity as a positive transactivator and represses the expression of gene encoding Hsps was cloned and characterized (NAKAI et al. 1997). HSF4 lacks the carboxyl-terminal hydrophobic repeat which is present among all vertebrate HSFs and is preferentially expressed in the heart, brain, skeletal muscles, and pancreas.

II. Regulation of DNA-Binding Activity of HSF1

An increase in the levels of denatured, unfolded or malfolded proteins triggers heat shock responses (GOFF and GOLDBERG 1985; ANANTHAN et al. 1986). It was also suggested by experimental evidence that intracellular levels of free Hsp70 contribute to the feedback regulation of Hsp70 expression (DiDOMENICO et al. 1982; CRAIG and GROSS 1991). Thus, in eukaryotic cells, a model for regulation of DNA-binding activity of HSF1 has been proposed (ABRAVAYA et al. 1992; BECKMANN et al. 1992; MORIMOTO 1993; MOSSER et al. 1993; MORIMOTO et al. 1994). In the model, Hsps themselves negatively regulate heat shock gene expression via an autoregulating loop. Under unstressed conditions, HSF1 is present in both the nucleus and the cytoplasm as a monomer that has no DNA-binding activity through transient interactions with Hsp70 or other Hsps. Under stressful conditions, such as heat shock or physiological stresses, an increase in denatured, misfolded or aggregated proteins creates a large pool of new protein substrates that compete with HSF1 for association with Hsp70. Thus, such stresses initiate the removal of the negative regulatory influence on the DNA-binding activity of HSF1. HSF1 assembles into a trimer, translocates into the nucleus, binds to HSE, undergoes a serine phosphorylation, and acquires transcriptional activity. This activation of HSF1 DNA-binding activity leads to increased transcription and synthesis of Hsps including Hsp70, which then interact with HSF1 and interfere with the DNA-binding activity of HSF1.

C. Damage by Ischemia and Reperfusion

I. Ischemia

"Ischemia" refers to the condition of a tissue in which blood flow is insufficient to meet metabolic demands. Although an ischemic tissue experiences hypoxia, the term "ischemia" is not synonymous with "hypoxia," and the effects are quite different (ALLEN and ORCHARD 1987). During ischemia, the cells are deprived of not only oxygen for mitochondrial respiration, but also of other substrates, such as glucose, that are essentially supplied by blood flow. In addition, physiologically toxic metabolites that are washed out with

Table 1. Metabolic and structural changes associated with
ischemia/reperfusion

ATP loss
Acidosis
Altered osmotic control
Accumulation of various lipid species
Reactive oxygen species
Calcium overload
Arachidonic acid and its metabolites
Loss of membrane integrity
Intracellular ionic derangement
Structural disorganization
Cell swelling

---

normal circulation accumulate in the tissue (Table 1; for review see Hillis
and Braunwald 1977; Reimer and Ideker 1987; Piper 1990; Karmazyn
1991).

As a result of oxygen deprivation, oxidative metabolism, electron transport, and adenosine triphosphate (ATP) production by oxidative phosphorylation in mitochondria rapidly decline. At the early stage, some ATP is still
produced with a compensatory increase in anaerobic glycolysis. Intracellular
acidosis develops due to the accumulation of lactate and hydrogen ions, and
anaerobic glycolysis is suppressed. These alterations contribute to damage to
the cell membrane and induce cell swelling, intracellular accumulation of
calcium, loss of intracellular potassium, and other disturbances of membrane
ion transport.

Cellular injury as a result of advanced ischemia is mediated by progressive
membrane damage involving several contributory factors (Reimer and Ideker
1987). Accumulation of calcium on other metabolic alterations induces
phospholipase activation, phospholipid degradation, and release of
lysophospholipids and free fatty acids. Reduced mitochondrial fatty acid metabolism induces the accumulation of various lipid species, such as acyl
CoA and acyl carnitine which can be incorporated into membranes and
impair their function. Reactive oxygen species, such as free radicals and toxic
oxygen species, are produced in ischemic cells and activated leukocytes. These
toxic chemicals result in peroxidative damage of fatty acids in membrane
phospholipids. The activation of protease probably induces damage of
cytoskeletal filaments and their stabilizing effect on the sarcolemma is lost.
These changes result in a progressive increase in nonspecific membrane permeability, further breakdown in the intracellular ionic environment, and ATP
depletion. Cellular metabolism and ATP generation nearly cease, and glycogen stores are depleted. As glycolysis and mitochondrial function are totally
lost, autolysis of cells begins, and the leakage of cellular contents increases
extensively.

II. Reperfusion

The term "reperfusion" refers to the period following the restoration of blood flow to the ischemic tissues. Although reperfusion is undoubtedly an important step in salvaging tissues subjected to ischemia, it may also induce additional injury (for review see KARMAZYN 1991; MAXWELL and LIP 1997). For instance, in the heart, although reperfusion procedure is an effective means of myocardial salvage, evidence obtained under clinical situations as well as under various experimental laboratory conditions has revealed a potential detrimental influence of reperfusion of the myocardium. Additional injury or further damage as a result of reperfusion after ischemia, is referred to as "reperfusion injury" and involves transient contractile dysfunction termed "myocardial stunning," increased deleterious arrhythmia, and cell death.

The effects of reperfusion are complex, and the pathogenesis and mechanism of reperfusion injury are still uncertain (MCCORD 1985; KARMAZYN 1991; HEARSE and BOLLI 1992; MAXWELL and LIP 1997). Reperfusion is associated with a variety of events including generation of oxygen free radicals, activation of neutrophils and platelets, increased synthesis of arachidonic acid and its metabolites, activation of the Na^+/H^+ exchanger, and abrupt intracellular calcium overload. Considering the variety of underlying mechanisms and manifestations, a single all-embracing theory may not account for reperfusion injury.

D. Regulation of Hsps by Ischemia/Reperfusion in the Brain and Heart

I. Induction of Hsps by Ischemia/Reperfusion in the Brain
(see also Chaps. 8 and 11, this volume)

Induction of Hsp70 in established models of global cerebral ischemia was first reported in gerbils (NOWAK 1985), and then in rats (DIENEL et al. 1986). The increased expression of Hsp70 mRNAs in the hippocampus after global ischemia was also shown by in situ hybridization using specific oligonucleotide probes (NOWAK 1991). Both in experimental animals and in humans, the neurons of the brain after brief global ischemia are discretely lost and involve selectively vulnerable cell populations (ITO et al. 1975; KIRINO 1982; PULSINELLI et al. 1982; SMITH et al. 1984; ROSS and GRAHAM 1993). The delayed death of CA1 pyramidal neurons in the hippocampus has been extensively studied and has been found to occur approximately 2–4 days after ischemia (KIRINO 1982; PULSINELLI et al. 1982). The expression of Hsp70 after global ischemia is also selectively distributed with relative vulnerability (NOWAK 1991; KAWAGOE et al. 1992b). During reperfusion after 5-min global ischemia, the accumulation of Hsp70 mRNA is initially expressed in all of the major hippocampal neuron populations but is sequentially lost from the less vulnerable dentate granule cells and CA3 pyramidal neurons, whereas the signal is persistently expressed in CA1 neurons for up to 48 h in gerbils. The duration of Hsp70 mRNA induction seems to directly correlate with relative vulnerability of the neuron populations. In the protein level, immunological localization of Hsp70 at 48 h of recirculation after 5-min ischemia in gerbils is observed only in CA3 but not in CA1 neurons that show prolonged expression of the Hsp70 mRNA (Vass et al. 1988). On the contrary, brief 2-min ischemia results in clear staining of Hsp70 in CA1 neurons. Similarly, in rats, CA1 neuronal cells are strongly stained with the anti-Hsp70 antibody 2 days after 5-min ischemia, while only CA3 and CA4 neurons are stained after 30-min ischemia (Nishi et al. 1993). The main factor contributing to this difference between Hsp70 mRNA and protein expression in vulnerable neurons seems to be the well documented prolonged deficit in translational activity (Dienel et al. 1980; Thilmann et al. 1986).

Focal ischemic insults for prolonged periods of time lead to infarction with injury of all cell types within the affected region. The border zone surrounding the ischemic core is a "penumbra" with intermediate flow in which a graded cellular response is observed (Astrup et al. 1981). The distribution of Hsp70 mRNA expression in the focal ischemia model seems to correlate well with this distribution of pathology. Relatively soon after ischemia, the expression of Hsp70 mRNA was reported to be concentrated in the penumbra region, with relatively little mRNA within the core region in rats (Welsh et al. 1992). The accumulation of Hsp70 mRNA has been more carefully examined and shown to be localized in the core region of the infarct in the early period of ischemia (2 h), and then to move to the penumbra region (4–8 h) in spontaneously hypertensive rats (Higashi et al. 1994). With reperfusion after prolonged focal ischemia, the level of Hsp70 mRNA apparently increased within this ischemic core (Kawagoe et al. 1992a; Welsh et al. 1992; Kinouchi et al. 1993).

Although relatively few studies have been reported on the induction of various stress proteins other than the Hsp70 family (Hsp70 and HSC70) after global ischemia, the induction of several other stress proteins was reported in a focal cerebral ischemia model (Higashi et al. 1994). Glucose-regulated protein 78 (GRP78), a member of the Hsp70 family located in the endoplasmic reticulum (ER), Hsp27, a cytoplasmic small stress protein, and Hsp47, a collagen-binding stress protein located in the ER (Hirayoshi et al. 1991) are differentially induced after focal cerebral ischemia in rats (Fig. 1). Hsp70 mRNA almost disappeared after 48 h of ischemia and, in contrast, marked accumulation of Hsp27 and Hsp47 mRNA was clearly observed after 48 h (Higashi et al. 1994). In situ hybridization shows different localization of Hsp70 and Hsp27 mRNA after focal ischemia (Higashi et al. 1994). Thus, the distinct induction kinetics of several stress proteins and the distinct localization of induced mRNAs suggest that they have distinct roles during the stress response and may cooperatively contribute to the protection of neuronal cells from stressful insults such as ischemia.

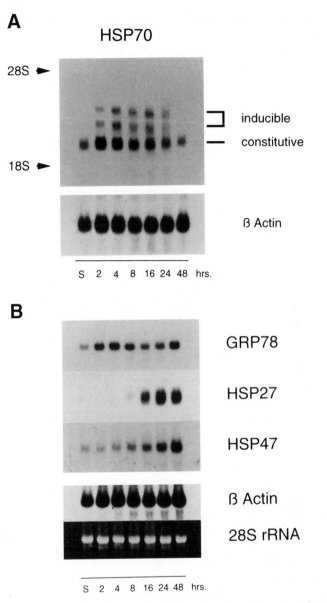

Fig. 1A,B. Time course of mRNA accumulation of various stress proteins after focal cerebral ischemia. Rats were sacrificed at indicated time points after focal ischemic treatment. Total RNA was extracted from ischemic cerebral neocortices of three animals at each time point and examined by Northern blot analysis using ^{32}P-labeled probes for each stress protein. Hybridized bands with β-actin probe and 28S rRNA stained with ethidium bromide were shown as internal controls. *S* indicates sham-operated controls

II. Induction of Hsps by Ischemia/Reperfusion in the Heart
(see also Chap. 12, this volume)

Several laboratories have demonstrated increased expression of Hsp70 mRNA and protein using different experimental systems in response to global or partial ischemia (Dillmann et al. 1986; Currie 1987; Mehta et al. 1988; Knowlton et al. 1991; Das et al. 1993; Myrmel et al. 1994). Dillmann et al. (1986) first demonstrated the induction of Hsp70 by ischemia. In the open chest dog model, ischemia of the heart caused by occluding the left anterior descending coronary artery was reported to induce high level expression of Hsp70 mRNA and protein by two-dimensional gel electrophoresis. In the open chest rabbit heart, 5-min coronary artery occlusion was reported to induce a twofold accumulation of Hsp70 mRNA in the ischemic area after 2-h reperfusion as measured by Northern blot analysis, whereas four repeated cycles of 5-min ischemia and 5-min reperfusion resulted in a threefold increase in Hsp70 mRNA (Knowlton et al. 1991). Moreover, after repeated 5-min global ischemia and 10-min reperfusion, the induction of Hsp27 and Hsp90 mRNA as well as Hsp70 mRNA was shown in isolated and perfused rat hearts by the Langendorff method (Das et al. 1993). We examined the time course of the accumulation of mRNAs for Hsp70 and Hsp90 during ischemia, postischemic reperfusion, or heat shock (Nishizawa et al. 1996). The levels of mRNAs were faint during global ischemia and clearly induced during heat shock or postischemic reperfusion. Although the expression of Hsp70 mRNA during heat shock was greater than the expression during ischemia/reperfusion, Hsp90 mRNA was significantly more strongly induced in ischemia/reperfusion than in heat shock indicating the presence of regulatory mechanisms other than HSF. Using in situ hybridization analysis of isolated rat hearts, Hsp70 mRNA was shown to be localized around the border of the necrotic region after ischemia/reperfusion (Plumier et al. 1996).

E. Regulation of HSF Activation by Hypoxia or Ischemia/Reperfusion

I. HSF Activation by Hypoxia

Benjamin et al. (1990) first reported the induction of binding activity of HSF by hypoxia. With the gel mobility shift assay using a double-stranded HSE oligonucleotide, they demonstrated that transient exposure of cultured mouse myogenic cells to a hypoxic atmosphere stimulates HSE binding activity of HSF through mechanisms that are independent of new protein synthesis. They also showed that the activation of HSF in hypoxic cells is temporally associated with induction of endogenous Hsp70 gene transcription and that induction of the human Hsp70 promoter requires an intact HSE. Thus, they indicated that hypoxia and heat shock induce expression of the Hsp70 gene by similar mechanisms. Mestril et al. (1994) also reported the HSE binding activity of HSF by hypoxia and found that HSF1 binds to HSE during hypoxia as well as during heat shock by supershift assays using specific antisera against HSF1 and HSF2 in the cell line derived from embryonic rat hearts, H9c2. Similarly, HSF1 was reported to be the main component of HSE-binding activity induced by hypoxia in a human carcinoma cell line or in murine cortical astrocytes (MIVECHI et al. 1994; BERGERON et al. 1996). MIVECHI et al. (1994) examined the HSF activation by a variety of stresses in a human lung carcinoma cell line and demonstrated the induction of HSE-binding activity and phosphorylation of HSF1 with increased synthesis of Hsps by heat shock, hypoxia, ethanol, or sodium arsenite. On the other hand, there were no detectable increases in HSF1 phosphorylation after treatment with X-irradiation or canavanine, an amino acid analogue. GIACCIA et al. (1992) reported the kinetics of HSE-binding activity of HSF by hypoxia in normal and tumor cell lines of murine and human origins, and their results suggested that the binding activity of HSF may be a useful marker for monitoring tumor hypoxia.

II. HSF Activation by Ischemia in the Brain

In rat brains, we first examined the activation of HSF during focal cerebral ischemia or after heat shock using spontaneously hypertensive rats (Fig. 2; HIGASHI et al. 1995). When the rats were subjected to whole body hyperthermia, the gel mobility shift assay using HSE-containing synthetic probes revealed increased HSE-binding activity in the brain. Similarly, the ischemic treatment, by occluding the right middle cerebral artery and the right common carotid artery, caused increased HSE-binding activity in the brain extract. During the time course of focal cerebral ischemia, high HSE-binding activity appeared 30–60 min after the ischemic treatment, and then gradually decreased. This induction of HSE-binding activity preceded the accumulation of Hsp70 mRNA which was detected by Northern blot analysis. HSE-binding activity was found to be higher in the extract from the ischemic center than from the peripheral regions 30 min after ischemic treatment, whereas the majority of HSE-binding activity moved to the border zone after 4 h of ischemia. This result is also consistent with the time-dependent location of Hsp70 mRNA during focal cerebral ischemia as described above (HIGASHI et al. 1994). Supershift assays using specific antisera against HSF1 and HSF2 showed that HSF activated by ischemia as well as by hyperthermia, is mainly composed of HSF1 (HIGASHI et al. 1995). The HSE-binding activity after whole body heat shock was higher in the cerebellum than in the cerebral cortex and hippocampus, which was consistent with the amount of HSF1 examined by Western blot analysis using specific antibodies.

III. HSF Activation by Ischemia/Reperfusion in the Heart

In the heart, we also systematically investigated the induction of HSE-binding activity of HSF by ischemia/reperfusion as well as by heat shock (NISHIZAWA et al. 1996). Isolated and perfused rat hearts by the Langendorff method were subjected to 2- to 60-min global ischemia by clamping the aortic cannula

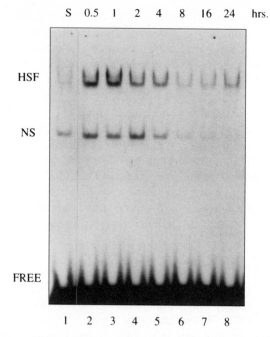

Fig. 2. Time course of HSE-binding activity of HSF during focal ischemia in rat brain. Whole cell extracts from neocortices were obtained at the indicated time points after ischemia (occlusion of the right middle cerebral artery and the right common carotid artery), and were analyzed by gel mobility shift assay using a radiolabeled HSE oligonucleotide. *S* indicates sham-operated controls. The position of the HSF/HSE complex (*HSF*), a non-specific complex (*NS*) and the unbound probe (*Free*) are indicated

followed by 10- to 60-min reperfusion or no reperfusion, and the HSE-binding activities of whole cell extracts were analyzed by gel mobility shift assay (Fig. 3). The activation of HSF during global ischemia began to be detected 3 min after clamping, reached a peak after 6 min and was then attenuated. After 20- or 40-min ischemia, reperfusion induced a burst of activation and the activity continued for more than 60 min. Thus, although the activation of HSF during global ischemia was weak and rapidly attenuated, postischemic reperfusion induced significant activation of HSF. At any time point, the HSE-binding activity was significantly weaker in comparison with that in heat-shocked hearts that were perfused with 42°C buffer. Recently, we demonstrated the HSE-binding activities of HSF in hearts repeatedly submitted to 10-min global ischemia and 10-min reperfusion (NISHIZAWA et al. 1997). The HSE binding activity during repetitive ischemia/reperfusion increased steadily, and after the third ischemia it reached a level equal to or higher than that in the heat-shocked hearts. By supershift assays using specific antisera against HSF1 and HSF2, we demonstrated that HSF1 is the primary component of HSE-binding activity induced by ischemia or reperfusion as well as by

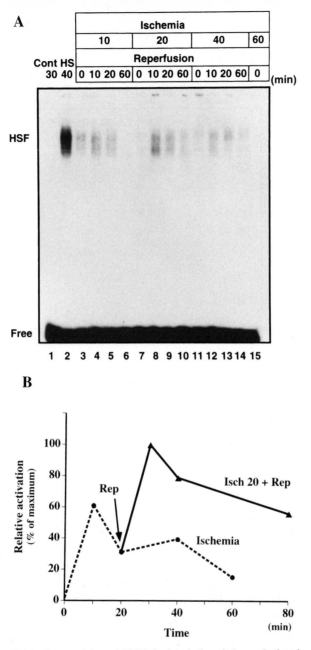

Fig. 3A,B. HSE-binding activity of HSF during ischemia/reperfusion in isolated and perfused rat heart. A gel mobility shift assay was performed with whole cell extracts prepared from control or ischemia-reperfused hearts. Reperfusion was carried out by perfusing with 37°C buffer for the time indicated after global ischemia. **A** Time course of HSE-binding activity during reperfusion after 10 (*lanes 4–6*), 20 (*lanes 8–10*), and 40 min (*lanes 12–14*) of global ischemia. Results for control (*Cont*) or heat-shocked (*HS*) hearts are also shown. **B** Relative HSE-binding activity during ischemia or reperfusion after 20-min ischemia (*Isch 20+Rep*) is shown. The levels of activated HSF in **A** were estimated using a bio-image analyzer and normalized to the level at 10-min reperfusion after 20-min ischemia. The *arrow* indicates the onset of reperfusion (*Rep*)

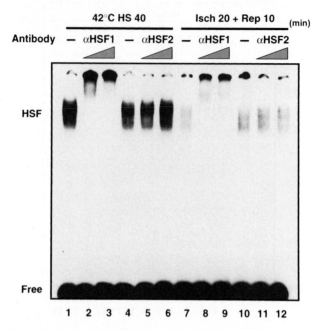

Fig. 4. Specific activation of HSF1 in heat-shocked or ischemia-reperfused heart. Whole cell extracts were prepared from heat-shocked (HS40, 42°C for 40 min) or ischemia-reperfused (20-min ischemia followed by 10-min reperfusion, Isch20+Rep10) hearts. Cell extracts were incubated with anti-HSF1 (*lanes 2, 3, 8 and 9*) or anti-HSF2 (*lanes 5, 6, 11 and 12*) antiserum before the DNA binding reaction, and gel mobility shift assay was performed. Cell extracts without antiserum (*lanes 1, 4, 7 and 10*) were similarly analyzed. In both extracts, supershifts or decreases in mobility of the complex were observed when antiserum against HSF1 was used (*lanes 2, 3, 8 and 9*), although anti-HSF2 antiserum had no effect (*lanes 5, 6, 11 and 12*). The volume of antiserum (diluted 1:10 with PBS) added was 0.5 μl (*lanes 2, 5, 8 and 11*) or 1.0 μl (*lanes 3, 6, 9 and 12*)

heat shock in the heart (Fig. 4; NISHIZAWA et al. 1996). The levels of mRNA for Hsp70 and Hsp90 were examined by Northern blot analysis to determine whether HSF activation resulted in the subsequent transcription of Hsps (NISHIZAWA et al. 1996). Although the induction of Hsp70 mRNA during heat shock was greater than that during ischemia/reperfusion, Hsp90 mRNA was significantly more strongly induced in ischemia/reperfusion than in heat shock. Individual Hsps may also be regulated by additional mechanisms other than HSF.

IV. HSF Activation by Ischemia/Reperfusion in Other Tissues

In the rat liver in vivo model, TACCHINI et al. (1993) examined the HSE-binding activity of HSF during ischemia/reperfusion by occlusion or reperfusion of the hilar pedicle of the left lateral and median lobes. They showed that HSF activity was present at the end of 60 min of ischemia, increased further after 30 min, but disappeared 60 min after reperfusion.

In the kidney, VAN WHY et al. (1994) demonstrated the HSE binding activity of HSF by ischemia/reperfusion using an in vivo rat model. They subjected the kidney to ischemia/reperfusion with a balloon-cuffed vascular occluder placed around the aorta, proximal to the renal artery, and inflated the cuff monitoring femoral artery pressure. HSF binding to HSE initially occurred after 15 min of ischemia, increased after 30 min of reflow after 45-min ischemia, and began to decline after 2 h of reflow. They examined the activation of HSF and cellular ATP levels in the renal cortex during graded vascular occlusion and it was indicated that the intensity of the activity was related to the degree of ischemia.

F. Ischemic Tolerance by Hsps in the Brain and Heart

I. Ischemic Tolerance by Hsps in the Brain

A large number of studies demonstrated a phenomenon of the ischemic tolerance as well as thermotolerance in various experimental systems after mild heat shock or mild ischemia (for review see Chap. 11, this volume, and MORIMOTO et al. 1994). With the previous brief ischemia, significant protection of CA1 neurons after subsequent prolonged ischemia was observed in gerbils (KITAGAWA et al. 1990; KATO et al. 1991; KIRINO et al. 1991) and then in rats (Y. LIU et al. 1992; NISHI et al. 1993). Such phenomena in other regions of the brain have also been reported (KITAGAWA et al. 1991). The induction of ischemic tolerance has been demonstrated after hyperthermia in vivo (CHOPP et al. 1989; KITAGAWA et al. 1991) or after systemic oxidative stress (OHTSUKI et al. 1992). Other stress proteins as well as Hsp70 are also induced simultaneously after the same treatment (HIGASHI et al. 1994). The cellular functions of these stress proteins in vivo remain to be elucidated. In addition to such a stress response, changes in apparent neuronal vulnerability appear to involve complex interactions of other mechanisms.

II. Myocardial Protection Against Ischemia by Hsps

Protective roles against ischemic injury have been the primary concern regarding Hsp for cardiologists and cardiovascular investigators (for review see Chap. 12, this volume, and YELLON and LATCHMAN 1992; MARBER 1994; YELLON and MARBER 1994; PLUMIER and CURRIE 1996; WILLIAMS 1997). The association of heat shock response with enhanced postischemic ventricular recovery was first reported by CURRIE et al. (1988): rats were exposed to 15 min of 42°C hyperthermia and 24 h later the hearts were isolated, perfused and submitted to global ischemia. In the heat-shocked hearts, recovery of contractile force, rate of contraction and rate of relaxation significantly improved, creatine kinase release associated with reperfusion injury decreased, and Hsp70 increased. They also showed a strong relationship between the accumulation of Hsp70 and the postischemic ventricular recovery 0, 24, 48, 96, and 192 h after hyperthermia (KARMAZYN et al. 1990). In focal ischemic rat hearts by occlusion and reperfusion of the left coronary artery, DONNELLY et al. (1992) demonstrated the reduction of the infarct size by prior whole-body hyperthermia. Similar protective effects of pretreatment by hyperthermia or mild ischemia on the functional recovery or the reduction of infarct size after heart ischemia, have been reported in various experimental systems or conditions (X. LIU et al. 1992; YELLON et al. 1992; CURRIE et al. 1993; MARBER et al. 1993; HUTTER et al. 1994; MARBER et al. 1994). Recently, it was clearly demonstrated that postischemic myocardial recoveries, including the size of infarction, contractile function and release of creatine kinase, significantly improved in transgenic mice that overexpressed inducible human or rat Hsp70 (MARBER et al. 1995; PLUMIER et al. 1995; HUTTER et al. 1996; RADFORD et al. 1996). In addition, the protective roles against ischemic insult of Hsps other than Hsp70 have been demonstrated. In both rat neonatal cardiomyocytes and the myogenic H9c2 cell line, cells infected with an adenoviral construct overexpressing both Hsp60 and Hsp10, which are known to form chaperonin complex in the mitochondria, were found to be protective against simulated ischemia (LAU et al. 1997). More recently, the overexpression of the small heat shock proteins Hsp27 and αB-crystalin was also demonstrated to protect against ischemic change in cardiomyocytes (MARTIN et al. 1997).

G. Mechanisms of HSF Activation by Hypoxia or Ischemia/Reperfusion

I. Specific Activation of HSF1 by Hypoxia or Ischemia/Reperfusion

As mentioned above, the primary component of HSF activated by hypoxia or ischemia/reperfusion was demonstrated to be HSF1 in various cells (MESTRIL et al. 1994; MIVECHI et al. 1994; BERGERON et al. 1996) and in the brain (HIGASHI et al. 1995) and heart of rats (NISHIZAWA et al. 1996). HSF1 acquires DNA-binding activity in response to heat shock or other physiological stresses and seems to be the general stress-response factor (for review see Chap. 10 and MORIMOTO et al. 1994). These findings suggest that the regulatory pathway of heat shock gene transcription activated by hypoxia or ischemia/reperfusion is essentially identical to that activated by heat shock.

II. Signals for the Activation of HSF1 by Hypoxia or Ischemia/Reperfusion

During hypoxia/reoxygenation or ischemia/reperfusion, multiple and complex pathological changes occur as mentioned above. These changes include accumulation of intracellular Ca^{2+}, loss of high-energy phosphate esters, membrane damage, mitochondrial failure, altered osmotic control, decreased intracellular pH, and the subsequent production of oxygen free radicals involving the arachidonic acid metabolic pathway. Although the detailed mechanisms remain to be determined, reactive oxygen species, arachidonic acid, and ATP depletion appear to be the major causes of HSF1 activation by hypoxia or ischemia/reperfusion. These changes likely disturb protein metabolism and then produce substrates for the molecular chaperones, resulting in activation of HSF1.

1. ATP Depletion

BENJAMIN et al. (1992) investigated the intracellular signals generated in hypoxic or ischemic cells and indicated that the effects of ATP depletion alone are sufficient to induce the DNA-binding of HSF when oxidative metabolism is impaired. They examined the effects of glucose deprivation and the metabolic inhibitor rotenone on DNA-binding activity of HSF in cultured C2 myogenic cells under normoxic conditions. Severe reductions in ATP concentrations (<30% of control levels) and intracellular pH (7.3 ~ 6.9) resulting from glucose deprivation plus rotenone were associated with HSE-binding activity of HSF. When the intracellular pH was reduced by administration of amiloride and sodium propionate to 6.7 with normal ATP levels, HSE-binding activity of HSF was not induced. In contrast, severe depletion of ATP remained inducing the activation of HSF even at normal pH levels in the presence of high K^+ and nigericin. VAN WHY et al. (1994) examined the activation of HSF and cellular ATP levels in the renal cortex of rats in vivo during graded ischemia and indicated that the intensity of the activity was related to the degree of ischemia indicated by cellular ATP levels.

On the other hand, IWAKI et al. (1993) examined the induction of Hsp70 mRNA and the amount of intracellular ATP during hypoxia/reoxygenation and metabolic stress using cultured neonatal cardiomyocytes. The appearance of Hsp70 mRNA preceded the intracellular ATP depletion caused by hypoxia. Moreover, in our study using isolated rat hearts, although HSF activation was very weak during global heart ischemia, during which ATP was reported to rapidly deplete (STEENBERGEN et al. 1990), prompt and significant activation was induced by postischemic reperfusion, during which the ATP level gradually increased (NISHIZAWA et al. 1996). Considering these findings, ATP depletion is likely not the main stimulus for heat shock response during hypoxia/reoxygenation or ischemia/reperfusion.

2. Reactive Oxygen Species

We showed that the induction of HSE-binding activity of HSF during ischemia was weak, the activity soon attenuated and that postischemic reperfusion significantly induced the binding activity of HSF in isolated and perfused rat hearts (NISHIZAWA et al. 1996). Postischemic reperfusion was reported to enhance the HSF activity induced by ischemia also in livers and kidneys of rats (TACCHINI et al. 1993; VAN WHY et al. 1994). Therefore, it was suggested that the primary signal that activated HSF during ischemia/reperfusion was produced mainly during postischemic reperfusion.

It has been established that a burst of production of reactive oxygen species including hydrogen peroxide (H_2O_2), superoxide radical (O_2^-), and hydroxyl radical (OH) occurs during the early moments of reperfusion of ischemic tissues (KLONER et al. 1989; KUKREJA and HESS 1992; for review see McCORD 1985; MAXWELL and LIP 1997). These reactive oxygen species are derived from a variety of sources, such as the xanthine oxidase system, activated neutrophils, the electron transport chain of mitochondria, and the arachidonic acid pathway.

KUKREJA et al. (1994) studied the accumulation of Hsp70 mRNA during exposure to exogenous ROS and during postischemic reperfusion in isolated rat hearts, and concluded that one of the potential mechanisms of expression of Hsp70 elicited by ischemia/reperfusion may involve oxygen radicals. In addition, induction of Hsp70 mRNA by ischemia/reperfusion was reported to be inhibited by intravenous infusion of recombinant human superoxide dismutase in pig livers (SCHOENIGER et al. 1994).

Recently, we demonstrated that ROS played an important role in the induction of the binding activity of HSF and the accumulation of mRNA for Hsp70 and Hsp90 in ischemia-reperfused hearts (NISHIZAWA et al. 1997). This conclusion is based on several observations in isolated and perfused rat heart. Firstly, we demonstrated the burst activation of HSF1 in hearts submitted to repetitive ischemia/reperfusion, which was reported to cause recurrent bursts of free radical generation. Secondly, this burst activation of HSF1 in the repetitive ischemia-reperfused heart was significantly reduced by treatment with either allopurinol, an inhibitor of xanthine oxidase, or catalase, a scavenger of H_2O_2. Thirdly, significant binding activity of HSF1 was observed upon perfusion with the buffer containing H_2O_2 or xanthine plus xanthine oxidase.

Despite the large number of studies, controversy continues as to the effects of oxidative stresses on the heat shock response. Several studies have shown the induction of Hsps by various oxidative stresses or reactive oxygen species (POLLA et al. 1987; JORNOT et al. 1991; HEUFELDER et al. 1992; LU et al. 1993). However, BRUCE et al. (1993) reported that although H_2O_2 or menadione induced HSE-binding activity of HSF, Hsps were not synthesized in NIH-3T3 cells. On the other hand, it was reported that phorbol esters did not cause HSF activation but induced Hsp synthesis in human monocytes, and that the increased mRNA stabilization was responsible for this induction of Hsp (JACQUIER-SARLIN et al. 1995). Recently, the involvement of the redox mechanism in the heat shock signal transduction pathway has been suggested (HUANG et al. 1994; JACQUIER-SARLIN and POLLA 1996). JACQUIER-SARLIN and POLLA (1996) reported that H_2O_2 exerted a dual effect in the human pre-monocytic cells: it reversibly inhibited DNA-binding activity of HSF as well as induced the binding activity. In addition, they proposed that the time required for thioredoxin induction provided an explanation for the lack of Hsp synthesis upon exposure to ROS, despite the activation of HSF. Thus, complex, multi-step regulation of the stress response to oxidative stress is suggested, and the differences among the studies may be due to the cell specificity, the type of oxidative stress and the subcellular location of the ROS generation (see also, Chap. 13, this volume).

3. Arachidonic Acid and Its Metabolites

Arachidonic acid and its metabolites, such as prostaglandins and thromboxane, are accumulated during ischemia and enhanced by reperfusion (for review see KARMAZYN 1991; VAN DER VUSSE et al. 1994; MAXWELL and LIP 1997). On the other hand, a relationship between heat shock gene expression and the metabolic pathway of arachidonic acid was suggested by a study that demonstrated treatment of human cells with prostaglandins A1, A2, and J2 induced heat shock protein synthesis (OHNO 1988; SANTORO et al. 1989). Moreover, arachidonate or antiproliferative prostaglandins were shown to induce the HSE-binding activity of HSF in human cells (AMICI et al. 1992; JURIVICH et al. 1994). It has also been observed that treatment with anti-inflammatory drugs, such as sodium salicylate (JURIVICH et al. 1992) and indomethacin (LEE et al. 1995), which inhibit arachidonate metabolism, leads to the activation of HSF1 DNA-binding activity. Furthermore, pretreatment with arachidonate (JURIVICH et al. 1994) or indomethacin (LEE et al. 1995) lowers the temperature threshold for induction of HSF1 activation. Although the detailed mechanisms remain to be determined, arachidonic acid metabolism likely contributes to the heat shock response induced by hypoxia or ischemia/reperfusion.

4. Decreased Intracellular pH

The main stimulus for HSF activation seems to be distinct from the decreased intracellular pH. Reduction of pH to nonphysiologic levels (<6.4) was reported to induce DNA-binding activity of HSF in HeLa cells (MOSSER et al. 1990). However, severe acidosis (pH 6.7) in the physiological range did not induce HSF binding activity in cultured myogenic cells (BENJAMIN et al. 1992), and the drop in intracellular pH to 6.8 failed to activate the characteristic puffs of polytene chromosomes seen during heat shock in *Drosophila* (DRUMMOND et al. 1986).

H. Clinical Application and Future Perspective

The tissue protective effect of Hsps from ischemia/reperfusion injury has been demonstrated. Moreover, recent interesting studies indicate that gene transfection with Hsps or pharmacological therapy that induces Hsps may result in new effective therapies although further investigation will be required to establish the general usefulness of these approaches (Maulik et al. 1994; Morris et al. 1996; Suzuki et al. 1997; Vigh et al. 1997). Suzuki et al. (1997) showed that in vivo gene transfection by intracoronary infusion of the hemagglutinating virus of Japan (HVJ)-liposome caused overexpression of Hsp70 in rat hearts, resulting in enhancement of myocardial tolerance to ischemia/reperfusion injury. In addition, a hydroxylamine derivative, Bimoclomol, demonstrated cytoprotection effects together with a lack of toxicity and side effects under several experimental conditions, including ischemia, by increasing synthesis of Hsps (Vigh et al. 1997).

Further research regarding the underlying mechanisms of Hsp induction, less noxious means to induce these proteins effectively, and improved biotechnology for gene regulation or gene delivery are necessary. We believe that further study will result in valuable clinical applications relevant to high-risk cardiovascular surgery, transplantation, or treatments for ischemic diseases such as myocardial infarction or stroke.

References

Abravaya K, Myers MP, Murphy SP, Morimoto RI (1992) The human heat shock protein hsp70 interacts with HSF, the transcription factor that regulates heat shock gene expression. Genes Dev 6:1153–1164

Allen DG, Orchard CH (1987) Myocardial contractile function during ischemia and hypoxia. Circ Res 60:153–168

Amici C, Sistonen L, Santoro MG, Morimoto RI (1992) Antiproliferative prostaglandins activate heat shock transcription factor. Proc Natl Acad Sci USA 89:6227–6231

Ananthan J, Goldberg AL, Voellmy R (1986) Abnormal proteins serve as eukaryotic stress signals and trigger the activation of heat shock genes. Science 232:522–524

Astrup J, Siesjo BK, Symon L (1981) Thresholds in cerebral ischemia – the ischemic penumbra. Stroke 12:723–725

Baler R, Dahl G, Voellmy R (1993) Activation of human heat shock genes is accompanied by oligomerization, modification, and rapid translocation of heat shock transcription factor HSF1. Mol Cell Biol 13:2486–2496

Beckmann RP, Lovett M, Welch WJ (1992) Examining the function and regulation of hsp 70 in cells subjected to metabolic stress. J Cell Biol 117:1137–1150

Benjamin IJ, Kroger B, Williams RS (1990) Activation of the heat shock transcription factor by hypoxia in mammalian cells. Proc Natl Acad Sci USA 87:6263–6267

Benjamin IJ, Horie S, Greenberg ML, Alpern RJ, Williams RS (1992) Induction of stress proteins in cultured myogenic cells. Molecular signals for the activation of heat shock transcription factor during ischemia. J Clin Invest 89:1685–1689

Bergeron M, Mivechi NF, Giaccia AJ, Giffard RG (1996) Mechanism of heat shock protein 72 induction in primary cultured astrocytes after oxygen-glucose deprivation. Neurol Res 18:64–72

Bruce JL, Price BD, Coleman CN, Calderwood SK (1993) Oxidative injury rapidly activates the heat shock transcription factor but fails to increase levels of heat shock proteins. Cancer Res 53:12–15

Chopp M, Chen H, Ho KL, Dereski MO, Brown E, Hetzel FW, Welch KM (1989) Transient hyperthermia protects against subsequent forebrain ischemic cell damage in the rat. Neurology 39:1396–1398

Clos J, Westwood JT, Becker PB, Wilson S, Lambert K, Wu C (1990) Molecular cloning and expression of a hexameric Drosophila heat shock factor subject to negative regulation. Cell 63:1085–1097

Craig EA, Gross CA (1991) Is hsp70 the cellular thermometer? Trends Biochem Sci 16:135–140

Currie RW (1987) Effects of ischemia and perfusion temperature on the synthesis of stress- induced (heat shock) proteins in isolated and perfused rat hearts. J Mol Cell Cardiol 19:795–808

Currie RW, Karmazyn M, Kloc M, Mailer K (1988) Heat-shock response is associated with enhanced postischemic ventricular recovery. Circ Res 63:543–549

Currie RW, Tanguay RM, Kingma J Jr (1993) Heat-shock response and limitation of tissue necrosis during occlusion/reperfusion in rabbit hearts. Circulation 87:963–971

Das DK, Engelman RM, Kimura Y (1993) Molecular adaptation of cellular defences following preconditioning of the heart by repeated ischaemia. Cardiovasc Res 27:578–584

DiDomenico BJ, Bugaisky GE, Lindquist S (1982) The heat shock response is self-regulated at both the transcriptional and posttranscriptional levels. Cell 31:593–603

Dienel GA, Pulsinelli WA, Duffy TE (1980) Regional protein synthesis in rat brain following acute hemispheric ischemia. J Neurochem 35:1216–1226

Dienel GA, Kiessling M, Jacewicz M, Pulsinelli WA (1986) Synthesis of heat shock proteins in rat brain cortex after transient ischemia. J Cereb Blood Flow Metab 6:505–510

Dillmann WH, Mehta HB, Barrieux A, Guth BD, Neeley WE, Ross J Jr (1986) Ischemia of the dog heart induces the appearance of a cardiac mRNA coding for a protein with migration characteristics similar to heat-shock/stress protein 71. Circ Res 59:110–114

Donnelly TJ, Sievers RE, Vissern FL, Welch WJ, Wolfe CL (1992) Heat shock protein induction in rat hearts. A role for improved myocardial salvage after ischemia and reperfusion? Circulation 85:769–778

Drummond IA, McClure SA, Poenie M, Tsien RY, Steinhardt RA (1986) Large changes in intracellular pH and calcium observed during heat shock are not responsible for the induction of heat shock proteins in Drosophila melanogaster. Mol Cell Biol 6:1767–1775

Fawcett TW, Sylvester SL, Sarge KD, Morimoto RI, Holbrook NJ (1994) Effects of neurohormonal stress and aging on the activation of mammalian heat shock factor 1. J Biol Chem 269:32272–32278

Giaccia AJ, Auger EA, Koong A, Terris DJ, Minchinton AI, Hahn GM, Brown JM (1992) Activation of the heat shock transcription factor by hypoxia in normal and tumor cell lines in vivo and in vitro. Int J Radiat Oncol Biol Phys 23:891–897

Goff SA, Goldberg AL (1985) Production of abnormal proteins in E. coli stimulates transcription of lon and other heat shock genes. Cell 41:587–595

GUSTO (1993) An international randomized trial comparing four thrombolytic strategies for acute myocardial infarction. The GUSTO investigators. N Engl J Med 329:673–682

Hearse DJ, Bolli R (1992) Reperfusion induced injury: manifestations, mechanisms, and clinical relevance. Cardiovasc Res 26:101–108

Heufelder AE, Wenzel BE, Bahn RS (1992) Methimazole and propylthiouracil inhibit the oxygen free radical-induced expression of a 72 kilodalton heat shock protein in Graves' retroocular fibroblasts (see comments). J Clin Endocrinol Metab 74:737–742

Higashi T, Takechi H, Uemura Y, Kikuchi H, Nagata K (1994) Differential induction of mRNA species encoding several classes of stress proteins following focal cerebral ischemia in rats. Brain Res 650:239–248

Higashi T, Nakai A, Uemura Y, Kikuchi H, Nagata K (1995) Activation of heat shock factor 1 in rat brain during cerebral ischemia or after heat shock. Mol Brain Res 34:262–270

Hillis LD, Braunwald E (1977) Myocardial ischemia. N Engl J Med 296:971–978, 1034–1041, 1093–1036

Hirayoshi K, Kudo H, Takechi H, Nakai A, Iwamatsu A, Yamada KM, Nagata K (1991) HSP47: a tissue-specific, transformation-sensitive, collagen-binding heat shock protein of chicken embryo fibroblasts. Mol Cell Biol 11:4036–4044

Huang LE, Zhang H, Bae SW, Liu AY (1994) Thiol reducing reagents inhibit the heat shock response. Involvement of a redox mechanism in the heat shock signal transduction pathway. J Biol Chem 269:30718–30725

Hutter JJ, Mestril R, Tam EK, Sievers RE, Dillmann WH, Wolfe CL (1996) Overexpression of heat shock protein 72 in transgenic mice decreases infarct size in vivo. Circulation 94:1408–1411

Hutter MM, Sievers RE, Barbosa V, Wolfe CL (1994) Heat-shock protein induction in rat hearts. A direct correlation between the amount of heat-shock protein induced and the degree of myocardial protection. Circulation 89:355–360

ISIS-2 (1988) Randomised trial of intravenous streptokinase, oral aspirin, both, or neither among 17,187 cases of suspected acute myocardial infarction: ISIS-2. ISIS-2 (Second International Study of Infarct Survival) Collaborative Group. Lancet 2:349–360

Ito U, Spatz M, Walker J Jr, Klatzo I (1975) Experimental cerebral ischemia in mongolian gerbils. I. Light microscopic observations. Acta Neuropathol (Berl) 32:209–223

Iwaki K, Chi SH, Dillmann WH, Mestril R (1993) Induction of HSP70 in cultured rat neonatal cardiomyocytes by hypoxia and metabolic stress. Circulation 87:2023–2032

Jacquier-Sarlin MR, Jornot L, Polla BS (1995) Differential expression and regulation of hsp70 and hsp90 by phorbol esters and heat shock. J Biol Chem 270:14094–14099

Jacquier-Sarlin MR, Polla BS (1996) Dual regulation of heat-shock transcription factor (HSF) activation and DNA-binding activity by H2O2: role of thioredoxin. Biochem J 318:187–193

Jornot L, Mirault ME, Junod AF (1991) Differential expression of hsp70 stress proteins in human endothelial cells exposed to heat shock and hydrogen peroxide. Am J Respir Cell Mol Biol 5:265–275

Jurivich DA, Sistonen L, Kroes RA, Morimoto RI (1992) Effect of sodium salicylate on the human heat shock response. Science 255:1243–1245

Jurivich DA, Sistonen L, Sarge KD, Morimoto RI (1994) Arachidonate is a potent modulator of human heat shock gene transcription. Proc Natl Acad Sci USA 91:2280–2284

Karmazyn M, Mailer K, Currie RW (1990) Acquisition and decay of heat-shock-enhanced postischemic ventricular recovery. Am J Physiol 259:H424–431

Karmazyn M (1991) Ischemic and reperfusion injury in the heart. Cellular mechanisms and pharmacological interventions. Can J Physiol Pharmacol 69:719–730

Kato H, Liu Y, Araki T, Kogure K (1991) Temporal profile of the effects of pretreatment with brief cerebral ischemia on the neuronal damage following secondary ischemic insult in the gerbil: cumulative damage and protective effects. Brain Res 553:238–242

Kawagoe J, Abe K, Sato S, Nagano I, Nakamura S, Kogure K (1992a) Distributions of heat shock protein (HSP) 70 and heat shock cognate protein (HSC) 70 mRNAs after transient focal ischemia in rat brain. Brain Res 587:195–202

Kawagoe J, Abe K, Sato S, Nagano I, Nakamura S, Kogure K (1992b) Distributions of heat shock protein-70 mRNAs and heat shock cognate protein-70 mRNAs after transient global ischemia in gerbil brain. J Cereb Blood Flow Metab 12:794–801

Kinouchi H, Sharp FR, Hill MP, Koistinaho J, Sagar SM, Chan PH (1993) Induction of 70-kDa heat shock protein and hsp70 mRNA following transient focal cerebral ischemia in the rat. J Cereb Blood Flow Metab 13:105–115

Kirino T (1982) Delayed neuronal death in the gerbil hippocampus following ischemia. Brain Res 239:57–69

Kirino T, Tsujita Y, Tamura A (1991) Induced tolerance to ischemia in gerbil hippocampal neurons. J Cereb Blood Flow Metab 11:299–307

Kitagawa K, Matsumoto M, Tagaya M, Hata R, Ueda H, Niinobe M, Handa N, Fukunaga R, Kimura K, Mikoshiba K, et al (1990) "Ischemic tolerance" phenomenon found in the brain. Brain Res 528:21–24

Kitagawa K, Matsumoto M, Kuwabara K, Tagaya M, Ohtsuki T, Hata R, Ueda H, Handa N, Kimura K, Kamada T (1991a) "Ischemic tolerance" phenomenon detected in various brain regions. Brain Res 561:203–211

Kitagawa K, Matsumoto M, Tagaya M, Kuwabara K, Hata R, Handa N, Fukunaga R, Kimura K, Kamada T (1991b) Hyperthermia-induced neuronal protection against ischemic injury in gerbils. J Cereb Blood Flow Metab 11:449–452

Kloner RA, Przyklenk K, Whittaker P (1989) Deleterious effects of oxygen radicals in ischemia/reperfusion. Resolved and unresolved issues. Circulation 80:1115–1127

Knowlton AA, Brecher P, Apstein CS (1991) Rapid expression of heat shock protein in the rabbit after brief cardiac ischemia. J Clin Invest 87:139–147

Kukreja RC, Hess ML (1992) The oxygen free radical system: from equations through membrane-protein interactions to cardiovascular injury and protection. Cardiovasc Res 26:641–655

Kukreja RC, Kontos MC, Loesser KE, Batra SK, Qian YZ, Gbur C Jr, Naseem SA, Jesse RL, Hess ML (1994) Oxidant stress increases heat shock protein 70 mRNA in isolated perfused rat heart. Am J Physiol 267:H2213–H2219

Lau S, Patnaik N, Sayen MR, Mestril R (1997) Simultaneous overexpression of two stress proteins in rat cardiomyocytes and myogenic cells confers protection against ischemia-induced injury. Circulation 96:2287–2294

Lee BS, Chen J, Angelidis C, Jurivich DA, Morimoto RI (1995) Pharmacological modulation of heat shock factor 1 by antiinflammatory drugs results in protection against stress-induced cellular damage. Proc Natl Acad Sci USA 92:7207–7211

Liu X, Engelman RM, Moraru II, Rousou JA, Flack Jd, Deaton DW, Maulik N, Das DK (1992) Heat shock. A new approach for myocardial preservation in cardiac surgery. Circulation 86:II358–II363

Liu Y, Kato H, Nakata N, Kogure K (1992) Protection of rat hippocampus against ischemic neuronal damage by pretreatment with sublethal ischemia. Brain Res 586:121–124

Lu D, Maulik N, Moraru II, Kreutzer DL, Das DK (1993) Molecular adaptation of vascular endothelial cells to oxidative stress. Am J Physiol 264:C715–C722

Marber MS, Latchman DS, Walker JM, Yellon DM (1993) Cardiac stress protein elevation 24 hours after brief ischemia or heat stress is associated with resistance to myocardial infarction. Circulation 88:1264–1272

Marber MS (1994) Stress proteins and myocardial protection. Clin Sci 86:375–381

Marber MS, Walker JM, Latchman DS, Yellon DM (1994) Myocardial protection after whole body heat stress in the rabbit is dependent on metabolic substrate and is related to the amount of the inducible 70-kD heat stress protein. J Clin Invest 93:1087–1094

Marber MS, Mestril R, Chi SH, Sayen MR, Yellon DM, Dillmann WH (1995) Overexpression of the rat inducible 70-kD heat stress protein in a transgenic mouse increases the resistance of the heart to ischemic injury. J Clin Invest 95:1446–1456

Martin JL, Mestril R, Hilal-Dandan R, Brunton LL, Dillmann WH (1997) Small heat shock proteins and protection against ischemic injury in cardiac myocytes (see comments). Circulation 96:4343–4348

Maulik N, Wei Z, Liu X, Engelman RM, Rousou JA, Das DK (1994) Improved postischemic ventricular functional recovery by amphetamine is linked with its ability to induce heat shock. Mol Cell Biochem 137:17–24

Maxwell SR, Lip GY (1997) Reperfusion injury: a review of the pathophysiology, clinical manifestations and therapeutic options. Int J Cardiol 58:95–117

McCord JM (1985) Oxygen-derived free radicals in postischemic tissue injury. N Engl J Med 312:159–163

Mehta HB, Popovich BK, Dillmann WH (1988) Ischemia induces changes in the level of mRNAs coding for stress protein 71 and creatine kinase M. Circ Res 63:512–517

Mestril R, Chi SH, Sayen MR, Dillmann WH (1994) Isolation of a novel inducible rat heat-shock protein (HSP70) gene and its expression during ischaemia/hypoxia and heat shock. Biochem J 298:561–569

Mivechi NF, Koong AC, Giaccia AJ, Hahn GM (1994) Analysis of HSF-1 phosphorylation in A549 cells treated with a variety of stresses. Int J Hyperthermia 10:371–379

Morimoto RI (1993) Cells in stress: transcriptional activation of heat shock genes. Science 259:1409–1410

Morimoto RI, Tissières A, Georgopoulus C (1994) The biology of heat shock proteins and molecular chaperones. Cold Spring Harbor Press, New York

Morris SD, Cumming DV, Latchman DS, Yellon DM (1996) Specific induction of the 70-kD heat stress proteins by the tyrosine kinase inhibitor herbimycin-A protects rat neonatal cardiomyocytes. A new pharmacological route to stress protein expression? J Clin Invest 97:706–712

Mosser DD, Kotzbauer PT, Sarge KD, Morimoto RI (1990) In vitro activation of heat shock transcription factor DNA-binding by calcium and biochemical conditions that affect protein conformation. Proc Natl Acad Sci USA 87:3748–3752

Mosser DD, Duchaine J, Massie B (1993) The DNA-binding activity of the human heat shock transcription factor is regulated in vivo by hsp70. Mol Cell Biol 13:5427–5438

Murphy SP, Gorzowski JJ, Sarge KD, Phillips B (1994) Characterization of constitutive HSF2 DNA-binding activity in mouse embryonal carcinoma cells. Mol Cell Biol 14:5309–5317

Myrmel T, McCully JD, Malikin L, Krukenkamp IB, Levitsky S (1994) Heat-shock protein 70 mRNA is induced by anaerobic metabolism in rat hearts. Circulation 90:II299–II305

Nakai A, Morimoto RI (1993) Characterization of a novel chicken heat shock transcription factor, heat shock factor 3, suggests a new regulatory pathway. Mol Cell Biol 13:1983–1997

Nakai A, Kawazoe Y, Tanabe M, Nagata K, Morimoto RI (1995) The DNA-binding properties of two heat shock factors, HSF1 and HSF3, are induced in the avian erythroblast cell line HD6. Mol Cell Biol 15:5168–5178

Nakai A, Tanabe M, Kawazoe Y, Inazawa J, Morimoto RI, Nagata K (1997) HSF4, a new member of the human heat shock factor family which lacks properties of a transcriptional activator. Mol Cell Biol 17:469–481

Nishi S, Taki W, Uemura Y, Higashi T, Kikuchi H, Kudoh H, Satoh M, Nagata K (1993) Ischemic tolerance due to the induction of HSP70 in a rat ischemic recirculation model. Brain Res 615:281–288

Nishizawa J, Nakai A, Higashi T, Tanabe M, Nomoto S, Matsuda K, Ban T, Nagata K (1996) Reperfusion causes significant activation of heat shock transcription factor 1 in ischemic rat heart. Circulation 94:2185–2192

Nishizawa J, Nakai A, Matsuda K, Ban T, Nagata K (1997) Reactive oxygen species play an important role in the activation of heat shock factor 1 in ischemia-reperfused heart. Circulation 96:I312

Nowak T Jr (1985) Synthesis of a stress protein following transient ischemia in the gerbil. J Neurochem 45:1635–1641

Nowak T Jr (1991) Localization of 70 kDa stress protein mRNA induction in gerbil brain after ischemia. J Cereb Blood Flow Metab 11:432–439

Ohno S (1988) Codon preference is but an illusion created by the construction principle of coding sequences. Proc Natl Acad Sci USA 85:4378–4382

Ohtsuki T, Matsumoto M, Kuwabara K, Kitagawa K, Suzuki K, Taniguchi N, Kamada T (1992) Influence of oxidative stress on induced tolerance to ischemia in gerbil hippocampal neurons. Brain Res 599:246–252

Perisic O, Xiao H, Lis JT (1989) Stable binding of Drosophila heat shock factor to head-to-head and tail-to-tail repeats of a conserved 5 bp recognition unit. Cell 59:797–806

Piper HM (1990) Pathophysiology of severe ischemic myocardial injury. Kluwer Academic, Dordrecht

Plumier JC, Ross BM, Currie RW, Angelidis CE, Kazlaris H, Kollias G, Pagoulatos GN (1995) Transgenic mice expressing the human heat shock protein 70 have improved post-ischemic myocardial recovery. J Clin Invest 95:1854–1860

Plumier JC, Currie RW (1996) Heat shock-induced myocardial protection against ischemic injury: a role for Hsp70? Cell Stress Chaperones 1:13–17

Plumier JC, Robertson HA, Currie RW (1996) Differential accumulation of mRNA for immediate early genes and heat shock genes in heart after ischaemic injury. J Mol Cell Cardiol 28:1251–1260

Polla BS, Healy AM, Wojno WC, Krane SM (1987) Hormone 1 alpha,25-dihydroxyvitamin D3 modulates heat shock response in monocytes. Am J Physiol 252:C640–C649

Pulsinelli WA, Brierley JB, Plum F (1982) Temporal profile of neuronal damage in a model of transient forebrain ischemia. Ann Neurol 11:491–498

Rabindran SK, Giorgi G, Clos J, Wu C (1991) Molecular cloning and expression of a human heat shock factor, HSF1. Proc Natl Acad Sci USA 88:6906–6910

Radford NB, Fina M, Benjamin IJ, Moreadith RW, Graves KH, Zhao P, Gavva S, Wiethoff A, Sherry AD, Malloy CR, Williams RS (1996) Cardioprotective effects of 70-kDa heat shock protein in transgenic mice. Proc Natl Acad Sci USA 93:2339–2342

Reimer KA, Jennings RB (1979) The "wavefront phenomenon" of myocardial ischemic cell death. II. Transmural progression of necrosis within the framework of ischemic bed size (myocardium at risk) and collateral flow. Lab Invest 40:633–644

Reimer KA, Ideker RE (1987) Myocardial ischemia and infarction: anatomic and biochemical substrates for ischemic cell death and ventricular arrhythmias. Hum Pathol 18:462–475

Ross DT, Graham DI (1993) Selective loss and selective sparing of neurons in the thalamic reticular nucleus following human cardiac arrest. J Cereb Blood Flow Metab 13:558–567

Santoro MG, Garaci E, Amici C (1989) Prostaglandins with antiproliferative activity induce the synthesis of a heat shock protein in human cells. Proc Natl Acad Sci USA 86:8407–8411

Sarge KD, Zimarino V, Holm K, Wu C, Morimoto RI (1991) Cloning and characterization of two mouse heat shock factors with distinct inducible and constitutive DNA-binding ability. Genes Dev 5:1902–1911

Sarge KD, Murphy SP, Morimoto RI (1993) Activation of heat shock gene transcription by heat shock factor 1 involves oligomerization, acquisition of DNA-binding activity, and nuclear localization and can occur in the absence of stress (published errata appear in Mol Cell Biol 1993, 13(5):3122–3123 and 1993, 13(6):3838–3839). Mol Cell Biol 13:1392–1407

Sarge KD, Park-Sarge OK, Kirby JD, Mayo KE, Morimoto RI (1994) Expression of heat shock factor 2 in mouse testis: potential role as a regulator of heat-shock protein gene expression during spermatogenesis. Biol Reprod 50:1334–1343

Schoeniger LO, Andreoni KA, Ott GR, Risby TH, Bulkley GB, Udelsman R, Burdick JF, Buchman TG (1994) Induction of heat-shock gene expression in postischemic pig liver depends on superoxide generation. Gastroenterology 106:177–184

Schuetz TJ, Gallo GJ, Sheldon L, Tempst P, Kingston RE (1991) Isolation of a cDNA for HSF2: evidence for two heat shock factor genes in humans. Proc Natl Acad Sci USA 88:6911–6915

Sistonen L, Sarge KD, Phillips B, Abravaya K, Morimoto RI (1992) Activation of heat shock factor 2 during hemin-induced differentiation of human erythroleukemia cells. Mol Cell Biol 12:4104–4111

Sistonen L, Sarge KD, Morimoto RI (1994) Human heat shock factors 1 and 2 are differentially activated and can synergistically induce hsp70 gene transcription. Mol Cell Biol 14:2087–2099

Smith ML, Auer RN, Siesjo BK (1984) The density and distribution of ischemic brain injury in the rat following 2–10 min of forebrain ischemia. Acta Neuropathol (Berl) 64:319–332

Steenbergen C, Murphy E, Watts JA, London RE (1990) Correlation between cytosolic free calcium, contracture, ATP, and irreversible ischemic injury in perfused rat heart. Circ Res 66:135–146

Suzuki K, Sawa Y, Kaneda Y, Ichikawa H, Shirakura R, Matsuda H (1997) In vivo gene transfection with heat shock protein 70 enhances myocardial tolerance to ischemia-reperfusion injury in rat. J Clin Invest 99:1645–1650

Tacchini L, Schiaffonati L, Pappalardo C, Gatti S, Bernelli-Zazzera A (1993) Expression of HSP 70, immediate-early response and heme oxygenase genes in ischemic-reperfused rat liver. Lab Invest 68:465–471

Tanabe M, Nakai A, Kawazoe Y, Nagata K (1997) Different thresholds in the responses of two heat shock transcription factors, HSF1 and HSF3. J Biol Chem 272:15389–15395

Thilmann R, Xie Y, Kleihues P, Kiessling M (1986) Persistent inhibition of protein synthesis precedes delayed neuronal death in postischemic gerbil hippocampus. Acta Neuropathol (Berl) 71:88–93

van der Vusse GJ, van Bilsen M, Reneman RS (1994) Ischemia and reperfusion induced alterations in membrane phospholipids: an overview. Ann NY Acad Sci 723:1–14

Van Why SK, Mann AS, Thulin G, Zhu XH, Kashgarian M, Siegel NJ (1994) Activation of heat-shock transcription factor by graded reductions in renal ATP, in vivo, in the rat. J Clin Invest 94:1518–1523

Vass K, Welch WJ, Nowak T Jr (1988) Localization of 70-kDa stress protein induction in gerbil brain after ischemia. Acta Neuropathol (Berl) 77:128–135

Vigh L, Literati PN, Horvath I, Torok Z, Balogh G, Glatz A, Kovacs E, Boros I, Ferdinandy P, Farkas B, Jaszlits L, Jednakovits A, Koranyi L, Maresca B (1997) Bimoclomol: a nontoxic, hydroxylamine derivative with stress protein-inducing activity and cytoprotective effects. Nat Med 3:1150–1154

Welsh FA, Moyer DJ, Harris VA (1992) Regional expression of heat shock protein-70 mRNA and c-fos mRNA following focal ischemia in rat brain. J Cereb Blood Flow Metab 12:204–212

Wiederrecht G, Seto D, Parker CS (1988) Isolation of the gene encoding the S. cerevisiae heat shock transcription factor. Cell 54:841–853

Williams RS (1997) Heat shock proteins and ischemic injury to the myocardium (editorial; comment). Circulation 96:4138–4140

Yellon DM, Latchman DS (1992) Stress proteins and myocardial protection. J Mol Cell Cardiol 24:113–124

Yellon DM, Pasini E, Cargnoni A, Marber MS, Latchman DS, Ferrari R (1992) The protective role of heat stress in the ischaemic and reperfused rabbit myocardium. J Mol Cell Cardiol 24:895–907

Yellon DM, Marber MS (1994) Hsp70 in myocardial ischaemia. Experientia 50:1075–1084

CHAPTER 10
Autoregulation of the Heat Shock Response

Y. Shi and R.I. Morimoto

A. Introduction

The heat shock response provides a homeostatic mechanism that enables cells to survive exposure to extreme environmental stress. The elevated synthesis of heat shock proteins and the biochemical properties of molecular chaperones is essential to prevent nascent polypeptides from premature non-productive interactions and to protect non-native proteins from misfolding and aggregation. Under conditions of normal cell growth, heat shock proteins are also essential for protein synthesis, protein folding and assembly, and protein degradation. Adaptation to stress requires the rapid, yet transient, inducible transcription of heat shock genes. The kinetics of induction and the magnitude of the heat shock response is proportional to the nature and duration of the stress (DiDomenico et al. 1982a,b; Mosser et al. 1988; Straus et al. 1990; Abravaya et al. 1991). Although overexpression of certain heat shock proteins prevents the appearance of protein aggregates and is cytoprotective to diverse forms of stress, chronic overproduction of heat shock proteins such as Hsp70 is deleterious for cell growth at normal growth temperatures (Bahl et al. 1987; Feder et al. 1992). The heat shock response, therefore, ensures that a critical balance is maintained between the levels of non-native proteins and chaperones as a sensor of stress and the regulated and inducible transcription of heat shock genes as the adaptation to stress.

The activation of heat shock genes in response to elevated temperatures and other environmental stresses has been studied extensively as a paradigm for inducible gene expression. In prokaryotes and eukaryotes, genetic and biochemical evidence supports a role for heat shock proteins in a feedback regulatory loop mediated by interactions between heat shock transcription factors (HSF1 in eukaryotes and $\sigma 32$ in prokaryotes) and heat shock proteins. This review will discuss the regulation of the heat shock transcription in *E. coli*, yeast, *Drosophila*, and vertebrates with an emphasis on the interplay between HSF/$\sigma 32$ and heat shock proteins in the normal and stressed cell (see also Chap. 3, this volume).

B. Regulation of the Heat Shock Response in Eukaryotes

I. Overview

In higher eukaryotes, stress-induced regulation of the heat shock response occurs principally by activation of HSFs from an inert non-DNA binding form to the transcriptionally competent, DNA binding state (MORIMOTO et al. 1990; LIS and WU 1993; MORIMOTO 1993; WU 1995). Among eukaryotes, the complexity of HSF regulation varies among species with yeast and *Drosophila* encoding a single HSF gene (WIEDERRECHT et al. 1988; SORGER and PELLHAM 1988; CLOS et al. 1990), whereas four HSF genes (HSF1–4) have been isolated and characterized in the mouse, chicken, and human genomes (RABINDRAN et al. 1991; SARGE et al. 1991; SCHUETZ et al. 1991; NAKAI and MORIMOTO 1993; NAKAI et al. 1997). Comparison of the various cloned HSF genes reveals an overall sequence identity of 40% with a high degree of conservation in the DNA binding domain containing a winged helix-turn-helix motif (HARRISON et al. 1994; MORIMOTO et al. 1994; VUISTER et al. 1994; WU et al. 1994, 1995), an extended hydrophobic repeat (HR-A/B) involved in trimerization (SORGER and NELSON 1989; CLOS et al. 1990; PETERANDERL and NELSON 1992), and a carboxyl-terminal localized transactivation domain (GREEN et al. 1995; SHI et al. 1995; ZUO et al. 1995; WISNIEWSKI et al. 1996; NAKAI et al. 1997). With the exception of the HSF in budding yeasts and human HSF4, another hydrophobic repeat (HR-C) is located adjacent to the transactivation domain, which has been suggested to function in suppression of trimer formation by interaction with HR-A/B (NAKAI and MORIMOTO 1993; RABINDRAN et al. 1993; ZUO et al. 1994; OROSZ et al. 1996; ZANDI et al. 1997; FARKAS et al. 1998). Other features unique to specific HSFs are the presence of an amino terminal transactivation domain in the *S. cerevisiae* HSF and the lack of a functional transactivation domain in human HSF4 (NAKAI et al. 1997). A common feature of all HSFs is the negative regulation of their activity, either DNA binding or transcriptional competence.

Of the HSFs co-expressed in vertebrates, HSF1 has the properties of the principal stress-induced transcriptional activator that acquires both DNA binding and transcriptional activity and is functionally homologous to yeast and *Drosophila* HSF (BALER et al. 1993; SARGE et al. 1993). In avian cells, HSF1 and HSF3 are coexpressed and both factors are activated by chemical and physiological stress (NAKAI et al. 1995; TANABE et al. 1997). The initial observations suggested that HSF3 was a redundant stress activator; however, cells deficient for HSF3, yet expressing HSF1, were severely compromised for the transcriptional activation of the endogenous heat shock genes (TANABE et al., 1998). Reintroduction of a normal HSF3 gene into HSF3 null cells restored a complete heat shock response revealing that HSF3 is essential for the heat shock response. HSF3 is also activated in a non-heat shock dependent manner by co-expression with the oncogene myb and is dependent on direct interaction between the DNA binding domains of HSF3 and Myb (KANEI-ISHII et al.

1997). As Myb functions as a growth regulated transcription factor, the interaction between Myb and HSF3 reveals a genetic crosstalk between cell growth and the stress response.

Additional evidence for a role of HSFs in cell growth and differentiation follows from evidence that HSF2 acquires DNA-binding activity during hemin treatment of human K562 cells (THEODORAKIS et al. 1989; SISTONEN et al. 1992, 1994), during murine spermatocyte differentiation (SARGE et al. 1994), and during murine embryogenesis (MEZGER et al. 1994; RALLU et al. 1997). HSF2 as for other HSFs is ubiquitously expressed and negatively regulated. Its DNA binding activity can be induced in tissue cultured cells by inhibitors of the ubiquitin-dependent proteasome (MATHEW et al., 1998). As the ubiquitin-dependent proteasome degrades short-lived and damaged proteins, these observations reveal that HSFs, by the activation of HSF2, provide the cell with a stress-regulated response to ensure degradation of misfolded proteins as a complement to the role of HSF1 which ensures that stress-induced non-native proteins do not misfold and aggregate.

Induction of HSF1 is a multi-step process distinguished by events of activation and attenuation. Upon heat shock, HSF1 translocates to the nucleus, oligomerizes to a DNA binding and transcriptionally inert state, undergoes inducible phosphorylation and activates the transcription of heat shock genes. During prolonged heat shock, the response attenuates as characterized by transcriptional repression through association of HSF1 with the molecular chaperones Hsp70 and Hdj1, dissociation of HSF1 trimers from DNA and refolding to the inert monomer. Translocation of HSF1 from the cytoplasm to the nucleus has been described in *Drosophila* and mammalian cells (LARSON et al. 1988; MOSSER et al. 1990; ZIMARINO et al. 1990; WESTWOOD et al. 1991; BALER et al. 1993; SARGE et al. 1993; SISTONEN et al. 1994; WU et al. 1994, 1995), yet others have observed that HSF1 can also be constitutively nuclear localized (WESTWOOD et al. 1991; WU et al. 1994). In part some of these differences may be cell-type specific due to differences in the recognition properties of anti-HSF1 monoclonal and polyclonal antisera (MORIMOTO et al. 1994; WU et al. 1994, 1995). Maintenance of HSF1 in a repressed state is a delicate matter and easily disrupted; for example, overexpression of HSF1 by transient transfection results in constitutively active trimers, and mutation of critical phosphoserines (S303 A/S308 A) causes constitutive derepression and appearance of active trimers (KNAUF et al. 1996; KLINE and MORIMOTO 1997). Acquisition of HSF1 DNA-binding activity can be uncoupled from transcriptional activity as demonstrated by the anti-inflammatory drugs sodium salicylate, indomethacin or ibuprofen. These drugs induce the appearance of HSF1 trimers which bind to DNA, yet lack inducible phosphorylation and are transcriptionally inert (JURIVICH et al. 1992; GIARDINA and LIS 1995; COTTO et al. 1996). Exposure of salicylate treated cells to heat shock, in the presence of cycloheximide to prevent de novo HSF1 synthesis, results in the appearance of the transcriptionally active state of HSF1 (COTTO et al. 1996).

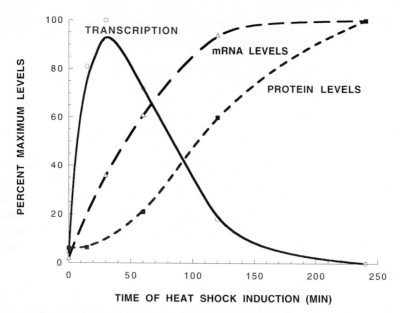

Fig. 1. Kinetics of the heat shock response. Comparison of the transcriptional rate of Hsp70 genes (*indicated by open circles*), the mRNA levels of Hsp70 (*open triangles*), and the synthesis and accumulation of Hsp70 proteins (*solid squares*) during a 4 h heat shock treatment of HeLa cells at 42°C. The percent maximum levels of each parameter are plotted against the time of heat shock treatment. The transcriptional rate was measured by nuclear run-on analysis; the mRNA levels were analyzed by S1 nuclease protection; Hsp70 protein levels were determined by western blot analysis with Hsp70 specific antibody

Among the key events that define attenuation of the heat shock response is the repression of inducible gene transcription during continued exposure to heat shock. Exposure of human cells to a 42°C heat shock results in the rapid and transient inducible transcription of heat shock genes with maximal rates of Hsp70 and Hsp90 transcription occurring within 30–60 min; thereafter the rate of heat shock gene transcription declines rapidly prior to the loss of HSF1 DNA binding activity (Abravaya et al. 1991; Kline and Morimoto 1997). During heat shock, the levels of heat shock messenger RNAs increase and are maintained at high levels due to the effects of heat shock on message stability (Mosser et al. 1988). This results in the elevated synthesis and accumulation of heat shock proteins (Fig. 1). There is a direct correlation between the arrest of heat shock gene transcription and the accumulation of heat shock proteins. Furthermore, during heat shock at 43°C, neither heat shock gene transcription attenuates nor does the synthesis of heat shock proteins occur.

II. Biochemical Study of Autoregulation in Higher Eukaryotes

Support for a role of molecular chaperones in the regulation of the heat shock response was obtained by biochemical studies which showed that HSF1 trimers induced during heat shock were associated with Hsp70 (ABRAVAYA et al. 1992; BALER et al. 1992; RABINDRAN et al. 1994; NUNES and CALDERWOOD 1995; SHI et al. 1998). The Hsp70-HSF complexes are ATP-sensitive, typical of chaperone-substrate interactions and can be reconstituted in vitro (ABRAVAYA et al. 1992; SHI et al. 1998). Through the use of HSF1 deletion mutants and direct in vitro binding assays, a site for Hsp70 binding was mapped to the transactivation domain of HSF1 (SHI et al. 1998). The consequence of interaction between Hsp70 and HSF1 was assessed using a stably transfected cell line conditionally expressing human Hsp70 under the control of a tetracycline-regulated promoter. Overexpression of Hsp70 selectively inhibited the induction of heat shock gene transcription with little or no effect on the formation of HSF1 trimers or on the inducible phosphorylation of HSF1 (SHI et al. 1998). These results argue strongly that Hsp70 is the negative regulator of HSF1 activity and that the repression of heat shock gene transcription which occurs during attenuation is due to the repressive effects of Hsp70 binding to the HSF1 transactivation domain. These results are supported by other studies where overexpression of Hsp70 in rat and *Drosophila* cells did not interfere with the activation of HSF1 DNA binding activity (RABINDRAN et al. 1994).

Identification of the transactivation domain of HSF1 as a chaperone binding site offers a number of intriguing possibilities including the role of Hsp70 as a competitor for HSF1 interaction with the basal transcriptional machinery (MASON and LIS 1997) or the role of Hsp70 in the conformational change of HSF1. As shown schematically, the consequence of Hsp70 binding is that the transactivation domain is rendered inaccessible to the transcriptional machinery, thus resulting in the transcriptional repression (Fig. 2). A related role for chaperones as regulators of transcriptional activators has also been described for the family of steroid aporeceptors, although in this case multiple chaperones are recruited to maintain the activator in a repressed state by formation of a stable chaperone-substrate complex (BOHEN and YAMAMOTO 1994). It is tempting to consider that the interactions observed between p53 and Hsp70 (Hsc70) could also reflect a form of chaperone-dependent regulation of a transcriptional activator (HUPP et al. 1992).

Although much of the emphasis on the role of heat shock proteins in the regulation of the heat shock response has centered on Hsp70, other molecular chaperones such as Hdj1 may also be important. Hdj1 interacts with HSF1 in higher eukaryotes and negatively regulates HSF1 transcriptional activity (SHI et al. 1998). A role for members of the DnaJ family in regulation of the heat shock response is supported by the observation in *S. cerevisiae* that the DnaJ homologue, SIS1, negatively regulates its own expression. However, SIS1 autoregulation requires the heat shock element (HSE) and other sequences, suggesting that additional regulatory molecules might be involved (ZHONG et al. 1996). HSF may also associate with Hsp90; however, this result seems variable as such associations have been detected for yeast, rat and rabbit Hsp90 (NADEAU et al. 1993; NAIR et al. 1996) but not with human Hsp90 (BALER et al. 1992; RABINDRAN et al. 1994; SHI et al. 1998).

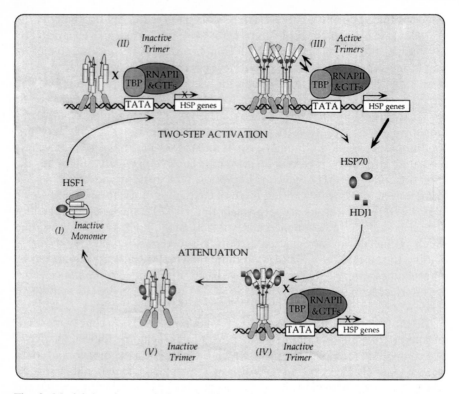

Fig. 2. Model for the regulation of eukaryotic heat shock response. Activation of heat shock transcription factor 1 (*HSF1*) involves conversion of HSF1 from a pre-existing inactive monomer (*I*) to a DNA binding yet transcriptionally incompetent trimer (*II*). Further activation of HSF1 (*III*) leads to the hyperphosphorylation of HSF1 and the transcription of heat shock genes. Synthesis and accumulation of heat shock proteins, in turn, results in the attenuation of the heat shock response. Attenuation is a two-step process, involves the association of Hsp70 and Hdj1 with the activation domain of HSF1 (*IV*) to make HSF1 transcriptionally inactive (*IV, V*), and subsequent conversion of HSF1 trimer to a monomer. The inability of HSF1 to interact with the basal transcriptional machinery and to transcribe heat shock genes is indicated by "*X*", while the interaction of HSF1 with TBP and RNAPII is *represented by double arrows*

III. Genetic Evidence for Autoregulation of the Heat Shock Response in Yeast and *Drosophila*

The initial evidence to indicate that the heat shock response is autoregulated was based on exposure of *Drosophila* cells to amino acid analogues. Rather than the expected transient stress response which occurs in response to heat shock, treatment with azetidine (a proline analogue) resulted in the continuous activation of heat shock gene expression (DiDomenico et al. 1982b). Amino acid analogues induce the heat shock response by their incorporation into nascent polypeptides which misfold. Amino acid analogue-containing nascent polypeptides associate stably with Hsp70 (BECKMANN et al. 1990) and their sequestration of Hsp70 is thought to result in activation of heat shock gene transcription. However, unlike the heat shock response where attenuation is linked to the de novo synthesis of heat shock proteins, the Hsp70 synthesized during amino acid analogue-induced stress is itself misfolded and therefore non-functional, consequently the heat shock response does not attenuate. These observations have led to the suggestion that functional heat shock proteins are required for autoregulation of the heat shock response.

Genetic evidence to support the autoregulation of the heat shock response in yeast has shown that overexpression of Ssa1p (cytosolic Hsp70) dampens the heat shock response from both the ssa1 and ssa4 promoters (STONE and CRAIG 1990). Likewise, deletion of ssa1ssa2 resulted in an unusually high level of expression of Ssa3p and other heat shock proteins which was mediated by activation of yHSF; these results strongly implicated HSF as a potential target for autoregulation by members of the Hsp70 family (BOORSTEIN and CRAIG 1990). The relationship between yHSF and Hsp70 was further supported by a search for extragenic suppressors of the temperature-sensitive phenotype of an ssa1ssa2 strain which also uncovered HSF as an interactive component of the regulatory response (NELSON et al. 1992). A spontaneous mutant EXA3 which could reverse the growth defect of ssa1ssa2 is very closely linked to the gene encoding HSF, and another mutation identified in the genetic screen maps to the HSF gene (HALLADAY and CRAIG 1995).

C. Regulation of the Heat Shock Response in Prokaryotes

I. Overview

The heat shock response in *E. coli* is under the control of σ32, the product of the rpoH (htpR) gene. σ32 associates with the RNAP core (E) to form the E σ32 holoenzyme which recognizes promoters of most heat shock genes (GROSSMAN et al. 1984; COWING et al. 1985; BLOOM et al. 1986). A second heat shock factor in *E. coli* is σE (σ24) which constitutes a complementary stress response that responds to misfolded proteins in the periplasm and outer membrane (MECSAS et al. 1993; ROUVIERE and GROSS 1996). Activation of σE leads to the induction of at least ten different proteins, four of which have been definitely identified: the periplasmic protease DegP, the other heat shock σ factor, σ32, the periplasmic peptidyl prolyl isomerase FkpA, and σE itself (ERICKSON et al. 1987; LIPINSKA et al. 1988; ERICKSON and GROSS 1989; WANG and KAGUNI 1989; RAINA et al. 1995; ROUVIERE et al. 1995; DANESE and SILHAVY 1997). Cells lacking σE are sensitive to elevated temperatures, sodium dodecyl sulfate (SDS)/ethylenediaminetetraacetic acid (EDTA), and crystal violet (Hiratsu et al. 1995; Raina et al. 1995; Rouviere et al. 1995). The temperature-sensitive phenotype could be restored by activation of a second signal transduction cascade, the Cpx pathway, revealing that *E. coli* has at least two partially overlapping stress signal cascades capable of relieving extracytoplasmic stress (Connolly et al. 1997).

Transcriptional activation of the principal heat shock genes htpG, dnaK, dnaJ, grpE, and groEL/S requires stress-induced changes in the levels of $\sigma32$. Under normal growth conditions the concentration of $\sigma32$ is very low, approximately 10–30 copies per cell at 30°C (Craig and Gross 1991) as a result of its short half life ($t_{1/2}=1$ min; Grossman et al. 1987; Tilly et al. 1989). Following heat shock, $\sigma32$ levels increase rapidly, in part due to the elevated synthesis of $\sigma32$ (Grossman et al. 1987; Straus et al. 1987; Kamath-Loeb and Gross 1991; Nagai et al. 1991) and its increased stability during heat shock. The consequence of elevated levels of $\sigma32$ is a burst of heat shock gene transcription (Lesley et al. 1987; Skelly et al. 1987; Straus et al. 1987).

II. Genetic Evidence for Autoregulation of the *E. coli* Heat Shock Response

Mutations in the heat shock proteins DnaK, DnaJ,GrpE and under-expression of GroEL/S result in the enhanced expression of heat shock genes in cells at normal growth temperatures and lead to an extended heat shock response after shift to high temperatures (Tilly et al. 1983; Straus et al. 1990; Kanemori et al. 1994). These cells are defective in the control of $\sigma32$ synthesis and exhibit increased stability of $\sigma32$ (Grossman et al. 1987; Tilly et al. 1989; Straus et al. 1990). The molecular chaperones DnaK, DnaJ and GrpE have been shown to be involved in $\sigma32$ degradation by targeting $\sigma32$ to several ATP dependent proteases FtsH, Lon, Clp, and HslVU (Tomoyasu et al. 1995; Kanemori et al. 1997). These proteases could synergistically affect the in vivo turnover of $\sigma32$ (Kanemori et al. 1997). The negative control of $\sigma32$ synthesis and stability is mediated by the interaction of region C of $\sigma32$ with the DnaK chaperone machinery. Deletion or mutation of region C spanning residues 122–144 of $\sigma32$ leads to sustained synthesis and prolonged half-life of $\sigma32$ (Nagai et al. 1994). The activity of $\sigma32$ is also subjected to negative regulation by heat shock proteins. This is supported by the observation that temperature downshift from 42°C to 30°C leads to a strong and rapid repression of heat shock gene transcription in *E. coli* cells independent of $\sigma32$ protein levels. In addition, the induction of heat shock proteins following overproduction of $\sigma32$ from a multicopy plasmid is only transient, despite the remaining elevated levels of $\sigma32$ (Tilly et al. 1983; Straus et al. 1989). These findings together indicate that heat shock gene expression is controlled by heat shock proteins modulating the synthesis, degradation and activity of $\sigma32$ (Gross et al. 1990; Bukau 1993; Yura et al. 1993; Georgopoulos et al. 1994; Yura 1996).

III. Biochemical Studies on Autoregulation of the *E. coli* Heat Shock Response

Consistent with the genetic evidence that DnaK, DnaJ and GrpE are directly implicated in the regulation of σ32 was the demonstration of specific interactions of these proteins with components of the σ32-dependent transcriptional machinery. Preparations of purified *E. coli* RNA polymerase holoenzyme exhibited cross-reactivity with an antibody against DnaK (SKELLY et al. 1988). This result was clarified by the demonstration that the DnaK chaperone machine (DnaK, DnaJ, GrpE) associated with σ32 (LIBEREK et al. 1992; GAMER et al. 1992). Features of the DnaK-σ32 complex were further characterized by gel filtration and glycerol gradient analysis (LIBEREK et al. 1992) and co-immunoprecipitation assay (LIBEREK and GEORGOPOULOS 1993).

The interaction of DnaK, DnaJ, and GrpE with σ32 is ATP dependent. DnaJ increases the efficiency of DnaK binding to σ32 in the presence of ATP and leads to the formation of DnaK-DnaJ-σ32 complexes containing ADP (LIBEREK and GEORGOPOULOS 1993; GAMER et al. 1996). GrpE functions to increase the rate of nucleotide release, thereby allowing subsequent ATP binding and complex dissociation (GAMER et al. 1996). The equilibrium between free active σ32 and DnaK/DnaJ-bound inactive σ32 constitutes an important element for homeostatic control of heat shock gene regulation. The equilibrium could be shifted to active σ32 by sequestration of the chaperones with heat-denatured proteins (GAMER et al. 1996). In addition, degradation of σ32 could also be counteracted by competition of chaperones with a segment of the lambda phage cIII, an inducer of the heat shock response (BAHL et al. 1987).

The DnaK binding sites within σ32 were established by screening a peptide library and revealed two high affinity binding sites located central and peripheral to the regulatory region C of σ32. These results are consistent with the genetic studies in which mutation of region C leads to the defective regulation of σ32 synthesis and stability (McCARTY et al. 1996). The identification of high affinity binding sites for DnaK within region C provides a basis for further understanding how DnaK regulates σ32 activity and protein levels. The interaction of DnaK and DnaJ with σ32 prevents σ32 from binding to RNAP, thus resulting in the arrest of heat shock gene transcription (LIBEREK et al. 1992; LIBEREK and GEORGOPOULOS 1993). In addition to the sequestration of σ32 away from RNAP, DnaK chaperone machinery autoregulates the *E. coli* heat shock response by simultaneous reactivation of heat-aggregated σ70 and switching σ70 and σ32 assembly with RNAP (BLASZCZAK et al. 1995). A schematic model for the regulation of *E. coli* heat shock response is presented in Fig. 3. In this model, aspects of DnaK chaperone machine mediated autoregulation: modulation of σ32 activity, presentation of σ32 to the proteases, and simultaneous reactivation of heat-aggregated σ70 to compete RNAP core from binding to σ32 are incorporated.

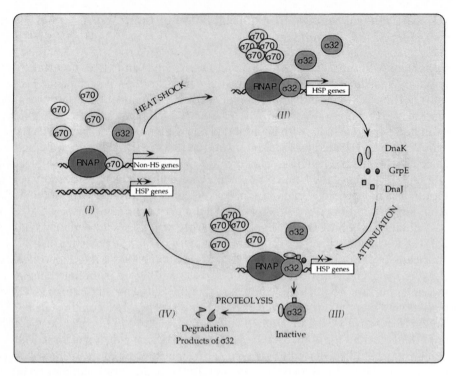

Fig. 3. Model for the regulation of the *E. coli* heat shock response. The activation of heat shock gene expressions in *E. coli* depends on the increased intracellular concentration of σ32 and the binding of σ32 to the RNAP core in place of σ70, which is heat-aggregated and inactivated (*I, II*). Association of σ32 with the RNAP core directs the RNAP to the upstream regions of heat shock genes and leads to the synthesis of heat shock proteins. Accumulated DnaK/DnaJ proteins bind to σ32, prevent its association with the RNAP core and target the σ32 to the proteolytic degradation pathway. Simultaneously, DnaK chaperone machinery disaggregates and reactivates σ70, which in turn competes the RNAP core back from σ32 and attenuates the heat shock response (*III, IV*)

D. Common Features of the Prokaryotic and Eukaryotic Heat Shock Response

The synthesis of heat shock proteins occurs rapidly, yet transiently, upon temperature upshift in cells of all organisms. It has long been speculated from studies in *Drosophila* and yeast that heat shock proteins function in an auotregulatory loop to modulate the intensity and duration of the heat shock response (Craig and Gross 1991). Indeed, this feature appears to be evolutionarily conserved as indicated by the role of heat shock proteins in the regulation of σ32 activity (Bukau 1993; Georgopoulos et al. 1994; Yura 1996). As described here, the regulatory pathway for the heat shock response, whether in *E. coli* or humans, involves inducible transcriptional activation by heat shock factors and negative regulation by heat shock proteins. The heat shock induced conversion of monomeric inactive HSF1 to a DNA-binding and transcriptional competent trimer leads to the interaction of HSF1 with basal transcriptional machinery, and subsequent activation of heat shock gene transcription. This is analogous to the consequence of association of σ32 to the RNAP core in *E. coli*. Components of the Hsp70 chaperone machine interact directly with HSF and σ32 respectively to control the levels or activity of the activators. The generality of regulatory strategies employed by the σ32 regulon and the eukaryotic heat shock genes and the well characterized regulatory mechanism in *E. coli* might help us to better understand the autoregulatory mechanism of the heat shock response in eukaryotes. The comparison from *E. coli* to human brings up an intriguing question: does eukaryotic Hsp70 dampen heat shock gene transcriptions by interfering with HSF binding to RNA polymerase II containing basal transcription machinery as DnaK controls σ32 activity in *E. coli*? Further understanding the mode and consequences of the interplay between molecular chaperones and a global regulatory factor HSF will shed more light on the mechanism of attenuation of the heat shock response.

References

Abravaya K, Phillips B, Morimoto RI (1991) Attenuation of the heat shock response in HeLa cells is mediated by the release of bound heat shock transcription factor and is modulated by changes in growth and in heat shock temperatures. Genes Dev 5:2117–2127

Abravaya K, Myers MP, Murphy SP, Morimoto RI (1992) The human heat shock protein Hsp70 interacts with HSF, the transcription factor that regulates heat shock gene expression. Genes Dev 6:1153–1164

Bahl H, Echols H, Straus DB, Court D, Crowl R, Georgopoulos CP (1987) Induction of the heat shock response of E. coli through stabilization of sigma 32 by the phage lambda cIII protein. Genes Dev 1:57–64

Baler R, Welch WJ, Voellmy R (1992) Heat shock gene regulation by nascent polypeptides and denatured proteins: Hsp70 as a potential autoregulatory factor. J Cell Biol 117:1151–1159

Baler R, Dahl G, Voellmy R (1993) Activation of human heat shock genes is accompanied by oligomerization, modification, and rapid translocation of heat shock transcription factor HSF1. Mol Cell Biol 13:2486–2496

Beckmann RP, Mizzen LA, Welch WJ (1990) Interaction of Hsp 70 with newly synthesized proteins: implications for protein folding and assembly. Science 248:850–854

Blaszczak A, Zylicz M, Georgopoulos C, Liberek K (1995) Both ambient temperature and the DnaK chaperone machine modulate the heat shock response in Escherichia coli by regulating the switch between σ70 and σ32 factors assembled with RNA polymerase. EMBO J 14:5085–5093

Bloom M, Skelly S, VanBogelen R, Neidhardt F, Brot N, Weissbach H (1986) In vitro effect of the Escherichia coli heat shock regulatory protein on expression of heat shock genes. J Bacteriol 166:380–384

Bohen SP, Yamamoto KR (1994) Modulation of steroid receptor signal transduction by heat shock proteins. In: Morimoto RI, Tissieres A, Georgopoulos C (eds) The biology of heat shock proteins and molecular chaperones. Cold Spring Harbor Laboratory Press, New York, pp 313–334

Boorstein WR, Craig EA (1990) Transcriptional regulation of ssa3, an hsp70 gene from Saccharomyces cerevisiae. Mol Cell Biol 10:3262–3267

Bukau B (1993) Regulation of the Escherichia coli heat-shock response. Mol Microbiol 9:671–680

Clos J, Westwood JT, Becker PB, Wilson S, Lambert K, Wu C (1990) Molecular cloning and expression of a hexameric Drosophila heat shock factor subject to negative regulation. Cell 63:1085–1097

Connolly L, De Las Penas A, Alba BM, Gross CA (1997) The response to extracytoplasmic stress in Escherichia coli is controlled by partially overlapping pathways. Genes Dev 11:2012–2021

Cotto JJ, Kline M, Morimoto RI (1996) Activation of heat shock factor 1 DNA binding precedes stress-induced serine phosphorylation. J Biol Chem 271:3355–3358

Cowing DW, Bardwell JC, Craig EA, Woolford C, Hendrix RW, Gross CA (1985) Consensus sequence for Escherichia coli heat shock gene promoters. Proc Natl Acad Sci USA 82:2679–2683

Craig EA, CA Gross (1991) Is Hsp70 the cellular thermometer? Trends Biochem Sci 16:135–140

Danese PN, Silhavy TJ (1997) The sigma(E) and the Cpx signal transduction systems control the synthesis of periplasmic protein-folding enzymes in Escherichia coli. Genes Dev 11:1183–1193

DiDomenico BJ, Bugaisky GE, Lindquist S (1982a) Heat shock and recovery are mediated by different translational mechanisms. Proc Natl Acad Sci USA 79:6181–6185

DiDomenico BJ, Bugaisky GE, Lindquist S (1982b) The heat shock response is self-regulated at both the transcriptional and posttranscriptional levels. Cell 31:593–603

Erickson JW, Vaughn V, Walter WA, Neidhardt FC, Gross CA (1987) Regulation of the promoters and transcripts of rpoH, the Escherichia coli heat shock regulatory gene. Genes Dev 1:419–432

Erickson JW, Gross CA (1989) Identification of the sigma E subunit of Escherichia coli RNA polymerase: a second alternate sigma factor involved in high-temperature gene expression. Genes Dev 3:1462–1471

Farkas T, Kutskova YA, Zimarino V (1998) Intramolecular repression of mouse heat shock factor 1. Mol Cell Biol 18:906–918

Feder JH, Rossi JM, Solomon J, Solomon N, Lindquist S (1992) The consequences of expressing Hsp70 in Drosophila cells at normal temperatures. Genes Dev 6:1402–1413

Gamer J, Bujard H, Bukau B (1992) Physical interaction between heat shock proteins DnaK, DnaJ, and GrpE and the bacterial heat shock transcription factor sigma 32. Cell 69:833–842

Gamer J, Multhaup G, Tomoyasu T, McCarty JS, Rudiger S, Schonfeld H-J, Schirra C, Bujard H, Bukau B (1996) A cycle of binding and release of the DnaK, DnaJ and GrpE chaperones regulates activity of the Escherichia coli heat shock transcription factor σ32. EMBO J 15:607–617

Georgopoulos C, Liberek K, Zylicz M, Ang D (1994) Properties of the heat shock proteins of Escherichia coli and the autoregulation of the heat shock response. In: Morimoto RI, Tissieres A, Georgopoulos C (eds) The biology of heat shock proteins and molecular chaperones. Cold Spring Harbor Laboratory Press, New York, pp 209–250

Giardina C, Lis JT (1995) Sodium salicylate and yeast heat shock gene transcription. J Biol Chem 270:10369–10372

Green M, Schuetz TJ, Sullivan EK, Kingston RE (1995) A heat shock-responsive domain of human HSF1 that regulates transcription activation domain function. Mol Cell Biol 15:3354–3362

Gross CA, Straus DB, Erickson JW, Yura T (1990) The function and regulation of heat shock proteins in E.coli. In: Morimoto RI, Tissieres A, Georgopoulos C (eds) Stress proteins in biology and medicine. Cold Spring Harbor Laboratory Press, New York, pp 167–189

Grossman AD, Erickson JW, Gross CA (1984) The htpR gene product of E. coli is a sigma factor for heat-shock promoters. Cell 38:383–390

Grossman AD, Straus DB, Walter WA, Gross CA (1987) σ32 synthesis can regulate the synthesis of heat shock proteins in Escherichia coli. Genes Dev 1:179–184

Halladay JT, Craig EA (1995) A heat shock transcription factor with reduced activity suppresses a yeast hsp70 mutant. Mol Cell Biol 15:4890–4897

Harrison CJ, Bohm AA, Nelson HCM (1994) Crystal structure of the DNA binding domain of the heat shock transcription factor. Science 263:224–227

Hiratsu K, Amemura M, Nashimoto H, Shinagawa H, Makino K (1995) The rpoE gene of Escherichia coli, which encodes sigma E, is essential for bacterial growth at high temperature. J Bacteriol 177:2918–2922

Hupp TR, Meek DW, Midgley CA, Lane DP (1992) Regulation of the specific DNA binding function of p53. Cell 71:875–886

Jurivich DA, Sistonen L, Kroes RA, Morimoto RI (1992) Effect of sodium salicylate on the human heat shock response. Science 255:1243–1245

Kamath-Loeb AS, Gross CA (1991) Translational regulation of sigma 32 synthesis: requirement for an internal control element. J Bacteriol 173:3904–3906

Kanei-Ishii C, Tanikawa J, Nakai A, Morimoto RI, Ishii S (1997) Activation of heat shock transcription factor 3 by c-Myb in the absence of cellular stress. Science 277:246–248

Kanemori M, Mori H, Yura T (1994) Effects of reduced levels of GroE chaperones on protein metabolism: enhanced synthesis of heat shock proteins during steady-state growth of Escherichia coli. J Bacteriol 176:4235–4242

Kanemori M, Nishihara K, Yanagi H, Yura T (1997) Synergistic roles of HslVU and other ATP-dependent proteases in controlling in vivo turnover of sigma32 and abnormal proteins in Escherichia coli. J Bacteriol 179:7219–7225

Kline MP, Morimoto RI (1997) Repression of the heat shock factor 1 transcriptional activation domain is modulated by constitutive phosphorylation. Mol Cell Biol 17:2107–2115

Knauf U, Newton EM, Kyriakis J, Kingston RE (1996) Repression of human heat shock factor 1 activity at control temperature by phosphorylation. Genes Dev 10: 2782–2793

Larson JS, Schuetz TJ, Kingston RE (1988) Activation in vitro of sequence-specific DNA binding by a human regulatory factor. Nature 335:372–375

Lesley SA, Thompson NE, Burgess RR (1987) Studies of the role of the Escherichia coli heat shock regulatory protein sigma 32 by the use of monoclonal antibodies. J Biol Chem 262:5404–5407

Liberek K, Galitski TP, Zylicz M, Georgopoulos C (1992) The DnaK chaperone modulates the heat shock response of Escherichia coli by binding to the σ32 transcription factor. Proc Natl Acad Sci USA 89:3516–3520

Liberek K, Georgopoulos C (1993) Autoregulation of the Escherichia coli heat shock response by the DnaK and DnaJ heat shock proteins. Proc Natl Acad Sci USA 90:11019–11023

Lipinska B, Sharma S, Georgopoulos C (1988) Sequence analysis and regulation of the htrA gene of Escherichia coli: a sigma 32-independent mechanism of heat-inducible transcription. Nucleic Acids Res 16:10053–10067

Lis JT, Wu C (1993) Protein traffic on the heat shock promoter: parking, stalling and trucking along. Cell 74:1–20

Mason PB Jr, Lis JT (1997) Cooperative and competitive protein interactions at the hsp70 promoter. J Biol Chem 272:33227–33233

Mathew A, Mathur SK, Morimoto RI (1998) The heat shock response and protein degradation: regulation of HSF2 by the ubiquitin-dependent proteasome. Mol Cell Biol 17:5091–5098

McCarty JS, Rudiger S, Schonfeld HJ, Schneider-Mergener J, Nakahigashi K, Yura T, Bukau B (1996) Regulatory region C of the E. coli heat shock transcription factor, sigma32, constitutes a DnaK binding site and is conserved among eubacteria. J Mol Biol 256:829–837

Mecsas J, Rouviere PE, Erickson JW, Donohue TJ, Gross CA (1993) The activity of sigma E, an Escherichia coli heat-inducible sigma-factor, is modulated by expression of outer membrane proteins. Genes Dev 7:2618–2628

Mezger V, Rallu M, Morimoto RI, Morange M, Renard JP (1994) Heat shock factor 2-like activity in mouse blastocytes. Dev Biol 166:819–822

Morimoto RI, Tissieres A, Georgopoulos C (1990) The stress response, function of the proteins, and perspectives. In: Morimoto RI, Tissieres A, Georgopoulos C (eds) Stress proteins in biology and medicine. Cold Spring Harbor Laboratory Press, New York, pp 1–36

Morimoto RI (1993) Cells in stress: transcriptional activation of heat shock genes. Science 269:1409–1410

Morimoto RI, Jurivich DA, Kroeger PE, Mathur SK, Murphy SP, Nakai A, Sarge K, Abravaya K, Sistonen LT (1994) Regulation of heat shock gene transcription by a family of heat shock factors. In: Morimoto RI, Tissieres A, Georgopoulos C (eds) The biology of heat shock proteins and molecular chaperones. Cold Spring Harbor Laboratory Press, New York, pp 417–455

Mosser DD, Theodorakis NG, Morimoto RI (1988) Coordinate changes in heat shock element binding activity and hsp70 gene transcription rates in human cells. Mol Cell Biol 8:4736–4744

Mosser DD, Kotzbauer PT, Sarge KD, Morimoto RI (1990) In vitro activation of heat shock transcription factor DNA-binding by calcium and biochemical conditions that affect protein conformation. Proc Natl Acad Sci USA 87:3748–3752

Nadeau K, Das A, Walsh CT (1993) Hsp90 chaperonins possess ATPase activity and bind heat shock transcription factors and peptidyl prolyl isomerase. J Biol Chem 268:1479–1487

Nagai H, Yuzawa H, Yura T (1991) Interplay of two cis-acting mRNA regions in translational control of sigma 32 synthesis during the heat shock response of Escherichia coli. Proc Natl Acad Sci USA 88:10515–10519

Nagai H, Yuzawa H, Kanemori M, Yura T (1994) A distinct segment of the sigma 32 polypeptide is involved in DnaK-mediated negative control of the heat shock response in Escherichia coli. Proc Natl Acad Sci USA 91:10280–10284

Nakai A, Morimoto RI (1993) Characterization of a novel chicken heat shock transcription factor, heat shock factor 3, suggests a new regulatory pathway. Mol Cell Biol 13:1983–1997

Nakai A, Kawazoe Y, Tanabe M, Nagata K, Morimoto RI (1995) The DNA-binding properties of two heat shock factors, HSF1 and HSF3, are induced in the avian erythroblast cell line HD6. Mol Cell Biol 15:5268–5278

Nakai A, Tanabe M, Kawazoe Y, Inazawa J, Morimoto RI, Nagata K (1997) HSF4, a new member of the human heat shock factor family which lacks properties of a transcriptional activator. Mol Cell Biol 17:469–481

Nair SC, Toran EJ, Rimerman RA, Hjermstad S, Smithgall TE, Smith DF (1996) A pathway of multi-chaperone interactions common to diverse regulatory proteins: estrogen receptor, Fes tyrosine kinase, heat shock transcription factor HSF1, and the aryl hydrocarbon receptor. Cell Stress Chaperones 1:237–250

Nelson RJ, Heschl MF, Craig EA (1992) Isolation and characterization of extragenic suppressors of mutations in the SSA hsp70 genes of Saccharomyces cerevisiae. Genetics 131:277–285

Nunes SL, Calderwood SK (1995) Heat shock factor-1 and the heat shock cognate 70 protein associate in high molecular weight complexes in the cytoplasm of NIH-3T3 cells. Biochem Biophy Res Commun 213:1–6

Orosz A, Wisniewski J, Wu C (1996) Regulation of Drosophila heat shock factor trimerization: global sequence requirements and independence of nuclear localization. Mol Cell Biol 16:7018–7030

Peteranderl R, Nelson HCM (1992) Trimerization of the heat shock transcription factor by a triple-stranded alpha-helical coiled-coil. Biochem 31:12272–12276

Rabindran SK, Giorgi G, Clos J, Wu C (1991) Molecular cloning and expression of a human heat shock factor, HSF1. Proc Natl Acad Sci USA 88:6906–6910

Rabindran SK, Haroun RI, Clos J, Wisniewski J, Wu C (1993) Regulation of heat shock factor trimer formation: role of a conserved leucine zipper. Science 259:230–234

Rabindran SK, Wisniewski J, Li L, Li GC, Wu C (1994) Interaction between heat shock factor and Hsp70 is insufficient to suppress induction of DNA-binding activity in vivo. Mol Cell Biol 14:6552–6560

Raina S, Missiakas D, Georgopoulos C (1995) The rpoE gene encoding the sigma E (sigma 24) heat shock sigma factor of Escherichia coli. EMBO J 14:1043–1055

Rallu M, Loones M, Lallemand Y, Morimoto RI, Morange M, Mezger V (1997) Function and regulation of heat shock factor 2 during mouse embryogenesis. Proc Natl Acad Sci USA 94:2392–2397

Rouviere PE, De Las Penas A, Mecsas J, Lu CZ, Rudd KE, Gross CA (1995) rpoE, the gene encoding the second heat-shock sigma factor, sigma E, in E. coli. EMBO J 14:1032–1042

Sarge KD, Zimarino V, Holm K, Wu C, Morimoto RI (1991) Cloning and characterization of two mouse heat shock factors with distinct inducible and constitutive DNA-binding ability. Genes Dev 5:1902–1911

Sarge K, SP Murphy, Morimoto RI (1993) Activation of heat shock transcription by HSF1 involves oligomerization, acquisition of DNA binding activity, and nuclear localization and can occur in the absence of stress. Mol Cell Biol 13:1392–1407

Sarge KD, Park-Sarge OK, Kirby JD, Mayo KE, Morimoto RI (1994) Expression of heat shock factor 2 in mouse testis: potential role as a regulator of heat-shock protein gene expression during spermatogenesis. Biol Reprod 50:1334–1343

Schuetz TJ, Gallo GJ, Sheldon L, Tempst P, Kingston RE (1991) Isolation of a cDNA for HSF2: evidence for two heat shock factor genes in humans. Proc Natl Acad Sci USA 88:6911–6915

Shi Y, Kroeger PE, Morimoto RI (1995) The carboxyl-terminal transactivation domain of heat shock factor 1 is negatively regulated and stress responsive. Mol Cell Biol 15:4309–4318

Shi Y, Mosser DD, Morimoto RI (1998) Molecular chaperones as HSF1 specific transcriptional repressors. Genes Dev 12:654–666

Sistonen L, Sarge KD, Phillips B, Abravaya K, Morimoto RI (1992) Activation of heat shock factor 2 during hemin-induced differentiation of human erythroleukemia cells. Mol Cell Biol 12:4104–4111

Sistonen L, Sarge KD, Morimoto RI (1994) Human heat shock factors 1 and 2 are differentially activated and can synergistically induce hsp70 gene transcription. Mol Cell Biol 14:2087–2099

Skelly S, Coleman T, Fu CF, Brot N, Weissbach H (1987) Correlation between the 32-kDa sigma factor levels and in vitro expression of Escherichia coli heat shock genes. Proc Natl Acad Sci USA 84:8365–8369

Skelly S, Fu CF, Dalie B, Redfield B, Coleman T, Brot N, Weissbach H (1988) Antibody to sigma 32 cross-reacts with DnaK: association of DnaK protein with Escherichia coli RNA polymerase. Proc Natl Acad Sci USA 85:5497–5501

Sorger PK, Pelham HR (1988) Yeast heat shock factor is an essential DNA-binding protein that exhibits temperature-dependent phosphorylation. Cell 54:855–864

Sorger PK, Nelson HCM (1989) Trimerization of a yeast transcriptional activator via a coiled-coil motif. Cell 59:807–813

Stone DE, Craig EA (1990) Self-regulation of 70-kilodalton heat shock proteins in Saccharomyces cerevisiae. Mol Cell Biol 10:1622–1632

Straus DB, Walter WA, Gross CA (1987) The heat shock response of E. coli is regulated by changes in the concentration of sigma 32. Nature 329:348–351

Straus DB, Walter WA, Gross CA (1989) The activity of sigma 32 is reduced under conditions of excess heat shock protein production in Escherichia coli. Genes Dev 3:2003–2010

Straus DB, Walter WA, Gross CA (1990) Dnak, DnaJ, and GrpE heat shock proteins negatively regulate heat shock gene expression by controlling the synthesis and stability of σ32. Genes Dev 4:2202–2209

Tanabe M, Nakai A, Kawazoe Y, Nagata K (1997) Different thresholds in the responses of two heat shock transcription factors, HSF1 and HSF3. J Biol Chem 272:15389–15395

Tanabe M, Kawazoe Y, Takeda S, Morimoto RI, Nagata K, Nakai A (1998) Disruption of the HSF3 gene results in the severe reduction of heat shock gene expression and loss of thermotolerance. EMBO J 17:1750–1758

Theodorakis NG, Zand DJ, Kotzbauer PT, Williams GT, Morimoto RI (1989) Hemin-induced transcriptional activation of the HSP70 gene during erythroid maturation in K562 cells is due to a heat shock factor-mediated stress response. Mol Cell Biol 9:3166–3173

Tilly K, McKittrick N, Zylicz M, Georgopoulos C (1983) The DnaK protein modulates the heat shock response of Escherichia coli. Cell 34:641–646

Tilly K, Spence J, Georgopoulos C (1989) Modulation of stability of the Escherichia coli heat shock regulatory factor sigma. J Bacteriol 171:1585–1589

Tomoyasu T, Gamer J, Bukau B, Kanemori M, Mori H, Rutman AJ, Oppenheim AB, Yura T, Yamanaka K, Niki H, Hiraga S, Ogura T (1995) Escherichia coli FtsH is a membrane-bound, ATP-dependent protease which degrades the heat-shock transcription factor sigma 32. EMBO J 14:2551–2560

Vuister GW, Kim SJ, Wu C, Bax A (1994) NMR evidence for similarities between the DNA-binding regions of Drosophila melanogaster heat shock factor and the helix-turn-helix and HNF-3/forkhead families of transcription factors. Biochemistry 33:10–16

Wang QP, Kaguni JM (1989) A novel sigma factor is involved in expression of the rpoH gene of Escherichia coli. J Bacteriol 171:4248–4253

Westwood JT, Clos J, Wu C (1991) Stress-induced oligomerization and chromosomal relocalization of heat-shock factor. Nature 353:822–827

Wiederrecht G, Seto D, Parker CS (1988) Isolation of the gene encoding the S. cerevisiae heat shock transcription factor. Cell 54:841–853

Wisniewski J, Orosz A, Allada R, Wu C (1996) The C-terminal region of Drosophila heat shock factor (HSF) contains a constitutively functional transactivation domain. Nucleic Acids Res 24:367–374

Wu C, Clos J, Giorgi G, Haroun RI, Kim S-J, Rabindran SK, Westwood T, Wisniewski J, Yim G (1994) Structure and regulation of heat shock transcription factor. In: Morimoto RI, Tissieres A, Georgopoulos C (eds) The biology of heat shock proteins and molecular chaperones. Cold Spring Harbor Laboratory Press, New York, pp 417–455

Wu C (1995) Heat shock transcription factors: structure and regulation. Annu Rev Cell Dev Biol 11:441–469

Yura T, Nagai H, Mori H (1993) Regulation of the heat-shock response in bacteria. Annu Rev Microbiol 47:321–350

Yura T (1996) Regulation and conservation of the heat-shock transcription factor sigma 32. Genes Cells 1:277–284

Zandi E, Tran TN, Chamberlain W, Parker CS (1997) Nuclear entry, oligomerization, and DNA binding of the Drosophila heat shock transcription factor are regulated by a unique nuclear localization sequence. Genes Dev 11:1299–1314

Zhong T, Luke MM, Arndt KT (1996) Tanscriptional regulation of the yeast DnaJ homologue SIS1. J Biol Chem 271:1349–1356

Zimarino V, Wilson S, Wu C (1990) Antibody-mediated activation of Drosophila heat shock factor in vitro. Science 249:546–549

Zuo J, Baler R, Dahl G, Voellmy R (1994) Activation of the DNA-binding ability of human heat shock transcription factor 1 may involve the transition from an intramolecular to an triple-stranded coiled-coil structure. Mol Cell Biol 14:7557–7568

Zuo J, Rungger D, Voellmy R (1995) Multiple layers of regulation of human heat shock transcription factor 1. Mol Cell Biol 15:4319–4330

CHAPTER 11

The Cellular Stress Gene Response in Brain

I.R. BROWN and F.R. SHARP

A. Introduction

Considerable advances in the molecular biology of the heat shock response and the role of heat shock proteins (hsps) in repair and protective mechanisms have been made using mammalian cells grown in tissue culture (MORIMOTO 1993; MORIMOTO et al. 1994; see also Chap. 3, this volume). Recently, an increasing amount of work has been carried out on intact thermoregulating animals. As will be shown in this article, the heat shock response is physiologically relevant since heat shock genes are turned on in the mammalian nervous system following stress treatments such as fever-like temperature, focal cerebral ischemia and subarachnoid hemorrhage. The brain is a complex structure composed of many cell types. As will become apparent, neuronal and glial cell types exhibit differences in constitutive expression of hsps and the type of brain cell that activates the heat shock response in vivo depends on the nature and the severity of the stress.

B. Response of the Brain to Physiologically Relevant Temperature Increase

I. Differential Induction of Heat Shock mRNA in Different Cell Types of the Hyperthermic Brain

Early studies demonstrated that increases in body temperature of 2–3°C in the rabbit resulted in a transient induction of hsp70 mRNA and protein in the brain during a period of overall inhibition of protein synthesis (for reviews see BROWN 1990, 1994). In situ hybridization has been used to map out the pattern of expression of hsp70 genes in the control and hyperthermic mammalian brain. This procedure enables one to identify the cells in a complex tissue which are expressing the gene of interest. One hour after hyperthermia, a marked induction of hsp70 mRNA was observed in fiber tracts throughout the rabbit forebrain and cerebellum, a pattern consistent with a glial response to fever-like temperature (SPRANG and BROWN 1987). Induction was also noted in the choroid plexus, microvasculature and granule cell layer of the cerebellum. Constitutive expression was observed in several neuronal enriched areas such as hippocampal regions CA1 to CA4 and the Purkinje layer of the cerebellum.

In 1 h hyperthermic animals, induction of hsp70 mRNA was not detected in neuronal cells other than granule neurons in the cerebellum.

These initial in situ hybridization studies were carried out using a 35S-labeled riboprobes which hybridized to both constitutively expressed and stress-inducible hsp70 transcripts. Subsequently, these studies were extended using riboprobes which discriminate hyperthermia-inducible 2.7 kb transcripts from constitutively expressed 2.5 kb transcripts (Brown and Rush 1990; Manzerra and Brown 1992a). A striking neuronal-glial difference in the localization of the two transcripts was noted in large Purkinje neurons in the cerebellum which are surrounded by smaller Bergmann glial cells. These neurons showed high levels of the hsc70 transcript and no induction of the hsp70 species in 1 h hyperthermic animals. In contrast, adjacent Bergmann glial cells demonstrated no detectable hsc70 mRNA and a very strong induction of hsp70 mRNA. A similar neuronal/glial difference in the expression of hsc70 and hsp70 transcripts was noted in the mammalian spinal cord (Manzerra and Brown 1992b).

To improve the localization of heat shock transcripts to individual neural cells, the in situ hybridization procedure was modified to utilize non-radioactive digoxigenin (DIG)-UTP labeled riboprobes (Foster et al. 1995). Neurons in the brainstem and in the molecular layer of the cerebellum showed expression of hsc70 mRNA while signal was not detected in adjacent glial cells. Hsc70 mRNA was highly localized to the cytoplasm of individual neurons which were identified by a neuron-specific enolase marker. A strong induction of hsp70 mRNA was noted in glial cells in cerebellar layers and in the brainstem of hyperthermic animals and not in adjacent large neurons.

Different types of glial cells exist in the mammalian brain and the question arises as to whether they show a differential hsp70 induction in response to a fever-like temperature shock. Oligodendrocytes synthesize and maintain the myelin sheath around axons to facilitate efficient transmission of nerve impulses. Astrocytes and microglia are reactive glial cell types which can be identified respectively by their expression of glial fibrillary acidic protein (GFAP) and the lectin GSA-B4. A protocol that combined DIG in situ hybridization and cytochemistry on the same tissue section was employed to identify glial cell types showing induction of heat shock transcripts. In response to a fever-like temperature increase, over 95% of the oligodendrocytes in four regions of the rabbit forebrain induced hsp70 mRNA by 2 h as did a subpopulation of the microglia while none of the GFAP positive astrocytes showed induction (Foster and Brown 1997).

These data suggest that specific glial cell populations exhibit distinct temperature thresholds for activation of the heat shock response. Studies with astrocytes grown in tissue culture have reported an induction of hsp70 mRNA; however, the temperature increment used in the vitro experiments (+8°C) was greater than that used in the in vivo experiment (+2.5°C; Nishimura et al. 1988,

1992). It is intriguing that virtually all the the oligodendrocytes in the forebrain induced hsp70 mRNA in response to a fever-like stress whereas the GFAP positive astrocyte population showed no response. Whether this reflects a sensitivity of components of the myelin sheath to physiologically relevant temperature increases is not known.

II. Intracellular Targeting of Neural Heat Shock mRNAs

There is growing interest in the phenomenon of intracellular targeting of specific mRNAs. Certain mRNA species are transported out of the cell body into cellular processes where they are locally translated into proteins at the appropriate time (LIPSHITZ 1995). This could provide a strategy for rapidly increasing levels of heat shock proteins in intracellular domains of neural cells that are remote from the cell body. High resolution in situ hybridization revealed that after hyperthermia, hsp70 mRNA was strongly induced in oligo-dendrocytes and transported into the cellular processes of these glial cells (FOSTER and BROWN 1996). Induction of hsp70 mRNA was not observed in several populations of large neurons. However, these neurons showed high levels of constitutive hsc70 mRNA and these transcripts showed more distal transport into dendritic processes following a fever-like temperature (FOSTER and BROWN 1996). Transport of heat shock messages in the processes of neural cells could provide a mechanism for rapidly increasing levels of hsps in cellular compartments which are remote from the cell body.

III. Cell Type Differences in Neural Heat Shock Proteins

Although the induction of hsp70 protein has been investigated in numerous systems by Western blotting and immunocytochemistry, comparatively few studies have characterized the specificity of the antibodies which are em-ployed. It has been demonstrated that certain antibodies which are specific to hsp70 in some mammal species, react with both hsp70 and hsc70 in related mammals (MANZERRA et al. 1997). Stress-inducible hsp70 and constitutive hsc70 protein show extensive amino acid sequence homology; however, they can be resolved by two-dimensional Western blotting in mammals (MANZERRA et al. 1997). This study demonstrated that basal levels of hsp70 isoforms were present in brain regions of the control rabbit and that these were elevated following hyperthermia whereas levels of hsc70 were similar in control and hyperthermic tissue. Multiple isoforms of hsp70 were detected but tissue-specific differences were not apparent in various organs of the rabbit. How-ever, species differences were observed as fewer hsp70 isoforms were noted in rat and mouse. In the control rabbit, higher levels of hsc70 protein were present in neural tissues compared to non-neural. Following a fever-like tem-perature, induction of hsp70 was greatest in non-neural tissues such as liver, heart, muscle, spleen and kidney compared to the nervous system. Preexisting hsc70 protein, which is high in nervous tissue, may dampen the level of induction of hsp70 in the stress response. Given this observation, caution is required in the employment of hsp70 induction as an index of cellular stress since endogenous levels of hsc70, and perhaps hsp70, may modulate the level of induction

Immunocytochemistry using antibodies which are specific to either hsp70 or hsc70 protein revealed pronounced differences in their cellular distribution in the mammalian nervous system (Manzerra et al. 1993). Large neurons, such as Purkinje neurons in the cerebellum, which show abundant levels of hsc70 protein in the cell body and apical dendrite, demonstrated no detectable induction of hsp70 protein after a fever-like temperature, whereas adjacent Bergmann glial cells show a robust induction of hsp70 protein. Does this reflect that these neurons are not stressed by the fever-like temperature perhaps due to the buffering action of their abundant endogenous hsc70?

Use of hsp70 protein induction alone as a marker of cellular stress may not be sufficient to predict whether neurons are perturbed following the fever-like temperature shock. In order to investigate the neuronal response to hyperthermia in more detail, an additional characteristic feature of the cellular stress response was employed namely nuclear translocation of heat shock proteins (Manzerra and Brown 1996). Following an increase in body temperature of $2.7°C$, nonneuronal cell types in the brain such as oligodendrocytes and ependymal cells induce hsp70 and rapidly translocate the protein to the nucleus. Several populations of neurons do not translocate their endogenous hsc70 protein to the nucleus unless the temperature is further increased to $+3.4°C$. Neuronal induction of hsp70 protein was not observed at either temperature.

It appears that different cell types in the mammalian brain have different set points for the induction of hsp70. Within the glial cell population, oligodendrocytes require a lower temperature increment for induction compared to astrocytes. Several populations of neurons appear to be buffered against induction of the heat shock response, perhaps due to their high constitutive levels of hsc70. The neuronal heat shock response may be activatible in stages depending on the level of stress on individual cell populations (Manzerra and Brown 1996). Following a fever-like temperature increase, a neuronal response is not observed, perhaps due to the buffering action of high levels of preexisting hsc70 in these cells. In response to a slightly higher hyperthermic stress, neuronal cells undergo a partial heat shock response, namely nuclear translocation of hsc70 protein. Severe stress, such as ischemia, results in a full heat shock response in neurons involving induction of hsp70. Neurons may be somewhat preprotected against minor stress and this enables them to avoid induction of the heat shock response and accompanying difficulties such as transient inhibition of protein synthesis which may hamper ongoing neurotransmission activity which is critical to the functioning of the organism.

IV. Expression of Heat Shock Proteins in the Developing Brain

Heat shock proteins are highly conserved proteins which are induced in cells upon exposure to elevated temperatures and other forms of cellular stress. In addition, most hsps are also present in the unstressed cells where they play vital roles in normal cellular function. Previous work has shown that neurons in the adult mammalian brain exhibit high levels of hsp90 and hsc70 mRNA and protein, as well as basal levels of hsp70 mRNA (MANZERRA et al. 1993; QURAISHI and BROWN 1995; FOSTER and BROWN 1996b). It was of interest to determine when these high levels are attained during postnatal neural development, a time of extensive neuronal differentiation.

Western blot analysis revealed that abundant levels of hsc70 and hsp90 were attained early in postnatal neural development and maintained in the adult (D'SOUZA and BROWN 1998b). In contrast, hsp60, a nuclear-encoded mitochondrial hsp, showed a major developmental increase, perhaps reflective of a developmental increase in mitochondrial content in the brain. A similar development increase was noted in another mitochondrial protein, cytochrome oxidase, subunit IV. Analysis of hsp70 protein showed low basal amounts in the unstressed brain and a developmental increase in the cerebral hemispheres. These observations on developmental expression suggest that these hsps are differentially regulated during postnatal development whereas hsps are coordinately induced following heat stress. Immunocytochemical studies demonstrated a neuronal localization of hsp90, hsc70, and hsp60 at all stages of postnatal development examined as well as in the adult, suggesting a role for these hsps in both the developing and fully differentiated neuron (D'SOUZA and BROWN 1998b).

V. Activation of Neural Heat Shock Transcription Factor HSF1

Molecular mechanisms which underlie the heat shock response have commonly been analyzed using tissue culture systems with less investigation of the intact mammal. In mammalian tissue culture systems, a supraphysiological temperature increment of 5°C is required to elicit a robust activation of the heat shock transcription factor HSF1 to a DNA-binding form and subsequent triggering of the induction of heat shock genes (JURIVICH et al. 1994). Such temperature increases are lethal to a mammal. How do intact, thermoregulating mammals react to lower temperature increments of 2–3°C which are similar to increases attained during fever and inflammation? In contrast to mammalian tissue culture experiments, a fever-like increase in body temperature of 2.5°C is sufficient to activate neural HSF1 to a DNA-binding form and induce hsp70 mRNA and protein in the rabbit (BROWN and RUSH 1996). Similar observations have been reported in rat brain (HIGASHI et al. 1995). It is likely that additional factors are present in vivo which modulate the threshold temperature for activation of HSF1 and facilitate the induction of the heat shock response in the nervous system at temperatures which are physiologically relevant. These factors are not present in tissue culture systems and increased temperature is required to induce the heat shock response in vitro.

It has been reported that exposure of cultured HeLa cells to low concentrations of arachidonate, which alone does not induce HSF1 DNA binding, can reduce the temperature threshold for HSF1 activation in these tissue culture cells to levels which are physiologically relevant (JURIVICH et al. 1994). These results indicate that molecules such as arachidonate, a central mediator of the inflammatory response, can influence the threshold temperature for induction of the heat shock response. Thus additional factors, which are present in the intact animal and not present in tissue culture cells, likely play roles in triggering the in vivo heat shock response at physiologically relevant increases in body temperature (BROWN and RUSH 1996). Restraint or immobilization stress has been shown to induce hsp70 in the adrenal gland of the intact rat through ACTH dependent activation of HSF1 (FAWCETT et al. 1994). However, attempts to reproduce the in vivo response to ACTH in cultured adrenocortical cells using added ACTH or other stimulators of cAMP have not been successful. Again, factors which are present in the intact animal appear to be required to impact on HSF1 and trigger the heat shock response.

VI. In Vivo Transcription Rate of Heat Shock Genes in the Brain

Previous studies in the nervous system which have examined the induction of hsp70 in the nervous system following hyperthermia have employed Northern blotting and in situ hybridization techniques (for reviews see BROWN 1990, 1994). These studies investigate the steady state levels of hsp70 mRNA transcripts and therefore do not distinguish between changes in mRNA stability/turnover or changes in the actual transcription rate of the genes. A convenient method for monitoring changes in transcription rates is the nuclear run-on transcription assay. Previous investigations have used this technique to study the transcription of heat shock genes in tissue culture systems (ABRAVAYA et al. 1991; VASQUEZ et al. 1993; SISTONEN et al. 1994; MATHUR et al. 1994), however the procedure has not been applied to nuclei isolated from neural tissue of hyperthermic animals.

Recently the nuclear run-on transcription assay has been modified for use in vivo in a neural system and fever-like temperatures has been shown to induce a major up-regulation in the transcription rate of hsp70 (D'SOUZA et al. 1998a). The transcription rate of several non-heat shock genes was also studied in the hyperthermic brain and little change was noted relative to the induction of hsp70. Gel mobility shift assays revealed a tight correlation between the kinetics of activation of neural HSF1 to a DNA-binding form and the time course of changes in the in vivo transcription rate of hsp70 in brain regions. These studies show that a fever-like temperature increase induces a major up-regulation in the in vivo transcription rate of hsp70 in the mammalian nervous system with little effect on the transcription rate of other genes.

VII. Neuroprotective Effect of Heat Shock Protein in the Retina

The protective effect of heat shock has been observed in the nervous system. For example, prior whole body hyperthermia in rats has been shown to protect retinal photoreceptors against degeneration induced by subsequent bright light exposure (BARBE et al. 1988). In collaboration with the laboratory of Micheal Tytell we have identified the cell types in the retina which induce hsp70 mRNA and protein in response to whole body hyperthermia (TYTELL et al. 1993, 1994). Photoreceptors are the main site of hsp70 induction and the time of maximal hsp70 protein accumulation in these cells corresponds to the time when photoreceptors are maximally protected against degeneration by bright light. Evidence that hsp70 may be neuroprotective is suggested by our observation that intraocular injection of purified hsp confers protection to photoreceptors against bright light damage whereas control protein injected into the other eye of the same animal does not (TYTELL et al. 1993).

VIII. Conclusions

In summary, different cell types in the mammalian brain exhibit different set points for induction of hsp70 in response to fever-like temperatures. An interesting question is whether the amount of preexisting hsc/hsps in a particular cell type influences the level of induction of hsp70 in the stress response. Certain classes of neurons may be "preprotected" against mild stress by their endogenous level of hsc70. The heat shock response may be activatible in stages depending on the level of stress with certain neurons first responding by translocating their endogenous hsc70 to the nucleus and inducing hsp70 if the stress is more intense. Translocation of heat shock mRNAs to the cellular processes of neural cells and resultant local protein synthesis may provide a mechanism for rapid delivery of hsps to intracellular domains which are remote from the cell body. Factors which are present in vivo and absent in tissue culture systems may modulate the threshold for activation of HSF1 to a DNA-binding form and permit neural induction of the heat shock response at temperatures which are physiologically relevant. Interestingly, oligodendrocytes in the mammalian brain are highly sensitive to fever-like temperatures in rapidly inducing hsp70. In the retinal system, hsp70 demonstrates neuroprotective effects since intraocular injection of the protein confers protection to photoreceptors against degeneration induced by subsequent bright light.

C. Cellular Stress Gene Response to Focal Cerebral Ischemia

I. Hsp70 and Delineation of the Penumbra

Hsp70 mRNA and protein are induced following focal brain ischemia (GONZALEZ et al. 1989) just as they are induced following global ischemia.

Although the pattern of induction is similar to that observed following global ischemia, there are some basic differences since focal ischemia can produce tissue infarction. In global ischemic brain injury there is selective neuronal death. Following infarction there is death of neurons and glia, or death of all cellular elements including neurons, glia and vascular endothelial cells.

Following permanent middle cerebral artery (MCA) occlusions, hsp70 mRNA is induced in the entire distribution of the MCA (KINOUCHI et al. 1993a,b, 1994a) and extends for some distance beyond the infarction (KINOUCHI et al. 1993b). The hsp70 mRNA is induced within 15 min of the ischemia (ABE et al. 1992) (KAMII et al. 1994; WANG et al. 1993) and is maximum at 3 h in caudate and 8–24 h in cortex (ABE et al. 1992; KAMII et al. 1994; WANG et al. 1993; KINOUCHI et al. 1993a,b, 1994a).

One day following MCA occlusions, hsp70 protein is expressed mainly in endothelial cells in the center of the infarction. At the margins of the infarction hsp70 protein is expressed in microglia and to a lesser extent in astrocytes. The induction occurs in cells inside the infarct and just outside the infarct (GONZALEZ et al. 1989, 1991; SHARP et al. 1991, 1993; SHARP and SAGAR 1994; SHARP 1995; GASPARY et al. 1995; PLANAS et al. 1997).

Hsp70 protein is also expressed in neurons outside regions of infarction (GONZALEZ et al. 1989; LI et al. 1992, 1993). Hsp70 mRNA is expressed in neurons outside areas of infarction (KINOUCHI et al. 1993a,b, 1994a) and it is these cells that also express hsp70 protein (KINOUCHI et al. 1993b). We have proposed that the region outside an infarction where neurons express both hsp70 mRNA and hsp70 protein can be viewed as the "ischemic penumbra" (KINOUCHI et al. 1993a,b). The volume of the penumbra can be quantified as the volume of brain outside an infarction where hsp70 protein is expressed in neurons.

Within areas of infarction many neurons and glia do not express hsp70 mRNA or protein, presumably due to insufficient ATP and other energy-related molecules. They sustain a transcriptional and translational block. Some neurons and glia, particularly at the margins of infarctions, do synthesize hsp70 mRNA. However, many of these cells do not synthesize hsp70 protein and probably die since they are within the infarct. These cells sustain a translational block, but not a transcriptional block for hsp70. Lastly, a fair number of glia, including microglia and some astrocytes, synthesize hsp70 mRNA and hsp70 protein at the margins of the infarction. In spite of making hsp70 protein, a number of these glial cells appear to die within the infarcts. Those outside the infarction may survive. The fate of the endothelial cells within an infarction that make hsp70 mRNA and hsp70 protein is unknown.

Most of the neurons containing hsp70 protein in the penumbra generally survive the focal ischemia. This is based upon the normal morphology of the cells. In addition, DNA nick end-labeling (TUNEL) and heat shock protein (hsp70) immunocytochemistry show that the cells labeled with the two methods are mainly separate populations of cells (STATES et al. 1996). In the cortex, hsp70 immunoreactive neurons were located outside areas of infarction and showed little evidence of DNA fragmentation. Hsp70-stained cortical neurons were intermingled with TUNEL stained cells near the infarct, but extended for greater distances away from the infarct. Cells that stained for either hsp70 protein or DNA fragmentation existed in close proximity to one another. Approximately 5–7% of hsp70-stained cells were TUNEL stained and 6% of TUNEL-positive cells also stained for hsp70. There was no hsp70 staining or DNA fragmentation in the brains of sham-operated controls or in the brains of animals 7 days following MCA occlusions.

This shows that ischemic cells capable of translating hsp70 protein generally do not undergo DNA fragmentation. Most hsp70 protein-containing neurons in the cortical "penumbra" and hippocampus survive the ischemic injury and are "reversibly injured." CA1 hippocampal pyramidal neurons die or are reversibly injured in approximately half of the animals following permanent MCA occlusions (STATES et al. 1996).

This study of hippocampus following MCA occlusion (STATES et al. 1996) and others (KINOUCHI et al. 1994a) show hsp70 stress gene expression in hippocampus, thalamus and substantia nigra (KINOUCHI et al. 1994a; STATES et al. 1996) following MCA occlusions. This is of interest since these regions are outside the MCA distribution, and therefore were not ischemic. We have speculated that ischemic depolarization of cortex and striatum may activate distant brain structures. If depolarization was sustained, this might lead to cellular injury and hsp70 induction in these structures (SHARP and SAGAR 1994; SHARP 1995; KINOUCHI et al. 1994b). This is supported by our finding that pre-administration of MK801 blocked hsp70 induction in thalamus and hippocampus following MCA occlusions (KINOUCHI et al. 1994b). This finding was interpreted to mean that cortical spreading depression activated cells in thalamus and hippocampus; and since MK801 blocks spreading depression in cortex this blocked hsp70 induction in thalamus and hippocampus (KINOUCHI et al. 1994b).

II. Hsp32 (HO-1) Spreading Depression Mediated Induction in Microglia

Hsp32 induction was compared to hsp70 heat shock protein induction following focal ischemia (NIMURA et al. 1996). One day following a short duration of focal ischemia, HO-1 and hsp70 staining in striatum occurred mainly in endothelial cells in infarcts and in glial cells around the areas of infarction. HO-1 protein was not induced in cortex by this short period of ischemia, whereas hsp70 was induced in cortical neurons in the MCA distribution. One day following prolonged MCA ischemia, both HO-1 and hsp70 were induced in neurons in cortex in the MCA distribution. *HO-1, however, was induced in glial cells throughout ipsilateral cortex, inside as well as outside the MCA distribution.* The results show that translation and/or transcription of the HO-1 and hsp70 genes are blocked in neurons and glia destined to die within infarcts, whereas translation of these stress genes continues in the endothelial cells. The duration of ischemia required to induce hsp70 in cortical neurons appears to be less than that required to induce HO-1 in cortical glia (Nimura et al. 1996).

HO-1 protein was induced in microglia throughout the ipsilateral cortex, inside as well as outside areas of infarction (Nimura et al. 1996; Bergeron et al. 1997; Geddes et al. 1996). *These results are of interest since they demonstrate that focal MCA ischemia can induce a stress gene in microglia throughout the cortex outside areas of infarction.* Glial cells outside areas of infarction mount a stress response to the infarction, and this response occurs at long distances from the infarction. Alhough the mechanism of this distant response is speculative, it could be related to spreading depression. Following MCA occlusion, waves of spreading depression occur throughout the entire rat cerebral hemisphere. It is likely that glia, as well as neurons, are depolarized by these waves of spreading depression. This may result in calcium entry into glia, activation of cAMP, or activation of other intracellular messengers that could induce c-fos or related immediate early genes in the glia. Once Fos and Jun family members were induced, these transcription factors could bind to the AP-1 site on the promoter of the HO-1 gene and induce HO-1 in microglia throughout the hemisphere. A similar mechanism could occur in astrocytes, resulting in the induction of genes with AP-1 sites in astrocytes. Such genes would include GFAP in astrocytes.

III. Hsp27 Spreading Depression Mediated Induction in Astrocytes

A large number of heat shock proteins are induced following focal ischemia including hsp70, grp78, hsp27, hsp90 and hsp47 (Higashi et al. 1994; Bergeron et al. 1997; Massa et al. 1996; Wang et al. 1993). Messenger RNAs of the hsp70 family proteins were induced within 4 h after ischemia and then rapidly decreased, whereas hsp27 and hsp47 mRNAs reach a maximum level of expression at 24 h and 48 h after ischemia, respectively. In situ hybridization showed that the expression of inducible hsp70 mRNA was observed predominantly in regions adjacent to the ischemic core. Hsp27 mRNA, however, was expressed over a broad area of the ipsilateral cerebral neocortex except for the ischemic center 24 h after ischemia (Higashi et al. 1994).

Hsp27 protein was not expressed in normal brain but was induced in microglia in the ischemic center at 4 h after the stroke. Hsp27 was induced in reactive astrocytes distributed widely in the ipsilateral hemisphere at 1–14 days following the stroke (Kato et al. 1995). This pattern of whole hemisphere induction of hsp27 in astrocytes is similar to the whole hemisphere induction of HO-1 in microglia. We propose that the mechanism of induction is similar for both genes. That is, MCA occlusion produces spreading depression. The spreading depression induces immediate early genes in both microglia and in astrocytes. Induction of c-fos in microglia results in induction of HO-1 in microglia. Induction of c-fos, or a related immediate early gene, in astrocytes results in induction of GFAP and perhaps genes like hsp27. This generalized microglial and astrocytic stress gene response around areas of ischemia likely contributes to the cortical neuronal and glial re-organization that occurs in cortex surrounding areas of infarction.

IV. Glucose Transporters/grp75/grp78: HIF Mediated Induction

Another group of stress proteins, called the glucose regulated proteins (grps), is also induced in ischemic brain. These proteins are induced by low glucose, hypoxia, and calcium ionophores (MASSA et al. 1996). Grp78 is found in the endoplasmic reticulum and may play a role in chaperoning proteins through the ER and in glycosylating proteins in the ER (LI et al. 1994; MASSA et al. 1996). Grp75, the mitochondrial hsp70, is found in mitochondrial membranes and chaperones proteins in and out of the mitochondria (MASSA et al. 1995, 1996). At least one of the glucose transporters, GLUT1, is also a glucose regulated protein that transports glucose in and out of the brain – across endothelial cell membranes and astrocytic end feet membranes (VANNUCCI et al. 1996).

Grp78 and grp75 are probably present in every cell. Immunocytochemistry for grp75, however, shows that it is highly expressed in neurons, and most highly expressed in very large neurons in the basal forebrain, paraventricular nucleus, motor neurons and others (MASSA et al. 1995). Following MCA occlusion, grp75 is highly induced in neurons in the cingulate and parietal cortex, regions outside the MCA infarction (MASSA et al. 1995, 1996). GLUT1 mRNA is also induced following focal and global ischemia (VANNUCCI et al. 1996; GERHART et al. 1994; LEE and BONDY 1993), presumably contributing to increased glucose transport into ischemic/hypoxic/hypoglycemic brain. The neuronal glucose transporter, GLUT3, decreases in areas where neurons are lost (MCCALL et al. 1995).

Although the mechanisms of GLUT1, grp75 and grp78 induction are not known, it is possible that tissue hypoxia contributes to the induction of these and related genes. GLUT1 is known to be a target gene for a transcription factor called HIF, hypoxia-inducible-factor (SEMENZA 1994). Once HIF is induced by hypoxia, HIF binds to target genes (WANG and SEMENZA 1993) and induces these genes. HIF target genes include the glycolytic enzymes, VEGF and GLUT1 (SEMENZA et al. 1994; EBERT et al. 1995; STEIN et al. 1995). It is possible that grp75 and other hypoxia/glucose inducible genes are induced by HIF. If true, then HIF and all of its target genes would be induced in the hypoxic regions around cerebral infarction.

V. Conclusions

It is tempting to propose three major classes of stress gene response around an area of infarction (Fig. 1). The first region is one where individual neurons are injured, have denatured proteins within them, and express hsp70 protein and related proteins that respond to protein denaturation. This region we have

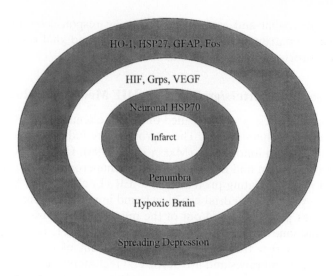

Fig. 1. A series of molecular penumbras can be described around an area of infarction. The first is described by a region where the hsp70 heat shock gene is induced in neurons, and likely corresponds to a limited region around an infarct classically described as the "penumbra." There is a broader region outside this which includes all brain tissue that was transiently hypoxic and where some degree of blood flow reduction occurred, and where hypoxia inducible factor (*HIF*) and its target genes are induced. There is an even broader region outside this where no reduction of blood flow occurs, but where spreading depression occurs which results in induction of a variety of genes including transcription factors like fos and zinc finger genes

named the "penumbra." The second region around an area of infarction may reflect a zone of hypoxia, where blood flow was reduced for a period of time that resulted in decreased oxygen delivery and induction of HIF and its target genes including the grps, GLUT1, and glycolytic enzymes. It is likely that the region of hypoxic brain is considerably larger than the penumbra defined by hsp70 induction (Fig. 1). Lastly, there is an extensive region around an infarct where spreading depression induces changes of gene expression, and results in induction of c-fos, c-jun and other immediate early genes in neurons and glia, and results in the induction of stress genes like HO-1 in microglia and hsp27 in astrocytes. The zone of spreading depression involves the entire hemisphere in a rat brain, but may only involve several sulci in human brain.

D. Cellular Stress Gene Response to Subarachnoid Hemorrhage

I. Clinical Syndrome of Subarachnoid Hemorrhage and the Role of HO

Following the rupture of arterial aneurysms, there is leakage of arterial blood into the subarachnoid space of brain (HALEY et al. 1993, 1994, 1997). Although some patients die immediately following this subarachnoid hemorrhage (SAH), many survive. Red blood cells within the cerebrospinal fluid space (CSF) lyse within a few hours, releasing hemoglobin and other intracellular components (MACDONALD and WEIR 1991). The fate of the iron containing hemoglobin is unknown, though it is postulated to lead to the delayed vasospasm and stroke following SAH (HALEY et al. 1993; KASSELL et al. 1996; MACDONALD and WEIR 1991, 1994; SASAKI and KASSELL 1990; FINDLAY et al. 1995).

The heme released from hemoglobin is metabolized by hemeoxygenase (HO) (MAINES 1988, 1992, 1996). HO is an intracellular protein that metabolizes heme, either derived from extracellular or intracellular sources, to carbon monoxide (CO), iron (Fe), and biliverdin. The biliverdin is then metabolized by biliverdin reductase to bilirubin (EWING et al. 1993; EWING and MAINES 1995). The bilirubin formed within cells must be cleared from the brain – either by diffusion into blood vessels or by bulk flow of the CSF.

Two different proteins account for the HO protein enzymatic activity: HO-1 and HO-2 (VREMAN and STEVENSON 1988). HO-2 is the product of one and perhaps several different genes (MAINES 1996). HO-2 is found in most neurons. It is generally not inducible, except by adrenal steroids (MAINES et al. 1996; WEBER et al. 1994). HO-2 accounts for most of the basal HO enzymatic activity in the brain (MAINES 1992, 1996). The possible role of HO-2 in the metabolism of heme following subarachnoid hemorrhage is described below.

HO-1 protein which is the product of a separate gene is expressed at low levels in normal, rodent adult brain. HO-1 is known to be the heat shock gene, hsp32 (SHIBAHARA et al. 1993). HO-1 has heat shock elements in its promoter, and is inducible by heat shock in all tissues examined (EWING et al. 1994; EWING and MAINES 1991). However, HO-1 is also inducible by a wide variety of factors, including c-fos (LAVROVSKY et al. 1994). Heme induces HO-1 by virtue of a heme/metal responsive element in the HO-1 promoter (CHOI and ALAM 1996). Hypoxia induces HO-1 by virtue of hypoxia-inducible-factor responsive element in its promoter (LEE et al. 1997). Oxidative stress also induces HO-1, likely due to the presence of a NFKB responsive element in the HO-1 promoter (CHOI and ALAM 1996). Hence, HO-1 is a very complex gene – crucial to the metabolism of extracellular heme, but also inducible by a wide variety of cellular stresses (CHOI and ALAM 1996).

II. Induction of HO-1 Following Experimental Subarachnoid Hemorrhage

To examine the role of HO following SAH, either lysed blood, whole blood, bovine serum albumin (BSA) or saline were injected into the cisterna magna of adult rats. HO-2 protein did not change on Western blots or on immunocytochemistry studies following these injections. This is consistent with current ideas that HO-2 mRNA and protein are not generally inducible, and do not appear to be inducible by the heme in hemoglobin (Matz et al. 1996a).

HO-1, however, was markedly induced after subarachnoid injections of both lysed blood and whole blood via the cisterna magna (Matz et al. 1996a–d, 1997). HO-1 protein was massively induced on Western blots of brain following lysed and whole blood injections into the cisterna magna. This induction occurred almost exclusively in microglia. HO-1 induction in microglia occurred in every structure examined, including cortex, hippocampus, striatum, thalamus, hypothalamus, brainstem and cerebellum. In addition, HO-1 was induced in astrocytes in specific layers of the hippocampus, and in the Bergmann glia of cerebellum (Matz et al. 1996a).

There was little HO-1 induction following cisternal injections of either BSA or saline. These injections produced modest HO-1 induction in astrocytes in specific layers of the hippocampus (Matz et al. 1996a). Thus the subarachnoid injections, themselves, produced a minimal stress response in hippocampal astrocytes. These results are important for showing that injection of the protein, BSA, does not induce HO-1 throughout brain. Hence, denatured proteins were not the stimulus for HO-1 induction in microglia.

III. Induction of HO-1 Following Subarachnoid Injections of Hemoglobin and Protoporphyrins

Based upon the results of the lysed and whole blood injections, we postulated that heme derived from hemoglobin was the major stimulus for HO-1 induction in microglia. To address this, the distribution of labeled hemoglobin was examined. Hemoglobin was biotinylated and injected into the cisterna magna. The biotinylated hemoglobin distributed throughout the ventricular system and subarachnoid spaces within a few hours. Biotinylated hemoglobin penetrated deeply into every structure in the brain. Cellular localization suggested a large amount of the labeled hemoglobin was found in neurons (Turner et al. 1997).

To assess whether heme from hemoglobin might be responsible for HO-1 induction, animals had purified hemoglobin (HbAo), bovine serum albumin (BSA) or saline injected into the cisterna magna. HbAo induced HO-1 in microglia throughout the entire brain, including the cortex, striatum, hippocampus, thalamus, hypothalamus, brainstem and cerebellum (Turner et al. 1997). The HO-1 induction by HbAo was identical to that observed following injections of lysed blood and injections of whole blood into the cisterna magna. This suggested that heme from hemoglobin, derived either from purified hemoglobin or from blood, was taken up into microglia where the heme induced HO-1 in microglia (Turner et al. 1997; Matz et al. 1996a).

To further examine the role of HO in metabolizing heme in the brain, the heme analogue tin protoporphyrin was also injected into the cisterna magna of adult rats. Tin protoporphyrin induced HO-1 protein in both neurons and in microglia, but not in astrocytes. Tin protoporphyrin is known to be a powerful inducer of HO-1. The protoporphyrin could directly stimulate the heme element on the HO-1 promoter, or inhibit heme metabolism, resulting in increased heme within cells and induction of HO-1 (MAINES and TRAKSHEL 1992; McCOUBREY and MAINES 1993; RUBLEVSKAYA and MAINES 1994; SMITH et al. 1993). The tin protoporphyrin results suggested that tin protoporphyrin was taken up into both neurons and microglia, where the porphyrin induced HO-1. Heme and tin protoporphyrin are not taken up into astrocytes, and hence do not induce HO-1 in astrocytes.

The mechanism of heme induction of HO-1 in microglia is unknown. As noted, it is thought that there is a heme-binding element in the HO-1 promoter that leads to induction of HO-1 by heme. Alternatively, heme may activate heat shock factors and induce HO-1 via heat shock elements. Hemin induces embryonic and fetal globins, and elevated expression of the stress genes hsp70, hsp90, and grp78/BiP in erythroid cells. Hemin activates the DNA-binding activity of HSF2, whereas heat shock induces predominantly the DNA-binding activity of a distinct factor, HSF1. Hemin activation of HSF2 might selectively activate HO-1 transcription in microglial cells (SISTONEN et al. 1992, 1994). It is also possible that heme itself is not responsible for HO-1 induction following SAH. Prostaglandins have been implicated in induction of HO-1, and could be released following SAH (AMICI et al. 1992). Induction of c-fos, NFKB and HIF following SAH could also play a role in inducing HO-1.

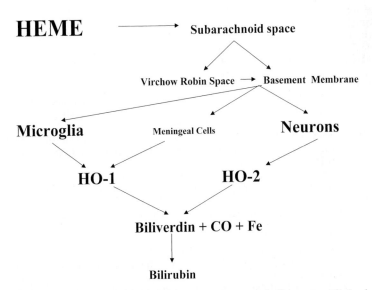

Fig. 2. Proposed cellular elements and enzymes that metabolize extracellular heme to biliverdin, carbon monoxide and iron

IV. Model for Metabolism of Heme by Microglia, Neurons and Meningeal Cells Following Subarachnoid Hemorrhage

A tentative hypothesis about how heme released from hemoglobin after subarachnoid hemorrhage is metabolized is shown in Fig. 2. Following subarachnoid hemorrhage, free hemoglobin can diffuse throughout the subarachnoid spaces. Some heme is taken up into meningeal cells, where it induces HO-1. The bulk of the heme, however, is taken up into both microglia and neurons. In microglia the heme induces HO-1 that then metabolizes the heme. In neurons, heme is taken up and metabolized by HO-2 to CO, Fe and biliverdin. Biliverdin reductase in microglia, neurons and meningeal cells then metabolizes biliverdin to bilirubin. The bilirubin then either diffuses back across the vessels or is carried to the venous system by bulk flow of CSF to the superior sagittal sinus.

References

Abe K, Kawagoe J, Araki T, Aoki M, Kogure K (1992) Differential expression of heat shock protein 70 gene between the cortex and caudate after transient focal cerebral ischaemia in rats. Neurol Res 14:381–385

Abravaya K, Phillips B, Morimoto RI (1991) Attentuation of the heat shock response in Hela cells is mediated by the release of bound heat shock transcription factor and is modulated by changes in growth and in heat shock temperatures. Genes Dev 5:2117–2127

Amici C, Sistonen L, Santoro MG, Morimoto RI (1992) Antiproliferative prostaglandins activate heat shock transcription factor. Proc Natl Acad Sci USA 89:6227–6231

Barbe MF, Tytell M, Gower DJ, Welch WJ (1988) Hyperthermia protects against light damage in the rat retina. Science 241:1817–1820

Bergeron M, Ferriero DM, Vreman HJ, Stevenson DK, Sharp FR (1997) Hypoxia-ischemia, but not hypoxia alone, induces the expression of heme oxygenase-1 (HSP32) in newborn rat brain. J Cereb Blood Flow Metab 17:647–658

Brown IR (1990) Induction of heat shock (stress) genes in the mammalian brain by hyperthermia and other traumatic events: a current perspective. J Neurosci Res 27:247–255

Brown IR (1994) Induction of heat shock genes in the mammaliam brain by hyperthermia and tissue injury. In: Mayer J, Brown IR (eds) Heat shock proteins in the nervous system. Academic, London, pp 31–53

Brown IR, Rush SJ (1990) Expression of heat shock genes (hsp70) in the mammaliam brain: distinguishing constitutively expressed and hyperthermia-inducible species. J Neurosci Res 25:14–19

Brown IR, Rush SJ (1996) In vivo activation of neural heat shock transcription factor HSF1 by a physiologically relevant increase in body temperature. J Neurosci Res 44:52–57

Choi AM, Alam J (1996) Heme oxygenase-1: function, regulation, and implication of a novel stress-inducible protein in oxidant-induced lung injury. Am J Respir Cell Mol Biol 15:9–19

D'Souza CA, Rush SJ, Brown IR (1998a) Effect of hyperthermia on the transcription rate of heat shock genes in the rabbit cerebellum and retina assayed by nuclear run-ons. J Neurosci Res 52:538–548

D'Souza SM, Brown IR (1998b) Constitutive expression of heat shock proteins hsp90, hsc70, hsp70 and hsp60 in neural and non-neural tissues of the rat during postnatal development. Cell Stress Chaperones 3:188–199

Ebert BL, Firth JD, Ratcliffe PJ (1995) Hypoxia and mitochondrial inhibitors regulate expression of glucose transporter-1 via distinct Cis-acting sequences. J Biol Chem 270:29083–29089

Ewing JF, Maines MD (1991) Rapid induction of heme oxygenase 1 mRNA and protein by hyperthermia in rat brain: heme oxygenase 2 is not a heat shock protein. Proc Natl Acad Sci USA 88:5364–5368

Ewing JF, Maines MD (1995) Immunohistochemical localization of biliverdin reductase in rat brain: age related expression of protein and transcript. Brain Res 672:29–41

Ewing JF, Raju VS, Maines MD (1994) Induction of heart heme oxygenase-1 (hsp32) by hyperthermia: possible role in stress-mediated elevation of cyclic $3':5'$-guanosine monophosphate. J Pharmacol Exp Ther 271:408–414

Ewing JF, Weber CM, Maines MD (1993) Biliverdin reductase is heat resistant and coexpressed with constitutive and heat shock forms of heme oxygenase in brain. J Neurochem 61:1015–1023

Fawcett TW, Sylvester SL, Sarge KD, Morimoto RI, Holbrook NJ (1994) Effects of neurohormonal stress and aging on the activation of mammalian heat shock factor 1. J Biol Chem 269:32272–32278

Findlay JM, Kassell NF, Weir BK, Haley EC Jr, Kongable G, Germanson T, Truskowski L, Alves WM, Holness RO, Knuckey NW (1995) A randomized trial of intraoperative, intracisternal tissue plasminogen activator for the prevention of vasospasm. Neurosurgery 37:168–176; discussion 177–168

Foster JA, Brown IR (1996a) Intracellular localization of heat shock mRNAs (hsc70 and hsp70) to neural cell bodies and processes in the control and hyperthermic rabbit brain. J Neurosci Res 46:652–665

Foster JA, Brown IR (1996b) Basal expression of stress-inducible hsp70 mRNA detected in hippocampal and cortical neurons of normal rabbit brain. Brain Res 724:73–83

Foster JA, Brown IR (1997) Differential induction of heat shock mRNA in oligodendrocytes, microglia, and astrocytes following hyperthermia. Mol Brain Res 45:207–218

Foster JA, Rush SJ, Brown IR (1995) Localization of constitutive-and hyperthermia-inducible heat shock mRNAs (hsc70 and hsp70) in the rabbit cerebellum and brainstem by non-radioactive in situ hybridization. J Neurosci Res 41:603–612

Gaspary H, Graham SH, Sagar SM, Sharp FR (1995) Hsp70 heat shock protein induction following global ischemia in the rat. Brain Res Mol Brain Res 34:327–332

Geddes JW, Pettigrew LC, Holtz ML, Craddock SD, Maines MD (1996) Permanent focal and transient global cerebral ischemia increase glial and neuronal expression of heme oxygenase-1, but not heme oxygenase-2, protein in rat brain. Neurosci Lett 210:205–208

Gerhart DZ, Leino RL, Taylor WE, Borson ND, Drewes LR (1994) GLUT1 and GLUT3 gene expression in gerbil brain following brief ischemia: an in situ hybridization study. Brain Res Mol Brain Res 25:313–322

Gonzalez MF, Lowenstein D, Fernyak S, Hisanaga K, Simon R, Sharp FR (1991) Induction of heat shock protein 72-like immunoreactivity in the hippocampal formation following transient global ischemia. Brain Res Bull 26:241–250

Gonzalez MF, Shiraishi K, Hisanaga K, Sagar SM, Mandabach M, Sharp FR (1989) Heat shock proteins as markers of neural injury. Brain Res Mol Brain Res 6:93–100

Haley EC Jr, Kassell NF, Apperson-Hansen C, Maile MH, Alves WM (1997) A randomized, double-blind, vehicle-controlled trial of tirilazad mesylate in patients with aneurysmal subarachnoid hemorrhage: a cooperative study in North America. J Neurosurg 86:467–474

Haley EC Jr, Kassell NF, Torner JC (1993) A randomized controlled trial of high-dose intravenous nicardipine in aneurysmal subarachnoid hemorrhage. A report of the Cooperative Aneurysm Study. J Neurosurg 78:537–547

Haley EC Jr, Kassell NF, Torner JC, Truskowski LL, Germanson TP (1994) A random-ized trial of two doses of nicardipine in aneurysmal subarachnoid hemorrhage. A report of the Cooperative Aneurysm Study. J Neurosurg 80:788–796

Higashi T, Takechi H, Uemura Y, Kikuchi H, Nagata K (1994) Differential induction of mRNA species encoding several classes of stress proteins following focal cere-bral ischemia in rats. Brain Res 650:239–248

Higashi T, Nakai A, Uemura Y, Kikuchi H, Nagata K (1995) Activation of heat shock factor 1 in rat brain during ischemia or after heat shock. Mol Brain Res 34:262–270

Jurivich DA, Sistonen L, Sarge KD, Morimoto RI (1994) Arachidonate is a potent modulator of human heat shock gene transcription. Proc Natl Acad Sci USA 91:2280–2284

Kamii H, Kinouchi H, Sharp FR, Koistinaho J, Epstein CJ, Chan PH (1994) Prolonged expression of hsp70 mRNA following transient focal cerebral ischemia in transgenic mice overexpressing CuZn-superoxide dismutase. J Cereb Blood Flow Metab 14:478–486

Kassell NF, Haley EC Jr, Apperson-Hansen C, Alves WM (1996) Randomized, double-blind, vehicle-controlled trial of tirilazad mesylate in patients with aneu-rysmal subarachnoid hemorrhage: a cooperative study in Europe, Australia, and New Zealand. J Neurosurg 84:221–228

Kato H, Kogure K, Liu XH, Araki T, Kato K, Itoyama Y (1995) Immunohistochemical localization of the low molecular weight stress protein hsp27 following focal cere-bral ischemia in the rat. Brain Res 679:1–7

Kinouchi H, Sharp FR, Chan PH, Koistinaho J, Sagar SM, Yoshimoto T (1994a) Induction of c-fos, junB, c-jun, and hsp70 mRNA in cortex, thalamus, basal gan-glia, and hippocampus following middle cerebral artery occlusion. J Cereb Blood Flow Metab 14:808–817

Kinouchi H, Sharp FR, Chan PH, Mikawa S, Kamii H, Arai S, Yoshimoto T (1994b) MK-801 inhibits the induction of immediate early genes in cerebral cortex, thala-mus, and hippocampus, but not in substantia nigra following middle cerebral artery occlusion. Neurosci Lett 179:111–114

Kinouchi H, Sharp FR, Hill MP, Koistinaho J, Sagar SM, Chan PH (1993a) Induction of 70-kDa heat shock protein and hsp70 mRNA following transient focal cerebral ischemia in the rat. J Cereb Blood Flow Metab 13:105–115

Kinouchi H, Sharp FR, Koistinaho J, Hicks K, Kamii H, Chan PH (1993b) Induction of heat shock hsp70 mRNA and hsp70 kDa protein in neurons in the 'penumbra' following focal cerebral ischemia in the rat. Brain Res 619:334–338

Lavrovsky Y, Schwartzman ML, Levere RD, Kappas A, Abraham NG (1994) Identi-fication of binding sites for transcription factors NF-kappa B and AP-2 in the promoter region of the human heme oxygenase 1 gene. Proc Natl Acad Sci USA 91:5987–5991

Lee PJ, Jiang B-H, Chin BY, Iyer NV, Alam J, Semenza GL, Choi AMK (1997) Hypoxia-inducible Factor-1 mediates transcriptional activation of the heme oxygenase-1 gene in response to hypoxia. J Biol Chem 272:5375–5381

Lee WH, Bondy CA (1993) Ischemic injury induces brain glucose transporter gene expression. Endocrinology 133:2540–2544

Li Y, Chopp M, Garcia JH, Yoshida Y, Zhang ZG, Levine SR (1992) Distribution of the 72-kd heat-shock protein as a function of transient focal cerebral ischemia in rats. Stroke 23:1292–1298

Li WW, Sistonen L, Morimoto RI, Lee AS (1994) Stress induction of the mammalian GRP78/BiP protein gene: in vivo genomic footprinting and identification of p70CORE from human nuclear extract as a DNA-binding component specific to the stress regulatory element. Mol Cell Biol 14:5533–5546

Li Y, Chopp M, Zhang ZG, Zhang RL, Garcia JH (1993) Neuronal survival is associ-ated with 72-kDa heat shock protein expression after transient middle cerebral artery occlusion in the rat. J Neurol Sci 120:187–194

Lipshitz HD (ed) (1995) Localized RNAs. Landes, Austin

Macdonald RL, Weir BK (1991) A review of hemoglobin and the pathogenesis of cerebral vasospasm. Stroke 22:971–982

Macdonald RL, Weir BK (1994) Cerebral vasospasm and free radicals. Free Radic Biol Med 16:633–643

Maines MD (1988) Heme oxygenase: function, multiplicity, regulatory mechanisms, and clinical applications. FASEB J 2:2557–2568

Maines MD (1992) Heme oxygenase: clinical applications and functions. CRC Press, Boca Raton

Maines M (1996) Carbon monoxide and nitric oxide homology: differential modulation of heme oxygenases in brain and detection of protein and activity. Methods Enzymol 268:473–488

Maines MD, Eke BC, Zhao X (1996) Corticosterone promotes increased heme oxygenase-2 protein and transcript expression in the newborn rat brain. Brain Res 722:83–94

Maines MD, Trakshel GM (1992) Differential regulation of heme oxygenase isozymes by Sn- and Zn-protoporphyrins: possible relevance to suppression of hyperbilirubinemia. Biochim Biophys Acta 1131:166–174

Manzerra P, Brown IR (1992a) Distribution of constitutive-and hyperthermia-inducible heat shock mRNA species (hsp70) in the Purkinje layer of the rabbit cerebellum. Neurochem Res 17:559–564

Manzerra P, Brown IR (1992b) Expression of heat shock genes (hsp70) in the rabbit spinal cord: localization of constitutive-and hyperthermia-inducible mRNA species. J Neurosci Res 31:606–615

Manzerra P, Brown IR (1996) The neuronal stress response: nuclear translocation of heat shock proteins as an indicator of hyperthermic stress. Exp Cell Res 229:35–47

Manzerra P, Rush SJ, Brown IR (1993) Temporal and spatial distribution of heat shock mRNA and protein (hsp70) in the rabbit cerebellum in response to hyperthermia. J Neurosci Res 36:480–490

Manzerra P, Rush SJ, Brown IR (1997) Tissue-specific differences in heat shock protein hsc70 and hsp70 in the control and hyperthermic rabbit. J Cell Physiol 170:130–137

Massa SM, Longo FM, Zuo J, Wang S, Chen J, Sharp FR (1995) Cloning of rat grp75, an hsp70-family member, and its expression in normal and ischemic brain. J Neurosci Res 40:807–819

Massa SM, Swanson RA, Sharp FR (1996) The stress gene response in brain. Cerebrovasc Brain Metab Rev 8:95–158

Mathur SK, Sistonen L, Brown IR, Murphy SP, Sarge, KD, Morimoto RI (1994) Deficient induction of human hsp70 heat shock gene transcription in Y79 retinoblastoma cells despite activation of heat shock factor 1. Proc Natl Acad Sci USA 91:8695–8699

Matz P, Turner C, Weinstein PR, Massa SM, Panter SS, Sharp FR (1996a) Heme-oxygenase-1 induction in glia throughout rat brain following experimental sub-arachnoid hemorrhage. Brain Res 713:211–222

Matz P, Weinstein P, States B, Honkaniemi J, Sharp FR (1996b) Subarachnoid injections of lysed blood induce the hsp70 stress gene and produce DNA fragmentation in focal areas of the rat brain. Stroke 27:504–512; discussion 513

Matz PG, Massa SM, Weinstein PR, Turner C, Panter SS, Sharp FR (1996c) Focal hyperexpression of hemeoxygenase-1 protein and messenger RNA in rat brain caused by cellular stress following subarachnoid injections of lysed blood. J Neurosurg 85: (in press)

Matz PG, Sundaresan S, Sharp FR, Weinstein PR (1996d) Induction of hsp70 in rat brain following subarachnoid hemorrhage produced by endovascular perforation. J Neurosurg 85:138–145

Matz PG, Weinstein PR, Sharp FR (1997) Heme oxygenase-1 and heat shock protein 70 induction in glia and neurons throughout rat brain after experimental intra-cerebral hemorrhage. Neurosurgery 40:152–160; discussion 160–152

McCall AL, Moholt-Siebert M, VanBueren A, Cherry NJ, Lessov N, Tiffany N, Thompson M, Downes H, Woodward WR (1995) Progressive hippocampal loss of immunoreactive GLUT3, the neuron-specific glucose transporter, after global forebrain ischemia in the rat. Brain Res 670:29–38

McCoubrey WK Jr, Maines MD (1993) Domains of rat heme oxygenase-2: the amino terminus and histidine 151 are required for heme oxidation. Arch Biochem Biophys 302:402–408

Morimoto R (1993) Cells in stress: transcriptional activation of heat shock genes. Science 259:1409–1410

Morimoto R, Tissieres A, Georgopoulos C (1994) The biology of the heat shock proteins and molecular chaperones. Cold Spring Harbor Laboratory Press, New York

Nimura T, Weinstein PR, Massa SM, Panter S, Sharp FR (1996) Heme oxygenase-1 (HO-1) protein induction in rat brain following focal ischemia. Brain Res Mol Brain Res 37:201–208

Nishimura RN, Dwyer BE, Welch W, Cole R, de Vellis J, Liotta K (1988) The induction of the major heat-stress protein in pruified rat glial cells. J Neurosci Res 20:12–18

Nishimura RN, Dwyer BE, de Vellis J, Clegg KB (1992) Characterization of the major 68 kDa heat shock protein in a rat transformed astroglial cell line. Mol Brain Res 12:203–208

Planas AM, Soriano MA, Estrada A, Sanz O, Martin F, Ferrer I (1997) The heat shock stress response after brain lesions: induction of 72 kDa heat shock protein (cell types involved, axonal transport, transcriptional regulation) and protein synthesis inhibition. Prog Neurobiol 51:607–636

Quraishi H, Brown IR (1995) Expression of heat shock protein 90 (hsp90) in neural and nonneural tissues of the control and hyperthermic rabbit. Exp Cell Res 219:358–363

Rublevskaya I, Maines MD (1994) Interaction of Fe-protoporphyrin IX and heme analogues with purified recombinant heme oxygenase-2, the constitutive isozyme of the brain and testes. J Biol Chem 269:26390–26395

Sasaki T, Kassell NF (1990) The role of endothelium in cerebral vasospasm. Neurosurg Clin North Am 1:451–463

Semenza GL (1994) Regulation of erythropoietin production. New insights into molecular mechanisms of oxygen homeostasis. Hematol Oncol Clin North Am 8:863–884

Semenza GL, Roth PH, Fang HM, Wang GL (1994) Transcriptional regulation of genes encoding glycolytic enzymes by hypoxia-inducible factor 1. J Biol Chem 269:23757–23763

Sharp FR (1995) Stress proteins are sensitive indicators of injury in the brain produced by ischemia and toxins. J Toxicol Sci 20:450–453

Sharp FR, Kinouchi H, Koistinaho J, Chan PH, Sagar SM (1993) Hsp70 heat shock gene regulation during ischemia. Stroke 24[Suppl 12]:I72–I75

Sharp FR, Lowenstein D, Simon R, Hisanaga K (1991) Heat shock protein hsp72 induction in cortical and striatal astrocytes and neurons following infarction. J Cereb Blood Flow Metab 11:621–627

Sharp FR, Sagar SM (1994) Alterations in gene expression as an index of neuronal injury: heat shock and the immediate early gene response. Neurotoxicology 15:51–59

Shibahara S, Yoshizawa M, Suzuki H, Takeda K, Meguro K, Endo K (1993) Functional analysis of cDNAs for two types of human heme oxygenase and evidence for their separate regulation. J Biochem (Tokyo) 113:214–218

Sistonen L, Sarge KD, Morimoto RI (1994) Human heat shock factors 1 and 2 are differentially activated and can synergistically induce hsp70 gene transcription. Mol Cell Biol 14:2087–2099

Sistonen L, Sarge KD, Phillips B, Abravaya K, Morimoto RI (1992) Activation of heat shock factor 2 during hemin-induced differentiation of human erythroleukemia cells. Mol Cell Biol 12:4104–4111

Smith A, Alam J, Escriba PV, Morgan WT (1993) Regulation of heme oxygenase and metallothionein gene expression by the heme analogs, cobalt-, and tin-protoporphyrin. J Biol Chem 268:7365–7371

Sprang GK, Brown IR (1987) Selective induction of a heat shock gene in fibre tracts and cerebellar neurons of the rabbit brain detected by in situ hybridization. Mol Brain Res 3:89–93

States BA, Honkaniemi J, Weinstein PR, Sharp FR (1996) DNA fragmentation and hsp70 protein induction in hippocampus and cortex occurs in separate neurons following permanent middle cerebral artery occlusions. J Cereb Blood Flow Metab 16:1165–1175

Stein I, Neeman M, Shweiki D, Itin A, Keshet E (1995) Stabilization of vascular endothelial growth factor mRNA by hypoxia and hypoglycemia and coregulation with other ischemia-induced genes. Mol Cell Biol 15:5363–5368

Tytell M, Barbe MF, Brown IR (1993) Stress (heat shock) protein accumulation in the central nervous system: its relationship to cell stress and damage. In: Seil FJ (ed) Advances in neurology. Raven, New York, pp 292–303

Tytell M, Barbe MF, Brown IR (1994) Induction of heat shock (stress) protein 70 and its mRNA in the normal and light-induced rat retina after whole body hyperthermia. J Neurosci Res 38:19–31

Turner C, Bergeron M, Matz P, Zegna A, Noble L, Panter S, Sharp FR (1997) Heme-oxygenase-1 (HO-1, hsp32) is induced in microglia throughout brain by subarachnoid hemoglobin. J Cereb Blood Flow Metab (in press)

Vannucci SJ, Seaman LB, Vannucci RC (1996) Effects of hypoxia-ischemia on GLUT1 and GLUT3 glucose transporters in immature rat brain. J Cereb Blood Flow Metab 16:77–81

Vazquez J, Pauli D, Tissieres A (1993) Transcriptional regulation in Drosophila during heat shock: a nuclear run-on analysis. Chromosoma 102:233–248.

Vreman HJ, Stevenson DK (1988) Heme oxygenase activity as measured by carbon monoxide production. Anal Biochem 168:31–38

Wang GL, Semenza GL (1993) Characterization of hypoxia-inducible factor 1 and regulation of DNA binding activity by hypoxia. J Biol Chem 268:21513–21518

Wang S, Longo FM, Chen J, Butman M, Graham SH, Haglid KG, Sharp FR (1993) Induction of glucose regulated protein (grp78) and inducible heat shock protein (hsp70) mRNAs in rat brain after kainic acid seizures and focal ischemia. Neurochem Int 23:575–582

Weber CM, Eke BC, Maines MD (1994) Corticosterone regulates heme oxygenase-2 and NO synthase transcription and protein expression in rat brain. J Neurochem 63:953–962

CHAPTER 12

Heat Stress Proteins and Their Relationship to Myocardial Protection

R. Carroll and D.M. Yellon

A. Introduction

Acute myocardial infarction is the most common cause of death in men in the Western world. The treatment of this condition is no longer simply supportive, awaiting the complications of ischaemic injury, but has entered a new era where the mortality of acute myocardial infarction can be approximately halved by thrombolytic agents and aspirin (ISIS-2 1988), with the greatest benefit seen in those treated soon after the onset of symptoms. The lack of a reduction in mortality when thrombolytic treatment is administered late is most likely due to the fact that the prolonged coronary occlusion has resulted in such severe necrosis of the myocardium that little benefit can be derived by restoring blood flow (Reimer and Jennings 1979). Therefore any intervention that could delay the onset of tissue necrosis could buy valuable time by extending the effective temporal window for thrombolysis. Attempts to limit myocardial infarct size over the last decade with exogenous pharmacological agents have been largely unsuccessful (Hearse and Yellon 1983), prompting us and others to explore the heart's endogenous protective mechanisms to ascertain if this route may provide us with the knowledge required to protect the myocardium from severe ischaemic injury.

B. Heat Stress and the Stress Response

All organisms respond at the cellular level to stresses such as heat by the preferential synthesis of a group of proteins known as the heat shock or stress induced proteins. This stress response relates to the process whereby general protein synthesis is inhibited whilst the synthesis of stress proteins is enhanced. A range of studies have now demonstrated the importance of these proteins in both the stressed and the unstressed cell, where they perform functions that seem essential to the maintenance of cellular integrity. Over the last decade interest in this group of proteins has grown with the recognition that manipulation of stress protein content in cells appears to be associated with enhanced cell survival following subsequent injury. Indeed studies suggest that stress proteins may be capable of delaying ischaemic injury to the myocardium (see editorials Yellon and Latchman 1992; Black and Lucchesi 1993; Yellon et al. 1993) and as such could be exploited to our advantage by offering a future approach to increase the effective time window for thrombolysis.

C. Are Stress Proteins Protective?

It is known that stress proteins have the ability to interact with other proteins by altering their conformation, e.g. hsp70 and mitochondrial protein import; protecting them from inappropriate and premature interactions, e.g. hsp90 and the steroid receptor; encouraging the correct refolding of denatured proteins, e.g. hsp90 and Fab or citrate synthase; aid in the degradation of denatured proteins, e.g. with the use of ubiquitin; detect subtle changes in the conformation of other proteins, e.g. hsp70 and clatharin triskelions; and stabilise the cytoskeleton, e.g. alphaB-crystallin interacting with actin and desmin. All these interactions suggest that stress proteins may be capable of increasing cell viability and integrity both during and following denaturing stresses. Therefore if unregulated these proteins may allow the cell to behave in a more efficient manner when it is subjected to subsequent potentially lethal ischaemic stress.

D. Evidence for the Ability of Stress Proteins To Protect the Cell

I. Thermotolerance

When prokaryotic, or eukaryotic cells, are acutely exposed to temperatures several degrees above their normal ambient temperature, the cells or animals will die within a predictable period of time. However, if the cells are initially exposed to the elevated temperature for a shorter time period and allowed to recover at normal temperatures, it has been shown that they can survive a subsequent period of otherwise lethal heat stress. This universal phenomenon is known as "acquired thermotolerance" (GERNER and SCHEIDER 1975). Stress proteins induced by the first period of sublethal thermal stress are attractive candidates to explain the acquisition of thermotolerance.

The evidence that stress proteins are involved in thermotolerance is strong. For example with regard to hsp70 in eukaryotic cell lines; thermal tolerance and stress protein induction are temporally related (SUGAHARA and SAITO 1988); the degree of stress protein expression following stable transfection in cells correlates with the degree of thermotolerance (LI et al. 1991); cell lines expressing abnormal hsp70 with a missing ATPase domain retain thermotolerance (LI et al. 1992); cells microinjected with a monoclonal antibody recognising a shared epitope on constitutive and inducible forms of hsp70 are thermosensitive (RIABOWAL et al. 1988); mutants selected for multiple copies of the hsp70 promoter region are thermosensitive (JOHNSTON and KUCEY 1988; in this study copies of the hsp70 regulatory region presumably competed with the endogenous hsp70 encoding genes for factors that activate hsp70 expression, since hsp70 induction was reduced by 90%); stress proteins incorporating amino acid analogues (i.e. non-functional stress proteins) result in thermal sensitisation (LI and LASZLO 1985). Other stress proteins may also be important since rodent cell lines stably transfected with hsp27 (homologous to alphaB crystallin) are "naturally" thermotolerant (LANDRY and CHRETIEN 1989), whilst cells with an hsp104 heat stress protein deletion are unable to acquire thermotolerance (SANCHEZ and LINDQUIST 1990). In addition transfection with antisense RNA for hsp90 (to specifically hybridise with, and prevent, the translation of native hsp90 mRNA) also prevents the acquisition of thermotolerance (BANSAL et al. 1991). For further discussion see Chap. 3, this volume.

However, it is not certain whether thermotolerance represents better recovery following heat stress or actual attenuation of injury during the thermal stress. LASZLO (1992) has demonstrated that protein synthesis and RNA synthesis are similarly depressed following heat stress of thermotolerant and naive fibroblasts; however, recovery of both processes was enhanced in cells overexpressing hsp70 (either by transfection or by prior treatment with sublethal thermal stress or sodium arsenite). In contrast other investigators (BURGMAN and KONINGS 1992) have found evidence for protection during thermal stress since conformational changes within proteins, detected by both SH group targeted electron spin resonance and mobility by thermal gel analysis, occur at a higher temperature in thermotolerant eukaryotic cell lines. Clearly protection may occur both during and following the thermal denaturing stress and these mechanisms are not mutually exclusive.

The arguments are further complicated by the fact that thermotolerance in some situations can occur in the absence of stress protein induction, leading to the speculation that two types of thermotolerance exist, stress proteins being responsible for the acquisition of long but not short-lasting thermotolerance.

II. Cross-tolerance

If sublethal heat stress induces stress proteins that protect against subsequent lethal heat shock, can other non-thermal stresses that also induce stress proteins give rise to similar thermal tolerance? Interestingly, this does seem to be the case. In addition, and of more relevance to experimental cardiology, thermally induced stress proteins seem capable of protecting against stresses which themselves induce stress protein synthesis. Such cross-protection is known as cross-tolerance.

There are numerous studies that support the concept of cross-tolerance first suggested by LI and HAHN (1978), who demonstrated that a hamster cell line could be rendered resistant to both adriamycin and heat toxicity by pretreatment with ethanol. Later studies have demonstrated similar findings but with very different stresses. For example; whole body heat stress in rats protects retinal pigment cells from light injury, protection being temporally dependent on hsp72 (Barbe et al. 1988); pre-treatment with hydrogen peroxide protects against subsequent oxidative stress (Christman et al. 1985); heat stress protects against subsequent oxidative stress in a number of models (for review, see Yellon and Latchman 1992; Yellon et al. 1993); heat stressed human breast cancer cells are rendered resistant to doxorubicin, an effect that seems related to hsp70 and hsp27 cell content (Ciocca et al. 1992) and heat stressed neuronal cells are resistant to the excito-toxic effects of glutamate, an effect dependent on protein synthesis and related to hsp70 (Lowenstein et al. 1991; Rordorf et al. 1991).

III. Stress Proteins and the Heart

Stress proteins have been detected in the myocardium of a variety of mammalian species. Their synthesis has been shown to be increased by whole body temperature elevation (Currie and White 1983; Currie 1987; Currie et al. 1990) and other stressful stimuli including ischaemia (Mehta et al. 1988), brief ischaemia and reperfusion (Knowlton et al. 1990), anoxia (Tuijl et al. 1991), hypoxia (Howard and Geoghegan 1986), pressure or volume overload (Delcayre et al. 1988), mechanical stretch (Knowlton et al. 1991), cytokines (Low Friedrich et al. 1992) and drugs such as vasopressin (Moalic et al. 1989), isoproterenol (White and White 1986), hydrogen peroxide (Low Friedrich et al. 1989), L-type (slow) calcium channel blockers (Low Friedrich and Schoeppe 1991) and other cardiotoxic drugs and heavy metals (Low Friedrich et al. 1991). Our group have recently shown the ability of the ansamycin antibiotic, Herbimycin-A, to protect cultured cardiomyocytes, with induction of hsp72 (Morris et al. 1996). The implication of these findings is that stress proteins may play an important role in the cardiac stress response. Although many of these stimuli induce members of the 70-kDa family of stress proteins, there appear to be differences in the pattern of stress protein induction when other families of stress proteins are considered. For example, cardiotoxic drugs preferentially induce a 30-kDa stress protein (Low Friedrich et al. 1991). It is possible that each stress induces its own subtly differing profile of stress proteins best suited to meet the cellular consequences of that particular stress.

IV. Heat Stress and Myocardial Protection

Following the suggestion that: (a) tissues with thermally pre-elevated stress proteins are resistant to stresses that normally induce stress proteins (cross-tolerance) and (b) myocardial ischaemia causes stress protein induction (Mehta et al. 1988; Knowlton et al. 1990), investigators have been directly interested in examining whether myocardial tissue entering ischaemia with a pre-elevation of stress proteins is resistant to infarction (CURRIE et al. 1988, 1993; DONNELLY et al. 1992; YELLON et al. 1992).

CURRIE et al. (1988) were the first investigators to show that temperature elevation to 42°C in rats resulted in concomitant cardiac stress protein and catalase induction, and an attenuation of ischaemia/reperfusion injury. Using an isolated heart model these investigators demonstrated that following ischaemia/reperfusion, contractile function is enhanced whilst creatine kinase release is dramatically reduced in heat stress compared with control hearts. These findings have been confirmed by our group in both the rat (PASINI et al. 1991) and the rabbit (YELLON et al. 1992). Moreover, these authors have observed improvements in additional parameters of protection in the heat stressed rabbit heart post-ischaemia (YELLON et al. 1992). These include preservation of high energy phosphates, a reduction in oxidative stress during reperfusion (as measured by lower levels of oxidised glutathione) and significant mitochondrial preservation following ischaemia. There does, however, appear to be some species variation in the metabolic changes associated with protection following heat stress. In the rabbit for example, higher levels of high energy phosphates mirror the enhanced contractile activity of heat stressed hearts during reperfusion (YELLON et al. 1992). In the rat, however, the enhanced contractile activity following ischaemia in the heat stressed groups is not associated with differences in high energy phosphate content between heat stressed and control hearts (CURRIE et al. 1988; CURRIE and KARMAZYN 1990; PASINI et al. 1991).

The protective effects of whole body heat stress have also been shown in the hypertrophied heart, which ordinarily has an increased susceptibility to ischaemic injury. In the hypertrophied rat heart 24 h after whole body heat stress preliminary evidence suggests that ischaemic changes are diminished whilst contractile function is enhanced (CORNELUSSEN et al. 1993).

Not all studies have found heat stress to be cardioprotective. For example, preliminary studies by one group (WALL et al. 1990) have been unable to demonstrate any protection against "no-flow" ischaemia following heat stress in the isolated unpaced working rat heart, with rate pressure product and cardiac output as endpoints. These investigators, although following an established protocol known to enhance cardiac stress protein synthesis, did not measure stress proteins directly.

The studies summarised to this point have demonstrated protection expressed in terms of myocardial contractility and metabolic state. The ability of whole body heat stress to reduce the extent of myocardial infarction as examined by triphenyl tetrazolium staining has also been examined. In a preliminary study in the rat model, infarct size is reduced following 37.5 min of regional ischaemia, with protection being temporally related to elevated stress protein levels at 24–96 h following whole body heat stress (LOESSER et al. 1992). In addition, our group (WALKER et al. 1993) have demonstrated a similar reduction in infarct size following 45 min of regional ischaemia and 2 h reperfusion in the buffer perfused rabbit heart removed 24 h after whole body heat stress.

Interestingly, in contrast to isolated heart studies, controversy surrounds the ability of whole body heat stress to reduce infarct size in vivo. In the rabbit, heat stress 24 h prior to ischaemia was unable to reduce infarct size following a 45-min coronary occlusion (YELLON et al. 1992) although protection was found following a 30-min occlusion by CURRIE et al. (1993). A similar dependence on the length of coronary occlusion is seen in the rat. DONNELLY and co-workers (1992) have demonstrated a reduction in infarct size in the rat following a 35-min, but not a 45-min, coronary occlusion initiated 24 h after whole body heat stress. Moreover in this model the reduction in infarct size, following a graded heat stress procedure, is related to the degree of stress protein induction (HUTTER et al. 1994). The apparent dependency of protection on the length of ischaemic insult is difficult to explain. One possibility is that the protection conferred by heat stress is only moderate, and that as the severity of the ischaemic insult increases the protection becomes less evident. A similar phenomenon occurs with ischaemic preconditioning in dogs where a marked reduction in infarct size occurs with 60 min of coronary occlusion, but not with 90 min (MIURA et al. 1992). Another apparent anomaly is the fact that although infarct size is reduced after a 30-min coronary occlusion 24 h following whole body heat stress in the rabbit, no protection is seen at 48 h after heat stress at a time when cardiac stress protein content is still increased (CURRIE et al. 1993).

Other investigators have been able to demonstrate in vivo protection following hot blood cardioplegia of the pig heart (LIU et al. 1992). In another rather "novel" approach (SCHOTT et al. 1990), a microwave diathermy probe was applied to the canine heart. Although such regional hyperthermia was shown to increase stress protein mRNA expression, there was no concomitant protection against infarction. However, the time course of this study was such that appreciable stress protein accumulation was unlikely to have occurred in the 1 h between thermal stress and coronary ligation. In addition, the spatial relationship between heat-treated myocardium and the subsequent area of hypoperfusion during coronary ligation was not documented.

The cause for the discrepancy between in vivo and in vitro studies is not clear. However, observations from our laboratory suggest that whole body heat stress may activate a blood borne component that overrides the beneficial effect of cardiac stress protein induction. This inference was made after noticing that blood from a heat stressed support rabbit; when used to perfuse an isolated rabbit heart, significantly increased infarct size (WALKER et al. 1993). It may be that whole body heat stress, although conferring myocardial protection, causes confounding physiological changes which have negative effects on infarct size. This is consistent with the finding that cytotoxic T-cells directed against myocardial heat shock proteins are induced in rats by stresses that elevate myocardial heat stress protein content (HUBER 1992), and that these cells are cytotoxic in vitro to heat stressed myocytes from the same species. In addition, the possibility remains that the duration of recovery after heat stress and length of the ischaemic insult may also influence the ultimate infarct size and appearance of cardioprotection.

V. Heat Stress Proteins and Ischaemic Preconditioning

Acquired thermotolerance, where sublethal hyperthermia protects against subsequent lethal hyperthermia, is similar in concept to ischaemic preconditioning, with sublethal ischaemia protecting against subsequent lethal ischaemia. One could speculate that stress proteins synthesised in response to the first brief episode of preconditioning ischaemia are involved in the protection against the subsequent injury. In agreement with this idea KNOWLTON et al. (1990) demonstrated that brief bursts of ischaemia, such as those used in preconditioning protocols, can induce hsp70 mRNA and protein accumulation. The mechanism by which stress proteins are induced by short episodes of ischaemia may be secondary to the free radical stress induced by reperfusion, since in the isolated rat heart stress protein induction following a 15 min infusion of xanthine plus xanthine oxidase is quantitatively similar to that induced by ischaemia with reperfusion (KUKREJA and HESS 1992). However, in both the study by KNOWLTON et al. (1990) and work from our laboratory (KUCUKOGKU et al. 1991), elevated levels of the hsp70 protein were only manifest at 2–24 h after the ischaemic insult. In contrast the protective effect of classical preconditioning is lost approximately 1 h after the initial brief ischaemic episode (VAN WINKLE et al. 1991).

The involvement of stress protein in ischaemic preconditioning has been further questioned by a study (THORNTON et al. 1990) which indicates that the protective effect of preconditioning can be observed under conditions where de novo protein synthesis has been almost entirely inhibited. Thus, it is unlikely that stress proteins are involved in the protection observed in early ischaemic preconditioning. However, the changes in mRNA coding for stress proteins indicate an adaptive response to ischaemia which may predict a delayed protection dependent on stress protein synthesis.

Indeed this is what we found when we (MARBER et al. 1993) either heat stressed or administered repetitive sublethal ischaemia to rabbits hearts and were able to demonstrate a significant limitation of infarct size 24 h later. Furthermore there was a significantly increased expression of the 70-kDa hsp in both the hearts that had the prior heat stress as well as the prior ischaemic insults (MARBER et al. 1993). Reports using the dog model (HOSHIDA et al. 1993; YAMASHITA et al. 1997) have also demonstrated effects of a second phase of protection existing at 24 h after preconditioning with 4 repeated, 5-min episodes of ischaemia. Interestingly, a similar phenomenon appears to occur within the brain, where ischaemic pre-treatment with two, repeated episodes of 2-min bilateral carotid occlusions is capable of limiting the neuronal cell loss that follows a subsequent more prolonged bilateral carotid occlusion (KITAGAWA et al. 1990, 1991; LIU et al. 1992). For this protective effect to be manifest, the short occlusions must precede the long occlusion by at least 24 h, a time interval known to result in cerebral heat stress protein accumulation in an identical model (Nowak 1990). It is interesting to speculate if myocardial adaptation to ischaemia may explain the apparent benefit of a 7 day or more history of angina prior to myocardial infarction (Muller et al. 1990), although likely collateral vessel formation and concomitant medication make any definite conclusions impossible. In contrast other attempts to induce myocardial protection by ischaemic pre-treatment have been unsuccessful. For example, Donelly et al. (1992) compared the protective benefit of heat stress with 24-h of recovery to 20 min of ischaemia with 8 h of reperfusion. Following a subsequent 35-min occlusion in the rat, heat stress pre-treatment reduced infarct size, whilst ischaemic pre-treatment did not. However, as heat stress resulted in a more marked stress protein accumulation, the authors concluded that ischaemic pre-treatment failed to protect because of insufficient stress protein accumulation.

Other methods of stress protein induction have included immobilisation stress, which has also been shown to protect against subsequent ischaemia/reperfusion injury assessed by CPK leakage, contractile function and reperfusion arrhythmias (Meerson and Malyshev 1989; Meerson et al. 1992).

VI. Heat Stress and Protection Against Non-ischaemic Injury

The observations that cardiac tissues synthesise stress proteins in response to a variety of stresses have encouraged investigators to explore the breadth of protection that follows heat stress. In this regard Meerson et al. (1991a,b) has demonstrated that stress protein induction by either heat or immobilisation protects the isolated rat and rabbit heart against a subsequent calcium paradox, a finding confirmed by ourselves in the isolated rabbit heart (Marber et al. 1993).

The fact that oxidant stress is capable of inducing cardiac stress proteins (Low Friedrich et al. 1989; Kukreja et al. 1994) has prompted Su et al. (1992) to examine the protective benefits of prior heat stress with exposure to H_2O_2 as the final stress. In a rat myocyte culture model, heat stress is capable of inducing acquired thermotolerance and limiting myocyte injury on subsequent H_2O_2 exposure.

Several lines of evidence suggest that hsp70i is a cytoprotective protein conferring tissue tolerance to ischaemia-reperfusion injury. For example, Hutter et al. (1994) showed that the rise in myocardial hsp70i content correlated with both the degree of hyperthermia and the extent of subsequent ischaemic tolerance. Our group showed that the post-hypoxic functional recovery of papillary muscles from rabbits pre-treated with hyperthermia correlated with the hsp70i content of an adjacent papillary muscle from the same animal (Marber et al. 1994). More recently, studies with transgenic mice that overexpress hsp70i (Marber et al. 1995; Plumier et al. 1995) and studies involving transfection of the gene encoding for hsp70 into isolated myogenic cells (HEADS et al. 1994, 1995; MESTRIL et al. 1994) and cardiac myocytes (CUMMING et al. 1996) show more convincingly that the protein directly confers protection against ischaemic injury. While there is good evidence then that the presence of hsp70 is associated with an ischaemia-tolerant phenotype, the extent to which this particular stress protein accounts for myocardial protection following heat stress is unclear.

VII. Mechanisms of Cardiac Protection by Elevated Temperature

All studies using heat to elevate stress protein synthesis result in a large number of physiological perturbations which may in themselves have cardioprotective properties. For example, Currie's group (CURRIE et al. 1988; KARMAZYN et al. 1990) have shown that heat stress also results in an increase in the endogenous levels of the anti-oxidant enzyme catalase. Moreover, they have demonstrated (KARMAZYN et al. 1990) that inactivating catalase with 3-AT (3-aminotriazole) results in an abolition of the protective effect normally observed at 24 and 48 h after heat stress. This protective role of catalase would be dependent upon its ability to minimise the damage caused by secondary free radical generation to sulphydryl containing enzymes, DNA and lipids (CECONI et al. 1988) by catalysing the conversion of H_2O_2 to water. Following a period of prolonged ischaemia the importance of catalase is increased since there is a marked reduction in the activity of SOD as well as in the ratio of reduced to oxidised glutathione (CECONI et al. 1988). In agreement with such a mechanism, we have observed a reduction in the levels of oxidised glutathione in the coronary effluent following ischaemia/reperfusion in heat stressed hearts, suggesting that a second, alternative line of antioxidant defence exists in these hearts (CURRIE and TANGUAY 1991). The picture, however, becomes more complicated since cardiac mRNA levels for catalase are not increased by heat stress (KUKREJA and HESS 1992). The increase in catalase may therefore result from post-translational mechanisms. An alternative suggestion is that stress proteins may modulate the catalytic activity of catalase by direct interaction with the enzyme (FERRARI et al. 1989).

Other evidence suggests that stress proteins may be able to limit myocardial damage independent of an antioxidant effect. Recent reports suggest that the injury occurring during the calcium paradox can be influenced by procedures that cause stress protein synthesis (MEERSON et al. 1991a,b; MARBER et al. 1993). The precise mode by which the calcium paradox damages the heart is a matter of controversy, but free-radical production is probably not involved (ALTSCHULD et al. 1991). It is thought that during the period of low calcium exposure changes occur in the structural proteins of the myocyte so as to increase fragility, and, on calcium repletion, the return of contractile activity causes myocyte mechanical disruption (STEENBERG et al. 1987). A similar process involving cytoskeletal disruption may also occur during ischaemia (BENNARDINI et al. 1992). Heat stress proteins are known to alter the physical properties of actin and desmin (GREEN and LIEM 1989) and may themselves form an integral part of the cytoskeleton (Havre and Hammond 1988). Yet another possible mechanism of protection is that during heat stress protein synthesis (apart from the stress proteins) is inhibited. A similar response has been noted during other forms of stress in cardiac tissue (Pauly et al. 1991) and it has been postulated that such a response allows the cell to redirect energy into more vital cell processes during and following times of stress (Yellon et al. 1993). A 17-kDa stress protein which inhibits protein translation has been isolated from cardiac tissue and is expressed in response to heat and pressure overload, providing yet another possible mechanism by which heat stress may confer myocardial protection.

Although the specific changes that result in myocyte death during ischaemia are poorly understood, alterations in the structural conformational of proteins will inevitably occur secondary to changes in pH, ionic concentration and free radical stress. The general protective properties of stress proteins may be able to attenuate or correct these changes.

E. Conclusions

Stress proteins are induced in the heart by various physiological and pathological conditions such as ischaemia; these findings together with other information have led to the hypothesis that stress proteins may have play an important role in adaptation to stress. This hypothesis has been examined by elevating the stress protein content of the heart by heat and other means prior to ischaemia. The protective effect of elevated temperature on the ability of the heart to survive a subsequent stress has now been demonstrated by a number of different laboratories using different animal species with various endpoints.

Stress proteins are induced by the brief periods of sub-lethal ischaemia known to cause ischaemic preconditioning. However, this induction of stress proteins has not as yet been shown to be directly responsible for the protection observed. However, studies in which direct tranfection of cells with hsp70 genes and/or transgenic mice in which hsps have been overexpressed do clearly indicate a direct protection role for these proteins.

The weight of evidence presented suggests that stress proteins increase the resistance of the heart to ischaemia and may offer an endogenous route to myocardial protection. Such a route represents an obvious pathway for therapeutic intervention. Future investigators, by using either pharmacological or genetic manipulations, will address the problem of cardiac stress protein induction independent of physical stress. It is hoped that by such methods it may be eventually possible to protect the heart by a specific but non-abusive stimulation of its own adaptive protective mechanisms.

Acknowledgements. Dr. Richard Carroll is sponsored by the British Heart Foundation. We also thank the Hatter Foundation for continued support.

References

Altschuld R, Ganote CE, Nayler WG, Piper HM (1991) What constitutes the calcuim paradox? J Mol Cell Cardiol 23:765–767

Bansal G, Norton PM, Latchman DS (1991) The 90kd heat shock protein protects cells from stress but not from viral infection. Exp Cell Res 195:303–306

Barbe M, Tytell M, Gower DJ, Welch WJ (1988) Htperthermia protects against light damage in the rat retina. Science 241:1871–1820

Bennardini F, Wrzosek A, Chiesi M (1992) Alpha B-crystallin in cardiac tissue. Association with actin and desmin filaments. Circ Res 71:288–294

Black A, Subjeck JR (1991) The biology and physiology of the heat shock and glucose-regulated stress protein systems. Karger, Basel

Black S, Lucchesi BR (1993) Heat shock proteins and the ischemic heart. An endogenous protective mechanism. Circulation 87:1048–1051

Burgman P, Konings AWT (1992) Heat induced protein denaturation in the particular fraction of HeLa S3 cells: effect of thermotolerance. J Cell Physiol 153:88–94

Ceconi C, Curello S, Cargooioni A, Ferrari R, Albertini A, Visioli O (1988) The role of glutathione status in the protection against ischaemic and reperfusion damage: effects of N-acetyl cysteine. J Mol Cell Cardiol 20:5–13

Christman M, Morgan RW, Jacobson FS, Ames BN (1985) Positive control of a regulon for defences against oxidative stress and some heat shock proteins in Salmonella typhimurium. Cell 41:753–762

Ciocca DR, Fuqua SA, Lock Lim S, Toft DO, Welch WJ (1992) Response of human breast cancer cells to heat shock and chemotherapeutic drugs. Cancer Res 52:3648–3654

Cornelussen R, Spiering W, Webers JH, de Bruin LG, Reneman RS, van der Vusse GJ, Snoeckx LH (1994) Heat shock improves ischaemic tolerance of hypertrophied rat hearts. Am J Physiol 267:H1941–H1947

Cumming, DVR, Heads J, Watson A, Latchman DJ, Yellon DM (1996) Differential protection of primary rat cardiocytes by transfection of specific heat stress proteins. J Mol Cell Cardiol 28:2343–2349

Currie R (1987) Effects of ischemia and perfusion temperature on the synthesis of stress induced (heat shock) proteins in isolated and perfused rat hearts. J Mol Cardiol 19:795–808

Currie R, Karmazyn M (1990) Improved post-ischemic ventricular recovery in the absence of changes in energy metabolism in working rat hearts following heat shock. J Mol Cell Cardiol 22:631–636

Currie RWM, Karmazyn M, Kloc M, Mailer K (1988) Heat-shock response is associated with enhanced postischemic ventricular recovery. Circ Res 63:543–549

Currie RW, BM, Davis TA (1990) Induction of the heat shock response in rats modulates heart rate, creatine kinase and protein synthesis after a subsequent hyperthermic treatment. Cardiovasc Res 24:87–93

Currie RW, Tanguay RM (1991) Analysis of RNA for transcripts for catalase and SP71 in rat hearts after in vivo hyperthermia. Biochem Cell Biol 69:375–382

Currie RW, Tanguay RM, Kingma JG (1993) Heat-shock response and limitation of tissue necrosis during occlusion/reperfusion in rabbit hearts. Circulation 87:963–971

Currie RW, White FP (1983) Characterization of the synthesis and accumulation of a 71-kilodalton protein induced in rat tissues after hyperthermia. Can J Biochem Cell Biol 61:438–446

Delcayre C, Samuel JL, Marotte F, Best Belpomene, Mercadier JJ, Rappaport, L (1988) Synthesis of stress proteins in rat cardiac myocytes 2–4 days after imposition of hemodynamic overload. J Clin Invest 82:460–468

Donnelly TJ, Sievers RE, Vissern FC, Welch WJ, Wolfe CL (1992) Heat shock protein induction in rat hearts. A role for improved myocardial salvage after ischemia and reperfusion? Circulation 85:769–778

Ferrari R, Ceconi C, Curello A, Ruigrok TJC (1989) No evidence of oxygen free radicals-mediated damage during the calcium paradox. Basic Res Cardiol 84:396–403

Gerner E, Scheider MJ (1975) Induced thermal resistance in HeLa cells. Nature 256:500–502

Green L, Liem RK (1989) Beta-internexin is a microtubule-associated protein identical to the heat-shock cognate protein and the clathrin uncoating ATPase. J Biol Chem 264:15210–15215

Havre PA, Hammond GL (1988) Isolation of a translation-inhibiting peptide from myocardium. Am J Physiol 255:H1024–H1031

Heads RJ, Latchman DS, Yellon DM (1994) Stable high level expression of a transfected human HSP70 gene protects a heart-derived muscle cell line against thermal stress. J Mol Cell Cardiol 26:695–699

Heads RJ, Yellon DM, Latchman DS (1995) Differential cytoprotection against heat stress or hypoxia following expression of specific stress protein genes in myogenic cells. J Mol Cell Cardiol 27:1669–1678

Hearse H, Yellon DM (1983) Why are we still in doubt about infarct size limitation? The experimental viewpoint. In: Hearse DJ, Yellon DM (eds) Therapeutic approaches to myocardial infarct size limitation. Raven, New York, pp 17–41

Hoshida S, Kuzuya T, Yamashita N, Fuji H, Oe H, Hori M, Suzuki K, Taniguchi Tada M (1992) Sublethal ischemia alters myocardial antioxidant activity in canine heart. Am J Physiol 264:H33–H39

Howard G, Geoghegan TE (1986) Altered cardiac tissue gene expression during acute hypoxia exposure. Mol Cell Biochem 69:155–160

Huber SA (1992) Heat-shock protein induction in adriamycin and picornavirus-infected cardiocytes. Lab Invest 67:218–224

Hutter M, Sievers RE, Wolfe CL (1994) Heat shock protein induction in rat hearts; a direct correlation between the amount of heat shock protein and the degree of myocardial protection. Circulation 89:355–360

ISIS-2 (International Study of Infarct Survival) study collaborative group (1988) Randomised trial of IV streptokinase,oral aspirin,both or neither among 17187 cases of suspected acute myocardial infarction. Lancet ii:349–360

Johnston R, Kucey BL (1988) Competitive inhibition of hsp70 gene expression causes thermosensitivity. Science 242:1551–1554

Karmazyn M, Mailer K, Currie RW (1990) Acquisition and decay of heat-shock-enhanced postischemic ventricular recovery. Am J Physiol 259:H424–H431

Kitagawa K, Matsumoto M, Tagaya M, Hata R, Ueda H, Niinobe M, Handa N, Fukunaga R, Kimura K, Mikoshiba K, Kamada T (1990) "Ischemic tolerance" phenomenon found in the brain. Brain Res 528:21–24

Kitagawa K, Matsumoto M, Kuwabara K, Tagaya M, Toshiho O, Hata R, Ureda H, Handa N, Kimura T (1991) Ischemia tolerance phenomenon detected in various brain regions. Brain Res 561:203–211

Knowlton A, Brecher P, Apstein CS (1990) Rapid expression of heat shock protein in the rabbit after brief cardiac ischemia. J Clin Invest 87:139–147

Knowlton AA, Eberli FR, Brecher P, Romo GM, Owen A (1991) A single myocardial stretch or decreased systolic fiber shortening stimulates the expression of heat shock protein 70 in the isolated, erythrocyte-perfused rabbit heart. J Clin Invest 88:2018–2025

Kucukogku S, Iliodromitis E, Van Winkle D, Downey J, Marber M, Heads R, Yellon DM (1991) Protection by ischemic preconditioning appears independent of stress protein synthesis. J Mol Cell Cardiol 23[Suppl V]:S73 (abstract)

Kukreja R, Kontos MC, Loesser KE, Batra SK, Qian, Y-Z, Geur CJ, Naseem SA, Jesse RL, Hess ML (1994) Oxidant Stress increases heat shock protein 70mRNA in isolated perfused rat heart. Am J Physiol 36:H2213–H2219

Kukreja RC, Hess ML (1992) The oxygen free radical system: from equations through membrane-protein interactions to cardiovascular injury and protection. Cardiovasc Res 26:641–655

Landry J, Chretien P, Lambert H, Hickey E, Weber LA (1989) Heat shock resistance conferred by expression of the human HSP27 gene in rodent cells. J Cell Biol 109:7–15

Laszlo A (1992) The thermoresistant state: protection from initial damage or better repair? Exp Cell Res 202:519–531

Li C, Laszlo A (1985) Amino acid analogs whilst inducing heat shock proteins sensitive cells to thermal damage. J Cell Physiol 115:116–122

Li G, Hahn GM (1978) Ethanol-Induced tolerance to heat and adriamycin. Nature 274:699–701

Li G, LI L, Liu Y, Mak JY, Chen L, Lee WMF (1991) Thermal response of rat fibroblasts stably transfected with the human 70KDa heat protein-encoding gene. Proc Natl Acad Sci USA 88:1681–1685

Li G, Li L, Liu RY, Rehman M, Lee WMF (1992) Heat shock protein hsp70 protects cells from thermal stress even after deletion of its ATP-binding domain. Proc Natl Acad Sci USA 89:2036–2040

Liu X, Engelman RM, Moraru LL, Rousou JA, Flach JE, Deaton DW, Maulik N, Das DU (1992) Heat shock. A new approach for myocardial preservation in cardiac surgery. Circulation 86:I358–I363

Liu Y, Kato H, Nakata N, Kogure V (1992) Protection of rat hippocampus against ischemic neuronal damage by pretreatment with sublethal ischemia. Brain Res 586:121–124

Loesser K, Vinnikova AK, Qian YK, Hess ML, Jesse RL, Kukreja RC (1992) Protection of ischaemia/reperfusion injury by stress in rats. Circulation 86[Suppl I]:I-557 (abstract)

Low I, Friedrich T, Schoeppe W (1991) Effects of calcium channel blockers on stress protein synthesis in cardiac myocytes. J Cardiovasc Pharmacol 17:800–806

Low I, Friedrich T, von Bredow F, Schoeppe W (1991) A cell culture assay for the detection of cardiotoxicity. J Pharmacol Methods 25:133–145

Low I, Friedrich T, Weisensee D, Mitrou P, Schoeppe W (1992) Cytokines induce stress protein formation in cultured cardiac myocytes. Basic Res Cardiol 87:12–18

Low I, Friedrich T, Schoeppe W (1989) Synthesis of shock proteins in cultured fetal mouse myocardial cells. Exp Cell Res 180:451–459

Lowenstein D, Chan PH, Miles MF (1991) The stress protein resonse in cultered neurones: characterization and evidence for a protective role in excitotoxicity. Neuron 7:1053–1060

Marber MS, Latchman DS, Walker JM, Yellon DM (1993) Cardiac stress protein elevation 24 hours after brief ischemia or heat stress is associated with resistance to myocardial infarction. Circulation 88:1264–1272

Marber MS, Mestril R, Chi SU, Sayer MR, Yellon DM, Dillmann WH (1995) Overexpression of the rat inducible 70-kD heat stress protein in a transgenic mouse increases the resistance of the heart to ischemic injury. J Clin Invest 95:1446–1456

Marber MS, Walker JM, Latchman DS, Yellon DM (1993) Attenuation by heat stress of a submaximal calcium paradox in the rabbit heart. J Mol Cell Cardiol 25:1119–1126

Marber MS, Walker JM, Latchman DS, Yellon DM (1994) Myocardial protection after whole body heat stress in the rabbit is dependent on metabolic substrate and is related to the amount of the inducible 70-kD heat stress protein. J Clin Invest 93:1087–1094

Meerson FZ, Malyshev IY (1989) Adaption to stress increases the heart resistance to ischemic and reperfusion arrhymias. J Mol Cell Cardiol 21:299–303

Meerson FZ, Malyshev I, Zamotrinsky AV (1991) Adaptive increase in the resistance of the heart to the calcium paradox. J Moll Cell Cardiol 23[Suppl V]:S162 (abstract)

Meerson FZ, Malyshev I, Zamotrinsky AV (1991) Phenomenon of the adaptive stabilization of sarcoplasmic and nuclear structures in myocardium. Basic Res Cardiol 3:205–214

Meerson FZ, Malyshev I, Zamtotrinsky AV (1992) Differences in adaptive stabilization of structures in response to stress and hypoxia relate with the accumulation of hsp70 isoforms. Mol Cell Biochem 111:87–89

Mehta HB, Popovich BK, Dillmann WH (1988) Ischemia induces changes in the level of mRNAs coding for stress protein 71 and creatine kinase M. Circ Res 63:512–517

Mestril R, Chi SH, Sayen MR, O'Reilly K, Dillman WH (1994) Expression of inducible stress protein 70 in rat heart myogenic cells confers protection against simulated ischemia-induced injury. J Clin Invest 93:759–767

Moalic JM, Bauters C, Himbert D, Bercovici J, Menas J, Guicheney P (1989) Phenylephrine, vasopressin and angiotensin II as determinants of proto-oncogene and heat-shock protein gene expression in adult rat heart and aorta. J Hypertens 7:195–201

Morris SB, Cumming DV, Latchmann DS, Yellon DM (1996) Specific induction of the 70 kD heat stress proteins by the tyrosine kinase inhibitor Herbimycin-A protects rat neonatal cardiomyocytes. A new pharmacological route to stress protein expression? J Clin Invest 97(3):244–251

Miura T, Adachi T, Ogawa T, Iwato T, Tsuchida A, Iimura O (1992) Myocardial infarct size limiting effect of ischemic preconditioning its natural decay and the effects of repetitive preconditioning. Cardiovasc Pathol 1:147–154

Muller D, Topol EJ, Califf RM, Sigmon KN, Gorman L, Gearge BS, Kereiakes DJ, Lee KL, Ellis SG (1990) Relationship between antcedent angina pectoris and short-term prognosis after thrombolytic therapy for acute myocardial infarction. Thrombolysis and angioplasty in myocardial infarction (TAMI) study group. Am Heart J 119:224–231

Nowak TS Jr (1990) Protein synthesis and the heart shock/stress response after ischemia. Cerebrovasc Brain Metab Rev 2:345–366

Pasini E, Cargnoni A, Ferrari R, Marber MS, Latchman DS, Yellon DM (1991) Heat stress and oxidative damage following iscemia and reperfusion in the isolated rat heart. Eur Heart J 12[Suppl]:306

Pauly D, Kirk KA, McMillan JB (1991) Carnitine palmitoyltransferase in cardiac ischemia. A potential site for altered fatty acid metabolism. Circ Res 68:1085–1094

Plumier JC, Ross BM, Currie RW, Angelidis CE, Kazlaris H, Kollias G, Pagoulatos GN (1995) Transgenic mice expressing the human heat shock protein 70 have improved post-ischemic myocardial recovery. J Clin Invest 95:1854–1860

Reimer K, Jennings R (1979) The "wave phenomenon" of myocardial ischemic cell death. II. Transmural progression of necrosis within the framework of ischemic bed size (myocardium at risk) and collateral flow. Lab Invest 40:633–644

Riabowal K, Mizzan LA, Welch WJ (1988) Heat shock is lethal to fibroblasts micro injected with antibodies against hsp70. Science 242:433–436

Rordorf G, Koroshetz WJ, Bonventre JV (1991) Heat shock protects cultured neurons from glutamate toxicity. Neuron 7:1043–1051

Sanchez Y, Lindquist SL (1990) Hsp 104 required for induced thermotolerance. Science 248:1112–1115

Schott R, Nao B, Strieter R, Groh M, Kunkel S, McClanahan T, Schaper W, Gallagher K (1990) Heat shock does not precondition canine myocardium. Circulation 82[Suppl III]:III-464 (abstract)

Steenberg C, Hill ML, Jennings RB (1987) Cytoskeletal damage during myocardial ischemia: changes in Vinculin immunofluorescence staining during total in vitro ischemia in canine heart. Circ Res 60:478–486

Su C-Y, Dillmann WH, Woods WT, Owen OE (1992) Heat shock induced oxidative tolerance in muscle cells. Circulation 84[Suppl I]:I-33 (abstract)

Sugahara T, Saito M (1988) Hyperthermic oncology. Taylor and Francis, London

Thornton J, Striplin S, Liu GS, Swafford A, Stanley AW, Van Winkle DM, Downey JM (1990) Inhibition of protein synthesis does not block myocardial protection afforded by preconditioning. Am J Physiol 259:H1822–H1825

Tuijl M, van Bergan en Henegouwen PM, van Wijk R, Verkleij AJ (1991) The isolated neonatal rat-cardiomyocyte used as an in vitro model for "iscemia". II. Induction of the 68 kDa heat shock protein. Biochim Biophys Acta 1091:278–284

Van Winkle DM, Thornton J, Downey JM (1991) Cardioprotection from ischaemic preconditioning is lost following prolonged reperfusion in the rabbit. Coron Art Dis 2:613–619

Walker D, Pasini E, Marber MS, Iliodromitis E, Ferrari R, Yellon DM (1993) Heat stress limits infarct size in the isolated rabbit heart. Cardiovasc Res 27:962–967

Walker DM, Pasini E, Kucukoglu S, Marber MS, Iliodromitis E, Ferari R, Yellon DM (1993) Heat stress limits infarct size in the isolated perfused rabbit heart. Cardiovasc Res 27:962–967

Wall SR, Fliss H, Korecky B. Role of catalase in myocardial protection against ischaemia in heat shocked rats. Mol Cell Biochem 1993 Dec 22;129(2):187–194

White FP, White SR (1986) Isoproterenol induced myocardial necrosis is associated with stress protein synthesis in rat heart and thoracic aorta. Cardiovasc Res 20:512–515

Yamashita N, Hoshida S, Nishida M, Igarashi J, Aoki K, Hori M, Kuzuya T, Tada M (1997) Time course of tolerance to ischaemia-reperfusion injury and induction of heat shock protein 72 by heat stress in the rat heart. J Mol Cell Cardiol 29:1815–1821

Yellon DM, Latchman DS (1991) Stress proteins and myocardial protection (editorial). Lancet 337:271–272

Yellon DM, Iliodromitis E, Latchman DS, Van Winkle DM, Downey JM, Williams FM, Williams TJ (1992) Whole body heat stress fails to limit infarct size in the reperfused rabbit heart. Cardiovasc Res 26:342–346

Yellon DM, Latchman DS (1992) Stress proteins and myocardial protection. J Mol Cell Cardiol 24:113–124

Yellon DM, Latchman DS, Marber MS (1993) Stress proteins-an endogenous route to myocardial protection: fact or fiction? Cardiovasc Res 27:158–161

Yellon DM, Pasini E, Cargnoni A, Marber MS, Latchman DS, Ferrari, R (1992) The protective role of heat stress in the ischaemic and reperfused rabbit myocardium. J Mol Cell Cardiol 24:895–907

CHAPTER 13

Heat Shock Proteins in Inflammation and Immunity

M. Bachelet, G. Multhoff, M. Vignola, K. Himeno, and B.S. Polla

A. Introduction: Multiple Roles of Heat Shock Proteins in Inflammation and Immunity

Inflammation represents a localized or systemic response against tissue or cell injury, which, on the one hand, is essential among cell defense mechanisms and, on the other, is involved in a broad spectrum of diseases. According to the initiating event, the inflammatory response may involve, or not, an antigen-specific immune response. In the first case, the initiating agent is generally a microorganism or an antigen of unknown origin, while in the latter, cells may respond to injurious physical agents (foreign bodies, burns, radiations, trauma) or toxic chemicals. Lymphocytes are classically involved in the specific immune response, and phagocytes in its non-specific counterpart.

The mechanisms resulting in these different types of inflammation all include a highly conserved common inducible response that involves the synthesis of heat shock/stress proteins (Hsps). Hsps may play various roles in multiple steps of the inflammatory response. Here we will present the current status of knowledge on the inflammation-related heat shock/stress responses and their anti-inflammatory or pro-inflammatory functions, respectively. Several members of the Hsp family function as molecular chaperones, each of them in discrete cellular compartments; along with these intracellular chaperoning tasks, cytosolic/lysosomal members of the most conserved and abundant Hsp70 family also contribute to antigen processing (Vanbuskirk et al. 1991; Marietthoz et al. 1994; Jacquier-Sarlin and Polla, submitted).

Hsp gene polymorphism, in particular of the genes coding the 70-kDa Hsp that are localized within the major histocompatibility complex (MHC), is a likely contributor to disease susceptibility or stress resistance (Favatier et al. 1997). Hsp can act as antigens, either self or non-self, and, as mentioned above, contribute to antigen processing and presentation, and thus to the efficiency of an immune response. Furthermore, they have the ability to protect cells and tissues from the deleterious effects of numerous mediators of inflammation, such as reactive oxygen species (ROS) or tumor necrosis factor (TNF)α (Jättelä et al. 1989; Kantengwa et al. 1991; Villar et al. 1993). Particular attention will be paid here to the role of the Hsp70 family, which is endowed with critical protective properties. It should, however, be mentioned that the protective (anti-inflammatory) effects of the stress response might relate more directly to the inhibition of activation of the transcription factor NFκ-B (ROSSI et al. 1997; WONG et al. 1997) – NF-κB being central to the inflammatory process (BAEUERLE et al. 1996; BALDWIN et al. 1996) – than to the Hsps themselves.

With respect to the direct protective effects of Hsp70 against ROS, mitochondria have been proposed as selective targets for these effects (POLLA et al. 1996). While mitochondria are considered as the cellular switchboard for cell survival or cell death, and the type of cell death, whether apoptosis or necrosis (RICHTER et al.1996), numerous publications indicate that Hsp70 has the ability to protect cells from apoptosis (SAMALI and COTTER 1996; DIX et al. 1996; MEHLEN et al. 1996a). Although at first sight this anti-apoptotic effect might appear protective, it actually can promote persistence rather than resolution of acute inflammation, by preventing the physiological removal of inflammatory cells by apoptosis. Thus, the possibility that the anti-apoptotic effects of Hsp70 contribute to the amplification and the chronicity of the inflammatory process will be considered here, in particular in the light of recent results obtained in asthma, a paradigm for chronic eosinophilic inflammation of the upper airways, where overexpression of Hsps could contribute both to antigen processing and presentation, and to chronic inflammation.

While the precise regulation and functions of Hsps in inflammation are not yet fully understood, future research and new experimental approaches in this field appear of great potential.

B. Role of Hsp Localization in the Induction of an Immune Response

By the end of the 1980s, it was found that stress proteins were among the dominant antigens recognized by the immune system in a number of different diseases. They are important players in host-parasite interactions (ZUGEL et al. 1995; JACQUIER-SARLIN et al. 1994), in autoimmune diseases (ANDERTON et al. 1995; HEUFELDER et al. 1992), in neurodegenerative diseases (CHOPP 1993), in virus infections (DICESARE et al. 1992; SANTORO et al. 1989) and in transplant rejection (MOLITERNO et al. 1995). Although at first sight, a specific immune response induced by evolutionary conserved proteins appears paradoxical, stress proteins are indeed major targets for the cellular and humoral immune responses (YOUNG and ELLIOTT 1989). The involvement of Hsps in autoimmune diseases and in anti-cancer immune responses has been described by several groups (WINFIELD and JARJOUR 1991; HEUFELDER et al. 1992; KAUFMANN 1994; SRIVASTAVA 1994).

In order to be recognized as antigens, the Hsps have to be somehow expressed at the cell membrane of antigen-presenting cells (CHOUCHANE et al. 1994), e.g., monocyte-macrophages or B cells, and thus be accessible to cells of the immune system. This is particularly intriguing in the case of self Hsps.

Although indeed members of the Hsp70 family are found on the plasma membrane of certain cell types, the mechanism(s) underlying transport and anchorage to the plasma membrane has not been identified yet. Since all isoforms of Hsp70 lack a hydrophobic leader sequence, it has been assumed that Hsps may be transported to the plasma membrane following cell death and disruption of the plasma membrane integrity (MUTHUKRISHNAN et al. 1991). A number of recent observations, however, challenge this view and suggest specific, though yet unraveled, mechanisms for Hsp70 membrane expression:

1. Hsp70 is found immunohistochemically and by selective cell surface biotinylation on the surface of certain tumor cells but not of normal cells (MULTHOFF et al. 1995a, 1997).
2. Soluble Hsp70 is not detectable in the culture medium of viable tumor cells expressing membrane Hsp70 as well as non-expressing normal cells (MULTHOFF et al. 1995a).
3. Hsp homologs to rat Hsp70 are among a selected group of proteins that are transferred from glial cells to giant axons, thus indicating that Hsps may be transported through the plasma membrane without disruption of membrane integrity.
4. Recent data suggest that the transport of Hsp70 is clearly distinct from the classical ER to Golgi pathway. Brefeldin A, monensin or colchicine that block ER-to-Golgi trafficking, block the secretion of polypeptides containing hydrophobic signal sequences while they have no influence on the membrane expression of Hsp70 (MISUMI et al. 1986; MULTHOFF et al. 1997). Therefore, one might speculate that transport of Hsp70 to the cell surface occurs post-translationally through a non-Golgi-dependent pathway. Rapidly released proteins such as interleukin-1 (IL-1) or acidic and basic fibroblast growth factor (FGF) also leave cells via non-classical non-Golgi pathways.
5. Data derived from electron microscopy indicate that tumor cells expressing Hsp70 on their plasma membrane also express Hsps in vesicles that are co-stained with cathepsin D, suggesting that lysosomal vesicles are involved in the transport of Hsp70 to the plasma membrane (MULTHOFF, unpublished observation).

Anchorage of Hsp70 within the plasma membrane might either be due to direct interaction of Hsp70 with fatty acids (HIGHTOWER and GUIDON 1989) or to the formation of a larger protein complex. Proteins of the Hsp70 family bind to a variety of other cellular proteins, including clathrin baskets and coated vesicles, the transformation related protein p53 (PINHASI-KIMHI et al. 1986) and cytoskeletal elements (OHTSUKA et al. 1986). The association of Hsp70 with the transferrin receptor during reticulocyte maturation could contribute to Hsp70 plasma membrane anchorage via the formation of a protein-receptor complex. The receptor-mediated endocytosis of *Chlamydia trachomatis* into host endometrial cells is another example of outer membrane association mediated by an Hsp70-related protein (RAULSTON et al. 1993). Preliminary data from our group indicate that Hsp70 is able to form complexes with other molecular chaperones that contain a classical transmembrane domain; these complexes could be detected on the plasma membrane and in lysosomes of certain tumor cells. Thus, interactions between members of the Hsp70 family and normal or aberrant proteins are emerging as a unifying theme for membrane anchorage.

C. Hsps and Cell Adhesion in the Initiation of Inflammation

In order to create an inflammatory site, cells – mainly leukocytes – have to escape circulation and migrate into the tissues. The migration of leukocytes from the blood and their accumulation in a given tissue or organ is fundamental to the inflammatory process, and characterizes numerous inflammatory diseases such as bronchial asthma, rheumatoid arthritis or the adult respiratory distress syndrome (ARDS). Several steps of the migration process can be distinguished, including various adhesion molecules differentially affected by inflammatory mediators.

The first step of this process involves leukocyte adhesion. Indeed, leukocytes initially interact with the endothelium via selectins, causing cells to deviate from the normal flow and marginate on the endothelium. Next, chemoattractants induce the activation of leukocytes' integrins that mediate leukocyte migration into the inflamed site (ROSALES et al. 1995). Different types of leukocytes leave the bloodstream in an orderly fashion. Polymorphonuclear leukocytes (PMN) are first recruited and generally account for organ or tissue damage by an acute inflammation whereas monocytes and lymphocytes arrive later and may be recruited for long periods leading thereby to chronic inflammation (SPRINGER 1994; WARD and MARKD 1989). So far, one study reported that adherence to plastic surfaces was associated with an increase in Hsp70 expression in myelomonocytic cells (FINCATO el al. 1991). However, additional observations suggest a correlation between the expression of Hsps and the activation of adhesion molecules. Cells with a higher capacity to spontaneously adhere on plastic surfaces such as monocytes-macrophages (mφ) show higher levels of Hsp as compared to other cells (POLLA et al. 1995b), and differentiation of the myelomonocytic line U937 with 1,25-(OH)$_2$ vitamin D$_3$ increases the cell adhesion capacity in parallel with the expression of Hsp70 (POLLA et al. 1987). Moreover, we observed that adherent human monocytes are more resistant to elevated temperatures (44.5°C) as compared to the same cells cultured under non-adherent conditions (Bachelet, unpublished). This latter observation may reflect overexpression of Hsps upon adherence, thus resulting in thermotolerance. Leukocyte migration during inflammation involves morphological changes that allow leukocytes to cross the endothelium through cell junctions by ameboid movements or diapedesis.

Hsps may protect cells against mechanical stresses caused by shape changes during diapedesis.

The physiological importance of cell adhesion proteins has emerged, among others, from studies on patients with leukocyte adhesion deficiency (LAD), a clinical syndrome secondary to a genetic deficiency in adhesion proteins (ARNAOUT 1990). Patients with LAD are characterized by recurrent bacterial infections and abnormalities in a wide spectrum of adherence-dependent functions of leukocytes, mainly attributable to deficiency (or absence) of cell surface expression of $\beta2$ integrins. We have examined the capacity of monocytes from one patient with LAD to synthesize Hsps upon thermal stress, using biometabolic labeling and sodium dodecylsulfate polyacrylamide gel electrophoresis (SDS-PAGE). As observed in Fig. 1, monocytes from this patient appeared to synthesize similar levels of Hsps as compared to monocytes from a normal control, suggesting that decreased expression of $\beta2$ integrins does not modulate Hsp synthesis.

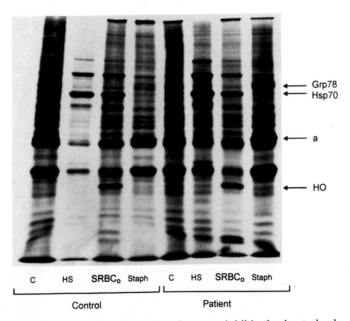

Fig. 1. Deficiency in adhesion molecules does not inhibit the heat shock response. Monocytes from a normal donor (control) or from a patient with LAD (patient) were exposed to heat shock, or allowed to phagocytose opsonized erythrocytes (SRBCo) or *Staphylococcus aureus* (*Staph*). Cells were labeled and processed for SDS-PAGE; aliquots corresponding to equal cell numbers were loaded onto each lane. Heat shock induced Hsp70; SRBCo, Hsp70 and heme oxygenase (HO); and *Staph*, predominantly glucose regulated protein (Grp)78. These proteins were induced both in control and in LAD cells, and rather more so in LAD than in control cells

D. Non-specific Immunity: Cells and Mediators Involved in the Induction of a Heat Shock/Stress Response

Non-specific inflammatory cells include mφ, PMN, eosinophils, and platelets, as well as many other cell types. Here we will concentrate on the first three, which are professional phagocytes involved in defense mechanisms. Activation of these phagocytes results in the release of a large repertoire of inflammatory mediators including ROS, lipid mediators and cytokines, that in turn participate in the inflammatory-related heat shock/stress response. We will focus in particular on mφ, as essential players in chronic inflammation, and as the highest producers of Hsps among the human cells examined. We will also examine specific differences between mφ and PMN. Many of the functions of mφ somehow relate to the heat shock/stress response, and we will consider both Hsp induction/regulation, and their functions in inflammation.

I. Monocytes-Macrophages

Monocytes-macrophages (mφ) display, among the circulating cells, a particularly interesting, most complex and diversified stress response. Heat shock induces in these cells high levels of synthesis and expression of Hsps of apparent molecular weight 110, 90, 68–73, 60–65, 58, 47, 27–30 kDa, while stresses such as phagocytosis might also induce glucose-regulated proteins (Grp), heme oxygenase (HO), ferritin, and possibly other as yet unidentified stress proteins. Furthermore, these mφ selectively respond to each stress: for example, during phagocytosis, the precise profile of Hsps will depend upon the type of phagocytic stimuli, the degree of activation of the respiratory burst and the type of ROS produced, and upon a balance between many potentially relevant second messengers, including calcium, kinases such as protein kinase C (PKC), mitogen activated protein kinases (MAPK), stress-activated protein kinases (SAPK), PKC, phosphatases, cyclic AMP, lipid mediators of inflammation, and proteases.

1. Reactive Oxygen Species

ROS generation by phagocytes occurs in response to multiple stressors including receptor-mediated phagocytosis, activation of PKC or the release of arachidonic acid metabolites, through the activation of the membrane associated NADPH-oxidase (TAUBER 1987; WATSON et al. 1990). Reduction of molecular oxygen to H_2O via NADPH-oxidase proceeds through a sequential one-electron transfer yielding superoxide anion (O_2^-), hydrogen peroxide (H_2O_2) and hydroxyl radicals ($\bullet OH$) (ROSEN et al. 1995). ROS play important roles in non-specific defense mechanisms such as killing of pathogenic microorganisms. At high levels, ROS can, however, be deleterious for the host as well, exerting proinflammatory effects and inducing cell or tissue injury. In contrast, at low levels, ROS are important signaling molecules. Selective ROS

(H_2O_2, ONOO⁻, •OH, but not O_2^- or nitric oxide [•NO]), induce Hsps as a protective mechanism against oxidative injury (JACQUIER-SARLIN et al. 1994; JACQUIER-SARLIN and POLLA 1996; POLLA et al. 1996). •OH in particular has been suggested to be involved in Hsp70 expression upon phagocytosis of xenogenic erythrocytes, or, in the presence of exogenous iron, of *Staphylococcus aureus* and *Pseudomonas aeruginosa* (KANTENGWA et al. 1993; BARAZZONE et al. 1996). Rodent peritoneal mφ activated with phagocytic stimuli also generate reactive nitrogen species such as •NO, that may be toxic to invading bacteria (MARLETTA 1989), while •NO does not by itself induce Hsp synthesis. •NO may interact with O_2^- to form peroxynitrite anions (ONOO⁻) that are strong pro-oxidants and induce Hsp70 in human monocytes (Richard et al., unpublished). Thus only specific ROS (H_2O_2, •OH and ONOO⁻) are inducers of Hsps, while O_2^- and •NO are not.

In terms of functions, Hsps clearly exert protective functions against the toxic effects of ROS. Mitochondria are selective targets of these protective effects (POLLA et al. 1996). Interestingly, mitochondria might also be central to the induction of Hsps by ROS, the difference between Hsp-inducing and non-inducing ROS being that they induce, or not, mitochondrial membrane depolarization (Polla et al., unpublished).

2. Lipid Mediators of Inflammation

Membrane phospholipids from leukocytes contain large amounts of arachidonic acid leading to the generation of numerous lipid mediators such as arachidonic acid (AA) derivatives and PAF-acether, upon activation of phospholipase (PL)A_2 and PLC activities (HOLTZMAN 1991). In human mφ, thromboxane (Tx)A_2 is the major AA metabolite produced, followed by leukotriene (LT)B_4, 5-hydroxyeicosatetraenoic acid (HETE), and prostaglandin (PG)E_2 (HOLTZMAN 1991). AA metabolism is associated with an influx of inflammatory cells to the site of inflammation: LTB_4, and to a lesser extent 5-HETE, have potent chemotactic activity for human PMN and eosinophils but not for mφ, while TxA_2 acts as a potent constrictor of vascular and airway smooth muscle. So far, major lipid mediators generated by mφ have not been demonstrated to interfere with the expression of Hsps in the same cells. This is not unexpected, since, on the one hand, mφ lack receptors for TxA_2 (BACHELET et al. 1992), and on the other, the synthesis of AA derivatives depends mainly upon PLA_2 activation in the presence of high extracellular Ca^{2+}, two factors shown not to affect Hsp expression in general (POLLA et al. 1995a). However, leukotrienes from the 5-lipoxygenase pathway exert several receptor-mediated biological activities on mφ that may account for the LTB_4-induced release of factors stimulating fibroblast proliferation (POLLA et al. 1985). Additional data from the group of KÖLLER describing 12-HETE-induced Hsp synthesis in human leukocytes (KÖLLER and KÖNIG 1990; KÖLLER et al. 1993) indicate that leukotrienes might indeed modulate Hsp expression in inflammation.

Moreover, mφ as well as PMN and eosinophils, bear receptors for PG. PGE_2 markedly increases intracellular cAMP levels in mφ and exert dual functions in inflammation, acting either as pro-inflammatory or anti-inflammatory factor (BONTA and PARNHAM 1982). PG of the type A and J (cyclopentenone PG) have been shown to exert antiproliferative and antiviral activities through a mechanism involving Hsp70, in several mammalian cell types (for review, SANTORO 1997). The antiviral activity of cyclopentenone PG, observed against a wide variety of DNA and RNA viruses, is always associated with their capacity to induce Hsp70.

3. Cytokines

Pro-inflammatory cytokines modulate Hsp synthesis via their pyrogenic activity as well as distinct mechanisms. IL-1 induces an ROS-dependent increase in Hsp70 in β cells of pancreas (HELQVIST et al. 1991), and IL-2 an IL-2 receptor-dependent accumulation of Hsp70 mRNA in lymphocytes (HAIRE et al. 1988). TNFα has been reported to induce Hsp70 in myelomonocytic cells (FINCATO et al. 1991), although other groups have been unable to reproduce these data (Polla et al., unpublished). In cultured chicken embryo cells, transforming growth factor (TGF)β increases the expression of Hsps, secondary to the stimulation of general protein synthesis and a subsequent increase in chaperone requirements (TAKENAKA and HIGHTOWER 1992). However, in many cells, cytokines do not appear to induce the synthesis of Hsps and whether or not cytokines induce a stress response seems to be tissue specific and to depend largely on the cytokine effects on the oxidant/anti-oxidant (im)balance.

Numerous studies on the interactions between Hsps and cytokines have highlighted the striking protection Hsps may provide against the toxic effects of cytokines, in particular, TNFα and IL-1 (for review, JACQUIER-SARLIN et al. 1994 and references therein). TNFα and IL-1 are cytokines with pleiotropic activities. These include the activation of the respiratory burst enzyme NADPH-oxidase, leading to a rapid rise in intracellular ROS originating in the mitochondria (HENNET et al. 1993), and provide a target for the protective effect of Hsps. Preexposure of the highly TNFα-sensitive mouse fibrosarcoma cell line WEHI 164, to temperatures ranging from 39°C to 42°C, prevents the cytotoxic effects of TNFα; the involvement of Hsp70 in this protection was confirmed in WEHI-transfected cells that overexpress Hsp70 (JÄÄTTELÄ et al. 1989; JÄÄTTELÄ 1993). In pancreatic β cells, Hsp70 also protects cells against a selective oxidative stress induced by IL-1 (MARGULIS et al. 1991).

In addition, TNFα and IL-1 induce phosphorylation of the low molecular weight Hsp27 resulting in its activation. Hsp27 also exerts specific protective functions: the protein is involved in the stabilization of the actin microfilament network (ARRIGO 1990), and counteracts TNFα-induced apoptosis (MEHLEN et al. 1996a), probably because of its ability to replenish intracellular reduced glutathione, thereby leading to increased levels of intracellular ROS (MEHLEN et al. 1996b), while oxidative stress has been proposed as a common final signal where several pathways associated with apoptosis converge (BUTTKE and SANDSTROM 1994).

4. Nuclear Factor κB (NF-κB)

Recent studies have suggested that an important mechanism by which the heat shock/stress response exerts protective effects during inflammation involves the inhibition of NF-κB nuclear translocation. NF-κB is found in the cytoplasm of cells in an inactive form associated with the inhibitor IκBα. Upon activation, IκBα is phosphorylated and undergoes proteolytic degradation to allow active NF-κB to translocate to the nucleus and stimulate transcription (BALDWIN 1996; BAEUERLE 1996). NF-κB plays a crucial role in regulating the transcription of several pro-inflammatory cytokines and chemokines including TNFα, IL-1 and IL-8, a potent chemoattractant for PMN (LEONARD and YOSHIMURA 1990). ROSSI et al. (1997) first reported that activation of HSF was associated with inhibition of NF-κB, through a mechanism involving inhibition of IκBα phosphorylation and degradation, an observation that has been confirmed since (WONG et al. 1997). Thus, inhibition of NF-κB activation represents a novel, Hsp-independent, anti-inflammatory effect of all HSF-activating compounds or factors, including heat shock and cyclopentenone PG (ROSSI et al. 1997).

II. Granulocytic Phagocytes

1. Polymorphonuclear Leukocytes (PMN)

PMN generally express Hsps to a lesser extent than mφ and their stress response appears to be differentially regulated (POLLA et al. 1995b). Phagocytic stimuli activate the generation of ROS in PMN as well as in mφ; however, Hsp synthesis is induced only in the latter. In addition, Hsp synthesis is known to occur in PMN, but not in mφ, stimulated with formyl methionyl leucyl phenylalanine (fMLP), the synthetic bacterial peptide analog that increases intracellular calcium and stimulates NADPH-oxidase, PLC, and PKC (TAUBER 1987).

These observed differences have been related to the inability of PMN to produce •OH, the putative key oxygen metabolite for the induction of Hsp. Indeed, lactoferrin, a typical PMN secretory product, is considered to prevent the generation of •OH through its ability to bind metal iron, a major promoter of •OH formation via the Fenton reaction (LIOCHEV and FRIDOVICH 1997). In the absence of •OH, the fMLP-induced Hsp expression may be mediated by the activation of PKC, via PLC activity (JACQUIER-SARLIN et al. 1995). Another hypothesis to explain these differences is that PMN are short-lived phagocytes that may not require significant protective mechanisms such as Hsp synthesis.

Among the secretory products of PMN stored in azurophilic granules (acid hydrolase, myeloperoxidase, lysozyme, and neutral proteases such as elastase and cathepsin G), cathepsin G appears most interesting, inducing in mφ the synthesis of a member of the Hsp70 family, the 78-kDa calcium-dependent Grp78 (PINOT et al., unpublished). Whether such induction also occurs in PMN and whether it relates to intracellular protein degradation remains to be determined.

2. Eosinophils

Eosinophils may exert beneficial roles in modulating immunoglobulin (Ig)E-mediated injury and in controlling parasitic infections but can also be harmful to the host in numerous clinical situations associated with eosinophilia, such as asthma. Eosinophil toxicity is not unexpected, since these cells secrete lipid mediators of inflammation, cytokines, ROS, and the granule constituents, major basic proteins (MBP), eosinophil cationic protein (ECP), and eosino-phil peroxidase (EPO), which are highly toxic for neighbor cells or parasites. While there is currently no available information about the capacity of eosino-phils to generate their own Hsps, Hsps have been shown to be overexpressed in alveolar macrophages and in the bronchial epithelium from patients with severe asthma and persistent airway eosinophilia (VIGNOLA et al. 1995; CHRISTIE et al. 1995). The in vitro exposure of human alveolar macrophages to purified eosinophil-derived proteins, however, does not induce Hsp synthesis, indicating that the mechanism by which activated eosinophils may induce Hsp synthesis in neighboring cells requires alternative events than eosinophilic toxic proteins (CHRISTIE et al. 1995). Among cytokines produced by eosino-phils, the Hsp70-inducing TGFβ (TAKENAKA and HIGHTOWER 1992) represents an interesting candidate. Another such candidate is LTC4, the major arachi-donic acid metabolite released by eosinophils, which exerts biological func-tions in mφ.

Figure 2 summarizes the interactions between inflammatory mediators in phagocytes, their own Hsps and the resulting proinflammatory/anti-inflammatory potential.

E. Cellular Immunity

Hsps are both extremely conserved and extremely immunogenic. These two characteristics appear quite divergent at first glance – and in order to reconcile them, Young and Cohen proposed the immunological homunculus theory, based on selection, classification and overrepresentation of certain conserved self-antigens, i.e., Hsps (COHEN and YOUNG 1991). These authors also make a clear distinction between autoimmunity and autoimmune disease, the former being actually protective against the latter. Thus, the fear for molecular mimicry of Hsps leading to autoimmune diseases is progressively vanish-ing, although Hsps have the ability to induce a broad immune response,

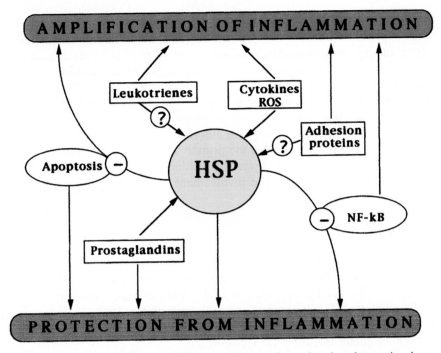

Fig. 2. Dual effects of Hsps and different inflammation-related pathways in phagocytes. Hsps can be induced in inflammation by various mediators, and might exert both
protective and potentially deleterious effects

with Hsp-reactive T cells, $\gamma\delta$ T cells, natural killer (NK) cells . . . and multiple
anti-Hsp antibodies.

I. T Cells

T cell mediated immune responses have been described predominantly for
members of the Hsp60–65, Hsp70 and Hsp90 families. Hsp-reactive T cells are
found in normal individuals and in cord-blood, indicating that these cells have
the ability to escape clonal deletion, which likely relates to the immunological
homunculus theory (MUNK et al. 1989; COHEN and YOUNG 1991). In terms of
pathology, however, emerging fields in which Hsps appear to play important
roles are, on the one hand, rheumatoid arthritis (discussed later in this book by
VAN EDEN) and the immune response against cancer. Indeed, human tumor-
infiltrating CD4+ cells (TIL) derived from melanomas, ovarian, lung, renal
cell and breast cancer have been shown to react specifically to Hsp70 expressing cell lines (YOSHINO et al. 1994). From these results it was concluded that
Hsp70-reactive T cells have to exist locally in certain tumor tissues and support
the local anti-tumor T cell response in tumors.

When purified from cells, Hsp70 and Hsp90 are associated with a broad range of peptides. Hsp70 associated peptides elicit a specific anti-cancer immunity in methylcholanthrene-induced sarcomas in mice (Udono et al. 1994). Vaccination of mice with tumor-derived Hsp70 preparation renders the mice immune to substantial challenge with the autologous tumor cells, while the Hsp90-related glycoprotein gp96 is able to prime CD8+ cells in vivo and purified tumor Gp96 fractions elicit a potent T cell response (Udono et al. 1994). Furthermore, vaccination with autologous tumor-derived Hsp-peptide complexes reduces the growth of the primary tumor as well as the metastatic burden, effects that can be abrogated by the depletion of CD4+ CD8+ T cells or NK cells, thus indicating the involvement of all three cell types in the protective immunity.

II. γδ T Cells

Evidence is accumulating that non-MHC restricted γδ TcR positive cells participate in the immune response to parasitic infections, autoimmune diseases, virus-induced diseases and also in the anti-cancer immune responses. In all these cases γδ T cells are involved in the recognition pathway of members of the Hsp60–65 families. Among the mycobacterial antigens Hsp65 is an immunodominant target for γδ T cells.

Several lines of evidence suggest that γδ T cells which recognize Hsp65 function in host defenses against pathogens (Born et al. 1990; O'Brien et al. 1992). Recently, however, Nagasawa et al. (1994) reported that γδ T cells play an essential role rather in the expression of Hsp65, in particular in host mφ of mice which acquired resistance against infection with *Toxoplasma gondii*. Hsp65 overexpressed in host mφ appears to play an essential role in host defenses by preventing apoptotic death of infected cells.

The proposed mechanisms for Hsp65 and its biological functions are illustrated in Fig. 3. In a first step, γδ T cells (Hisaeda et al. 1995), especially extrathymic γδ T cells (Hisaeda et al. 1996a), recognize *Toxoplasma*-associated antigens, a prime candidate of which is *Toxoplasma*-derived Hsp65, presenting on the surface of mφ as well as cytoplasma and mitochondria (Nagasawa et al. 1992). γδ T cells then secrete cytokines such as IFNγ and TNFα, which in turn activate mφ (Hisaeda et al. 1996b). The activated mφ exhibit an enhanced respiratory burst, releasing high levels of ROS and NO intermediates, which contribute to the killing of intracellular pathogens. The mφ then synthesize self Hsp, which is effective in protecting infected mφ from apoptotic cell death (Hisaeda et al. 1997). This programmed cell death appears to be caused by apoptosis-inducing factor(s) produced by *Toxoplasma protozoan*, especially high-virulent *Toxoplasma*. Although the biological role and biochemical characteristics of this product still are under investigation, its targets appear to be mφ and not γδ T cells. In any case, a synergistic effect between Hsp65 expression preventing apoptotic death of host cells and NO production, both mediated by IFNγ and TNFα which are generated by

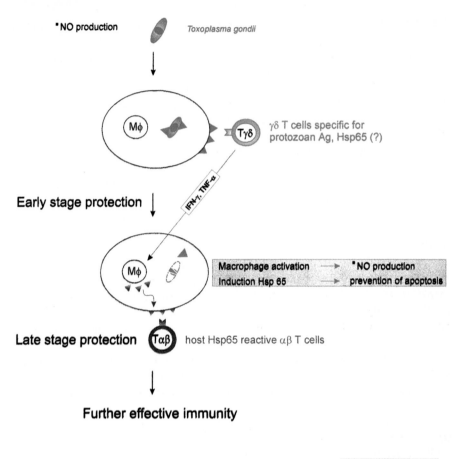

Fig. 3. Induction of an Hsp-dependent immune response by *Toxoplasma gondii*. *Toxoplasma*-derived Hsp65 is expressed on the surface of mφ, activating host γδ T cells. These cells then secrete IFNγ and TNFα, thus activating mφ and leading to host Hsp65 induction and membrane expression, thereby amplifying the immune response

γδ T cells, is required for the host defense. Furthermore, it should be noted that signals provided via receptors for cytokines toward Hsp65 expression and NO production are different from each other (HISAEDA and HIMENO 1997).

This contribution of Hsp65 to host defenses against *Toxoplasma* may actually apply broadly to host-pathogen interactions (ISHIKAWA et al. 1997). For example, in infections with *Leishmania major* and *Trypanosoma cruzi*, which like *Toxoplasma gondi*, are obligate intracellular parasites, the role of Hsp65 in protective immunity is quite similar (HIMENO and HISAEDA 1996).

III. Hsp, NK Cells and Cancer Immunity

Although a large number of Hsp-related tumor antigens have been identified so far (LURQUIN et al. 1989; VAN-DEN EYNDE et al. 1991), the direct interaction of tumor antigens with effector cells is not fully understood, especially in the case of non-MHC restricted effector mechanisms. NK cells were functionally defined in mediating the host's antitumor immune response for a long period of time.

More recently, several groups (MORETTA et al. 1994; LONG et al. 1996) provided direct evidence for the existence of distinct NK subclones with defined specificities against certain HLA alloantigens. TAMURA and colleagues (1993) investigated the role of the 70-kDa constitutive/cognate Hsp (Hsc70) as a possible tumor antigen, and suggested that Hsc70 expressed on the surface of tumor cells acts as recognition structure for non-MHC restricted CD4/CD8-double negative T cells.

Non MHC-restricted, TcR/CD3 negative NK-like effector cells are also to be considered in the immune response against Hsps. Indeed, the inducible Hsp70 is expressed on the surface of certain tumor cells, where it acts as a positive recognition signal for TcR/CD3 negative NK cells. In sarcoma cells, the cell surface expression of Hsp70 is induced by non-lethal heat shock (MULTHOFF et al. 1995a,b; MULTHOFF and HIGHTOWER 1996) or by treatment with the membrane-reactive alkyl-lysophospholipid derivative ET18-OCH3 (BOTZLER et al. 1996). This stress-inducible Hsp70 expression correlates with an enhanced sensitivity to lysis mediated by non-MHC restricted NK cells (MULTHOFF et al. 1995b).

Certain carcinoma cell lines exhibit Hsp70 cell surface expression under physiological conditions: the colon carcinoma line CX2 is known to stably express Hsp70 on about 60% of cells. Following cell sorting by Hsp70 expression, two sublines, CX+ and CX-, were generated. CX+ shows a stable high expression level of Hsp70 whereas CX- shows Hsp70 cell surface expression only on a minor population. Interestingly, HLA antigens and adhesion molecules were not different among CX+ and CX- cells. These two sublines provide an autologous tumor cell system for studying the role of Hsp70 as a recognition site on tumor cells.

Stress-independent plasma membrane expression of Hsp70 occurs in parallel with an increased sensitivity of NK-mediated tumor cells lysis (MULTHOFF et al. 1997). Hsp70 might act as one possible recognition structure for a distinct TcR/CD3 negative NK subpopulation. Although the complete Hsp70 protein could be immunoprecipitated from the membrane fraction of Hsp70-expressing tumor cells, the C-terminal part of Hsp70 is particularly immunogenic for NK cells, as determined by antibody binding studies using different Hsp70 specific antibodies with mapped recognition epitopes (Multhoff et al., unpublished). The positive signal for Hsp70-mediated lysis appears to dominate the negative regulatory signal for inhibition of lysis mediated by MHC alleles. Indeed, MHC class I expression is identical in both tumor cell types

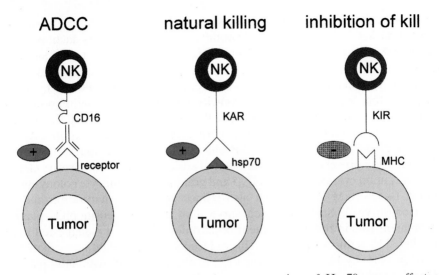

Fig. 4. Stress-independent plasma membrane expression of Hsp70 as an effector mechanism for the NK-mediated tumor cell lysis. NK cells contribute to antitumor immunity by either ADCC, or recognition of tumor cell membrane expressed Hsp70, while killer cell inhibitory/activatory receptors (KIR/KAR) modulate killing

(MULTHOFF et al. 1997), and even following treatment with IFNγ that enhances MHC expression on both tumor lines, the Hsp70 expressing CX+ is lysed better by NK cells. Furthermore, MHC class I-specific antibodies are unable to increase the lysis of Hsp70 expressing tumor cells and NK cells, the specificity of which against Hsp70 expressing tumor cells can be generated from MHC divergent donors.

A schematic illustration of different NK effector mechanisms is shown in Fig. 4. Besides CD16-antibody-mediated antibody-dependent cellular cytotoxicity (ADCC), Hsp70 might be considered as an immunogenic target structure on tumor cells for a specific TcR/CD3 negative NK subpopulation. Blocking studies, using purified recombinant Hsp70, revealed that these NK cells might express an Hsp70 receptor, the molecular characterization of which is currently under investigation. In this context the role of killer cell inhibitory/ activatory receptors (KIR/KAR) with specificity against distinct MHC alleles is also being analyzed.

F. The Paradigm of Asthma

Airway inflammation is considered the main cause of asthma. Acute asthma attacks induced by allergen challenge in allergic patients lead to an early inflammatory response characterized by a specific IgE-mediated activation of mast cells, alveolar macrophages and bronchial epithelial cells. IgE-mediated activation involves the release of proinflammatory mediators such as histamine, AA metabolites, PAF-acether, ROS, neuropeptides and cytokines that altogether lead to the constriction of airway and smooth muscle, mucus secretion and vasodilation. This acute phase response is often followed by a late response that contributes to the chronicity and to the exacerbation of inflammation which are typical features of the illness. It is suggested that the recruitment of peripheral blood cells, particularly eosinophils, plays a central role in the late phase asthmatic response. Besides cell-cell and cell-extracellular matrix adhesion processes, the recruitment of eosinophils into the airways clearly depends upon cytokines such as IL-5 and granulocyte-monocyte colony stimulating factor (GM-CSF), which also induce eosinophil maturation and increased survival. Such increased survival of activated eosinophils plays a prominent role in the development of chronic asthma and is suggested to be mediated by reduced apoptosis. This reduced apoptosis might be mediated, at least in part, by Hsp70 overexpression in the airways of asthmatic patients (Samali and Cotter 1996; Vignola et al. 1995). Indeed, Hsp70 expression in asthma significantly correlates with clinical severity (Vignola et al. 1995; Fajac et al. 1997), indicating that Hsp parallels the development of chronic airway inflammation (Vignola et al. 1995).

There are several possible biological consequences of the increased expression of Hsp70 in asthma. Firstly, Hsp70 may provide an effective protective mechanism against ROS and proinflammatory cytokines, widely released in the inflamed airways. The theoretical protective role of Hsp70 in asthma is supported by the evidence that anti-Hsp70 immunoreactivity of epithelial cells or alveolar macrophages significantly correlates with the number of eosinophils recovered in the bronchoalveolar lavage of asthmatics (Vignola et al. 1995). The toxicity of eosinophils mainly results in epithelium shedding caused by extracellular ECP; overexpression of Hsp70 by bronchial epithelial cells may protect these cells against the deleterious effects of the latter mediator.

Secondly, Hsp70 may contribute to the amplification of the inflammatory response. Indeed, Hsp70 may participate to antigen processing and/or presentation (Corrigan and Kay 1992) as well as in class II expression, or act itself as an antigen. Since the airway epithelium of patients with severe asthma is infiltrated by an increased number of T lymphocytes bearing both the $\alpha\beta$ and the $\gamma\delta$ receptors, it is conceivable that Hsp expressed by bronchial epithelial cells may activate Hsp-specific T cells, contributing to the perpetuation of the airway's inflammation.

Thirdly, the increased expression of Hsp70 (and Hsp27) in asthma may play a role in the regulation of cell apoptosis in epithelial cells and in inflammatory cells. With regard to the bronchial epithelium, recent evidence shows that very few epithelial cells are apoptotic in asthmatic patients (Vignola, unpublished). When apoptosis is detected, it localizes to the superficial layer of metaplastic lesions or to desquamated ciliated bronchial epithelial cells. This localization suggests that apoptosis contributes to tissue turnover by elimination of the epithelial cells after terminal differentiation or cellular damage (MOUNTZ et al. 1994), as well as to the maintenance of a balance between the rate of cell proliferation and death (SAVILL 1994; SCHULER et al. 1994). Thus, in asthma, despite the release of a wide range of cytotoxic or pro-apoptotic mediators, a low number of bronchial epithelial cells are apoptotic: these cells may efficiently protect themselves against noxious stimuli. Interestingly, the increased expression of Hsp70 in the epithelium of asthmatics is paralleled by an increased expression of bcl-2, and a low level of p53 and proliferating cell nuclear antigen (PCNA) expression, suggesting that the regeneration of the epithelial layer in asthma may be related more to the survival of basal epithelial cells than to their replication. By contrast, the increased expression of Hsp70 in inflammatory cells infiltrating the bronchial mucosa of asthmatic patients may have deleterious consequences. By their ability to reduce cell apoptosis and increase cell survival, the overexpression of Hsp70 by inflammatory cells may play a crucial role in the persistence of these cells in the inflamed tissues and in the pathogenesis of the chronicity of airway inflammation. Hence, the potential biological consequences of the expression of Hsp70 seem to be different according to the different cell types expressing these molecules.

In contrast, the low expression of Hsp70 observed in the airways of chronic bronchitis patients suggests that the mechanisms underlying the airway inflammation in this distinct disease differ from those involved in asthma. The presence of non-degranulated eosinophils observed in biopsies of patients with chronic bronchitis suggests a lower "aggressivity" of eosinophils in this clinical situation as compared to asthma. Airway inflammation in asthma and chronic bronchitis may differ in terms of cell recruitment, activation, and mediator release. These differences may be relevant to the regulation of Hsp70 expression, which may be specific for asthma rather than a general feature of chronic inflammation of the airways.

Finally, Hsps have been described to associate with cytosolic steroid receptors, suggesting a fundamental anti-inflammatory role for these proteins, particularly relevant to glucocorticoid therapy. Glucocorticoids, which suppress the release of arachidonic acid derivatives by inhibiting PLA_2, possess cytosolic receptors which bind Hsp90 molecules and form an inactive complex unable to bind DNA. This inactive form of the receptor is a multiprotein complex that also includes Hsp70. In the absence of steroids, it appears that the Hsp90/Hsp70 chaperoning system is required to maintain a proper receptor conformation for high steroid binding affinity. Upon steroid binding, Hsp90, but not Hsp70, dissociates, triggering receptor transformation from the inactive form to a steroid-activated state that binds to the appropriate response element in the promoters of glucocorticoid responsive genes (GRE), to bring about the final response in target cells (PRATT 1993), including the anti-inflammatory effects.

G. Conclusions and Perspectives

Though the primary function of Hsps is to rescue other proteins from denaturation, the fields of application of these proteins have expanded during recent years to broader areas, and particularly to the biomedical field, which includes infection, cancer and inflammatory diseases, as illustrated in several chapters of this book. Besides the evidence that Hsps exert protective, anti-inflammatory effects – though eventually, pro-inflammatory effects as well – a novel role for these proteins has been suggested, as a promising prognostic/ diagnostic marker in inflammatory and (auto)immune diseases. To promote this specific research area, we recently developed a new test that allows the rapid evaluation Hsp70 in human peripheral blood monocytes with an increased sensitivity and accuracy (Bachelet et al., 1998). The use of such new tools may allow in the near future the definition of clues to a better understanding of the influence of Hsps in immune and inflammatory-related diseases.

Acknowledgements. We are grateful to Dr. P. Daniel LEW for providing the cells from the LAD patient and to OM Pharma for partial support. M.B. and B.S.P. are supported by INSERM.

References

Anderton SM, van der Zee R, Prakken B, Noordzij A, van Eden W (1995) Activation of T cells recognizing self 60-kD heat shock protein can protect against experimental arthritis. J Exp Med 181:943–952

Arnaout MA (1990) Leukocyte adhesion molecules deficiency: its structural basis, pathophysiology and implications for modulating the inflammatory response. Immunol Rev 114:145–180

Arrigo AP (1990) TNF induces the rapid phosphorylation of the mammalian hsp28. Mol Cell Biol 10:1276–1280

Bachelet M, Chouaid C, Havet N, Masliah J, Barre A, Housset B, Vargaftig B (1992) Modulation of arachidonic acid metabolism and cyclic AMP content of human alveolar macrophages. Eicosanoids 5:185–190

Bachelet M, Mariéthoz E, Banzet N, Souil E, Pinot F, Polla CZ, Durand P, Bouchaert I, Polla BS (1998) Flow cytometry is a rapid and reliable method for evaluating heat shock protein 70 expression in human monocytes. Cell Stress & Chap 3:168–176

Baeuerle PA, Baltimore D (1996) NF-κB: ten years later. Cell 87:13–20

Baldwin AS Jr (1996) The NF-κB and IκB proteins: new discoveries and insights. Annu Rev Immunol 14:649–681

Barazzone C, Kantengwa S, Suter S, Polla BS (1996). Phagocytosis of *Pseudomonas aeruginosa* fails to elicit heat shock protein expression in human monocytes. Inflammation 20:243–262

Bonta IL, Parnham MJ (1982) Immunomodulatory antiinflammatory functions of E-type prostaglandins. Minireview with emphasis on macrophage-mediated effects. Int J Immunopharmacol 4:103–109

Born W, Hall L, Dallas A, Boymel J, Shinnick T, Young D, Brennan P, O'Brien R (1990) Recognition of a peptide antigen by heat shock reactive γδ T lymphocytes. Science 249:67–69

Botzler C, Kolb H-J, Issels RD, Multhoff G (1996) Noncytotoxic alkyl-lysophospholipid treatment increases sensitivity of leukemic K562 cells to lysis by natural killer (NK) cells. Int J Cancer 65:633–638

Buttke TM, Sandstrom PA (1994) Oxidative stress as a mediator of apoptosis. Immunol Today 15:7–10

Chopp M (1993) The roles of heat shock proteins and immediate early genes in central nervous system normal function and pathology. Curr Opin Neurol Neurosurg 6:6–10

Chouchane L, Bowers S, Sawasdikosol S, Simpson RM, Kindt TJ (1994) Heat shock proteins expressed on the surface of human T cell leukemia virus Type I-infected cell lines induce autoantibodies in rabbits. J Infect Dis 169:253–259

Christie P, Jacquier-Sarlin M, Janin A, Bousquet J, Polla BS (1995) Heat Shock proteins in eosinophilic inflammation. In: Van Eden W, Young DB (eds) Stress proteins in medicine. Dekker, New York, pp 479–493

Cohen IR, Young DB (1991) Autoimmunity, microbial immunity and the immunological homunculus. Immunol Today 12:105–110

Corrigan CJ, Kay AB (1992) Asthma. Role of T-lymphocytes and lymphokines. Br Med Bull 48:72–84

DiCesare S, Poccia F, Mastino A, Colizzi V (1992) Surface expressed heat shock proteins by stressed or human immunodeficiency virus (HIV)-infected lymphoid cells represent the target for antibody-dependent cellular cytotoxicity. Immunol 76:341–343

Dix DJ, Allen JW, Collins BW (1996) Targeted gene disruption of Hsp70–2 results in failed meiosis, germ cell apoptosis and male infertility. Proc Natl Acad Sci USA 93:3264–3268

Favatier F, Bornman L, Hightower LE, Günther E, Polla BS (1997). Variation in hsp gene expression and Hsp polymorphism: do they contribute to differential disease susceptibility and stress tolerance? Cell Stress Chap 2:141–155

Fajac I, Roisman GL, Lacronique J, Polla BS, Dusser DJ (1997) Bronchial gamma-delta T-lymphocytes and expression of heat shock proteins in mild asthma. Eur Respir J 10:633–638

Fincato G, Polentarutti N, Sica A, Mantovani A, Colotta F (1991) Expression of a heat-shock inducible gene of the HSP70 family in human myelomonocytic cells: regulation by bacterial products and cytokines. Blood 77:579–586

Haire RN, Peterson MS, O'Leary JJ (1988) Mitogen activation induces the enhanced synthesis of two heat-shock proteins in human lymphocytes. J Cell Biol 106:883–891

Helqvist S, Polla BS, Johannasen J, Nerup J (1991) Heat shock proteins induction in rat pancreatic islets by recombinant human interleukin 1 β. Diabetologia 34:150–156

Hennet T, Richter C, Peterhans E (1993) Tumor necrosis factor-alpha induces superoxide anion generation in mitochondria of L929 cells. Biochem J 289:587–592

Heufelder AE, Wenzel BE, Bahn RS (1992) Cell surface localization of a 72 kilodalton heat shock protein in retroocular fibroblasts from patients with Graves' ophthalmopathy. J Clin Endocrinol Metab 74:732–736

Hightower LE, Guidon PT (1989) Selective release from cultured mammalian cells of heat shock (stress) proteins that resemble glia-axon transfer proteins. J Cell Physiol 138:257–266

Himeno K, Hisaeda H (1996) Contribution of 65-kDa heat shock protein induced by gamma and delta T cells to protection against Toxoplasma gondii infection. Immunol Res 15:258–264

Hisaeda H, Nagasawa H, Maeda K, Maekawa Y, Ishikawa H, Ito Y, Good RA, Himeno K (1995) $\gamma\delta$ T cells play an important role in expression of hsp65 and in acquiring protective immune responses against infection with Toxoplasma gondii. J Immunol 155:244–249

Hisaeda H, Sakai T, Nagasawa H, Ishikawa H, Yasutomo K, Maekawa Y, Himeno K (1996a) Contribution of extrathymic $\gamma\delta$ T cells to the expression of heat shock protein and to protective immunity in mice infected with *Toxoplasma gondii*. Immunol 88:551–557

Hisaeda H, Sakai T, Ishikawa H, Maekawa Y, Yasutomo K, Nagasawa H, Himeno K (1996b) Mechanisms of Hsp65 expression induced by $\gamma\delta$ T cells in murine *Toxoplasma gondii* infection. Pathobiology 64:198–203

Hisaeda H, Sakai T, Ishikawa H, Maekawa Y, Yasumoto K, Good RA, Himeno K (1997) Heat shock protein 65 induced by $\gamma\delta$ T cells prevents apoptosis of macrophages and contributes to host defense in mice infected with *Toxoplasma gondii*. J Immunol 159:2375–2381

Hisaeda H, Himeno K (1997) The role of host-derived heat shock protein in immunity against *Toxoplasma gondii* infection. Parasitology Today 13:465–468

Holtzman M (1991) Arachidonic acid metabolism, Am Rev Resp Dis 143:188–203

Ishikawa H, Hisaeda Y, Maekawa Y, Himeno K (1997) Expression of heat shock proteins in host macrophages correlates with a protective potential against infection with *Leishmania major* in mice. Parasitology International 46:263–270

Jäättelä M, Saksela K, Saksela E (1989) Heat shock protects WEHI-164 target cells from the cytolysis by tumor necrosis factor a and β. Eur J Immunol 19:1413–1417

Jäättelä M (1993) Overexpression of major heat shock protein hsp70 inhibits tumor necrosis factor-induced activation of phospholipase A2. J Immunol 151:4286–4294

Jacquier-Sarlin MR, Fuller K, Dinh-Xuan AT, Richard M-J, Polla BS (1994) Protective effects of hsp70 in inflammation. Experientia 50:1031–1038

Jacquier-Sarlin MR, Jornot L, Polla BS (1995) Differential expression and regulation of hsp70 and hsp90 by phorbol esters and heat shock. J Biol Chem 270:14094–14099

Jacquier-Sarlin MR, Polla BS (1996) Dual regulation of heat-shock transcription factor (HSF) activation and DNA-binding activity by H2O2: role of thioredoxin. Biochem J 318:187–19

Kantengwa S, Donati YRA, Clerget M, Maridonneau-Parini I, Sinclair F, Mariéthoz E, Perin M, Rees ADM, Slosman DO, Polla BS (1991) Heat shock proteins: an autoprotective mechanism for inflammatory cells? Semin Immunol 3:49–56

Kantengwa S, Polla BS (1993) Phagocytosis of *Staphylococcus aureus* induces a selective stress response in human monocytes-macrophages (mφ): modulation by mφ differentiation and by iron. Infect Immun 61:1281–1287

Kaufmann SH (1994) Heat shock proteins and autoimmunity: a critical appraisal. Int Arch Allergy Immunol 103:317–322

Köller M, König W (1990) Arachidonic acid metabolism in heat-shock treated human leukocytes. Immunology 70:458–464

Köller M, Hensler T, Konig B,Prevost G, Alouf J, Konig W (1993) Induction of heat shock proteins by bacterial toxins, lipid mediators and cytokines in human leukocytes. Int J Med Microbiol Virol Parasitol Infect Dis 278:365–376

Leonard EJ, Yoshimura T (1990) Neutrophil attractant/activation protein-1 (NAP-1 [interleukin 8]). Am J Res Cell Mol Biol 2:479–481

Liochev SI, Fridovich I (1997) How does superoxide dismutase protect against tumor necrosis factor: a hypothesis informed by effect of superoxide on "free" iron. Free Radiat Biol Med 23:668–671

Long EO, Colonna M, Lanier LL (1996) Inhibitory MHC class I receptors on NK and T cells: a standard nomenclature. Immunol Today 17:100–111

Lurquin C, Van Pel A, Mariamé B, De Plaen E, Szikora J-P, Janssens J, Reddehase MJ, Lejeune J, Boon T (1989) Structure of the gene tum-transplantation antigen P91 A: the mutated exon encodes a peptide recognized with Ld by cytolytic cells. Cell 58:293–303

Margulis BA, Sandler S, Eizirik DL, Welsh N, Welsh M (1991) Liposomal delivery of purified heat shock protein hsp70 into rat pancreatic islets as protection against interleukin 1β-induced impaired β-cell function. Diabetes 40:1418–1422

Mariéthoz E, Tacchini-Cottier F, Jacquier-Sarlin M, Sinclair F, Polla B S (1994) Exposure of monocytes to heat shock does not increase class II expression but modulates antigen-dependent T cell responses. Int Immunol 6:925–930

Marletta MA (1989) Nitric oxide: biosynthesis and biological significance. Trends Biochem Sci 14:488–492

Mehlen P, Schulze-Osthoff K, Arrigo A-P (1996a) Small stress proteins as novel regulators of apoptosis. J Biol Chem 271:16510–16514

Mehlen P, Kretz-Romy C, Preville X, Arrigo A-P (1996b) Human hsp27, Drosophila hsp27 and human $\alpha\beta$-crystallin expression-mediated increase in glutathione is essential for the protection activity of these proteins against TNFα-induced cell death. EMBO J 15:2695–2706

Misumi Y, Miki K, Takatsuki A, Tamura G, Ikehara Y (1986) Novel blockade by brefeldin A of intracellular transport of secretory proteins in cultured rat hepatocytes. J Biol Chem 261:11398–11403

Moliterno R, Valdivia L, Pan F, Duquesnoy RJ (1995) Heat shock protein reactivity of lymphocytes isolated from heterotopic rat cardiac allografts. Transplantation 59:598–604

Moretta L, Ciccone E, Poggi A, Mingari MC, Moretta A (1994). Ontogeny, specific functions and receptors of human natural killer cells. Immunol Lett 40:83–88

Mountz JD, Wu J, Cheng J, Zhou T (1994) Autoimmune disease. A problem of defective apoptosis. Arthritis Rheum 37:1415–1420

Multhoff G, Botzler C, Wiesnet M, Müller E, Meier T, Wilmanns W, Issels RD (1995a) A stress inducible 72-kDa heat shock protein (HSP72) is expressed on the surface of human tumor cells, but not in normal cells. Int J Cancer 61:272–279

Multhoff G, Botzler C, Wiesnet M, Eissner G, Issels RD (1995b) CD3-large granular lymphocytes recognize a heat-inducible immunogenic determinant associated with the 72-kD heat shock protein on human sarcoma cells. Blood 86:1374–1382

Multhoff G, Hightower LE (1996) Cell surface expression of heat shock proteins and the immune response. Cell Stress Chap 1:167–176

Multhoff G, Botzler C, Jennen L, Schmidt J, Ellwart J, Issels R (1997) Heat shock protein 72 on tumor cells. A recognition structure for natural killer cells. J Immunol 158:4341–4350

Munk ME, Schoel B, Modrow S, Karr RW, Young RA, Kaufmann SHE (1989) T lymphocytes from healthy individuals with specificity to self epitopes shared by the mycobacterial and human 65-kilodalton heat shock protein. J Immunol 143:2844–2879

Muthukrishnan L, Warder E, McNeil PL (1991) Basic fibroblast growth factor is efficiently released from a cytosolic storage site through plasma membrane disruptions of endothelial cells. J Cell Physiol 148:1–16

Nagasawa H, Oka M, Maeda K, Chai J-G, Hisaeda H, Ito Y, Good RA, Himeno K (1992) Induction of heat shock protein closely correlates with protection against Toxoplasma gondii infection. Proc Natl Acad Sci USA 89:3155–3158

Nagasawa H, Hisaeda H, Maekawa Y, Fujioka H, Ito Y, Aikawa M, Himeno K (1994) $\gamma\delta$ T cells play a crucial role in expression of 65000 MW heat shock protein in mice immunized with Toxoplasma antigen. Immunol 83:347–352

O'Brien RL, Fu Y, Cranfill R, Dallas A, Ellis C, Reardon C, Lang J, Carding S, Kubo R, Born W (1992) Heat shock protein 60-reactive $\gamma\delta$ cells: a large, diversified T-lymphocyte subset with highly focused specificity. Proc Natl Acad Sci USA 89:4348–4352

Ohtsuka K, Nakamura H, Sato C (1986) Intracellular distribution of 73,000 and 72,000 dalton heat shock protein in HeLa cells. Int J Hyperthermia 2:267–75I

Pinhasi-Kimhi O, Michalovitz D, Ben-Zeev A, Oren M (1986) Specific interaction between the p53 cellular tumor antigen and major heat shock proteins. Nature 320:182–184

Polla BS, De Rochemonteix B, Junod AF, Dayer J-M (1985) Effects of LTB4 and Ca++ ionophore A23187 on the release by human alveolar macrophages of factors controlling fibroblast functions. Biochem Biophys Res Commun 129:560–567

Polla BS, Healy AM, Wojno WC, Krane SM (1987) Hormone 1a,25-dihydroxyvitamin D3 modulates heat shock response in monocytes. Am J Physiol 252:C640–C649

Polla BS (1988) A role for heat shock proteins in inflammation, Immunol Today 9:134–137

Polla BS, Mariéthoz E, Hubert D, Barazzone C (1995a) Heat-shock proteins in host-pathogen interactions: implications for cystic fibrosis. Trends Microbiol 10:392–396

Polla BS, Stubbe H, Kantengwa S, Maridonneau-Parini I, Jacquier-Sarlin MR (1995b) Differential induction of stress proteins and functional effects of heat shock in human phagocytes. Inflammation 19:323–377

Polla BS, Kantengwa S, François D, Salvioli S, Franceschi C, Marsac C, Cossarizza A (1996) Mitochondria are selective targets for the protective effects of heat shock against oxidative injury. Proc Natl Acad Sci USA 93:6458–6463

Pratt WB (1993) The role of heat shock proteins in regulating the function, folding and trafficking of the glucocorticoid receptor. J Biol Chem 268:21455–21458

Rauslton JE, Davis CH, Schmiel DH, Morgan MW, Wyrik PB (1993) Molecular characterization and outer membrane association of a *Chlamydia trachomatis* protein related to the hsp70 family of proteins. J Biol Chem 268:23139–23147

Richter C, Schweitzer M, Cossarizza A, Franceschi C (1996) Control of apoptosis by cellular ATP levels. FEBS Lett 378:107–110

Rosales C, Juliano RL (1995) Signal transduction by cell adhesion receptors in leuko-cytes. J Leukocyte Biol 57:189–198

Rosen GM, Pou S, Ramos CL, Cohen MS, Britigan BE (1995) Free radicals and phagocytic cells. FASEB J 9:200–209

Rossi A, Elia G, Santoro MG (1997) Inhibition of nuclear factor kB by prostaglandin A1: an effect associated with heat shock transcription factor activation. Proc Natl Acad Sci USA 94:746–750

Samali A, Cotter TG (1996) Heat shock proteins increase resistance to apoptosis. Exp Cell Res 223:163–170

Santoro MG (1997) Antiviral activity of cyclopentenone prostanoids. Trends Microbiol 5:276–281

Santoro MG, Garaci E, Amici C (1989) Prostaglandins with antiproliferative activity induce the synthesis of a heat shock protein in human cells. Proc Natl Acad Sci USA 86:8407

Savill J (1994) Apoptosis in disease. Eur J Clin Invest 24:715–73

Schuler D, Szende B, Borsi JD, Marton T, Bocsi J, Magyarossy E, Koos R, Csoka M (1994) Apoptosis as a possible way of destruction of lymphoblasts after glucocor-ticoid treatment of children with acute lymphoblastic leukemia. Pediatr Hematol Oncol 11:641–64

Springer TA (1994) Traffic signals for lymphocyte recirculation and leukocyte emigra-tion: the multistep paradigm. Cell 76:301–314

Srivastava PK (1994) Heat shock proteins in immune response to cancer: the fourth paradigm. Experientia 50:1054–1060

Takenaka IM, Hightower LE (1992) Transforming growth factor $\beta 1$ rapidly induces Hsp70 and Hsp90 molecular chaperones in cultured chicken embryo cells. J Cell Physiol 152:568–577

Tamura Y, Tsuboi N, Sato N, Kikuchi K (1993) 70 kDa heat shock cognate protein is a transformation-associated antigen and a possible target for the host's anti-tumor immunity. J Immunol 151:5516–5524

Tauber AI (1987) Protein knase C and the activation of the human neutrophil NADPH-oxidase. Blood 69:711–720

Vanbuskirk AM, DeNagel DC, Guagliardi LE, Brodsky FM, Pierce CK (1991) Cellular and subcellular distribution of PBP72/74, a peptide-binding protein that plays a role in antigen processing. J Immunol 146:500–506

Van-den Eynde B, Lethe B, Van-Pel A, De-Plaen E, Boon T (1991) The gene coding for a major tumor rejection antigen of tumor P815 is identical to the normal gene of syngeneic DBA/2 mice. J Exp Med 173:1373–1384

Vignola AM, Chanez P, Polla BS, Vic P, Godard P, Bousquet J (1995) Increased expression of heat shock protein 70 on airway cells in asthma and chronic bronchitis. Am J Res Cell Mol Biol 13:683–694

Villar J, Edelson JD, Post M, Mullen BM, Slutsky AS (1993) Induction of heat stress proteins is associated with decreased mortality in an animal model of acute lung injury. Am Rev Respir Dis 147:177–181

Udono H, Levey DL, Srivastava PK (1994) Cellular requirements for tumor-specific immunity elicited by heat shock proteins: tumor rejection antigen gp96 primes CD8+ T cells in vivo. Proc Natl Acad Sci USA 91:3077–3081

Ward PA, Markd RM (1989) The acute inflammatory reaction. Curr Opin Immunol 2:5–9

Watson F, Robinson J, Edwards SW (1990) Protein kinase C-dependent and -independent activation of the NADPH oxidase of human neutrophils. J Biol Chem 256:7432–7439

Winfield J, Jarjour W (1991) Do stress proteins play a role in arthritis and autoimmunity? Immunol Rev 121:193–220

Wong HR, Ryan M, Wispé JR (1997) Stress response decrease NF-κB nuclear translocation and increases IκBα expression in A549 cells. J Clin Invest 99:2433–2428

Yoshino I, Goedegebuure PS, Peoples GE, Lee KY, Eberlein TJ (1994) Human tumor-infiltrating CD4+ T cells react to B cell lines expressing heat shock protein 70. J Immunol 153:4149–4158

Young RA, Elliott TJ (1989) Stress proteins, infection, and immune surveillance. Cell 59:5–8

Zugel U, Schoel B, Yamamoto S, Hengel H, Morein B, Kaufmann SH (1995) Crossrecognition by CD8 T cell receptor alpha beta cytotoxic T lymphocytes of peptides in the self and the mycobacterial hsp60 which share intermediate sequence homology. Eur J Immunol 25:451–458

CHAPTER 14

Heat Shock Proteins in Embryonic Development

M. MORANGE

A. Introduction

Very early, even before they were characterized as proteins the synthesis of which is increased by heat shock and other proteotoxic treatments, the heat shock proteins (Hsps) were described as active partners in cell differentiation and organism morphogenesis. Let us mention two such observations. As early as 1982, the variations in small Hsp expression during *Drosophila* development were described (SIROTKIN and DAVIDSON 1982). As we shall see, the expression of small Hsps during the transition between proliferation and differentiation remains the center of numerous investigations. In 1982, JOHNSON's group described the transient expression during the early transcriptional activation of the mouse zygotic genome of two proteins of 70 kDa (FLACH et al. 1982), which were shown later to be two members of the Hsp70 family (BENSAUDE et al. 1983).

Interest in the spontaneous developmental expression of Hsps somehow vanished when the functions of these proteins were discovered in 1986. Chaperoning of nascent proteins and/or degradation of misfolded or denatured proteins are ubiquitous functions, required in every living cell, not specifically related to cell differentiation and development. The general feeling was that the characteristic expression of Hsps during development was simply reflecting the dramatic variations in protein synthesis and degradation occurring during development.

While it is likely that some of the developmental expression patterns of Hsps can be explained by the concomitant variations in protein turnover, many observations of the specific expression of Hsps in a limited group of cells and at specific differentiation stages cannot be reconciled with such a simple interpretative scheme. Recent data have demonstrated the crucial requirements for Hsps at specific developmental stages. As we shall see, this crucial requirement is linked with the fundamental role which these proteins play in the control of cell division, intracellular signaling pathways and cell death. By virtue of their involvement in such fundamental processes, Hsps are active partners in cell differentiation and development.

Our intent in this review is not to list all the observations which have accumulated on the characteristic expression of Hsps during differentiation and development. Some recent reviews have compiled many of these data (Heikkila 1993; Morange 1997). In addition, the diversity of differentiation processes in which Hsps have been implicated and of the organisms in which such observations have been made would transform this description into one long list of heteroclite observations. For instance, we will not cover in this review the numerous observations on the characteristic expression of Hsps in plants. A complete review was published last year (Boston and Vierling 1996). This review, as well as more recently published articles such as that by Wehmeyer et al. (1996), fully confirm that the expression of the different Hsps during plant development is too specific – too narrowly controlled – to be solely linked to the general metabolism of proteins. However, the Hsps of plants are less well characterized than those of the animal kingdom – with the exception of small Hsps, whose complex structure and evolution have been extensively studied in plants (Waters et al. 1996) as has their chaperone function (Lee et al. 1995, 1997). In addition, no data demonstrating the developmental role of Hsps have been obtained in plants, whereas, as we shall see, data in animals have demonstrated such a role in a limited number of cases and provided a description of the mechanisms of action of Hsps.

In the first section of this review, we will document this very characteristic developmental expression of Hsps in *Drosophila,* where it has been most intensively studied. Next, we will describe different situations where, in addition to the documentation of Hsp expression, additional experiments have been performed which have shown that this expression is required for the fulfillment of this developmental step. Explanatory models have been proposed: some of them, however, still await confirmation.

We will then turn our attention to the mechanisms involved in the developmental expression of these Hsps. The question is whether the factors which are responsible for the stress inducibility of these genes are also responsible for the regulation of Hsp expression during development.

This review intends more to open pathways of research than to provide definitive answers on the role of Hsps in development. Therefore, we will spend some time in describing the characteristics of some recently described chaperones (and co-chaperones) which are good candidates for playing a fundamental role in development and cell differentiation. We will also consider chaperones and co-chaperones in general: the fact that some of these chaperones have been recruited as Hsps, the synthesis of which is induced by proteotoxic treatments, is of minor importance for their developmental function.

Finally, we will devote the last part of this review to the role of Hsps in aging. Aging may be considered as the last stage of development. Not only does Hsp expression vary during aging, but recent data obtained by the screening of genes involved in aging both in *Drosophila* and nematodes have shown a close link between the capacity to mount a stress and heat shock response, and life expectancy. Favored models of the aging process propose that the mechanisms of defense – including the heat shock response and heat shock proteins – play a crucial role in the definition of life-span.

B. Specific Expression of Hsps During *Drosophila* Development

Probably the most recent, abundant and clear-cut observations concerning the differential expression of Hsps during development and differentiation have been collected in *Drosophila* for Hsp83 and the small Hsps (MICHAUD et al. 1997a; see also an earlier article by PERKINS et al. 1990 for HSC70).

Hsp83 is ubiquitously expressed during embryogenesis. However, studies have revealed an interesting pattern of expression and localization of the *hsp83* mRNAs during oogenesis and early embryogenesis (DING et al. 1993). The gene for Hsp83 is highly transcribed in nurse cells from stages 9 to 11, and from stage 10 the mRNA is transferred to the oocyte. During the first cell divisions, the maternal *hsp83* mRNA is concentrated at the posterior pole of the embryo through its selective stabilization. *hsp83* mRNA is then taken up by the pole cells. The gene is also transcribed from the zygotic genome at the syncitial blastoderm stage, but this latter expression is limited to the anterior third of the embryo. This localized expression is under the control of early developmental genes such as *bicoid*.

The differential expression of small Hsps is most obvious in the brain during development, and in testes and ovaries during postnatal differentiation. It concerns three of the four small Hsps: Hsp23, Hsp26 and Hsp27, Hsp22 not being expressed in early development (MICHAUD et al. 1997a). In general, two different experimental strategies have been used: the direct observation of the proteins with specific antibodies, or the use of promoter-driven reporter genes in transgenic flies.

Hsp23 is synthesized during development in a subset of glial cells called the middle glial cells (TANGUAY 1989) and it is also expressed in the salivary glands (ARRIGO and AHMED-ZADEH 1981). It is also highly expressed in late third instar larvae and during pupation, in response to an increased level of β-ecdysone. The amount of protein decreases at the adult stage to reach a minimum level in 1-week-old flies, except in the neurocytes and glial cells of the central nervous system and in the gonads (MARIN et al. 1993). In the male gonads Hsp23 is expressed constitutively in specific cells of the somatic lineage, such as the cyst cells. Hsp23 might also be associated with the spermatid bundles.

Hsp26 is also highly expressed in gonads in larvae, pupae and adults. It is expressed in spermatocytes as well as in nurse cells and developing oocytes (GLASER et al. 1986). During development, Hsp26 is also expressed in the epithelium, proventriculus, larval brain and ventral ganglion.

Hsp27 shares with Hsp23 some characteristics of expression: it is induced in the imaginal disks of third instar larvae and throughout the larval stages its expression is restricted to the CNS and gonads. However, Hsp27 also exhibits a specific pattern of expression. In late pupae, it is present at the top of the eye ommatidial unit. In the adult, it is present in a few cluster of cells dispersed in the organism, whereas no mRNAs for Hsp27 can be detected in the same structures, suggesting that the protein is stabilized in these cells (Pauli et al. 1990). In testes, in contrast with Hsp23, which is expressed in cells of the somatic lineage, Hsp27 is highly expressed in the spermatocytes. In ovaries, Hsp27 is expressed in nurse cells, but also in a small group of somatic cells at stages 8–10 of oogenesis. Remarkably, the intracellular localization of Hsp27 changes during development: it is localized in the nucleus of nurse cells during the first stages of oogenesis, whereas it shows a perinuclear and cytoplasmic localization from stage 8 (Marin and Tanguay 1996).

In summary, small Hsps exhibit a very specific pattern of expression during *Drosophila* development. Furthermore, the intracellular localization of these different small Hsps is not identical. Hsp22 is localized in mitochondria, whereas Hsp23 and Hsp26 are mainly found in the cytoplasm. As previously mentioned, Hsp27 localization changes during development, being mainly cytoplasmic, but nuclear in some cells at specific developmental stages. In addition, the small Hsps can be modified by phosphorylation and the pattern of phosphorylation is different from one tissue to another (Marin et al. 1996). As described in another chapter of this book, variations in the state of phosphorylation have been shown to be important for small Hsp functions. Another level of complexity, which we do not intend to describe in this review, is the inducibility by heat treatment of these different Hsps, which varies between cells and between different developmental stages (Michaud et al. 1997b).

What might be the role of small Hsps during *Drosophila* development? Early results showed that the expression of Hsps is dependent on the energy source used by the cells (Lanks et al. 1986). During development, the changes in energy metabolism are frequent and modifications in Hsp synthesis might reflect these changes. In addition, two "housekeeping" functions of Hsps might be important during development:

First, a protective effect: the accumulation of Hsps during development might be the way to protect cells and tissues against proteotoxic aggressions at specific stages of development (Hunter and Dix 1996). The demonstrated action of exogenously added HSC70 in preventing neural death after peripheral sensory nerve damage is a good example of this potential protective and repair effect of Hsps (Houenou et al. 1996). Therefore, a description of developmental changes of Hsps in mammalian brain might be useful in understanding the pathological consequences of a maternal hyperthermia on brain development (Brown 1994; Kato et al. 1995; Walsh et al. 1997).

Second, a chaperone function linked with the increase in protein synthesis and cytomorphological reorganization which occurs during development: a parallel variation in the level of the spontaneous expression of Hsps and in the rate of protein synthesis is well illustrated by the developmental study of

Hsp47, a chaperone specialized in the formation of collagen. Tissue-specific differences in the expression of Hsp47 during zebrafish development correlate with the requirement for its chaperone function, although some discrepancies between the expression of the chaperone and the various collagen types suggest that other chaperones can probably replace Hsp47 in its specialized function (LELE and KRONE 1997). A good correlation has also been found in different developmental studies between the activity of mitochondria and the level of expression of Hsp60, which is a chaperone involved in the import into and folding of proteins inside this organelle (PARANKO et al. 1996; D'SOUZA and BROWN 1998).

The data reported above indicate a precisely regulated pattern of expression of small Hsps during *Drosophila* development and provide compelling evidence in favor of a developmental role of these small Hsps, distinct from their general chaperone and protective function. However, those observations, while tantalizing, remain frustrating because there are no real clues on what might be the different developmental functions of these proteins.

In view of this situation, we felt compelled to turn towards situations where the expression of these proteins has been clearly linked with a specific developmental process and where testable hypotheses on the functions of Hsps during this process have been proposed.

C. Essential Roles of Hsps During Development

We have chosen three different Hsps and three developmental situations which are currently the best documented: the role of the small Hsp in the proliferation/developmental-transition in mammalian cells; the role of Hsp90 in vertebrate muscle formation and the role of Hsp70–2 in mouse meiosis.

I. Mammalian Small Hsp: A "Checkpoint" Between Proliferation, Differentiation and Cell Death

Frequently, proliferation and differentiation are alternative pathways between which cells must choose during the development of an organism. Many studies have demonstrated a transient increased expression of the mammalian small Hsp (called Hsp25, Hsp27 and Hsp28, depending on the authors and on the organisms) during the transition between these two developmental behaviors. In the case of T lymphocytes, induction of Hsp27 expression followed mitotic stimulation (HANASH et al. 1993). In addition, Hsp27 is highly abundant in immature thymocytes versus mature non-cycling T-cells. In all the other published cases, the small Hsp is transiently induced in the hours following the induction in vitro of the transition from the proliferative state to the differentiation pathway. For instance, in B lymphocytes the transient increase in Hsp28 coincides precisely with the peak of cellular proliferation and the onset of growth arrest (SPECTOR et al. 1992). Differentiation in vitro of fetal rat calvarial osteoblasts takes place during a 30-day period. The small Hsp transiently increases to a maximal value of 2.5 times control level between day 7 and day 18. This increase parallels the onset of growth arrest and precedes the synthesis of differentiation markers, such as osteocalcin (SHAKOORI et al. 1992). The link between induction of the small Hsp and growth arrest has also been demonstrated in the case of the human promyelocytic leukemia HL60 cells (SHAKOORI et al. 1992). These cells can be induced to differentiate towards the granulocytic phenotype by retinoic acid treatment or, alternatively, towards the monocyte/macrophage phenotype by treatment with PMA. Interestingly, the mechanisms involved in the transient increase of the levels of the small Hsps are different in the two differentiation pathways. Retinoic acid acts solely by stabilizing the protein (SPECTOR et al. 1994; see, however, MINOWADA and WELCH 1995 for a different result), whereas PMA increases the transcription of the gene (SPECTOR et al. 1993). This suggests that the increase in Hsp28 level is an important prerequisite for the commitment towards the differentiation pathway.

The final experimental systems we will describe are embryonal carcinoma and embryonic stem cells (STAHL et al. 1992; MEHLEN et al. 1997). The basal level of Hsp25 is variable from one cell line to another. However, differentiation induced by addition of retinoic acid, embryoid body formation or serum deprivation in LIF – a factor inhibiting differentiation of ES cells – all lead to a transient increase in the expression of small Hsp.

To these observations in vitro, one must add the observations made in the embryo which show a strong labeling of isolated cells or structures with antibodies targeted against small Hsp. Hsp25 is expressed at the onset of myogenic differentiation in the myocardium of the mouse cardiac primordium at day 9.5 or in the differentiating myotomes slightly later in development (GERNOLD et al. 1993; LOONES et al. 1997). It is also expressed in some cells or nuclei of the central nervous system. At mid-gestation, Hsp25 is expressed in ganglia such as the fifth ganglion or in nuclei such as the hypoglossal nucleus (LOONES et al. 1997). Hsp25 is also heavily expressed in isolated neurons in the myelencephalon and spinal cord – the labeled neurons probably correspond to motoneurons – but also in the medulla and thalamus. However, this high level of expression is not transient and persists in the adult (PLUMIER et al. 1997).

In nearly all differentiation systems which have been studied, the increase in the level of small Hsp is preceded by hyperphosphorylation of this protein (SPECTOR et al. 1993, 1994; but see MINOWADA and WELCH 1995 for contradictory results). However, in the case of the HL-60 cells, phosphorylation can be inhibited without any change in the PMA-induced differentiation (SCHULTZ et al. 1997). The increase in small Hsp is concomitant with an increased amount of the high molecular mass, aggregated form, of the protein (CHAUFOUR et al. 1996), which has been shown to be the active form for at least two functions of the small Hsp – the chaperone function, and the control of the intracellular glutathione level. In fact, the transient increase in Hsp27 level in ES cells is accompanied by a transient increase in the glutathione level (MEHLEN et al. 1997).

It is not clear whether small Hsp is essential for cell growth. Neoplastic B cell lines which do not synthesize Hsp28 grow as rapidly as the Hsp28-producing cells (SPECTOR et al. 1992). A reduction in the level of Hsp27 by antisense mRNAs does not affect the rate of growth of undifferentiated ES cells (MEHLEN et al. 1997), but it affects the growth of the breast cancer cell line MCF-7 (MAIRESSE et al. 1996) and of osteosarcoma cell lines (RONDEAUX et al. 1997). However, the transient increase in small Hsp seems to be essential for growth arrest and commitment towards cell differentiation. The same B cell lines which do not synthesize Hsp28 during normal growth express it when induced to differentiate by phorbol ester treatment. Moreover, a heat treatment induces both the synthesis of small Hsp and cell differentiation. In contrast, reduction in the level of small Hsp by the expression of antisense constructs fully prevents differentiation of ES cells after LIF deprivation. Cells go on dividing and die by apoptosis instead of differentiating (MEHLEN et al. 1997). Evidence in favor of an essential role of small Hsp in differentiation has also been obtained in HL60 cells. A decrease in the level of Hsp27 by the addition of antisense oligonucleotides does not block, but alters, the morphological changes induced by retinoic acid treatment, and it also reduces the extent to which proliferation is inhibited (CHAUFOUR et al. 1996). It has been claimed that Hsp28 is a substrate of a growth regulatory protease called myeloblastin expressed in the NB4 promyelocytic leukemia cells (SPECTOR et al. 1995). Myeloblastin is a serine protease which is downregulated in HL60 cells by inducers of both monocytic and granulocytic differentiation (BORIES et al. 1989). Therefore, downregulation of myeloblastin might be partly responsible for the increase in Hsp28 level. Inhibition of myeloblastin expression by antisense oligodeoxynucleotides is sufficient to inhibit proliferation and to induce differentiation of HL60 leukemia cells. Since a decrease in myeloblastin levels leads to an increase in small Hsp level, one might speculate that this increase is a major step in the induction of the differentiation pathway.

What might be the functional relationship between small Hsp level and differentiation? Small Hsp is involved in microfilament organization (LAVOIE et al. 1993; WELSH et al. 1996). Hsp27 is associated with the I band and the M-line regions of myofibrils in myocytes of heart and skeletal muscle (HOCH et al. 1996). Modifications of the cytoskeleton induced by the action of small Hsp might alter cell growth and differentiation. Other functions attributed to small Hsp might also interfere with cell growth and differentiation. For instance, small Hsp appears to control the red/ox state, through regulating glutathione abundance, and this may be linked with the anti-apoptotic role of small Hsp (MEHLEN et al. 1996a,b). Interestingly, all results point to small Hsp as being a crucial cellular checkpoint between proliferation, apoptosis and differentiation (ARRIGO 1997). Future studies will probably reveal by which mechanisms small Hsp (and αB-crystallin, which is highly similar in structure and function) plays such a fundamental role.

II. Hsp90 and the Control of Muscle Cell Differentiation Through the Regulation of Myogenic Transcription Factors

As in other vertebrates, two Hsp90 genes exist in zebrafish, *hsp90α* and *hsp90β* (Krone and Sass 1994). Whereas *hsp90β* is expressed at relatively high levels in the developing central nervous system and does not show a segmental pattern of expression within the somites, *hsp90α* mRNA is detectable by whole-mount in situ hybridization in only a subset of somitic cells along the notochord and in a region along the posterior part of the somite. *hsp90α* is also expressed in the pectoral fin primordia.

Hsp90 has been shown to act as a chaperone in vitro (Wiech et al. 1992) although its chaperone action in vivo appears limited (Nathan et al. 1997). It also interacts with denatured proteins to protect them against irreversible aggregation and degradation (Freeman and Morimoto 1996; Schneider et al. 1996). But Hsp90 has additional, more specific functions, which might be more relevant to an involvement in cell differentiation and development. Hsp90 interacts with newly synthesized steroid receptors (Dalman et al. 1989) and this interaction is required for a proper folding of these receptors (Picard et al. 1990). Hsp90 also interacts with other transcription factors such as the dioxin receptor and some other members of the PAS domain-containing HLH transcription factors (Perdew 1988; McGuire et al. 1995), as well as with MyoD (Shaknovich et al. 1992). The latter interaction is required in vitro for proper folding of MyoD as well as for its interaction with E12 (Shue and Kotz 1994).

It was therefore interesting to compare the expression of *hsp90α* and *myoD* during zebrafish development. The result shows that the overall expression pattern of *myoD* parallels that of *hsp90α*: both in the somites and in fin buds, even if the expression of *myoD* precedes the expression of *hsp90α* in the somites (Sass et al. 1996). The expression of both *hsp90α* and *myoD* decreases later in development, whereas the expression of differentiation markers (such as α-tropomyosin) remains high.

A similar expression of *hsp90α* has also been observed in the myogenic cells of the chicken somites (Sass and Krone 1997). However, different observations have been reported in mammals (Loones et al. 1997): no spatially restricted expression of *hsp90α* is observed in developing muscles, whereas at day 15.5, Hsp90α is expressed in some specific areas of the brain, at the border between the mesencephalon and the metencephalon, and between the telencephalon and the diencephalon, in the thalamus and in the amygdaloid neuroepithelium, a part of the hippocampal neuroepithelium (Loones et al., manuscript in preparation). All these data suggest that Hsp90α might have specific functions during development. However, the observations are different depending upon the organisms. In addition, the fact that it is co-expressed with transcription factors does not prove that Hsp90α plays a crucial role in the control of these factors. Work remains to be done to attribute to Hsp90 precise developmental functions.

III. Hsp70–2: A Specialized Chaperone Essential for Meiosis

Specific forms of the Hsp70 chaperone – and of the Hsp110 proteins (see later) – are synthesized during spermatogenesis. HSC70 t is expressed at the spermatid stage whereas Hsp70–2 is expressed earlier, at the spermatocyte stage (ZAKERI and WOLGEMUTH 1987; ZAKERI et al. 1988; ALLEN et al. 1988; DIX 1997). The expression of Hsp70–2 is not limited to testes – the protein is also expressed in skeletal muscle, brain and other tissues, but at a lower level than in the testes (BONNYCASTLE et al. 1994). Hsp70–2 has a high level of amino acid similarity (more than 80%) with the other members of the Hsp70 family. The first clue to its functions came from the knockout of the gene (DIX et al. 1996). The mice are viable and apparently normal except that whereas female mice with non-expressed *hsp70–2* gene are fertile, males are not. Spermatogenesis is blocked at the first meiotic division, at the pachytene stage of spermatocyte. Instead of progressing along the differentiation pathway, cells die by apoptosis (MORI et al. 1997).

The characteristics of the mutated animals suggest two possible roles for the Hsp70–2 protein: a role in the progression of the cell cycle from the G2 to the M phase of the first meiotic division, or a role in the assembly or disassembly of the synaptonemal complex which forms at this stage and is essential for the repair and recombination events occurring during the first meiotic division. Recent publications show that both hypotheses are correct and that Hsp70–2 plays a dual role at this stage of development. This protein, which has a cytoplasmic and a nuclear distribution, interacts with CDC2 (ZHU et al. 1997). This interaction is required for CDC2 kinase to acquire an active conformation, enabling it to interact with cyclin B1. The active CDC2/cyclin B1 complex allows the transition from the G2 to M phase. Addition of exogenous recombinant Hsp70–2 to extracts of *hsp70–2$^{-/-}$* testes containing inactive CDC2 is sufficient to activate CDC2 kinase by facilitating its association with cyclin B1.

This direct action of Hsp70–2 on cell cycle progression is not its only function at this stage of development. Hsp70–2 also interacts with the synaptonemal complex in spermatocytes, though not in oocytes (ALLEN et al. 1996). This interaction seems necessary, not for the formation of this complex, but for its disassembly prior to the diplotene stage (DIX et al. 1997). Many proteins associating with the synaptonemal complex have been described during the last few years: RAD51 (HAAF et al. 1995), UBC9 (KOVALENKO et al. 1996), and BRCA1 (SCULLY et al. 1997). Hsp70–2 might be required to chaperone one or another of these proteins during synaptonemal complex disassembly. Alternatively, the chaperone function of Hsp70–2 might be indirect, through its activation of CDC2 kinase. Indeed, some of the proteins present in the synaptonemal complex have potential sites of phosphorylation by CDC2.

In summary, it appears that Hsps do not participate in development directly as developmental genes do. Rather, they are active players at a higher level in the hierarchy of structures present in living organisms, at the cellular level where they control the balance between division, differentiation and death.

D. Mechanisms Regulating the Expression of Hsps During Differentiation and Development

In *Drosophila*, it has been demonstrated that the mechanisms controlling the expression of the small Hsps during development are different from those involved in the heat induction of the same proteins. Regulatory sequences, distinct from the heat shock element (HSE) sequences to which the heat shock transcription factor (HSF) binds, are present in each of the promoters and are responsible for the developmental control of these genes (DUBROVSKY et al. 1996; MICHAUD et al. 1997a). This also appears to be true for the homologous genes in mammals. If relatively little is known about the control of small Hsps, many data have been collected by the group of J. Piatigorsky on α-A and α-B crystallins (GOPAL-SRIVASTAVA and PIATIGORSKY 1993, 1994; GOPAL-SRIVASTAVA et al. 1995). In particular, PAX-6 has been shown to be required for the expression of these proteins in the lens (CVEKL et al. 1995; CVEKL and PIATIGORSKY 1996). Other factors and target sequences are required for their expression in the lung or muscle (HAYNES et al. 1995).

Little is known about the regulation of the developmental expression of the other Hsps. In three different experimental situations, the HSEs and the HSFs have been implicated in this developmental expression. In *Xenopus*, a constitutive HSE-binding activity distinct from the stress inducible HSF1 factor is expressed in stage I and II oocytes in parallel with a transient high expression of the *hsp70* gene (GORDON et al. 1997). In the murine testes, the expression of *hsp70–2* is concomitant with the presence of a high amount of a second HSF, HSF2. HSF2 is able to bind to the promoter of *hsp70–2*, even though no canonical HSE sequences are found in the *hsp70–2* promoter (SARGE et al. 1994). Finally, in mouse, the transient expression of the *hsp70* genes at the onset of zygotic genome transcription is concomitant with the presence of a high amount of maternal HSF1 in the nuclei of the embryo (CHRISTIANS et al. 1997). Transgenic mice harboring a construct containing the luciferase reporter gene under the control of the *hsp70–1* promoter reproduce this transient expression of the *hsp70* genes. Injection of large amounts of the *hsp70–1* promoter fragment in the early embryos of these transgenic mice abolishes the transient expression of luciferase, presumably by "squelching" the factors required for the expression of the endogenous genes. In similar experiments performed with promoter fragments in which the HSE sequences have been selectively mutated, there is only a limited inhibitory effect (CHRISTIANS et al. 1997). These experiments suggest that maternal HSF1 might be involved in this early expression of the *hsp70* genes in the mouse embryo. However, direct microinjection of hybrid constructs containing the *hsp70–3* promoter (which is quite similar to *hsp70–1*) show that the Sp1 sequence and the Sp1 transcription factor are required for its expression, not the HSE sequences or the HSF (BEVILACQUA et al. 1997). These conflicting data might be reconciled if the effect of HSFs on *hsp* gene expression is highly dependent on the chromatin structure. This has been shown to be the case in amphibian oocytes (LANDSBERGER and WOLFFE 1995; LANDSBERGER et al. 1995), and the role of chromatin structure in the stress inducibility of heat-shock genes is also well documented in *Drosophila* (see KINGSTON et al. 1996 for a review). Whatever the mechanisms involved in *hsp70* gene expression at this early developmental stage, this expression is essential: recent data show that transfection with antisense oligonucleotides complementary to Hsp70 blocks development before the blastocyst stage (DIX 1997b).

Among the different HSFs expressed in mammals, some might have a role in the developmental expression of Hsps. HSF2 was shown to be present and active during in vitro differentiation of the human K562 cell line (SISTONEN et al. 1992, 1994), in the testes (see earlier), in embryonal carcinoma cells (MURPHY et al. 1994) and in early embryos (MEZGER et al. 1994). It was hypothesized that HSF2 might be the HSF responsible for the developmental control of Hsp expression. The other transcription factors found in vertebrates, HSF3 and HSF4, have different functions. Chicken HSF3 is activated at higher temperatures than HSF1 and might be responsible for a response to severe stresses (TANABE et al. 1997). No transcriptional activator domain has been found in human HSF4 and this factor might act as a repressor (NAKAI et al. 1997).

Despite the recent demonstration that human HSF2 can efficiently substitute yeast HSF (LIU et al. 1997), the role of HSF2 in the expression of Hsps during development remains enigmatic. In uterus, HSF2 is activated in parallel with estrogen stimulation of Hsp90 expression, suggesting a role for HSF2 in Hsp90 expression (YANG et al. 1995). In fact, two spliced forms of HSF2 exist, only one of which (HSF2-α) is significantly active (FIORENZA et al. 1995; GOODSON et al. 1995) and promotes *hsp* gene expression during in vitro K562 differentiation (LEPPA et al. 1997a). In contrast, overexpression of HSF2-β has an inhibitory effect. Throughout most of mouse embryogenesis, HSF2-β is expressed, but no correlations have been found between the level and DNA-binding activity of HSF2 and the expression of Hsps (RALLU et al. 1997). Similarly, no correlation could be detected during rat spermatogenesis (ALASTALO et al. 1997). Furthermore, in addition to its role in the nucleus, HSF2 is found in the cytoplasm during late embryogenesis and during spermatogenesis, where its putative function remains unknown (ALASTALO et al. 1997; Rallu, personal communication).

One possible explanation to reconcile the different results is that HSFs – and HSF2 in particular – require the cooperation of other transcription factors to be active in transcription during development. For instance, HSF3 has been shown to be activated by c-Myb in the absence of stress (KANEI-ISHII et al. 1997). Perhaps a factor such as Sp1 has a similar action at the onset of the mouse zygotic genome transcription.

Another possibility is that HSFs are developmental transcription factors, but at least some of their developmental targets are distinct from the heat shock genes. Some observations of *Drosophila* HSF and mammalian HSF2 are in favor of such a hypothesis. Despite *Drosophila* HSF being required for oogenesis and early larval development (JEDLICKA et al. 1997), the inactivation of this HSF is not correlated with any changes in the level of expression of Hsps at these stages of development, and the developmental functions of *Drosophila* HSF are genetically separable from its functions in stress response. HSF1 and HSF2 have slightly different sequence specificities in DNA recognition (KROEGER and MORIMOTO 1994) and thioredoxin gene transcription is activated by HSF2, but not by HSF1 (LEPPA et al. 1997b). Clearly, further studies will be necessary to understand the mechanisms of the developmental expression of Hsps.

E. In Search of Additional Developmental Chaperones

The pleiotropic role of chaperones makes characterization of their developmental functions difficult. The most interesting results described thus far have been obtained with respect to the specialized, rather than generalist chaperones (Hsp25, Hsp90), as well as with a minor member of a large chaperone family (Hsp70–2).

The functional role of HSC70 t in spermatogenesis still awaits discovery. Another Hsp family, the Hsp110 family, harbors a member (APG-1) specifically expressed during spermatogenesis, the function of which has not yet been defined (KANEKO et al. 1997a,b). A protein, distantly related to Hsp90 and able to interact with the retinoblastoma protein, has been described, but its developmental expression has not been studied (CHEN et al. 1996).

Co-chaperones are required for the correct functioning of chaperones. They might also confer specificity upon the action of chaperones, as demonstrated for auxilin, a co-chaperone of HSC70, in uncoating clathrin-coated pits (RASSOW et al. 1995; UNGEWICKELL et al. 1996). Therefore co-chaperones are good candidates for controlling a developmental function of Hsps. Tissue-specific isoforms of co-chaperones have already been described in the case of Hsp40 which is, by itself, a chaperone, but also acts as a co-chaperone for Hsp70 (CYR et al. 1994). Two co-chaperones have been described for Hsp90: HOP, also called Sti1 or p60 (HONORE et al. 1992; LASSLE et al. 1997), which is required for the interaction of Hsp90 with HSC70 on one hand and with steroid receptors or v-Src on the other (CHANG et al. 1997); and CDC37 (p50), which is essential for the activity of various protein kinases, such as Cdk4, Raf, and sevenless (STEPANOVA et al. 1996; KIMURA et al. 1997; HUNTER and POON 1997). The variations in the level of these two proteins during development have not been reported. Nothing is known about the developmental variations in the level of the other co-chaperones of Hsp70, HIP and NM23 (LEUNG and HIGHTOWER 1997). Finally, it would be interesting to have data on the expression of ubiquitin and ubiquitin-conjugating enzymes during development. Degradation is the last step in protein chaperoning and ubiquitin-protein conjugates are selectively present in cells undergoing major cytomorphological reorganization (SCOTTING et al. 1991) as well as during cell apoptosis (SCHWARTZ et al. 1990; DELIC et al. 1993).

The choice of which chaperones to examine in developmental studies depends on the nature of their targets and of the processes in which these targets are active players. The experimental situations in which the developmental role of Hsps has been precisely characterized show that these proteins probably do not act directly on the differentiation and developmental programs but indirectly, by regulating the cell cycle, the balance between proliferation and differentiation and cell death. We have already described data concerning the interaction of different Hsps (Hsp90, Hsp70–2) with CDKs and therefore their involvement in the control of the cell cycle. Hsp70 overexpression is able to release the cell cycle block induced by doxorubicin (KARLSEDER et al. 1996). Hsp70 is also known to interact with mutant p53, whereas Rb is associated with a distant relative of Hsp90 (CHEN et al. 1996). Recently, interesting data were obtained on the involvement of Hsp70 and its co-chaperones in the control of apoptosis: Hsp70 overexpression blocks apoptosis (GABAI et al. 1997; ANDERSON et al. 1997), maybe by preventing stress kinase activation (GABAI et al. 1997), but certainly also at a more downstream level by the direct or indirect inhibition of caspases (MOSSER et al. 1997; ANDERSON et al. 1997). In a symmetric way, Bcl-2 protects against heat-induced cell death (STRASSER and ANDERSON 1995). Hsp70 interacts directly with BAG-1 (TAKAYAMA et al. 1997), which is a Bcl-2-binding protein with anti-apoptotic activity (TAKAYAMA et al. 1995). BAG-1 binds to the ATPase domain of Hsp70 and, in cooperation with Hsp40, stimulates the Hsp70's ATP hydrolysis activity (HOHFELD and JENTSCH 1997).

F. The Place of Hsps in Aging

Two alterations in the expression of Hsps linked with aging have been described: an attenuation of the Hsp synthesis induced by stress and a higher spontaneous expression of some of the Hsps. The decrease in the extent of Hsp induction following stress can be most easily studied by looking at the major inducible Hsp, Hsp70. In cells from aging rats, the level of induction of Hsp70 is reduced (FARGNOLI et al. 1990) whereas the kinetics of the stress response and the induction of the other Hsps remain unaltered. Similar observations have been made in different mammalian species (rat and human), in isolated animal tissues as well as in vivo (for a review, see HEYDARI et al. 1994). A similar decrease in the inducibility of Hsp70 is observed during cell senescence in in vitro cultures (LIU et al. 1989). The response to other proteotoxic stress is similarly altered. It is appealing to correlate these observations with well-known physiological observations concerning aging: the dramatic increase in the incidence of heat stroke with age (Heydari et al. 1994), or the increase in ischemic accidents in the brain and heart. In the latter organ, a mild heat shock or an increased expression of Hsp70 obtained by transgenesis (Marber et al. 1995; Plumier et al. 1995) has a protective effect against ischemia. In aged animals, the heat inducibility of Hsp70 is reduced in the heart in parallel with a decrease in the protective effect against ischemia conferred by a mild hyperthermic treatment (Locke and Tanguay 1996). The decrease in the inducibility of Hsp70 parallels a decrease in the inducibility of HSF1 binding. However, the level of HSF1 protein itself is unaltered by age, suggesting that the modifications associated with age concern the mechanisms of heat-induced activation (Liu et al. 1991; Heydari et al. 1993).

In *Drosophila*, in contrast, aging leads to the spontaneous expression of Hsp70 in flight and leg muscles, and of Hsp22 and Hsp23 in a broader range of tissues (Wheeler et al. 1995). The mechanism of this increase is transcriptional for the small Hsps but post-transcriptional for Hsp70. Since the expression of HSP70 is also increased in young flies mutant for catalase and superoxide dismutase, it was suggested that the induction of Hsp70 is a response to the accumulation in muscle tissues – where the oxygen metabolism is high – of oxidized, damaged proteins. To our knowledge, no such overexpression of Hsps during aging has been described so far in mammals.

The observations reported in *Drosophila* suggest the hypothesis that alterations in the heat shock response and in Hsp synthesis occurring during aging might be related to the accumulation of protein damage. This damage results from the action of reactive oxygen species formed during O_2 consumption. Caloric restriction is known to increase life-span in rodents (and probably also in humans) by decreasing the production of reactive oxygen species (Sohal and Weindruch 1996). Interestingly, the heat inducibility of Hsp70 synthesis in rat hepatocytes is less altered by age in animals which have been submitted to caloric restriction (Heydari et al. 1993, 1996).

The previous experiment suggests the existence not only of a correlation, but of a direct link between the alterations in the heat shock response and synthesis of HSPs on one part, and aging on the other. Many different models have been proposed to explain aging. At either extreme, aging might be either the result of a genetic program, or the mere consequence of the accumulation of molecular alterations during life. The recent isolation of genes from humans, *Drosophila* and, most efficiently, nematodes whose alterations can change life-span has favored the hypothesis that aging results from an accumulation of damage over time. However, these experiments demonstrated that the organisms have mechanisms for combating the alterations linked with age: life-span results from a balance between the amount of damage due to age and the efficiency of the protective and repair processes.

All the stress responses existing in living organisms – against UV, oxidative stress, or, more generally, proteotoxic stress – are therefore part of the "anti-aging program." Different data show that the heat shock response is not a minor player in this fight. Mutants selected for an increased life-span in nematodes have an increased intrinsic thermotolerance (LITHGOW et al. 1995). The same is true in *Saccharomyces cerevisiae* (KENNEDY et al. 1995). This increased thermotolerance is associated with an increased level of the small Hsp, Hsp16, in nematodes (LITHGOW 1996). Furthermore, repeated treatments of nematodes at sublethal temperatures increase thermotolerance and, in parallel, induce a low but significant increase in life expectancy (LITHGOW et al. 1995). The same result has been obtained in *Drosophila* (MAYNARD SMITH 1958; KHAZAELI et al. 1997): to demonstrate that the increase in life expectancy is due to Hsps and not to the heat treatment itself, the result has been confirmed in transgenic *Drosophila*, which differ only in the number of copies of the inducible Hsp70 gene (TATAR et al. 1997). An increase in the expression of Hsp70 of only 10% above normal levels increases the subsequent life-span. The latter plateaus at a value of about twice the normal value. Although the superinduction of Hsp70 is transient, the effect on age-specific survival is persistent.

Therefore an increased expression of Hsps is linked with organism survival, whereas a weakening of the stress response accompanies aging and the increased rate of mortality associated with it. In some sense, Hsps are "chaperoning extended life" (TATAR et al. 1997).

G. Conclusion

Hsps and chaperones show specific expression patterns during development and cell differentiation as well as during aging. Some of these variations are clearly associated with changes in protein turnover or with specific developmental requirements for the protective effect of these proteins. Some of the developmental variations are relatively transient and linked with more specific functions of these chaperones. A paradigm for these developmental functions has been revealed by the study of Hsp70–2 in spermatogenesis. It is likely that specific developmental functions will soon be attributed to small Hsp and Hsp90 as well. *hsp* genes do not act as developmental genes to direct development along one pathway or another. Rather, as important partners in differentiation, cell division, and apoptosis, they act at the cellular level to regulate development.

Acknowledgements. We are indebted to Dr. Sean Davidson for careful reading of this manuscript.

References

Alastalo T-P, Lönnström M, Leppä S, Kaarniranta K, Pelto-Huikko M, Sistonen L, Parvinen M (1998) Stage-specific expression and cellular localization of the heat shock factor 2 isoforms in the rat seminiferous epithelium. Exp Cell Res 240:16–27
Allen RL, O'Brien DA, Jones CC, Rockett DL, Eddy EM (1988) Expression of heat shock proteins by isolated mouse spermatogenic cells. Mol Cell Biol 8:3260–3266

Allen JW, Dix DJ, Collins BW, Merrick BA, He C, Selkirk JK, Poorman-Allen P, Dresser ME, Eddy EM (1996) Hsp70–2 is part of the synaptonemal complex in mouse and hamster spermatocytes. Chromosoma 104:414–421

Anderson RL, Buzzard KA, Giaccia A (1997) HSP72 modulates pathways of stress-induced apoptosis. 5th International Federation of Teratology Societies Conference, Sydney

Arrigo A-P (1998) Small stress proteins: chaperones that act as regulators of intracellular redox state and programmed cell death. Biol Chem 379:19–26

Arrigo AP, Ahmed-Zadeh C (1981) Immunofluorescence localization of a small heat shock protein (hsp23) in salivary gland cells of Drosophila melanogaster. Mol Gen Genet 185:73–79

Bensaude O, Babinet C, Morange M, Jacob F (1983) Heat shock proteins, first major products of zygotic gene activity in mouse embryo. Nature 305:331–333

Bevilacqua A, Fiorenza MT, Mangia F (1997) Developmental activation of an episomic hsp70 gene promoter in two-cell mouse embryos by transcription factor Sp1. Nucleic Acids Res 25:1333–1338

Bonnycastle LLC, Yu C-E, Hunt CR, Trask BJ, Clancy KP, Weber JL, Patterson D, Schellenberg GD (1994) Cloning, sequencing, and mapping of the human chromosome 14 heat shock protein gene (HSPA2). Genomics 23:85–93

Bories D, Raynal M-C, Solomon DH, Darzynkiewicz Z, Cayre YE (1989) Down-regulation of a serine protease, myeloblastin, causes growth arrest and differentiation of promyelocytic leukemia cells. Cell 59:959–968

Boston RS, Viitanen PV, Vierling E (1996) Molecular chaperones and protein folding in plants. Plant Mol Biol 32:191–222

Brown IR (1994) Induction of heat shock genes in the mammalian brain by hyperthermia and tissue injury. In: Mayer J, Brown I (eds) Heat shock proteins in the nervous system. Academic, London, pp 32–54

Chaufour S, Mehlen P, Arrigo A-P (1996) Transient accumulation, phosphorylation and changes in the oligomerization of Hsp27 during retinoic acid-induced differentiation of HL-60 cells: possible role in the control of cellular growth and differentiation. Cell Stress Chaperones 1:225–235

Chang H-CJ, Nathan DF, Lindquist S (1997) In vivo analysis of the Hsp90 Cochaperone Sti1 (p60). Mol Cell Biol 17:318–325

Chen C-F, Chen Y, Dai K, Chen P-L, Riley DJ, Lee W-H (1996) A new member of the hsp90 family of molecular chaperones interacts with the retinoblastoma protein during mitosis and after heat shock. Mol Cell Biol 16:4691–4699

Christians E, Michel E, Adenot P, Mezger V, Rallu M, Morange M, Renard J-P (1997) Evidence for the involvement of mouse heat shock factor 1 in the atypical expression of the HSP70.1 heat shock gene during mouse zygotic genome activation. Mol Cell Biol 17:778–788

Cvekl A, Kashanchi F, Sax CM, Brady JN, Piatigorsky J (1995) Transcriptional regulation of the mouse αA-crystallin gene: activation dependent on a cyclic AMP-responsive element (DE1/CRE) and a Pax-6-binding site. Mol Cell Biol 15:653–660

Cvekl A, Piatigorsky J (1996) Lens development and crystallin gene expression: many roles for Pax-6. Bioessays 18:621–630

Cyr DM, Langer T, Douglas MG (1994) DnaJ-like proteins: molecular chaperones and specific regulators of Hsp70. Trends Biochem Sci 19:176–181

Dalman FC, Bresnick EH, Patel PD, Perdew GH, Watson SJ Jr, Pratt WB (1989) Direct evidence that the glucocorticoid receptor binds to hsp90 at or near the termination of receptor translation in vitro. J Biol Chem 264:19815–19821

Delic J, Morange M, Magdelenat H (1993) Ubiquitin pathway involvement in human lymphocyte γ-irradiation-induced apoptosis. Mol Cell Biol 13:4875–4883

Ding D, Parkhurst SM, Halsell SR, Lipshitz HD (1993) Dynamic Hsp83 RNA localization during Drosophila oogenesis and embryogenesis. Mol Cell Biol 13:3773–3781

Dix DJ, Allen JW, Collins BW, Mori C, Nakamura N, Poorman-Allen P, Goulding EH, Eddy EM (1996) Targeted gene disruption of Hsp70–2 results in failed meiosis, germ cell apoptosis, and male infertility. Proc Natl Acad Sci USA 93:3264–3268

Dix DJ (1997a) Hsp70 expression and function during gametogenesis. Cell Stress Chaperones 2:73–77

Dix DJ (1997b) Expression and function of the stress inducible HSP70s during pre-implantation embryogenesis. Teratology 55:44

Dix DJ, Allen JW, Collins BW, Poorman-Allen P, Mori C, Blizard DR, Brown PR, Goulding EH, Strong BD, Eddy EM (1997) HSP70–2 is required for desynapsis of synaptonemal complexes during meiotic prophase in juvenile and adult mouse spermatocytes. Development 124:4595–4603

D'Souza S, Brown IR (1998) Constitutive expression of heat shock proteins hsp90, hsc70, hsp70 and hsp60 in neural and non-neural tissues of the rat during postnatal development. Cell Stress Chaperones 3:188–199

Dubrovsky EB, Dretzen G, Berger EM (1996) The Broad-Complex gene is a tissue-specific modulator of the ecdysone response of the Drosophila hsp23 gene. Mol Cell Biol 16:6542–6552

Fargnoli J, Kunisada T, Fornace AJ Jr, Schneider EL, Holbrook NJ (1990) Decreased expression of heat shock protein 70 mRNA and protein after heat treatment in cells of aged rats. Proc Natl Acad Sci USA 87:846–850

Fiorenza MT, Farkas T, Dissing M, Kolding D, Zimarino V (1995) Complex expression of murine heat shock transcription factors. Nucleic Acids Res 23:467–474

Flach G, Johnson MH, Braude PR, Taylor RAS, Bolton VN (1982) The transition from maternal to embryonic control in the 2-cell mouse embryo. The EMBO J 1:681–686

Freeman BC, Morimoto RI (1996) The human cytosolic molecular chaperones hsp90, hsp70 (hsc70) and hdj-1 have distinct roles in recognition of a non-native protein and protein refolding. EMBO J 15:2969–2979

Gabai VL, Meriin AB, Mosser DD, Caron AW, Rits S, Shifrin VI, Sherman MY (1997) Hsp70 prevents activation of stress kinases. J Biol Chem 272:18033–18037

Gernold M, Knauf U, Gaestel M, Stahl J, Kloetzel P-M (1993) Development and tissue-specific distribution of mouse small heat shock protein hsp25. Dev Genet 14:103–111

Glaser RL, Wolfner MF, Lis JT (1986) Spatial and temporal pattern of hsp26 expression during normal development. EMBO J 5:747–754

Goodson ML, Park-Sarge O-K, Sarge KD (1995) Tissue-dependent expression of heat shock factor 2 isoforms with distinct transcriptional activities. Mol Cell Biol 15:5288–5293

Gopal-Srivastava R, Piatigorsky J (1993) The murine αB crystallin/small heat shock protein enhancer: identification of αBE-1, αBE-2, αBE-3, and MRF control elements. Mol Cell Biol 13:7144–7152

Gopal-Srivastava R, Piatigorsky J (1994) Identification of a lens specific regulatory region (LSR) of the murine αB-crystallin gene. Nucleic Acids Res 22:1281–1286

Gopal-Srivastava R, Haynes JI II, Piatigorsky J (1995) Regulation of the murine αB-crystallin/small heat shock protein gene in cardiac muscle. Mol Cell Biol 15:7081–7090

Gordon S, Bharadwaj S, Hnatov A, Ali A, Ovsenek N (1997) Distinct stress-inducible and developmentally regulated heat shock transcription factors in Xenopus oocytes. Dev Biol 181:47–63

Haaf T, Golub EI, Reddy G, Radding CM, Ward DC (1995) Nuclear foci of mammalian Rad51 recombination protein in somatic cells after DNA damage and its localization in synaptonemal complexes. Proc Natl Acad Sci USA 92:2298–2302

Hanash SM, Strahler JR, Chan Y, Kuick R, Teichroew D, Neel JV, Hailat N, Keim DR, Gratiot-Deans J, Ungar D, Melhem R, Zhu XX, Andrews P, Lottspeich F, Eckerskorn C, Chu E, Ali I, Fox DA, Richardson BC, Turka LA (1993) Data base analysis of protein expression patterns during T-cell ontogeny and activation. Proc Natl Acad Sci USA 90:3314–3318

Haynes JI II, Gopal-Srivastava R, Frederikse PH, Piatigorsky J (1995) Differential use of the regulatory elements of the αB-crystallin enhancer in cultured murine lung (MLg), lens (αTN4–1) and muscle (C2C12) cells. Gene 155:151–158

Heikkila JJ (ed) (1993) Heat shock gene expression and development. Dev Genet 14:1–158

Heydari AR, Wu B, Takahashi R, Strong R, Richardson A (1993) Expression of heat shock protein 70 is altered by age and diet at the level of transcription. Mol Cell Biol 13:2909–2918

Heydari AR, Takahashi R, Gutsmann A, You S, Richardson A (1994) Hsp70 and aging. Experientia 50:1092–1098

Heydari AR, You S, Takahashi R, Gutsmann A, Sarge KD, Richardson A (1996) Effect of caloric restriction on the expression of heat shock protein 70 and the activation of heat shock transcription factor 1. Dev Genet 18:114–124

Hoch B, Lutsch G, Schlegel WP, Stahl J, Wallukat G, Bartel S, Krause EG, Benndorf R, Karczewski P (1996) HSP25 in isolated perfused rat hearts: localization and response to hyperthermia. Mol Cell Biochem 160/161:231–239

Höhfeld J, Jentsch S (1997) GrpE-like regulation of the Hsc70 chaperone by the anti-apoptotic protein BAG-1. EMBO J 16:6209–6216

Honoré B, Leffers H, Madsen P, Rasmussen HH, Vandekerckhove J, Celis JE (1992) Molecular cloning and expression of a transformation-sensitive human protein containing the TPR motif and sharing identity to the stress-inducible yeast protein STI1. J Biol Chem 267:8485–8491

Houenou LJ, Li L, Lei M, Kent CR, Tytell M (1996) Exogenous heat shock cognate protein Hsc70 prevents axotomy-induced death of spinal sensory neurons. Cell Stress Chaperones 1:161–166

Hunter ES, Dix DJ (1996) Antisense oligonucleotides against Hsp70-1 and 70-3 increase mouse embryonic sensitivity to arsenite-induced dysmorphogenesis in vitro. Teratology 53:86

Hunter T, Poon RYC (1997) Cdc37: a protein kinase chaperone? Trends Cell Biol 7:157–161

Jedlicka P, Mortin MA, Wu C (1997) Multiple functions of Drosophila heat shock transcription factor in vivo. EMBO J 16:2452–2462

Kanei-Ishii C, Tanikawa J, Nakai A, Morimoto RI, Ishii S (1997) Activation of heat shock transcription factor 3 by c-Myb in the absence of cellular stress. Science 277:246–248

Kaneko Y, Nishiyama H, Nonoguchi K, Higashitsuji H, Kishishita M, Fujita J (1997a) A novel hsp110-related gene, apg-1, that is abundantly expressed in the testis responds to a low temperature heat shock rather than the traditional elevated temperatures. J Biol Chem 272:2640–2645

Kaneko Y, Kimura T, Nishiyama H, Noda Y, Fujita J (1997b) Developmentally regulated expression of APG-1, a member of heat shock protein 110 family in murine male germ cells. Biochem Biophys Res Commun 233:113–116

Karlseder J, Wissing D, Holzer G, Orel L, Sliutz G, Auer H, Jäättelä M, Simon MM (1996) HSP70 overexpression mediates the escape of a doxorubicin-induced G2 cell cycle arrest. Biochem Biophys Res Commun 220:153–159

Kato M, Mizuguchi M, Takashima S (1995) Developmental changes of heat shock protein 73 in human brain. Dev Brain Res 86:180–186

Kennedy BK, Austriaco NR Jr, Zhang J, Guarente L (1995) Mutation in the silencing gene SIR4 can delay aging in S. cerevisiae. Cell 80:485–496

Khazaeli AA, Tatar M, Pletcher SD, Curtsinger JW (1997) Heat-induced longevity extension in Drosophila. I. Heat treatment, mortality and thermotolerance. J Gerontol A52:B48–B52

Kimura Y, Rutherford SL, Miyata Y, Yahara I, Freeman BC, Yue L, Morimoto RI, Lindquist S (1997) Cdc37 is a molecular chaperone with specific functions in signal transduction. Genes Dev 11:1775–1785

Kingston RE, Bunker CA, Imbalzano AN (1996) Repression and activation by multiprotein complexes that alter chromatin structure. Genes Dev 10:905–920

Kovalenko OV, Plug AW, Haaf T, Gonda DK, Ashley T, Ward DC, Radding CM, Golub EI (1996) Mammalian ubiquitin-conjugating enzyme Ubc9 interacts with Rad51 recombination protein and localizes in synaptonemal complexes. Proc Natl Acad Sci USA 93:2958–2963

Kroeger PE, Morimoto RI (1994) Selection of new HSF1 and HSF2 DNA-binding sites reveals difference in trimer cooperativity. Mol Cell Biol 14:7592–7603

Krone PH, Sass JB (1994) Hsp90α and hsp90β genes are present in the zebrafish and are differentially regulated in developing embryos. Biochem Biophys Res Commun 204:746–752

Landsberger N, Ranjan M, Almouzni G, Stump D, Wolffe AP (1995) The heat shock response in Xenopus oocytes, embryos, and somatic cells: a regulatory role for chromatin. Dev Biol 170:62–74

Landsberger N, Wolffe AP (1995) Role of chromatin and Xenopus laevis heat shock transcription factor in regulation of transcription from the X. laevis hsp70 promoter in vivo. Mol Cell Biol 15:6013–6024

Lanks KW (1986) Modulators of the eukaryotic heat shock response. Exp Cell Res 165:1–10

Lässle M, Blatch GL, Kundra V, Takatori T, Zetter BR (1997) Stress-inducible, murine protein mSTI1. J Biol Chem 272:1876–1884

Lavoie JN, Hickey E, Weber LA, Landry J (1993) Modulation of actin microfilament dynamics and fluid phase pinocytosis by phosphorylation of heat shock protein-27. J Biol Chem 268:24210–24214

Lee GJ, Pokala N, Vierling E (1995) Structure and in vitro molecular chaperone activity of cytosolic small heat shock proteins from pea. J Biol Chem 270:10432–10438

Lee GJ, Roseman AM, Saibil HR, Vierling E (1997) A small heat shock protein stably binds heat-denatured model substrates and can maintain a substrate in a folding-competent state. EMBO J 16:659–671

Lele Z, Krone PH (1997) Expression of genes encoding the collagen-binding heat shock protein (Hsp47) and type II collagen in developing zebrafish embryos. Mech Dev 61:89–98

Leppä S, Pirkalla L, Saarento H, Sarge KD, Sistonen L (1997a) Overexpresion of HSF2-β inhibits hemin-induced heat shock gene expression and erythroid differentiation in K562 cells. J Biol Chem 272:15293–15298

Leppä S, Pirkkala L, Chow SC, Eriksson JE, Sistonen L (1997b) Thioredoxin is transcriptionally induced upon activation of heat shock factor 2. J Biol Chem 272:30400–30404

Leung S-M, Hightower LE (1997) A 16-kDa protein functions as a new regulatory protein for Hsc70 molecular chaperone and is identified as a member of the Nm23/nucleoside diphosphate kinase family. J Biol Chem 272:2607–2614

Lithgow GJ, White TM, Melov S, Johnson TE (1995) Thermotolerance and extended life-span conferred by single-gene mutations and induced by thermal stress. Proc Natl Acad Sci USA 92:7540–7544

Lithgow GJ (1996) Invertebrate gerontology: the age mutations of Caenorhabditis elegans. Bioessays 18:809–815

Liu AY, Lin Z, Choi H, Sorhage F, Li B (1989) Attenuated induction of heat shock gene expression in aging diploid fibroblasts. J Biol Chem 264:12037–12045

Liu AY, Choi H, Lee Y, Chen KY (1991) Molecular events involved in transcriptional activation of heat shock genes become progressively refractory to heat stimulation during aging of human diploid fibroblasts. J Cell Physiol 149:360–366

Liu X-D, Liu PCC, Santoro N, Thiele DJ (1997) Conservation of a stress response: human heat shock transcription factors functionally substitute for yeast HSF. EMBO J 16:6466–6477

Locke M, Tanguay RM (1996) Diminished heat shock response in the aged myocardium. Cell Stress Chaperones 1:251–260

Loones M-T, Rallu M, Mezger V, Morange M (1997) HSP gene expression and HSF2 in mouse development. CMLS 53:179–190

Mairesse N, Horman S, Mosselmans R, Galand P (1996) Antisense inhibition of the 27 kDa heat shock protein production affects growth rate and cytoskeletal organization in MCF-7 cells. Cell Biol Int 20:205–212

Marber MS, Mestril R, Chi S-H, Sayen MR, Yellon DM, Dillmann WH (1995) Overexpression of the rat inducible 70-kD heat stress protein in a transgenic mouse increases the resistance of the heart to ischemic injury. J Clin Invest 95:1446–1456

Marin R, Valet JP, Tanguay RM (1993) hsp23 and hsp26 exhibit distinct spatial and temporal patterns of constitutive expression in Drosophila adults. Dev Genet 14:69–77

Marin R, Landry J, Tanguay RM (1996) Tissue-specific posttranslational modification of the small heat shock protein HSP27 in Drosophila. Exp Cell Res 223:1–8

Marin R, Tanguay RM (1996) Stage-specific localization of the small heat shock protein Hsp27 during oogenesis in Drosophila melanogaster. Chromosoma 105:142–149

Maynard Smith J (1958) Prolongation of the life of Drosophila subobscura by brief exposure of adults to a high temperature. Nature 181:496–497

McGuire J, Coumailleau P, Whitelaw ML, Gustafsson J-A, Poellinger L (1995) The basic helix-loop-helix/PAS factor Sim is associated with hsp90. J Biol Chem 270:31353–31357

Mehlen P, Kretz-Remy C, Préville X, Arrigo A-P (1996) Human hsp27, Drosophila hsp27 and human αB-crystallin expression-mediated increase in glutathione is essential for the protective activity of these proteins against TNFα-induced cell death. EMBO J 15:2695–2706

Mehlen P, Schulze-Osthoff K, Arrigo A-P (1996) Small stress proteins as novel regulators of apoptosis. J Biol Chem 271:16510–16514

Mehlen P, Mehlen A, Godet J, Arrigo A-P (1997) hsp27 as a switch between differentiation and apoptosis in murine embryonic stem cells. J Biol Chem 272:31657–31665

Mezger V, Rallu M, Morimoto RI, Morange M, Renard J-P (1994) Heat shock factor 2-like activity in mouse blastocysts. Dev Biol 166:819–822

Michaud S, Marin R, Tanguay RM (1997a) Regulation of heat shock gene induction and expression during Drosophila development. CMLS 53:104–113

Michaud S, Marin R, Westwood JT, Tanguay RM (1997b) Cell-specific expression and heat-shock induction of Hsps during spermatogenesis in Drosophila melanogaster. J Cell Sci 110:1989–1997

Minowada G, Welch W (1995) Variation in the expression and/or phosphorylation of the human low molecular weight stress protein during in vitro cell differentiation. J Biol Chem 270:7047–7054

Morange M (ed) (1997) Developmental control of heat shock and chaperone gene expression. CMLS 53:78–129 and 168–213

Mori C, Nakamura N, Dix DJ, Fujioka M, Nakagawa S, Shiota K, Eddy EM (1997) Morphological analysis of germ cell apoptosis during postnatal testis development in normal and hsp70–2 knockout mice. Dev Dyn 208:125–136

Mosser DD, Caron AW, Bourget L, Denis-Larose C, Massie B (1997) Role of the human heat shock protein hsp70 in protection against stress-induced apoptosis. Mol Cell Biol 17:5317–5327

Murphy SP, Gorzowski JJ, Sarge KD, Phillips B (1994) Characterization of constitutive HSF2 DNA-binding activity in mouse Embryonal Carcinoma Cells. Mol Cell Biol 14:5309–5317

Nakai A, Tanabe M, Kawazoe Y, Inazawa J, Morimoto RI, Nagata K (1997) HSF4, a new member of the human heat shock factor family which lacks properties of a transcriptional activator. Mol Cell Biol 17:469–481

Nathan DF, Vos MH, Lindquist S (1997) In vivo functions of the Saccharomyces cerevisiae Hsp90 chaperone. Proc Natl Acad Sci USA 94:12949–12956

Pauli D, Tonka C-H, Tissières A, Arrigi A-P (1990) Tissue-specific expression of the heat shock protein HSP27 during Drosophila melanogaster development. J Cell Biol 111:817–828

Paranko J, Seitz J, Meinhardt A (1996) Developmental expresion of heat shock protein 60 (HSP60) in the rat testis and ovary. Differentiation 60:159–167

Perdew GH (1988) Association of the Ah receptor with the 90kD heat shock protein. J Biol Chem 263:13802–13805

Perkins LA, Doctor JS, Zhang K, Stinson L, Perrimon N, Craig EA (1990) Molecular and developmental characterization of the heat shock cognate 4 gene of Drosophila melanogaster. Mol Cell Biol 10:3232–3238

Picard D, Khursheed B, Garabedian MJ, Fortin MG, Lindquist S, Yamamoto KR (1990) Reduced levels of hsp90 compromise steroid receptor action in vivo. Nature 348:166–168

Plumier J-C, Ross BM, Currie RW, Angelidis CE, Kazlaris H, Kollias G, Pagoulatos GN (1995) Transgenic mice expressing the human heat shock protein 70 have improved postischemic myocardial recovery. J Clin Invest 95:1854–1860

Plumier J-CL, Hopkins DA, Robertson HA, Currie RW (1997) Constitutive expression of the 27-kDa heat shock protein (hsp27) in sensory and motor neurons of the rat nervous system. J Comp Neurol 384:409–428

Rallu M, Loones M-T, Lallemand Y, Morimoto R, Morange M, Mezger V (1997) Function and regulation of heat shock factor 2 during mouse embryogenesis. Proc Natl Acad Sci USA 94:2392–2397

Rassow J, Voos W, Pfanner N (1995) Partner proteins determine multiple functions of Hsp70. Trends Cell Biol 5:207–212

Rondeaux P, Galand P, Horman S, Mairesse N (1997) Effects of antisense HSP27 gene expression in osteosarcoma cells. In Vitro Cell Dev Biol 33:655–658

Sarge KD, Park-Sarge O-K, Kirby JD, Mayo KE, Morimoto RI (1994) Expression of heat shock factor 2 in mouse testis: potential role as a regulator of heat-shock protein gene expression during spermatogenesis. Biol Reprod 50:1334–1343

Sass JB, Weinberg ES, Krone PH (1996) Specific localization of zebrafish hsp90α mRNA to myoD-expressing cells suggests a role for hsp90α during normal muscle development. Mech Dev 54:195–204

Sass JB, Krone PH (1997) HSP90α gene expression may be a conserved feature of vertebrate somitogenesis. Exp Cell Res 233:391–394

Schneider C, Sepp-Lorenzino L, Nimmesgern E, Ouerfelli O, Danishefsky S, Rosen N, Hartl FU (1996) Pharmacologic shifting of a balance between protein refolding and degradation mediated by Hsp90. Proc Natl Acad Sci USA 93:14536–14541

Schultz H, Rogalla T, Engel K, Lee JC, Gaestel M (1997) The protein kinase inhibitor SB203580 uncouples PMA-induced differentiation of HL-60 cells from phosphorylation of Hsp27. Cell Stress Chaperones 2:41–49

Schwartz LM, Myer A, Kosz L, Engelstein M, Maier C (1990) Activation of polyubiquitin gene expression during developmentally programmed cell death. Neuron 5:411–419

Scotting P, McDermott H, Mayer RJ (1991) Ubiquitin-protein conjugates and αB crystallin are selectively present in cells undergoing major cytomorphological reorganisation in early chicken embryos. FEBS Lett 285:75–79

Scully R, Chen J, Plug A, Xiao Y, Weaver D, Feunteun J, Ashley T, Livingston DM (1997) Association of BRCA1 with Rad51 in mitotic and meiotic cells. Cell 88:265–275

Shaknovich R, Shue G, Kohtz DS (1992) Conformational activation of a basic helix-loop-helix protein (myoD1) by the C-terminal region of murine hsp90 (hsp84). Mol Cell Biol 12:5059–5068

Shakoori AR, Oberdorf AM, Owen TA, Weber LA, Hickey E, Stein JL, Lian JB, Stein GS (1992) Expression of heat shock genes during differentiation of mammalian osteoblasts and promyelocytic leukemia cells. J Cell Biochem 48:277–287

Shue G, Kohtz DS (1994) Structural and functional aspects of basic helix-loop-helix protein folding by heat-shock protein 90. J Biol Chem 269:2707–2711

Sirotkin K, Davidson N (1982) Developmentally regulated transcription from Drosophila melanogaster site 67B. Dev Biol 89:196–210

Sistonen L., Sarge KD, Phillips B, Abravaya K, Morimoto RI (1992) Activation of heat shock factor 2 during hemin-induced differentiation of human erythroleukemia cells. Mol Cell Biol 12:4104–4111

Sistonen L, Sarge KD, Morimoto RI (1994) Human heat shock factors 1 and 2 are differentially activated and can synergistically induce hsp70 gene transcription. Mol Cell Biol 14:2087–2099

Sohal RS, Weindruch R (1996) Oxidative stress, caloric restriction and aging. Science 273:59–63

Spector NL, Samson W, Ryan C, Gribben J, Urba W, Welch WJ, Nadler LM (1992) Growth arrest of human B lymphocytes is accompanied by induction of the low molecular weight mammalian heat shock protein (hsp28). J Immunol 148:1668–1673

Spector NL, Ryan C, Samson W, Levine H, Nadler LM, Arrigo A-P (1993) Heat shock protein is a unique marker of growth arrest during macrophage differentiation of HL-60 cells. J Cell Physiol 156:619–625

Spector NL, Mehlen P, Ryan C, Hardy L, Samson W, Levine H, Nadler LM, Fabre N, Arrigo A-P (1994) Regulation of the 28 kDa heat shock protein by retinoic acid during differentiation of human leukemic HL-60 cells. FEBS Lett 337:184–188

Spector NL, Hardy L, Ryan C, Miller WH Jr, Humes JL, Nadler LM, Luedke E (1995) 28-kDa mammalian heat shock protein, a novel substrate of a growth regulatory protease involved in differentiation of human leukemia cells. J Biol Chem 270:1003–1006

Stahl J, Wobus AM, Ihrig S, Lutsch G, Bielka H (1992) The small heat shock protein hsp25 is accumulated in P19 embryonal carcinoma cells and embryonic stem cells of line BLC6 during differentiation. Differentiation 51:33–37

Stepanova L, Leng X, Parker SB, Harper JW (1996) Mammalian p50^{Cdc37} is a protein kinase-targeting subunit of Hsp90 that binds and stabilizes Cdk4. Genes Dev 10:1491–1502

Strasser A, Anderson RL (1995) Bcl-2 and thermotolerance cooperate in cell survival. Cell Growth Differ 6:799–805

Takayama S, Sato T, Krajewski S, Kochel K, Irie S, Millan JA, Reed JC (1995) Cloning and functional analysis of BAG-1: A novel Bcl-2-binding protein with anti-cell death activity. Cell 80:279–284

Takayama S, Bimston DN, Matsuzawa S-I, Freeman BC, Aime-Sempe C, Xie Z, Morimoto RI, Reed JC (1997) BAG-1 modulates the chaperone activity of HSP70/Hsc70. EMBO J 16:4887–4896

Tanabe M, Nakai A, Kawazoe Y, Nagata K (1997) Different thresholds in the responses of two heat shock transcription factors, HSF1 and HSF3. J Biol Chem 272:15389–15395

Tanguay RM (1989) Localized expression of a small heat shock protein, Hsp23, in specific cells of the central nervous system during early embryogenesis in Drosophila. J Cell Biol 109:155a

Tatar M, Khazaeli AA, Curtsinger JW (1997) Chaperoning extended life. Nature 390:30

Ungewickell E, Ungewickell H, Holstein SEH, Lindner R, Prasad K, Barouch W, Martin B, Greene LE, Eisenberg E (1995) Role of auxilin in uncoating clathrin-coated vesicles. Nature 378:632–635

Walsh D, Li Z, Wu Y, Nagata K (1997) Heat shock and the role of the HSPs during neural plate induction in early mammalian CNS and brain development. CMLS 53:198–211

Waters ER, Lee GJ, Vierling E (1996) Evolution, structure and function of the small heat shock proteins in plants. J Exp Bot 47:325–338

Wehmeyer N, Hernandez LD, Finkelstein RR, Vierling E (1996) Synthesis of small heat-shock proteins is part of the developmental program of late seed maturation. Plant Physiol 112:747–757

Welsh MJ, Wu W, Parvinen M, Gilmont RR (1996) Variation in expression of hsp27 messenger ribonucleic acid during the cycle of the seminiferous epithelium and co-localization of hsp27 and microfilaments in Sertoli cells of the rat. Biol Reprod 55:141–151

Wheeler JC, Bieschke ET, Tower J (1995) Muscle-specific expression of Drosophila hsp70 in response to aging and oxidative stress. Proc Natl Acad Sci USA 92:10408–10412

Wiech H, Buchner J, Zimmermann R, Jakob U (1992) Hsp90 chaperones protein folding in vitro. Nature 358:169–170

Yang X, Dale EC, Diaz J, Shyamala G (1995) Estrogen dependent expression of heat shock transcription factor: implications for uterine synthesis of heat shock proteins. J Steroid Biochem 52:415–419

Zakeri ZF, Wolgemuth DJ (1987) Developmental-stage-specific expression of the hsp70 gene family during differentiation of the mammalian male germ line. Mol Cell Biol 7:1791–1796

Zakeri ZF, Wolgemuth DJ, Hunt CR (1988) Identification and sequence analysis of a new member of the mouse HSP70 gene family and characterization of its unique cellular and developmental pattern of expression in the male germ line. Mol Cell Biol 8:2925–2932

Zhu D, Dix DJ, Eddy EM (1997) HSP70-2 is required for CDC2 kinase activity in meiosis I of mouse spermatocytes. Development 124:3007–3014

CHAPTER 15

Heat Shock Proteins in Rheumatoid Arthritis

W. van Eden

A. Introduction

In view of the fundamental role of stress proteins in the maintenance of protein homeostasis, it seems likely that malfunctions associated with members of stress protein families would have pathological effects. Such effects might be minimal under normal physiological conditions, but could be exacerbated at times when other disease stimuli trigger the requirement for local alterations in stress protein function in particular afflicted cells or tissues. During infection, it can be anticipated that the requirement of stress proteins for cell viability will be equally essential both for the pathogen and for the infected host. Just as stress proteins are essential in "normal" as well as stressed cells, it is clear that changes in stress protein expression will be associated with physiologically normal events accompanying infection as well as with any subsequent pathological events. In addition to the direct role of stress proteins in cell physiology, their potential medical influence is compounded by their ability to act as potent immunogens. Responses to microbial stress proteins are a prominent feature of the immune repertoire in patients and in experimental animals, and there has been wide discussion of the possibility that recognition of conserved, self-like, epitopes on such antigens could influence infectious and other diseases. Three broad hypotheses have been put forward concerning the relevance of immunological reactivity to stress proteins:

In some instances, reactivity to stress proteins can be seen as an example of "mimicry" in which an initial response against a pathogen component cross-reacts with a self protein, triggering autoimmune pathology. On the other hand such recognition of self could reflect normal regulatory events contributing to the maintenance of peripheral tolerance. In autoimmune arthritis, in fact there is, as we will argue in this paper, evidence in support of the latter possibility.

The detection of autoreactive immune responses to stress proteins in individuals without disease suggests they may be integral to normal immune function – possibly with a role in "immune surveillance," acting as a generalized recognition system for cells stressed by infection, malignancy, or other causes, or again with a role in maintaining peripheral tolerance.

Finally, immune responses to stress proteins can be viewed as a secondary event, reflecting tissue breakdown and release of intracellular proteins following any pathological disturbance.

B. Autoimmune Arthritis and Immunity to Bacterial Antigens

An intriguing and unique aspect of acquired imunity is its antigen specificity. The immune system is equipped with antigen-specific cellular elements that control the system in such a way that responses are precisely tailored to reactivity that helps to maintain the self-integrity of the host under a variety of circumstances. Of crucial importance to this would seem to be a proper discrimination between "self" and "nonself." And even beyond that, the immune system is faced with decision-making in order to mount responses exclusively to those encountered substances against which the generation of immune responses will have beneficial consequences. The system has evolved to avoid harmful responses by mechanisms such as tolerance. However, specificity by itself seems not to be a likely mechanism to ensure "self-tolerance."

The antigenic composition of "self" is so immensely diverse that avoidance of reactivity with "self" by the mere absence of cells having specificity for "self" would badly compromise the leftover repertoire of cells meant to cover the huge array of "non-self" antigens. Therefore, it seems appropriate to define "self-non-self" discrimination in a more functional manner: the immune system has developed, depending on the circumstances, distinct patterns of reactivity to different antigens. Such distinction seems determined by matters such as previous experience with the same or related antigens, by the molecular contexts in which epitopes are presented, by antigenic load, persistence, antigenic competition, etc. Such a functional definition of "self-non-self" discrimination is in accordance with a large body of experimental findings, showing that self-reactive elements such as disease-inducing autoreactive T cells are persistently contained within the healthy immune system (COHEN 1986). Moreover, at the molecular level no intrinsic differences have been shown to exist between the composition of "self" and "non-self" antigens. One may even argue that some "non-self" antigens are so frequently encountered by the immune system that, for the sake of efficiency, they would be better regarded as "self." This, for instance, may be true for many antigens present within the bacterial microflora, such as hsps. Our immune system is subject to the daily challenge by an environment replete with innumerable potentially harmful invaders. Fortunately, most of us successfully meet this challenge without jamming the immune system by continuously mounting responses to the antigens of such invaders. Most probably, some form of functional tolerance is controlling specific responsiveness with regard to such antigens, while mechanisms implicated in so-called natural resistance will take care of eliminating unwanted intruders. Usually, natural resistance is effective enough to prevent specific immunity from coming into action. When, however, natural resistance fails and infection is established, the necessity of distinguishing the "self" from the "non-self" appears. The prevalent functional tolerance is then broken by the need to respond to the exogenous invader. In this situation it seems that the immune response is forced to act, despite the risk of simultaneously responding to the "self." Especially in the case of arthritis, such a situation seems to develop relatively frequently (Table 1).

The classic example is acute rheumatic fever, where an infection of the throat with group A streptococci leads to arthritis and, in addition, to rheumatic heart disease – probably due to an antigenic mimicry between streptococcal M-protein and some heart muscle antigen. Furthermore, various forms of reactive arthritis have been identified which all, by definition, are the result of immune responses elicited by bacterial infection. Similary, Lyme arthritis appears to be a chronic complication of infection with *Borrelia burgdorferi* (STEERE et al. 1977). More direct proof of the principle that bacterial priming may evoke T-cell-dependent autoimmune arthritis has been obtained in experimental animals. Models have been studied in rodents where arthritis is induced by bacterial antigens, such as cell walls from streptococci (CROMARTIE et al. 1977) and anaerobes of the gastrointestinal flora (SEVERIJNEN et al. 1988), gram-negative bacteria such as *Yersinia* and *Salmonella* (HILL and YU 1987), or heat-killed mycobacteria.

Whether in humans the same principle may be held responsible for the origin of, for instance, rheumatoid arthritis is questionable. One may,

Table 1. Bacteria and arthritis

Human disease	Bacterial organisms	References
Acute rheumatic fever	Streptococci	BEACHY and STOLLERMAN 1973
Reactive arthritis	*Yersinia, Shigella, Salmonella, Gonococcus, Campylobacter*	AHO et al. 1985
Lyme arthritis	*Borrelia burgdorferi*	STEERE et al. 1977
Ankylosing spondylitis	*Klebsiella*, other gram-negatives?	KEAT 1986
Arthritis as side-effect of cancer immunotherapy	*M. bovis* BCG (mycobacteria)	TORISU et al. 1978
Rodent models (rat)		
Adjuvant arthritis	Mycobacteria	PEARSON 1956
SCW arthritis	*Streptococcus pyogenes* (group A)	CROMARTIE et al. 1977
Peptidoglycan polysaccharide induced arthritis	Anaerobic intestinal bacteria	SEVERIJNEN et al. 1988
	Bacterial organisms	References
Viable gram-negatives	*Yersinia, Salmonella*	HILL and YU 1987

however, assume that in the case of infection, situations may occur where normal regulatory control similarly is jeopardized. Alternatively, one may also envisage that under a certain constellation of events, antigenic stimulation in itself, without infection, may suffice to trigger reactivity directed against our "self." A possible example of such a situation in humans may be the arthritis that is seen to result as a side-effect of BCG immunotherapy (Torisu et al. 1978; Ochsenkuhn et al. 1990). In this therapy, for instance, attenuated mycobacteria are instilled in the bladder as a treatment for bladder cancer. Although transient, the resulting arthritis can be severe and is histologically characterized by a typical inflammatory synovitis with a T-cell infiltrate in the absence of mycobacterial antigen (Lamm et al. 1986; Hughes et al. 1989). The possible impact of the commensal intestinal bacterial flora on the maintenance of tolerance for critical antigens may be illustrated by the fact that arthritis is sometimes seen as a consequence of changes in the intestinal flora composition, for instance, after jejunal bypass surgery. Although in the latter case arthritis is thought to depend on circulating immune complexes (Utsinger 1980), the origin of this arthritis may also be due to a critical break of tolerance, as has been suggested to be the explanation for arthritis seen in association with Crohn's disease (Vanden Broek et al. 1988). Experimental evidence for the role of the intestinal flora in maintaining tolerance has been obtained in the rat model of adjuvant arthritis.

Adjuvant arthritis is easily induced in rats by intracutaneous immunization of heat-killed mycobacteria suspended in mineral oil (a substance known as complete Freund's adjuvant). Inbred rat strains differ in their susceptibility to adjuvant arthritis. Fisher rats are known to be resistant. However, when these rats were being bred under germ-free conditions they were shown to be susceptible; whereas, upon colonization with *E. coli* bacteria, these rats again became resistant to adjuvant arthritis (Kohasi et al. 1986). Thus, herewith, the presence of intestinal colonization was clearly shown to be a crucial factor in the maintenance of tolerance for "self" antigens critically involved in arthritis. Probably the phenomenon is more general and not unique for Fisher rats.

Among researchers working with models of experimental arthritis, it is known from experience that animals from colonies that are kept under stringent housing conditions tend to develop more severe disease than their conventionally reared counterparts. In other words, the cleaner the environment the less resistant animals are to the induction of disease. Moreover, in diabetes prone NOD mice, having a disease incidence of about 50–80%, the incidence may reach almost 100% as soon as the animals are kept under germ-free or certain specific pathogen-free conditions (Todd 1990). Thus, on the one hand it seems that immune stimulation by antigens or adjuvants or by infection with bacteria or even viruses, such as shown in the case of diabetes in the NOD mouse (Oldstone 1988), may protect against the development of autoimmune diseases. On the other hand, there is proof that immune stimulation by non-self antigens, as we have seen above, may induce such diseases. Therefore, it seems that the tolerance for "self" in our functional definition is an active principle that needs to be maintained by a controlled and continuous challenge with the "non-self." Since disease seems to be a consequence of a failure to maintain such tolerance, one may ask oneself whether it will be possible to restore such tolerance by the use of defined non-self antigens. In the experimental situation, prevention of disease by reinforcement of tolerance through preimmunization of the disease-susceptible animals seems to be relatively easy to achieve. For such purposes, at least in arthritis, we seem to have already defined a critical antigen. In the context of the present discussion it may come as no surprise that this antigen is a bacterial heat-shock protein. Invariably present in all bacterial organisms constituting our "non-self" environment and being prominently immunogenic, heat-shock proteins are attractive and, due to their evolutionary conservation, reliable target antigens for our immune response. Already in the naive non-immune individual the immune system seems to have set a rather selective focus on groups of conserved molecules, including hsp (COHEN and YOUNG 1991). In various arthritis models the evidence now has been collected that preimmunization with mycobacterial hsp60 induces protection against disease. Furthermore, although originally this mycobacterial hsp60 had been defined to be critically involved only in mycobacteria-induced adjuvant arthritis, it has now been established that also in other experimental models, including models without any bacterial involvement, immunity to hsp60 is closely associated with development of disease.

C. Hsp60 Is the Critical Antigen in Rat Adjuvant Arthritis

In adjuvant arthritis, disease is induced by substances that have no obvious antigenic relationships with joint tissues. Originally this experimental model was discovered by PEARSON (1956) when he observed the development of arthritis in rats after experimental immunizations using complete Freund's adjuvant containing heat-killed mycobacteria. More recently, other oily adjuvantia, such as pristane (THOMPSON et al. 1990) and the synthetic lipoidal amine CP 20961 (CHANG et al. 1980), have been shown to induce a similar form of arthritis. What these models have in common is an arthritis produced in the absence of immunization with a "self-antigen." Although the pathogenic mechanisms involved are probably not identical, they may well be related, since in all three models immunity to hsp antigens is involved, as we will discuss later.

Because of similarities seen in the histology of adjuvant arthritis and human rheumatoid arthritis, the model of mycobacteria-induced adjuvant arthritis has been explored in great detail. The critical role of cellular immunity in the model was established by the successful adoptive transfer of disease by thoracic duct lymphocytes (WHITEHOUSE et al. 1969). More definite proof of the role of T cells was obtained by HOLOSHITZ et al. (1983), when they were able to select an arthritogenic T-cell line from cells obtained from mycobacteria-immunized Lewis rats. When administered in sublethally irradi-ated syngeneic Lewis rats, this T-cell line, called A2, caused an arthritis which was indistinguishable from the actively, i.e., with mycobacteria, induced dis-ease. Moreover, in contrast to other non-arthritogenic T-cell lines, this same A2 T-cell line was found to induce protection against subsequent AA induc-tion in non-irradiated rats. From these and other observations it was con-cluded that the A2 cell line comprised T cells which had a central regulatory role in adjuvant arthritis. The fact that this line had been raised by in vitro selection with whole mycobacteria, following the exact protocol that had been successfully used earlier for the generation of an encephalitogenic T-cell line with specificity for basic protein of myelin (BEN-NUN and COHEN 1982), indi-cated that development of T-cell immunity with specificity for some mycobac-terial antigen was an essential feature of adjuvant arthritis.

Upon cellular cloning of A2, several CD4+ T-helper subclones were obtained. One of these subclones, called A2b, was found to be virulently arthritogenic (HOLOSHITZ et al. 1984). This particular subclone A2b was found to proliferate not only in the presence of mycobacterial antigens, but also in the presence of semi-purified preparations of cartilage proteoglycans (VAN EDEN et al. 1985). These findings led to the concept of molecular mimicry between a mycobacterial antigen and a cartilage-associated "self-antigen" as the critical pathogenic mechanism explaining mycobacteria-induced arthritis (VAN EDEN et al. 1987). With cell-line A2 and its cloned A2b we had identified some of the cells which seemed to play a central role in adjuvant arthritis. All cells had in common their unexpected double specificity for mycobacteria and proteoglycans. Therefore, identification of the antigens seen by these cells was expected to clarify the exact pathogenic mechanism of AA. While the original cell-line A2 had been found to show some minor proliferation in the presence of collagen type II (HOLOSHITZ et al. 1984), the even more virulently arthritogenic A2b did not. A2b, however, proliferated in the presence of crude proteoglycans. In particular, preparations enriched for link or core proteins of proteoglycans were found to simulate A2b, suggesting that the epitope mim-icked by mycobacteria was present in one of these cartilage-associated pro-teins. This mycobacterial antigen we have now defined exactly. Molecular cloning of M. bovis BCG genes in E. coli by THOLE et al. (1985) resulted in the expression of several mycobacterial proteins, amongst which was a 65-kDa protein. This protein turned out to be the antigen we were looking for. A2 and A2b turned out to respond vigorously, while none of the other clones tested showed any responsiveness (VAN EDEN et al. 1988). Furthermore, by Western blotting it became evident that polyclonal or monoclonal antibodies raised against mycobacteria were frequently reactive with this 65-kDa molecule, and not exclusively in mycobacteria but also in many other bacterial organisms (THOLE et al. 1988).

From the sequence homologies of the M. bovis 65-kDa protein with known heat shock proteins such as the E. coli GroEL, the 65 kDa of mycobac-teria was identified as a heat-shock protein, to be called hsp65 of mycobacteria

(YOUNG et al.1987). By the study of deletion mutants of the 65-kDa gene expressing only parts of the molecule and pEX2 expression products consisting of fusion proteins composed of *E.coli* betagalactosidase and various truncated derivatives of the 65-kDa protein, a provisional mapping of the epitopes of A2b was obtained in the area located from positions 171 to 234. Finally, by means of several overlapping synthetic peptides, an epitope was defined that was located at positions 180 through 188 that stimulated A2b (VAN EDEN et al. 1988). The amino acid sequence of this epitope had some limited sequence homology with a rat proteoglycan link protein sequence. Although supportive evidence for the relevance of this homology was obtained, for instance, in human T-cell responses, it is uncertain whether this link protein sequence is the target structure in adjuvant arthritis. The homology could also be with an as yet unidentified or sequenced cartilage-associated protein or with another protein type which is present in all cells and therefore all tissues.

D. Nasal Tolerance to hsp Peptides Suppresses Antigen and Non-Antigen Induced Arthritis

Recently, the unique relationship of the 180–188 sequence with the arthritic process was further substantiated by tolerizing rats for this particular sequence by administering this peptide in the nose (nasal tolerance) or by giving it subcutaneously in PBS at high dosages (high dose tolerance). This procedure was seen to protect the animals from the subsequent induction of arthritis by either mycobacteria in oil or by the non-antigenic synthetic adjuvant avridine (CP20961) (PRAKKEN et al. 1997). Apparently, this single bacterial epitope, which resembled a (so far not identified) self epitope at the site of inflammation, was capable of inducing regulatory mechanisms of peripheral tolerance. The success of this regimen in suppressing disease, and especially in the case of non-microbially induced disease, seemed to support the possibility that this microbial epitope, indeed, had a unique relationship with a disease critical self-antigen in the joint. Apparently, the exposition of such antigen at the mucosal surface of the nose is already sufficient to set reactive T cells in a regulatory mode with the capacity to enforce peripheral tolerance, leading to disease resistance.

E. Conserved hsp60 Epitopes Induce Arthritis Suppressive T Cells

The identification of mycobacterial hsp60 as a critical antigen in arthritis has led to many studies on the potential of hsp60 in modifying arthritis development. Immunization experiments in mice and rats using hsp60 proteins or its derivative peptides have never led to induction of arthritis. On the contrary, resistance to arthritis was seen to develop. Prior immunization of experimental animals with mycobacterial hsp60 has been found to protect against subsequent induction of AA (van Eden et al. 1988; Billingham et al. 1990), streptococcal cell wall arthritis (van den Broek et al. 1989) and avridine (a non-antigenic lipoidal amine) arthritis in rats (Billingham et al. 1990) and also pristane (Thompson et al. 1990) and collagen arthritis (Ito et al. 1991) in mice. Careful analyses of differential T cell responsiveness of whole mycobacteria (AA induction protocol) versus mycobacterial hsp60 (protection protocol) immunized rats have now indicated that the arthritis inductive capacity of whole mycobacteria does coincide with dominant responses directed to the mimicry epitope 180–188 (Anderton et al. 1994). This is not the case when the hsp60 protein is used for immunization. Cellular cloning of responsive T cells in the latter protective protocol has revealed clones that recognize sequences conserved between mycobacterial and mammalian hsp60. Adoptive transfer experiments have shown that such T cells, recognizing conserved sequences and in particular the mycobacterial 256–265 sequence, were capable of transferring protection against the disease (Anderton et al. 1995). Furthermore, it was demonstrated that such cells had the capacity of responding to heat-shocked spleen cells. Immunizations with conserved peptide 256–265 were also inducing arthritis protection and none of several other peptides containing non-conserved dominant T cell epitopes tested was capable of inducing any protection. Testing of the same 256–265 peptide in a very different arthritis model in Lewis rats, in this case avridine arthritis, revealed that also in a non-bacterially induced model the protective potential of the peptide was present. Protection was also found using the homologous rat (self) peptide. Recent experiments have shown similar protective effects to result from immunization with mycobacterial hsp70 both in AA and avridine arthritis. Also for the smaller subunit of the hsp60 complex (GroES of 10-kDa *E. coli*) its arthritis protective potential has been documented (Ragno et al. 1996).

F. Suppression in Arthritis Models Is Specific for Heat Shock Proteins

Given the observation that conserved epitopes of hsps induced resistance to arthritis by setting a focus of T cell reactivity directed to recognition of self homolog (human) hsp, one could easily ask whether other conserved bacterial proteins would have similar arthritis protective capacities. However, in striking contrast with the protection in autoimmune disease models obtained with hsps such as hsp60, 70 and 10, we have seen recently that a set of other conserved, but not stress inducible, bacterial proteins, such as superoxide dismutase, glyceraldehyde-3-phosphate-dehydrogenase and aldolase, did not protect in experimental arthritis, despite their good immunogenicity and sequence homologies with their mammalian homologs (Prakken et al., unpublished). Therefore, it seems that the protective quality of hsps is a unique aspect of hsps, possibly caused by their exquisite behavior of being upregulated locally and expression under conditions of stress such as existing at sites of inflammation.

G. Immune Mediated Diseases

Most diseases are associated with tissue or cell damage. In many cases of such damage, the liberation of (intracellular) antigens will lead to activation of the immune system. This situation is potentially harmful for our body, since it may lead to an inflammatory response. In the case of infectious diseases such inflammatory responses are needed in order to create an environment which is hostile to the invading microorganism. In most infectious diseases the inflammatory process leads to accompanying immunopathology, which is normally kept within bounds by the self-regulating capacity of the immune response. The mechanisms operative in this self-regulation are far from being understood at present. It is generally accepted, however, that the self-regulatory mechanisms operative in peripheral tolerance (see below) can become compromised, which then may result in the development of a so-called autoimmune disease.

The immunopathology seen in autoimmune diseases, diseases in which the immune system primarily attacks "self-tissues," resembles the immunopathology seen in inflammation resulting from infection. Autoimmune diseases are usually chronic and cause major socioeconomic losses. Examples of autoimmune diseases are rheumatoid arthritis, multiple sclerosis, type I diabetes, lupus, thyroiditis and myasthenia gravis. Since in these diseases "self-proteins" are being recognized by the immune system, the origin of autoimmune diseases is being thought to reside in a defective self-regulation, or incomplete tolerance to such self-proteins.

H. Rheumatoid Arthritis as a Model
   Autoimmune Disease

Rheumatoid arthritis (RA) is the most common inflammatory cause of disability in the western world. Although the underlaying cellular and molecular mechanisms have remained unclear until now, RA seems to provide a good example of how the interaction between genetic and environmental factors may lead to autoimmunity. RA is a most destructive joint disease and its clinical course is characterized by involvement of the small joints of the hands and feet followed by centripetal progression to larger joints and finally even cervical spine. The histology of the disease is characterized by hyperplastic synovial tissue which is heavily infiltrated by various types of leukocytes. It is very likely that the constant supply of new cells of the immune system is necessary to induce and subsequently maintain the inflammatory process.

The current treatments for RA are only symptomatic and can be divided into three lines. The first line of therapy consists of treatment with nonsteroidal anti-inflammatory drugs (NSAIDs). These drugs can control pain and swelling of the joints, but do not halt the progressive joint destruction associated with the disease. Furthermore, NSAIDs can cause upper gastrointestinal tract bleeding upon prolonged usage. In the case that RA remains active despite treatment with NSAIDs (which is usually the case), the second line of therapy can be applied, which consists of treatment with disease modifying antirheumatic drugs (DMARDs). These DMARDs, such as penicillamine, chloroquine, gold compounds and sulfasalazine, generally show some beneficial effect after a period of 3–6 months. However, due to severe side effects, treatment with DMARDs has to be stopped in about 25% of patients. The belief that the immune system is actively involved in the onset and pathogenesis of RA has led to the development of a third line of treatment with strong broad-acting immunosuppressive drugs such as cyclosporin A and methotrexate. These strong immunosuppressive drugs generally have severe side effects, such as nephrotoxicity and in the long-term cancer. Besides the adverse side effects with the currently used antirheumatic drugs, the long-term outcome of sequential monotherapy based on the therapeutic pyramid described above has been disappointing. No substantial evidence has been found proving that the drugs employed actually arrest the progression of joint destruction. Therefore, alternative strategies are clearly needed.

I. Hsps in Autoimmune and Other Inflammatory Diseases

In the past the most immunogenic protein of bacteria was known as the "common antigen of gram-negatives." This protein now has been recognized as being the hsp60 (GroEL of *E. coli*) molecule. Similarly, in parasitic infections, hsp70 and sometimes hsp90 are found to represent other major targets for the humoral immune response. This in itself predicts that the presence of antibodies directed at the major families of heat shock proteins will be a frequent and normal situation in most individuals and also indicates that immune responses to hsps are compatible with normal health. Despite the frequent occurrence of hsp antibodies in normal individuals it seems that in the majority of inflammatory diseases raised levels of hsp antibodies can be found (see van Eden and Young 1996). This has been reported for rheumatoid arthritis (RA), juvenile RA, reactive arthritis, Behçet's disease, SLE, Crohn's disease, insulin dependent diabetes mellitus (IDDM) and multiple sclerosis (MS). Significant hsp specific T-cell responses have been observed in RA, JRA (see below), Behçet's disease, MS and in graft infiltrating lymphocytes during transplant rejection episodes. Raised expression of hsps in diseased tissues has been documented in sarcoidosis, SLE, inflammatory liver diseases, chronic gastritis (gastric ulcer), celiac disease, MS, IDDM and atherosclerosis. Despite the general perception that hsps seem to play a role in different autoimmune diseases, there is no consensus on cause and effect relationships. Evidence in favor of hsps being a trigger leading to autoimmunity, because of their conserved nature, is essentially lacking. A more plausible possibility is that inflammation in general causes raised tissue expression, leading to the generation of hsp specific T- and B-cell responses.

J. Hsps in Human Arthritic Diseases

Similar to what has been documented in many other autoimmune conditions, rheumatic diseases have also been seen to feature the sequelae of immune responses to hsps. This is most evident from serology studies, which have shown the presence of raised levels of hsp60 specific antibodies in patients (BAHR et al. 1989). Raised expression of hsp in inflamed tissues has been documented for arthritic synovium in both RA and JRA (BOOG et al. 1992). In children this raised expression has been seen to be an event that occurred early in the development of disease. Despite earlier reports claiming also prominent T cell responses to mycobacterial hsp60 in advanced RA, more recent studies have suggested that proliferative T cell responses are, however, more confined to early RA and that in advanced RA such responses are less prominent. Alternatively, responses have been detected in functional assays sensitive to cytokine production, such as in assays measuring the effect of mononuclear cells on cartilage proteoglycan turnover in vitro (WILBRINK et al., 1993). In reactive arthritis and in oligoarticular juvenile rheumatoid arthritis (OA-JRA) T cell responses to hsp60 were again more prominent. In contrast to adult RA, these conditions are characterized by a remitting course of disease development.

In children with JCA, however, proliferative responses have been obtained in T lymphocytes taken from both the peripheral blood and the synovial compartment. Most significant and reliable responses were seen by stimulating the cells with self-antigen human hsp60 (DE GRAEFF-MEEDER et al. 1995; PRAKKEN et al. 1996). Since most of the patients, however, did respond to mycobacterial hsp60 as well, it seems that also in this case, conserved epitopes, equally present in both the human and mycobacterial hsp60, are recognized by patient T cells. By comparing patients of distinct clinical subgroups, it became evident that responders had oligoarticular (OA) forms of JCA, whereas non-responders had polyarticular or systemic JCA. In other words, those with a remitting form of disease responded, whereas those having a non-remitting form did not. Longitudinal studies indicated that (temporary) remission of OA-JCA coincided with disappearance of responses, and that a clinical relapse coincided with reappearance of responses. Altogether, the data obtained in JCA patients have shown that responses to human hsp60 as a self-antigen do occur and that they are associated with relatively benign forms of arthritis. The presence of responses during the active phase of disease, preceding remission, suggests, in line with the observations made in the AA model, that responses to self-hsp60 may positively contribute to mechanisms leading to disease remission. If so, possibilities for immunological intervention in JCA, and possibly also RA, may be found in strategies aimed at manipulating peripheral tolerance through the vaccination with hsp60, or peptides containing defined epitopes. For these purposes hsp60 may well serve as a useful ancillary autoantigen.

K. Hsp60 T Cell Responses in RA and JRA Are Associated with Suppressive Cytokine Production

In adult RA patients T cell responses to human hsp60 were detectable by culturing the cells in the presence of added IL4. Bacterial hsp60 was found to stimulate cells without added IL4. This suggested a Th2 nature of the T cells responding to the human molecule in particular (van Roon et al. 1997). In children with JRA (OA), responses to human hsp60 coincided with raised expression of IL4 (RT-PCR) in the synovial cells. Furthermore, stimulation of both synovial and peripheral blood T cells resulted in a raised CD30 (a possible marker for Th2 cells) expression in activated (CD45RO+) CD4+ and CD8+ T cells. In addition OA-JRA patient derived hsp60 specific T cell lines were shown to produce IL4 and TGFβ (Prakken et al. 1996). Altogether the data as obtained in human arthritis patients have shown the potential of hsps to trigger the release of Th2 associated and suppressive cytokines.

Hsps are an intrinsic component of the slow acting antirheumatic drug OM89. In various models of experimental autoimmunity it is known that exposure of the animals to exogenous bacterial flora contributes to resistance against disease induction. Also in adjuvant arthritis, germ-free animals were more susceptible than their conventionally reared counterparts (Kohashi et al. 1986). Furthermore, in Fisher rats it was shown that the susceptible germ-free animals became resistant upon gut re-colonization with E. coli bacteria.

In other words, exposure of the immune system to cross-reactive bacterial antigens, such as hsps, might well stimulate the immune system to resume control over unwanted self-reactive clones. In line with the known contribution of bacterial gut flora to tolerance, it seems best to effect such exposure through oral (or nasal) administration of bacterial antigens. Laboratoires-OM (Geneva) is producing a glycoprotein rich extract of E.coli (OM-89), which is marketed and used as a slow acting drug for the treatment of RA. It is administered orally and has shown a therapeutic efficacy comparable with that of gold in trials in RA patients (Rosenthal et al. 1991). Recent analyses have revealed that E.coli hsp60 (GroEL) and hsp70 (DnaK) are the dominant immunogens present in this material (Vischer and Van Eden 1994; Bloemendal et al. 1997). Furthermore, both in AA and in avridine arthritis in Lewis rats, OM-89 was found to protect against arthritis. Therefore, E.coli hsps, when administered orally, may trigger a T cell regulatory event that contributes to the control of RA, in a way similar to the effect of mycobacterial hsp60 in models of arthritis. It is therefore possible that the therapeutic potential of hsps in RA is already exemplified to some extent by the mode of action of OM-89.

L. Mechanisms by Which hsps Produce Protection in Autoimmune Arthritis

The early expression of hsp60 in inflamed synovial tissues (stress response) and the findings of immunological recognition of the protein both at the level of antibodies and T cells especially in remitting arthritic diseases seem to tie in very well with the data obtained in the experimental animal models. The animal experiments have shown that irrespective of the trigger that led to synovial inflammation, prior stimulation of immunity to self-hsp60 using bacterial hsp60 or its conserved peptide raised resistance to subsequent disease induction. Taken together these findings have indicated that T cell recognition of hsp60 (and also hsp10 or hsp70) at the site of inflammation does contribute to the control of the ongoing inflammatory response. The stimulation of such responses by prior immunization using bacterial hsp or its relevant peptides, therefore, is expected to facilitate such T cell mediated regulatory control (VAN EDEN et al., 1998).

Heat shock proteins are, despite their conserved nature and therefore antigenic relationship with "self," immunogenic to an exceptionally high degree. Although the reasons for this are as yet unclear, this phenomenon can be understood in terms of repertoire selection. Hsps are well expressed in the thymic medulla, at the sites where positive thymic selection occurs. This will lead to selection of a repertoire of T cells having a receptor that allows low affinity (partial agonist) interactions with self hsp also when it becomes overexpressed at sites of inflammation. High affinity receptors were negatively selected and therefore deleted. During infection with an hsp expressing microorganism (any bacterium, parasite, etc.), the same repertoire will be expanded by recognition of the microbial hsp homolog. As the homolog will be antigenically close to the self hsp, but not identical with it, this recognition will include high affinity interactions. By its nature these high affinity interactions may lead to an aggressive (Th1) anti-infectious response. At the same time this repertoire is numerically expanded. On re-encounter of these T cells with self-hsp as expressed in the inflammatory site, the interaction will again be of low affinity and lead to T cell regulation due to "partial agonist" driven T cell responses. In other words exposure of the immune system to microbial hsps is expanding a repertoire of potentially anti-inflammatory T cells. The same repertoire expansion can be effected by directed stimulation of the immune system with hsp proteins or peptides thereof. Given the existing evidence that responses to hsps comply with peripheral tolerance mechanisms of immune deviation, it is likely that bystander suppressive mechanisms (secretion of IL4, TGFβ) are operational in hsp induced anti-inflammatory mechanisms.

On the other hand it may be that under normal conditions low level expression of self-hsp epitopes by non-professional antigen presenting cells in the periphery or of conserved microbial hsp epitopes in the "tolerizing" gut environment is continuously noted by T cells. This recognition, however, in the absence of co-stimulation or in the otherwise tolerizing mode, will drive such T cells into a regulatory phenotype or anergy. Upon subsequent involvement of such cells in autoimmune inflammation, they will exert regulatory activity, also when they recognize their overexpressed antigen presented by the professional APC. Recently, we have demonstrated that anergic T cells suppressed, in co-culture, proliferative responses of other T cells as soon as the antigen that was used to induce anergy was added to the co-culture (Taams et al., submitted). In the case of heat shock proteins, thus anergized cells could focus their regulatory activity on sites of inflammation where heat shock proteins become temporarily overexpressed (bystander suppression). In infection the activity of such anergic regulators would be outweighed by a dominant frequency of T cells (vigorously) responding to non-conserved microbial hsp epitopes. Depending on the circumstances the pro-inflammatory response will (temporarily) break the peripheral tolerance. And indeed, upon immunization or infection, microbe specific epitopes are more dominantly recognized. The hsp induced protection against autoimmune inflammation is now explained by either the expansion of the self-hsp cross-reactive T cells which assume the anergic phenotype when encountering their antigen on non-professional APCs, or by further imprinting or spreading of the preexistent anergic state.

M. Lessons for the Development of Specific Immunotherapy in Autoimmunity

As argued above, immunity to bacterial antigens, such as hsps, may contribute to maintenance of self-tolerance as a hedge against autoimmunity. To achieve a lasting restoration of such tolerance in the case of disease, it seems most adequate to target immunotherapy at the reinforcement of natural mechanisms that contribute to such maintenance of self-tolerance. In other words, exposure of the immune system to bacterial antigens, such as hsps, might well stimulate the immune system to resume control over unwanted self-reactive clones. In line with the known contribution of bacterial gut flora to tolerance, it seems best to effect such exposure through oral administration of bacterial antigens. Although little support for the effectivity of such an approach can be obtained from work in experimental disease models, so far from experience in human medicine such support can be obtained. As mentioned above, Laboratories-OM (Geneva) has been producing *E. coli* bacterial lysates, containing bacterial hsps, which are used among others for the treatment of RA. For obvious reasons it would be of great interest to analyze such mechanisms at the level of T cell responses in patients treated with this material.

Positive findings of such an analysis would then possibly lead to attempts to develop better defined pharmaceutical compounds such as synthetic peptides. Such peptides could be composed of conserved sequences of bacterial hsps and be used to stimulate the frequency or activity of T cells with the potential to recognize self-hsp molecules, expressed at sites of inflammation.

On a limited scale initial experience with oral administration of a defined hsp peptide has been obtained by Bonnin and Albani (1998). Based on the sequence homology of an *E. coli* dnaJ sequence with the so-called "shared-epitope," a consensus sequence shared by RA associated HLA-DR molecules, the peptide QKRAA was administered to RA patients in an open label clinical trial. Upon analysis of the functional phenotype of responding T cells in these patients it appears that the oral treatment did effect a relative shift from Th1 to Th2 in cells with specificity for the dnaJ peptide.

The manipulation of peripheral tolerance with hsp immunization may work through the re-inforcing of natural protective mechanisms of T cell regulatory events in inflammation. The approach is broad and can be useful for the treatment of inflammation as seen in various autoimmune diseases and other inflammatory diseases such as allergy. Furthermore, spin-off may be expected for infection (including vaccines) and cancer. The ubiquitous nature of hsp expression in diseased tissues may well ensure that antigenic spreading as a possible tolerance escape mechanism will not easily take place. This is in contrast to alternative more conventional antigen specific approaches.

Effective treatment leading to reduced hsp expression in the tissues will lead to a gradual loss of therapeutic impact as a useful built-in feedback mechanism. In other words there is no therapeutic overshoot.

N. Conclusion

In the exploration of mechanisms of peripheral tolerance it has become evident that besides control through specific elimination of cells, interactive regulatory effects of antigens do play a role. Work on antagonistic peptides, anergic T cells or cells displaying a regulatory cytokine profile has provided an experimental basis for such effects. Heat shock proteins can provide an example of how such interactive regulatory effects can be targeted to sites of cellular stress. Immunological recognition of heat shock proteins as seen in association with virtually every inflammatory condition, including autoimmune conditions, such as rheumatoid arthritis, can be central to peripheral tolerance mechanisms meant to control inflammation beyond tissue specific antigenic manipulation.

Acknowledgement. I thank Ms. Ydwine van den Oosterkamp for expert editorial assistance.

References

Aho K, Leirisalo-Repo M, Repo H (1985) Reactive arthritis. Clin Rheum Dis 11:25

Anderton SM, Van der Zee R, Noordzij A, Van Eden W (1994) Differential hsp65 T cell epitope recognition following Adjuvant Arthritis inducing or protective immunisation protocols. J Immunol 152:3656–3664

Anderton SM, Van der Zee R, Prakken B, Noordzij A, Van Eden W (1995) Activation of T cells recognizing self 60 kDa heat shock protein can protect against experimental arthritis. J Exp Med 181:943–952

Bahr GM, Rook GAW, Al-Saffar M, Van Embden JDA, Stanford JL, Behbehani K (1989) An analysis of antibody levels to mycobacteria in relation to HLA type: evidence for non-HLA-linked high levels of antibody to the 65kD heat shock protein of M. tuberculosis in rheumatoid arthritis. Clin Exp Immunol 74:211–215

Beachy EN, Stollerman GH (1973) Mediation of cytotoxic effects of streptococcal M protein by non-type-specific antibody in human sera. J Clin Invest 52:2563

Ben-Nun A, Cohen IR (1982) Experimental autoimmune encephalomeyelitis(EAE) mediated by T cell lines: process of selection of lines and characterisation of the cells. J Immunol 129:303

Billingham MEJ, Butler R, Colston MJ (1990) A mycobacterial 65-kD heat shock protein induces antigen-specific suppression of adjuvant arthritis, but is not itself arthritogenic. J Exp Med 171:339–344

Bloemendal A, Van der Zee R, Rutten VPMG, Van Kooten PJS, Farine JC, Van Eden W (1997) Experimental immunisation with anti-rheumatic bacterial extract OM-89 induces T cell responses to heat-shock protein 60 and 70; modulation of peripheral immunological tolerance as its possible mode of action in the treatment of rheumatoid arthritis. Clin Exp Immunol 110:72–78

Bonnin D, Albani S (1998) Mucosal modulation of immune responses to heat-shock proteins in autoimmune arthritis. Biotherapy 10:213–221

Boog CJP, de Graeff-Meeder ER, Lucassen MA, van der Zee R, Voorhorst-Ogink, van Kooten PJS, Geuze HJ, van Eden W (1992) Two monoclonal antibodies generated against human hsp60 show reactivity with synovial membranes of patients with juvenile chronic arthritis. J Exp Med 175:1805

Chang Y-H, Pearson CM, Abe C (1980) Adjuvant polyarthritis IV. Induction by a synthetic adjuvant: immunologic, histopathogic and other studies. Arthritis Rheum 23:62

Cohen IR (1986) Regulation of autoimmune disease: physiological and therapeutic. Immunol Rev 94:5

Cohen IR, Young DB (1991) The immune system's view of invading microorganisms, autoimmunity and the immunological homunculus. Immunol Today 12:105

Cromartie WJ, Craddock JG, Schwab JH, Anderle SK, Yang C (1977) Arthritis in rats after systemic injection of streptococcal cells or cell walls. J Exp Med 146:1585

De Graeff-Meeder ER, Van Eden W, Rijkers GT, Prakken ABJ, Kuis W, Voorhorst-Ogink MM, Van der Zee R, Schuurman HJ, Helders PJM, Zegers BJM (1995) Juvenile chronic arthritis: T cell reactivity to human hsp60 in patients with a favourable course of arthritis. J Clin Invest 95:934–940

Hill JL, Yu DTY (1987) Development of an experimental animal model for reactive arthritis induced by Yersinia enterocolitica infection. Infect Immun 55:721

Holoshitz J, Naparstek Y, Ben-Nun A, Cohen IR (1983) Lines of T lymphocytes induce or vaccinate against autoimmune arthritis. Science 219:56–58

Holoshitz J, Matitiau A, Cohen IR(1984) Arthritis induced in rats by clones of T lymphocytes responsive to mycobacteria but not to collagen type II. J Clin Invest 73:211–215

Hughes RA, Allard SA, Maini RN (1989)Arthritis associated wtih adjuvant mycobacterial treatment for carcinoma of the bladder. Ann Rheum Dis 48:432

Ito J, Krco C, Yu D, Luthra HS, David CS (1991) Preadministration of a 65kD heat-shock protein GroEL, inhibits collagen induced arthritis in mice (abstract). J Cell Biochem 15A:284

Keat A (1986) Is spondylitis caused by Klebsiella? Immunol Today 7:144

Kohashi O, Kohashi Y, Takahashi T, Ozawa A, Shigematsu N (1986) Suppressive effect of E. coli on adjuvant induced arthritis in germ-free rats. Arthritis Rheum 29:547–555

Lamm DL, Stogdill VD, Stogdill B, Crispen RG (1986) Complications of bacille Calmette-Guerin immunotherapy in patients with bladder cancer. J Urol 135: 274

Ochsenkuhn T, Weber MW, Caselmann WH (1990) Arthritis after M. bovis immunotherapy for bladder cancer (letter). Ann Intern Med 112:882

Oldstone MBA (1988) Prevention of type I diabetes in nonobese diabetic mice by virus infection. Science 239:500

Pearson CM (1956) Development of arthritis, periarthritis and periostitis in rats given adjuvant. Proc Soc Exp Biol Med 91:95–101

Prakken ABJ, Van Eden W, Rijkers GT, Kuis W, Toebes EA, de Graeff-Meeder ER, Van der Zee R, Zegers BJM (1996) Autoreactivity to human hsp60 predicts disease remission in oligoarticular juvenile chronic arthritis. Arthritis Rheum 39:1826–1832

Prakken ABJ, Van der Zee R, Anderton SM, van Kooten PJS, Kuis W, Van Eden W (1997) Peptide induced nasal tolerance for a mycobacterial hsp60 T cell epitope in rats suppresses both adjuvant arthritis and non-microbially induced experimental arthritis. Proc Natl Acad Sci USA 94:3284–3289

Ragno S, Winrow VR et al (1996) A synthetic 10 kD heat-shock protein (hsp10) from Mycobacterium tuberculosis modulates adjuvant arthritis. Clin Exp Immunol 103:384–390

Severijnen AJ, Hazenberg MP, Van de Merwe JP (1988) Induction of chronic arthritis in rats by cell-wall fragments of anaerobic coccoid rods isolated from the faecal flora of patients with Crohn's disease. Digestion 39:118

Steere AC, Malawista SE, Suydman DR, Shope RE, Andiman WA, Ross MR, Steele FM (1977) Lyme arthritis: an epidemic of oligoarticular arthritis in children and adults in three connecticut communities. Arthritis Rheum 20:7–17

Thole JER, Dauwerse HG, Das PK, Groothuis DG, Schouls LM, Van Embden JDA (1985)Cloning of the Mycobacterium bovis BCG DNA and expression of antigens in Escherichia coli. Infect Immun 50:800–806

Thole JER, Hindersson P, De Bruyn J, Cremers F, Van der Zee J, De Cock H, Tommassen J, Van Eden W, Van Embden JDA (1988) Antigenic relatedness of a strongly immunogenic 65kD mycobacterial protein antigen with a similarly sized ubiquitous bacterial common antigen. Microb Pathog 4:71–83

Thompson SJ, Rook GA, Brearley RJ, Van der Zee R, Elson CJ (1990) Autoimmune reactions to heat-shock proteins in pristane induced arthritis. Eur J Immunol 20:2479

Todd JA (1990) Genetic control of autoimmunity in type I diabetes. Immunol Today 11:122

Torisu M, Miyahara T, Shinohara N, Ohsato K, Sonozaki H (1978) A new side effect of BCG immunotherapy: BCG-induced arthritis in man. Cancer Immunol Immunother 5:77–83

Utsinger PD (1980) Bypass disease: a bacterial antigen-antibody systemic immune complex disease. Arthritis Rheum 23:758

Van den Broek MF, Van de Putte LBA, Van den Berg WB (1988) Crohn's disease associated with arthritis: a possible role for crossrectivity between gut bacteria and cartilage in the pathogenesis of arthritis. Arthritis Rheum 31:1077

Van den Broek MF, Hogervorst EMJ, Van der Bruggen MCJ, Van Eden W, Van der Zee R, Van den Berg WB (1989) Protection against streptococcal cell wall induced arthritis by pretreatment with 65kD mycobacterial heat-shock protein. J Exp Med 170:449–466

Van Eden W, Holoshitz J, Nevo Z, Frenkel A, Klajman A, Cohen IR (1985) Arthritis induced by a T lymphocyte clone that responds to Mycobacterium tuberculosis and to cartilage proteoglycans. Proc Natl Acad Sci USA 82:5064–5067

Van Eden W, Holoshitz J, Cohen IR (1987) Antigenic mimicry between mycobacteria and cartilage proteoglycans: the model of adjuvant arthritis. In: Cruse TM (ed) Concepts immunopathol 4. Karger, Basel, pp 144–170

Van Eden W, Thole JER, Van der Zee R, Noordzij A, van Embden JDA, Hensen EJ, Cohen IR (1988) Cloning of the mycobacterial epitope recognized by T lymphocytes in adjuvant arthritis, Nature 331:171–173

Van Eden W, Young DB (1996) Stress proteins in medicine. Dekker, New York

Van Eden W, van der Zee R, Paul AGA, Prakken BJ, Wendling U, Anderton SM, Wanben MHM (1998) Do heat-shock proteins control the balance of T cell regulation in inflammatory diseases? Immunol Today 19:303–307

van Roon J, van Eden W, van Roy J, Lafeber F, Bijlsma JWJ (1997) Stimulation of suppressive T cell responses by human but not bacterial 60 kDa heat-shock protein in synovial fluid of patients with rheumatoid arthritis. J Clin Invest 100:459–463

Vischer TL, Van Eden W (1994) Oral desensibilization in rheumatoid arthritis (RA). Ann Rheum Dis 53:708–710

Whitehouse DJ, Whitehouse MW, Pearson CM (1969) Passive transfer of adjuvant induced arthritis and allergic encephalomyelitis in rats using thoracic duct lymphocytes. Nature 224:1322

Wilbrink B, Holewijn M, Bijlsma JWJ, van Roy JLAM, den Otter W, van Eden W (1993) Suppression of human cartilage proteoglycan synthesis by rheumatoid synovial fluid mononuclear cells activated with mycobacterial 60kD heat-shock protein. Arthritis Rheum 36:514

Young DB, Ivanyi J, Cox JH, Lamb JR (1987) The 65kDa antigen of mycobacteria – a common bacterial protein? Immunol Today 8:215–219

CHAPTER 16

Heat Shock Protein 60 and Type I Diabetes

S.G. NEWTON and D.M. ALTMANN

A. Introduction

Type I diabetes is characterised by the destruction of pancreatic insulin secreting beta cells and is commonly believed to have an autoimmune aetiology. During the clinically silent, early stages of disease the pancreas is infiltrated by inflammatory cells which mediate beta cell damage. The failure of glucose homeostasis observed in patients is a direct consequence of immunologically mediated beta cell destruction. While there is general agreement that diabetes is caused by autoreactive T cells, there is considerable debate about the antigen specificity of the diabetogenic T cells. As in other autoimmune diseases, identification of the target autoantigen(s) has become a major challenge with potential rewards in new therapies. However, there are no T cell mediated autoimmune diseases in which it has been possible to establish with certainty which self peptides trigger the initial pathology and which major histocompatibility complex (MHC) products present them.

Hsp60 has been implicated as a target self-antigen in a number of different autoimmune diseases including clinical and experimental arthritis, multiple sclerosis and diabetes. Several features make Hsp60 an attractive candidate autoantigen. It is strongly immunogenic, including the ability to elicit strong anti-self responses, even in healthy individuals (KAUFMANN et al. 1991). This is so despite the abundant expression of Hsp60 in mammalian thymus, normally considered to lead to tolerance by deletion of any reactive T cells (BIRK et al. 1996a). Such findings led to the analogy of an immunological homunculus: the notion that some immunogenic self antigens may not induce self-tolerance but rather may be required to shape the responses to exogenous self antigens (COHEN and YOUNG 1991). Another important feature of the Hsp60/65 family is their very high degree of evolutionary conservation. The extensive homology between these stress proteins in prokaryotic and eukaryotic species offers a potential source of molecular mimicry in the contribution of infection to autoimmunity. Lastly, as a stress protein, the expression of Hsp60 is frequently found to be up-regulated at inflammatory sites, offering the possibility of positive inflammatory feedback in autoimmune tissues. Against these arguments is the fact that these diseases are highly organ and tissue specific, an observation not easily explained by immune recognition of a more or less ubiquitously expressed protein.

The aim of this chapter will be to review the credentials of Hsp60 as a candidate autoantigen in these diseases before reviewing in detail the immunopathogenesis of type I diabetes in humans and non-obese diabetic (NOD) mice and considering the role of Hsp60 in the context of the various other beta cell targets which have been investigated.

B. Hsp60 and Autoimmune Diseases

The first evidence for the involvement of a heat shock protein in the pathogenesis of autoimmune disease came from an investigation of T cell specificities in adjuvant arthritis; this disease model has been prototypic in considering the role of Hsp60 in autoimmunity (see also Chap. 15, this volume). Adjuvant arthritis is inducible in some strains of rats by immunisation with complete Freund's adjuvant containing killed *Mycobacteria tuberculosis*. Holoshitz and colleagues selected arthritogenic T cells and used them to probe for antigens important for the development of disease (HOLOSHITZ et al. 1983). One clone, A2b, was capable of transferring arthritis to irradiated rats and shown by van Eden to proliferate in response to both whole *Mycobacteria tuberculosis* and to a protein moiety in the rat joint cartilage proteoglycan (HOLOSHITZ et al. 1983; VAN EDEN et al. 1985). The A2b clone was used to screen a *Mycobacteria tuberculosis* expression library and was found to respond specifically to mycobacterial Hsp65 (VAN EDEN et al. 1988). Recent work has shown that T cell epitopes contained within the carboxy terminal portion of mycobacterial Hsp65 are recognised early in the disease process and that determinant spread can then be observed in the late phases of disease. The observation that the anti-Hsp T cell response increases and diversifies with the onset of clinical disease has led to a reinforcement of the hypothesis that spreading of the immune response from the mycobacterial Hsp65 to the rat Hsp60 homologue is important for pathogenesis. Further credence is given to this hypothesis by the additional observation that divergence of the immune response does not occur in non-susceptible rat strains that share the same major histocompatibility class II locus (MOUDGIL et al. 1997).

The function of immunity to the Hsp60 family in the development of rheumatoid arthritis in humans is less clear. Early reports showed that T cells purified from patients' synovial joints are able to respond to *Mycobacteria tuberculosis* Hsp65 and that the T cells can respond to mycobacterial epitopes and to epitopes shared with human Hsp60 (VAN EDEN et al. 1989). There is also evidence of increased patient antibody responses to *Mycobacteria tuberculosis* and *Escherichia coli* Hsp65 (DE GRAEFF MEEDER et al. 1993; HANDLEY et al. 1996). This evidence, however, is balanced by the observation that T cell responses to bacterial and self heat shock protein 60 family members can be detected in many individuals with and without autoimmune disease and are not necessarily pathogenic (MUNK et al. 1989). Furthermore, T cell responses to Hsp60 have in some cases been correlated with remission. Human Hsp60 responsive T cells purified from the synovial fluid of rheumatoid arthritis patients proliferate in response to interleukin (IL-) 4 stimulation and produce greater amounts of IL-4 and less interferon gamma (IFNγ) than cells stimulated with mycobacterial Hsp65 (VAN ROON et al. 1997). This cytokine profile is associated with protection or remission in a number of autoimmune disorders. These human Hsp60 responsive T cell lines were also capable of inhibiting the production of tumour necrosis factor alpha (TNFα). A similar regulatory role for T cell responses against human Hsp60 has been postulated to exist in oligoarticular juvenile rheumatoid arthritis and juvenile chronic arthritis patients (PRAKKEN et al. 1996). T cell autoimmunity against human Hsp60 has been shown to be correlated with an entry into remission in patients with both of these disorders.

C. Hsp60 Reactivity and the NOD Mouse Model of Type I Diabetes

The NOD mouse constitutes a spontaneous animal model of human type I diabetes showing many parallels with the human disease. Like human type I diabetes patients, NOD mice show insulitis prior to disease, humoral and cellular immunity to beta cell antigens and carry multiple diabetes associated genetic susceptibility loci. Genetic studies have shown that the strongest NOD mouse susceptibility locus is the unique NOD major histocompatibility complex class II allele H2-A^{g7} (WICKER et al. 1995). This allele is unusual in having serine at position 57 of the beta chain rather than the more common aspartamine. Mutation of this serine to an aspartamine molecule prevents the onset of type I diabetes in class II transgenic NOD mice (O'REILLY et al. 1994). Todd and coworkers showed a strong correlation between the amino acid at residue 57 of the DQβ chain and susceptibility to type I diabetes, with serine and alanine bearing alleles over-represented (TODD et al. 1987). Biochemical studies comparing the major histocompatibility complex class II allele H2-A^{g7} from the NOD mouse and the diabetes associated human allele HLA-DQ8 (DQA1*0301/B1*0302) have shown clear similarities between the two molecules. These molecules are the most likely candidates for the MHC molecules responsible, in humans and mice, for presenting the islet cell peptides to diabetogenic T cells. Both molecules have similar peptide binding grooves with a rare preference for an acidic residue in the P9 anchor position. In addition both alleles are relatively SDS unstable (REIZIS et al. 1997a,b; CARRASCO-MARIN et al. 1996). This tendency for class II molecules, which would normally be seen on SDS-PAGE as a stable $\alpha\beta$ heterodimer complexed with self peptide, to dissociate into monomers is taken to indicate a low affinity for endogenous self peptides. Other studies implicate low affinity interactions as a possible route to autoimmunity through failure of self tolerance (LIU et al. 1995). An immunodominant epitope of Hsp60, p277, induces strong T cell responses in NOD mice yet has immeasurably low affinity for the class II molecule which presents it, H2-A^{g7} (REIZIS et al. 1997a; CARRASCO-MARIN et al. 1996).

Histological examination of pancreata from pre-diabetic NOD mice and patients reveals large numbers of infiltrating mononuclear cells, a process termed insulitis. In NOD mice, as in patients, inflammatory beta cell destruction results in the breakdown of glucose homeostasis due to insufficient insulin production. During the 1980's there was much interest in characterising the cellular infiltrate in order to define the effector cells responsible for beta cell destruction. Immunohistochemistry of the lesions demonstrated that the majority of the cells were $CD4^+$ or $CD8^+$ T cells but that macrophages, B cells and monocytes were also present (Miyazaki et al. 1985).

Evidence for T cell involvement in diabetogenesis is provided by three major observations. Neonatal thymectomy, T cell specific immunosuppression and antibody mediated T cell suppression all prevent disease (reviewed by Bach 1994). Additionally, NOD-scid mice, which fail to produce mature T and B lymphocytes due to a nonsense mutation in the DNA-dependent protein kinase catalytic subunit gene required for V(D)J recombination, show neither insulitis nor diabetes (Danska et al. 1996). Direct evidence for T cell involvement comes from adoptive transfer experiments in which neonatal or sublethally irradiated adult mice receive cells. The earliest studies of this sort showed that splenocytes from diabetic NOD mice were capable of transferring diabetes to recipient pre-diabetic mice within days of transfer (Bedossa et al. 1989; Wicker et al. 1986). Further investigations have characterised the splenocyte population that is responsible for disease transfer. Purified populations of T cells from splenocytes containing both $CD4^+$ and $CD8^+$ cells were capable of transferring disease. T cell populations from diabetic mice depleted of $CD8^+$ cells are capable of transferring diabetes to NOD-scid mice but with lower incidence and increased onset time when compared with mixed $CD4^+$ and $CD8^+$ T cell populations. This result was not replicated when cells from young, pre-diabetic donor mice were used; there was a requirement for both $CD4^+$ and $CD8^+$ cells. This led to the hypothesis that $CD8^+$ cells are necessary for the development of a diabetogenic T cell response, but that once the response is developed, these cells are not required for destruction of the pancreas (Rohane et al. 1995). Further evidence to support this hypothesis has come from numerous experiments utilising T cell clones. There have been many different adoptive transfer experiments with T cell clones utilising various different antigen specificities both known and unknown. Some $CD4^+$ clones recognising an array of different beta cell antigens are capable of transferring diabetes. However, no $CD8^+$ clones capable of transferring diabetes without $CD4^+$ T cell help have been reported. In any case, the requirement for $CD8^+$ cells cannot be absolute as T cell receptor transgenic mice expressing only $CD4^+$ T cells of a single specificity can develop spontaneous, accelerated diabetes in the absence of any other rearranged T cell receptors (Verdaguer et al. 1997).

The mechanism by which this damage is mediated is unclear. Beta cells are known to be very sensitive to inflammatory cytokine production and it maybe that this alone is sufficient to cause cell death by apoptosis. Another potential mechanism is offered by the observation that pancreatic beta cells can be driven to apoptose after signalling through Fas. Expression of Fas by beta cells is up-regulated by the inflammatory cytokine interleukin 1 beta. Beta cells from newly diagnosed type I diabetes patients express Fas but those from normal controls do not. These Fas expressing beta cells are in close proximity to Fas ligand expressing T cells. In support of this, NOD$^{lpr/lpr}$ mice, which express a mutated Fas receptor, are resistant to both spontaneous and transferred diabetes (CHERVONSKY et al. 1997).

Activated CD4+ T cells kept in long term culture differentiate into two broad groups known as T helper 1 and T helper 2 cells. This differentiation is known also to occur in vivo. T helper 1 cells secrete large amounts of IFNγ and IL-2 and are primarily associated with cellular immunity whilst T helper 2 cells secrete interleukins 4 and 10 and are mainly involved in humoral immunity (MOSMANN and COFFMAN 1989). There is evidence from many animal models of autoimmune disease, studies of patient T cell responses and immunohistochemistry of patient biopsies to support the view that T helper 1 cells promote autoimmune disease and that T helper 2 responses can be protective. Infiltrating islet cells are predominantly of a T helper 1 phenotype in both NOD mice and human patients (HEALEY et al. 1995; RABINOVITCH et al. 1996). A transgenic system has been used to generate diabetogenic CD4+ T cells bearing the same TCR but with different effector phenotypes. This was achieved by T cell activation in different cytokine milieus. Adoptive transfer of T helper 1 or T helper 2 cells resulted in insulitis but only T helper 1 cells were able to transfer diabetes to recipient mice (KATZ et al. 1995). This data, in conjunction with the indirect evidence from animal studies and patients, provides evidence that T helper 1 cells actively promote disease but provides no evidence that T helper 2 responses are protective.

D. Cell Types Required for Diabetogenesis in Patients

The poor availability of clinical specimens, coupled with the fact that insulitis is clinically silent, means that less is known about the cells needed for diabetogenesis in patients than in animal models of diabetes. The observations that have been made are similar to observations from studies in the NOD mouse suggesting that the mechanism of diabetogenesis is similar. Histological examination of a post-mortem pancreas obtained from a recently diagnosed diabetic patient revealed large numbers of infiltrating mononuclear cells including T and B lymphocytes. Pancreas biopsies from newly diagnosed patients show insulitis in the majority of patients comprised predominately of CD8+ T cells and macrophages with some CD4+ T cells and B lymphocytes (ITOH et al. 1993). Further evidence for lymphocyte involvement in insulitis and diabetogenesis is provided by observations following rejection of a HLA identical pancreas allograft by a diabetic recipient. The infiltrate was a mixed cell population comprised predominantly of CD8+ alpha/beta TCR positive cells. In vitro stimulation with anti-CD3 antibody resulted in the growth of CD8+ alpha/beta TCR positive cells, CD4+ alpha/beta TCR positive cells and CD4-/CD8- gamma/delta TCR positive cells (Santamaria et al. 1992). Similar cell populations could not be purified from insulitis free biopsies.

Flow cytometric analysis of patient peripheral blood samples has shown that pre-diabetic and newly diagnosed individuals show higher levels of interleukin-2 receptor positive T cells than disease free first degree relatives (Peakman et al. 1994). A proportion of T cell clones isolated from recently diagnosed diabetes patients are able to proliferate in a dose dependent manner to islet cell extracts. These potentially autoreactive T cells are absent from diabetes free first degree relatives.

E. Islet Cell Antibody Responses

In the mid 1970s many workers focused on the humoral immunity observed in diabetic patients, Bio-breeding (BB) rats and NOD mice in an attempt to define pancreatic beta cell antigens involved in the development of type I diabetes. Typically experiments were performed by purifying immunoglobulin from diabetic and control patient or animal sera and comparing the proteins immunoprecipitated or immunoblotted by the two groups. This experimental approach was successful and has led to the characterisation of a number of autoantigens in patients and animal models.

Spontaneous antibody responses are found in NOD mice to a number of different proteins including *Mycobacterium tuberculosis* heat shock protein 60, the p277 peptide from human heat shock protein 60, glutamic acid decarboxylase 65 and insulin (De-Aizpurua et al. 1994; Elias et al. 1990). The role that autoantibodies play in the development of disease is unclear, but it is known that they are not necessary for pathogenesis. A proportion of NOD mice made B cell deficient with a targeted mutation develop insulitis and diabetes (Yang et al. 1997).

Glutamic acid decarboxylase (GAD) is an enzyme with two isoforms, molecular weights 65 and 67 kDa, which result from differential splicing. GAD 65 and 67 both catalyse the production of the neurotransmitter gamma-aminobutyric acid. GAD protein expression is not ubiquitous and the two isoforms show differential expressions patterns suggesting that there are as yet unknown functional differences between them. Studies on murine message RNA show GAD 65 expression in testis, epithelial cells of the fallopian tube, pancreatic beta cells and brain. GAD 67 expression was observed in the same tissues as GAD 65 but was additionally observed in the spleen and liver. GAD 67 message is more abundant than GAD 65 message in the pancreas (Faulkner-Jones et al. 1993). There have been multiple reports demonstrating that sera from diabetic patients can immunoprecipitate or be used to immunoblot a 64 kDa antigen. After some initial confusion the protein was shown to be glutamic acid decarboxylase (reviewed by Cooke 1994). The incidence of anti-glutamic acid decarboxylase antibodies in diabetic patients is elevated when compared with non-diabetic first degree relatives suggesting that the response is a result of diabetogenesis rather than environmental stimulus. It has been observed that there are regions of glutamic acid decarboxylase which have sequence homology with the 2 C protein from coxsackievirus B4; however, no cross reactive spontaneous antibody or T cell responses have been observed (Hou et al. 1994). There is, therefore, no evidence that this phenomena is significant in the development of type I diabetes. Examination of the humoral response produced by a newly diagnosed diabetic patient has shown that the majority of the antibody response to glutamic acid decarboxylase is centred around the carboxy terminus of the molecule between amino acid residues 450–570 with a few determinants being found in other parts of the molecule. High titres of anti-glutamic acid decarboxylase antibodies are also observed in patients with Stiff man syndrome but are centred around epitopes at the N-terminus of the protein. It is as yet unknown whether these antibodies play a role in the pathogenesis of either disease (Daw et al. 1996).

There are a number of other proteins which have been suggested as targets of anti-islet cell antibodies. These include insulin as well as a novel chymotrypsinogen related protein and a 115 kDa protein tyrosine phosphatase family protein. The 37/40 kDa tryptic fragment which is frequently recognised by islet cell antigen positive sera has been shown to be derived from IA-2, a 105 kDa transmembrane protein belonging to the protein tyrosine phosphatase family (Kawasaki et al. 1996).

F. Autoreactive T Cell Responses

NOD mice show spontaneous T cell autoreactivity to a number of different antigens. Hsp65 autoimmunity in the NOD mouse was first observed by Elias and her colleagues (1990). It came as a surprise to many that an antigen clearly implicated in the pathogenesis of arthritis models was also involved in the development of the diabetes. The evidence does, however, strongly suggest that Hsp60/65 is a target of autoimmune attack. There are many other antigens with credentials as autoantigens and it is likely that diabetes can result from a loss of immunological tolerance to a number of different self-antigens. Data from NOD mice and diabetic patients suggests that the autoimmune T cell response is diverse and centres on both known and currently undefined antigens.

The first data suggesting that T cell responses to members of the heat shock protein 60 family could be diabetogenic came from studies of a pancreatic beta cell antigen, as then unknown, which was immunologically cross reactive with *Mycobacterium tuberculosis* heat shock protein 65. It was observed that antibody responses to *Mycobacterium tuberculosis* heat shock protein 65, insulin and anti-insulin antibodies preceded the clinical onset of disease. This antibody production was shown to be supported by spontaneous T cell reactivity to *Mycobacterium tuberculosis* heat shock protein 65 which could be observed prior to detectable serum antibodies (Elias et al. 1990). Experimental confirmation of autoimmune T cell responses directed against mammalian heat shock protein 60 rather than *Mycobacterium tuberculosis* heat shock protein 65 followed this report. Spontaneously derived T cell clones, from NOD splenocytes, proliferated more vigorously in response to human heat shock protein 60 than when stimulated with *Mycobacterium tuberculosis* heat shock protein 65. Following this finding a large effort was made to determine which portions of murine heat shock protein 60 were recognised by NOD T cells. Murine heat shock protein 60 was cloned from a pancreatic beta cell tumour, expression constructs created containing portions of the protein and T cell responsiveness was examined (Elias et al. 1991; Birk et al. 1996b). All the detected T cell responses observed were distal to position 437. Peptides were synthesised to cover the region from residue 437 to the carboxy terminal of the molecule using sequence from human heat shock protein 60. When NOD splenic T cell responses to these peptides were analysed it became evident that the response centred around the 24 amino acids between residues 437 and 460. This peptide, encoding at least one T cell epitope, has been named p277 (Elias et al. 1991). Recent work has shown that NOD T cells can recognise epitopes other than p277; a second peptide named p12 can induce strong proliferative responses. There are no known spontaneous T cell responses to p12 and currently no evidence exists that T cells specific for epitopes within p12 are capable of inducing diabetes.

Spontaneous T cell responses have been observed to a number of antigens other than heat shock protein 60, both characterised and uncharacterised. The relevance of these targets to the disease process is often assessed either by the ability of cell transfer to accelerate disease or by the ability of specific tolerance to prevent disease. T cell clones recognising epitopes within *Mycobacterium tuberculosis* Hsp65 and murine Hsp60 are capable of inducing diabetes in adoptive transfer experiments. There are two reported CD4+ T cell clones generated by stimulation with *Mycobacterium tuberculosis* heat shock protein 65 capable of transferring diabetes. It is not known whether the cells recognise a conserved part of the molecule which is cross-reactive with murine heat shock protein 60 (Elias et al. 1990). Three clones, C9, C7 and C27 which proliferate strongly when stimulated with human Hsp60 have been shown to be diabetogenic when transferred into NOD mice. The C9 clone has been extensively characterised; the epitope it recognises lies within the p277 peptide derived from human heat shock protein 60 (Elias et al. 1991). This peptide is conserved between mice and humans with only one amino acid residue difference and the clone is stimulated by either peptide.

Insulin is a beta cell specific protein and constitutes approximately eighty percent of total beta cell protein. This fact in itself makes it an attractive autoantigen. One study has shown that the majority of spontaneously arising early infiltrating T cells in young NOD mice proliferate in response to stimulation with insulin (WEGMANN et al. 1994). The key region of the insulin molecule recognised by the majority of CD4+ T cell clones encompasses residues 9 to 23 of the B chain (DANIEL et al. 1995). CD4+ T cell clones specific for this epitope are capable of transferring disease during adoptive transfer experiments with young NOD mice or NOD *scid* mice.

Spontaneous T helper 1 responses to GAD 65 are detected in some but not all studies of NOD mice. The spontaneous T cell response is initially centred around the carboxy terminal of glutamic acid decarboxylase 65, residues 509 to 543. As the mice age there is evidence of some epitope spread to other parts of the protein with strong responses being seen to residues 247 to 266 by the time the mice reach seven weeks of age (KAUFMAN et al. 1993). Longitudinal analysis has shown that these responses are first observed at the same time as the onset of insulitis. A study analysing the spontaneous appearance of T cell responses against GAD 65, GAD 67 and Hsp60 suggested that GAD 65 T cell responses precede Hsp60 T cell responses by approximately 14 days (TISCH et al. 1993).

G. T Cell Clones of Unknown Antigen Specificity

There are a large number of clones that are diabetogenic in adoptive transfer experiments for which the antigen specificity is unknown. The first reported diabetogenic T cell clone, BDC-2.5, was a CD4+ T cell clone derived from splenocytes from a newly diabetic NOD mouse cultured in the presence of NOD islet cells and irradiated splenocytes as antigen presenting cells. In vivo the clone was shown to be able to induce islet but not pituitary graft destruction (HASKINS et al. 1988). After adoptive transfer into young NOD mice BDC-2.5 and another islet antigen specific clone BDC-6.9, were shown to be able to produce extensive insulitis and diabetes (HASKINS and McDUFFIE 1990).

Many other groups isolated T cell clones from pre-diabetic NOD mice. The majority of workers used infiltrating cells as a starting cell population. Many of these T cell clones have been shown to induce insulitis; some also induce diabetes after adoptive transfer. The antigen specificity for these clones remains elusive. Extensive characterisation of the BDC series of clones by HASKINS and coworkers found that much of the activity is localised to the beta cell granule membrane (BERGMAN and HASKINS 1994).

H. Evidence from Suppression of Specific T Cell Responses in NOD Mice

There are a number of experimental protocols which have been used to induce T cell tolerance in NOD mice. All of these protocols expose the immune system to antigen in a non-inflammatory environment. It has been known for many years that the delivery of soluble antigen directly to the spleen by intravenous or intra-peritoneal injection favours tolerance induction. The same is true of immunisation across the nasal mucosa. The precise mechanism by which these protocols act are unknown. Most current hypothesis suggest that differences in the site and manner of antigen presentation induces an alteration in cytokine production, or the stimulation of regulatory cells, resulting in suppression of the immune response.

Manipulation of the immune response to Hsp60 has been clearly shown to inhibit the onset of diabetes in NOD mice. Pre-diabetic NOD mice immunised intra-peritoneally with the p277 peptide emulsified in incomplete Freund's adjuvant show significant protection from spontaneous disease. Thirty percent of p277 treated mice were diabetic at 8 months of age compared with 80% of irrelevant peptide treated mice. The protective effect was also evident in adoptive transfer experiments. The C9 clone, which recognises an epitope within the p277 peptide, is capable of transferring diabetes in adoptive transfer experiments. Mice pre-treated with p277, as in the previous experiment, were protected from diabetes (none was observed) whilst non-treated mice showed eighty percent diabetes after T cell transfer (ELIAS et al. 1991). Fragments of the p277 peptide are capable of partially protecting NOD mice but optimal therapy requires the whole molecule. Successful therapy is associated with a down-regulation of T cell responses to p277 and a burst of T helper 2 cytokines; splenic T cells from treated mice produce interleukins 4 and 10, whilst T cells from mice prior to treatment or from untreated controls produce IFNγ. This change in cytokine production is associated with the production p277 specific antibodies of the IgG1 subclass, presumably induced by the T helper 2 cytokines (ELIAS and COHEN 1995; ELIAS et al. 1997).

Mice have been generated which express murine Hsp60 under the control of a major histocompatibility class II promoter. This would be expected to increase protein expression in antigen presenting cells, including those of the thymus, resulting in increased negative selection and T cell tolerance. However, these mice do not manifest total T cell tolerance; anti-Hsp60 T cell responses are readily observed. There is, nevertheless, a delay in diabetes onset and insulitis. Fewer than 20% of transgenic mice show elevated blood glucose at 9 months of age compared with 80% of non-transgenic littermates (BIRK et al. 1996a).

There are a number of different reports showing that the administration of insulin orally or subcutaneously leads to a decrease in the incidence of diabetes in NOD mice and BB rats. Insulin B chain alone in incomplete Freund's adjuvant confers equal protection to whole insulin. The observation that a region of the B chain, residues 9–23, stimulates the majority of spontaneously arising insulin reactive T cell clone from NOD mice has led to analysis of the effects of treatment with this peptide (WEGMANN et al. 1994). Subcutaneous injection with B-(9–23) in incomplete Freund's adjuvant decreases diabetes incidence in NOD mice from 90% at 40 weeks of age to 10%. This protection was not dependent on the route of peptide administration; intranasal dosing also resulted in delayed diabetes onset (DANIEL and WEGMANN 1996).

Transgenic antigen expression of insulin has been used to explore the role of insulin auto-reactivity in NOD mouse diabetogenesis. As with the heat shock protein 60 transgenic mice, proinsulin II expression was driven by a major histocompatibility class II promoter. The data from this experiment were similar to the data from the Hsp60 transgenic mouse; total T cell tolerance was not observed but reduced insulitis was nevertheless present (FRENCH et al. 1997).

There are a number of reports demonstrating GAD 65 specific protection from diabetes although there is as yet no direct evidence that glutamic acid decarboxylase restricted T cells are capable of transferring diabetes to recipient mice. Intravenous injection of 3 week old NOD mice with GAD 65 in phosphate buffered saline reduces insulitis and results in loss of the spontaneous T cell responses to GAD 65 and Hsp60. These T cell responses were present in mice manipulated with a control antigen. Intrathymic injection with GAD 65 results in a significant delay in the onset of diabetes demonstrating that treatment can inhibit diabetes as well as insulitis (TISCH et al. 1993). Inhibition of the T cell response to GAD 65 by injection of GAD 65 in incomplete Freund's adjuvant causes a decrease in diabetes incidence and a switching of the T cell response from a T helper 1 to a T helper 2 response. Additionally, modulation of the immune response in this manner allows increased survival of syngeneic islet grafts in diabetic NOD mice (TIAN et al. 1996).

I. Concluding Remarks

Workers searching for the perceived holy grail of a single diabetes related autoantigen have had some good news and some bad news. The bad news is that we have progressed from a lack of candidate antigens to an excess. It is difficult to rationalise that a heterogeneous burst of T cell response to insulin, Hsp60, IA-2 and GAD 65 is necessary for disease. A requirement for multiple target antigens is further contradicted by the known high disease risk associated with HLA with DR3,4 alleles. This is often interpreted as evidence that a specific peptide/MHC complex is necessary to trigger disease. It has been suggested that the earliest detectable T cell response (usually believed to be centred on GAD 65) must be the real trigger and that the other observed responses are epitope spread that occurs as pathology develops. This argument does not, however, take account of the fact that the epitopes recognised by many potently diabetogenic T cell clones recognise currently unknown antigens making it impossible to place responses to these autoantigens longitudinally relative to diabetes onset. Additionally, ever more sensitive assays are being developed which may allow the characterisation of earlier responses to different antigens. The good news is that this debate may be of no therapeutic importance. Antigen specific suppression of T cell responses to Hsp60, GAD 65 and insulin have each been sufficient to inhibit diabetogenesis in

NOD mice. Clinical trails are currently underway to determine whether the same is true for diabetic patients. "Linked suppression" or tolerance to one target peptide inhibiting T cell responses to other antigens in the same tissue is well characterised. The prospect of Hsp60 immunotherapy of patients is thus a real one.

References

Bach J (1994) Insulin-dependent diabetes mellitus as an autoimmune disease. Endocr Rev 15:516–542

Bedossa P, Bendelac A, Bach J, Carnaud C (1989) Syngeneic T cell transfer of diabetes into NOD newborn mice: in situ studies of the autoimmune steps leading to insulin-producing cell destruction. Eur J Immunol 19:1947–1951

Bergman B, Haskins K (1994) Islet-specific T-cell clones from the NOD mouse respond to beta-granule antigen. Diabetes 43:197–203

Birk O, Douek D, Elias D, Takacs K, Dewchand H, Gur S, Walker M., van der Zee R, Cohen I, Altmann D (1996a) A role of Hsp60 in autoimmune diabetes: analysis in a transgenic model. Proc Natl Acad Sci USA 93:1032–1037

Birk O, Elias D, Weiss A, Rosen A, van der Zee R, Walker M, and Cohen I (1996b) NOD mouse diabetes: the ubiquitous mouse Hsp60 is a beta-cell target antigen of autoimmune T cells. J Autoimmun 9:159–166

Carrasco-Marin EJ, Kanagawa O, Unanue E (1996) The class II MHC I-Ag7 molecules from non-obese diabetic mice are poor peptide binders. J Immunol 156:450–458

Chervonsky A, Wang Y, Wong F, Visintin I, Flavell R, Janeway C, Mathis A (1997) The role of Fas in autoimmune diabetes. Cell 89:17–24

Cohen IR, Young DB (1991) Autoimmunity, microbial immunity and the immunological homunculus. Immunol Today 12:100–110

Cooke A (1994) Autoimmune disease. Gadding around the beta cell. Curr Biol 4:158–160

Daniel D, Gill R, Schloot N, Wegmann D (1995) Epitope specificity, cytokine production profile and diabetogenic activity of insulin-specific T cell clones isolated from NOD mice. Eur J Immunol 25:1056–1062

Daniel D, Wegmann D (1996) Protection of nonobese diabetic mice from diabetes by intranasal or subcutaneous administration of insulin peptide B-(9–23). Proc Natl Acad Sci USA 93:956–960

Danska J, Holland D, Mariathasan S, Williams K, Guidos C (1996) Biochemical and genetic defects in the DNA-dependent protein kinase in murine scid lymphocytes. Mol Cell Biol 16:5507–5517

Daw K, Ujihara N, Atkinson M, Powers A (1996) Glutamic acid decarboxylase autoantibodies in stiff-man syndrome and insulin-dependent diabetes mellitus exhibit similarities and differences in epitope recognition. J Immunol 156:818–825

de Graeff Meeder E, Rijkers G, Voorhorst M, Kuis W, van der Zee R, van Eden W, Zegers B (1993) Antibodies to human HSP60 in patients with juvenile chronic arthritis, diabetes mellitus, and cystic fibrosis. Pediatr Res 34:424–428

De-Aizpurua H, Chosich N, Harrison L (1994) Natural history of immunity to glutamic acid decarboxylase in non-obese diabetic (NOD) mice. J Autoimmun 7:643–653

Elias D, Markovits D, Reshef T, van der Zee R, Cohen IR (1990) Induction and therapy of autoimmune diabetes in the non-obese diabetic (NOD/Lt) mouse by a 65-kDa heat shock protein. Proc Natl Acad Sci USA 87:1576–1580

Elias D, Reshef T, Birk O, van der Zee R, Walker M, Cohen I (1991) Vaccination against autoimmune mouse diabetes with a T-cell epitope of the human 65-kDa heat shock protein. Proc Natl Acad Sci USA 88:3088–3091

Elias D, Cohen IR (1995) Treatment of autoimmune diabetes and insulitis in NOD mice with heat shock protein 60 peptide p277. Diabetes 44:1132–1138

Elias D, Meilin A, Ablamunits V, Birk O, Carmi P, Konen Waisman S, Cohen IR (1997) Hsp60 peptide therapy of NOD mouse diabetes induces a Th2 cytokine burst and downregulates autoimmunity to various beta-cell antigens. Diabetes 46:758–764

Faulkner-Jones B, Kun J, Harrison L (1993) Localization and quantitation of expression of two glutamate decarboxylase genes in pancreatic beta-cells and other peripheral tissues of mouse and rat. Endocrinology 133:2962–2972

French M, Allison J, Cram D, Thomas H, Dempsey Collier M, Silva A, Georgiou H, Kay T, Harrison L, Lew A (1997) Transgenic expression of mouse proinsulin II prevents diabetes in nonobese diabetic mice. Diabetes 46:34–39

Handley H, Yu J, Yu D, Singh B, Gupta R, Vaughan J (1996) Autoantibodies to human heat shock protein (hsp) 60 may be induced by Escherichia coli groEL. Clin Exp Immunol 103:429–435

Haskins K, McDuffie M (1990) Acceleration of diabetes in young NOD mice with a CD4+ islet-specific T cell clone. Science 249:1433–1436

Haskins K, Portas M, Bradley B, Wegmann D, Lafferty K (1988) T-lymphocyte clone specific for pancreatic islet antigen. Diabetes 37:1444–1448

Healey D, Ozegbe P, Arden S, Chandler P, Hutton J, Cooke A (1995) In vivo activity and in vitro specificity of CD4+ Th1 and Th2 cells derived from the spleens of diabetic NOD mice. J Clin Invest 95:2979–2985

Holoshitz J, Naparstek Y, Ben-Nun A, Cohen IR (1983) Lines of T lymphocytes induce or vaccinate against autoimmune arthritis. Science 219:56–58

Hou J, Said C, Franchi D, Dockstader P, Chatterjee N (1994) Antibodies to glutamic acid decarboxylase and P2-C peptides in sera from coxsackie virus B4-infected mice and IDDM patients. Diabetes 43:1260–1266

Itoh N, Hanafusa T, Miyazaki A, Miyagawa J, Yamagata K, Yamamoto K, Waguri M, Imagawa A, Tamura S, Inada M et al (1993) Mononuclear cell infiltration and its relation to the expression of major histocompatibility complex antigens and adhesion molecules in pancreas biopsy specimens from newly diagnosed insulin-dependent diabetes mellitus patients. J Clin Invest 92:2313–2322

Katz J, Benoist C, Mathis D (1995) T helper cell subsets in insulin-dependent diabetes. Science 268:1185–1188

Kaufman D, Tian J, Forsthuber T, Ting G, Robinson P, Atkinson M, Sercarz E, Tobin M, Lehmann P (1993) Spontaneous loss of T cell tolerance to glutamic acid decarboxylase in murine insulin dependent diabetes mellitus. Nature 366:69–72

Kaufmann S, Schoel B. van Embden J, Koga T, Wand-Wurttenberger A, Munk M, Steinhoff U (1991) Heat-shock protein 60: implications for pathogenesis of and protection against bacterial infections. Immunol Rev 121:67–90

Kawasaki E, Eisenbarth G, Wasmeier C, Hutton J (1996) Autoantibodies to protein tyrosine phosphatase-like proteins in type I diabetes. Overlapping specificities to phogrin and ICA512/IA-2. Diabetes 45:1344–1349

Liu GY, Fairchild PJ, Smith RM, Prowle JR, Kioussis D, Wraith DC (1995) Low avidity recognition of self-antigen by T cells permits escape from central tolerance. Immunity 3:407–415

Miyazaki A, Yamada K, Miyagawa J, Fujino-Kurihara H, Nakajima H, Nonaka K, Tarui S (1985) Predominance of T lymphocytes in pancreatic islets and spleen of pre-diabetic non-obese diabetic (NOD) mice: a longitudinal study. Clin Exp Immunol 60:622–630

Mosmann T, Coffman RL (1989) Th1 and Th2 cells: different patterns of lymphokine secretion lead to different functional properties. Annu Rev Immunol 7:145–173

Moudgil K, Chang T, Eradat H, Chen A, Gupta R, Brahn E, Sercarz (1997) Diversification of T cell responses to carboxy-terminal determinants within the 65-kD heat-shock protein is involved in regulation of autoimmune arthritis. J Exp Med 185:1307–1316

Munk M, Schoel B, Modrow S, Karr R, Young R, Kaufmann S (1989) T lymphocytes from healthy individuals with specificity to self-epitopes shared by the mycobacterial and human 65-kilodalton heat shock protein. J Immunol 143:2844–2849

O'Reilly LA, Healey D, Simpson E, Chandler P, Lund T, Ritter MA, Cooke A (1994) Studies on the thymus of non-obese diabetic (NOD) mice: effect of transgene expression. Immunology 82:275–286

Peakman M, Wen L, McNab GL, Watkins PJ, Tan KC, Vergani D (1994) T cell clones generated from patients with type 1 diabetes using interleukin-2 proliferate to human islet antigens. Autoimmunity 17:31–39

Prakken AB, van Eden W, Rijkers GT, Kuis W, Toebes EA, de Graeff Meeder ER, van der Zee R, Zegers BJ (1996) Autoreactivity to human heat-shock protein 60 predicts disease remission in oligoarticular juvenile rheumatoid arthritis. Arthritis Rheum 39:1826–1832

Rabinovitch A, Suarez Pinzon WL, Strynadka K, Lakey JR, Rajotte RV (1996) Human pancreatic islet beta-cell destruction by cytokines involves oxygen free radicals and aldehyde production. J Clin Endocrinol Metab 81:3197–3202

Reizis B, Bockova J, Konen-Waisman S, Mor F, Cohen I (1997a) Molecular characterization of the diabetes-associated mouse MHC class II protein, I-Ag7. Int Immunol 9:43–51

Reizis B, Altmann DM, Cohen IR (1997b) Biochemical characterization of the human diabetes-associated HLA-DQ8 allelic product: similarity to the major histocompatibility complex class II I-A^{g7} protein of non-obese diabetic mice. Eur J Immunol 27:2478–2484

Rohane PW, Shimada A, Kim DT, Edwards CT, Charlton B, Shultz LD, Fathman CG (1995) Islet-infiltrating lymphocytes from prediabetic NOD mice rapidly transfer diabetes to NOD-scid/scid mice. Diabetes 44:550–555

Santamaria P, Nakhleh RE, Sutherland DE, Barbosa JJ (1992) Characterization of T lymphocytes infiltrating human pancreas allograft affected by isletitis and recurrent diabetes. Diabetes 41:53–61

Tian J, M C-S, Herschenfeld A, Middleton A, Newman D, Mueller R, Arita S, Evans C, Atkinson M, Mullen Y, Sarvetnick N, Tobin A, Lehmann P, Kaufamn D (1996) Modulating autoimmune responses to GAD inhibits disease progression and prolongs islet graft survival in diabetes-prone mice. Nature Medicine 2:1348–1353

Tisch R, Singer S, Liblau R, Fugger L, McDevitt H (1993) Immune response to glutamic acid decarboxylase correlates with insulitis in non-obese diabetic mice. Nature 366:72–75

Todd JA, Bell JI, McDevitt HO (1987) HLA-DQ beta gene contributes to susceptibility and resistance to insulin dependent diabetes mellitus. Nature 329:599–604

van Eden W, Hogervorst EJ, van der Zee R, van Embden JD, Hensen EJ, Cohen IR (1989) The mycobacterial 65 kD heat-shock protein and autoimmune arthritis. Rheumatol Int 9:187–191

van Eden W, Holoshitz J, Nevo Z, Frenkel A, Klajman A, Cohen IR (1985) Arthritis induced by a T-lymphocyte clone that responds to Mycobacterium tuberculosis and to cartilage proteoglycans. Proc Natl Acad Sci USA 82:5117–5120

van Eden W, Thole JE, van der Zee R, Noordzij A, van Embden JD, Hensen EJ, Cohen IR (1988) Cloning of the mycobacterial epitope recognized by T lymphocytes in adjuvant arthritis. Nature 331:171–173

van Roon JA, van Eden W, van Roy JL, Lafeber FJ, Bijlsma JW (1997) Stimulation of suppressive T cell responses by human but not bacterial 60-kD heat-shock protein in synovial fluid of patients with rheumatoid arthritis. J Clin Invest 100:459–463

Verdaguer JSD, Amrani A, Anderson B, Averill N, Santamaria P (1997) Spontaneous autoimmune diabetes in monoclonal T cell nonobese diabetic mice. J Exp Med 186:1663–1677

Wegmann DR, Norbury-Glaser M, Daniel D (1994) Insulin-specific T cells are a predominant component of islet infiltrates in pre-diabetic NOD mice. Eur J Immunol 24:1853–1858

Wicker LS, Miller BJ, Mullen Y (1986) Transfer of autoimmune diabetes mellitus with splenocytes from nonobese diabetic (NOD) mice. Diabetes 35:855–860

Wicker LS, Todd JA, Peterson LB (1995) Genetic control of autoimmune diabetes in the NOD mouse. Annu Rev Immunol 13:179–200

Yang M CB, Gautam AM (1997) Development of insulitis and diabetes in B cell deficient NOD mice. J Autoimmun 10:257–260

CHAPTER 17
Heat Shock Proteins and Multiple Sclerosis

G. Ristori, C. Montesperelli, D. Kovacs, G. Borsellino, L. Battistini,
C. Buttinelli, C. Pozzilli, C. Mattei, and M. Salvetti

A. Introduction

Multiple sclerosis is an inflammatory disease of the central nervous system
(CNS) with myelin damage (loss of myelin sheaths and eventually secondary
axon damage). In most of cases it evolves from a relapsing-remitting phase to
a chronic progressive one which brings about stable disability and handicap. In
spite of the scanty certainties about its etiology, the pathogenesis of MS is
widely accepted to be autoimmune based on:

1. A potentially disregulated immune response in patients compared to
   controls (the presence of immunocompetent cells not only inside the MS
   lesions but also as autoreactive, "in-vivo"-activated effectors in the peripheral blood or in the cerebrospinal fluid of patients; the presence of immunoglobulin oligoclonal bands in the cerebrospinal fluid of patients; the
   association with "at risk" MHC haplotypes in patients).
2. The pathogenetic role of such autoreactive effectors in studies on the
   animal model of MS, experimental allergic encephalomyelitis (EAE); in
   rodents and in primates, the encephalitogenicity of myelin-specific T
   lymphocytes (alone or in combination with myelin-specific antibodies) has
   been demonstrated.
3. Finally, and most importantly, the therapeutic efficacy of strategies aimed
   at deleting or antagonizing the pathogenic autoaggressive effectors; most of
   the treatments, recently implemented in MS with encouraging results, tend
   to modulate, more or less specifically, the disregulated immune response of
   the patients. Examples of such therapies include interferon β, copolymer-1
   and intravenous high-dose immunoglobulins.

As in other organ-specific autoimmune diseases, the identification of
potential autoantigens began with the most abundant and most immunogenic
proteins associated with the target tissue. Among CNS potential autoantigens,
myelin basic protein (MBP), proteolipid protein (PLP) and, more recently,
myelin oligodendrocyte glycoprotein (MOG) have been extensively investigated and allowed, in part, to verify points 1 and 2 above. Nonetheless, studies
on the pathogenetic relevance of the above antigens are far from having
provided definitive answers: no clear cut differences have been described
between patients and controls with respect to precursor frequency, fine

specificity or other functional aspects of the T lymphocyte response to the above autoantigens. The research on the immune response to non-myelin- and non-CNS-specific (auto-) antigens may contribute to understanding the mechanisms underlying the chronicity and progression of the immunopathological process. The monophasic course of acute, immune-mediated encephalomyelitides has recently found a biological explanation in the great number of protective mechanisms that may halt the progression of the autoimmune process: suppressor cytokines and immune deviation processes, apoptosis of dangerous immunocompetent cells, the idiotypic-antiidiotypic network are all examples of defensive mechanisms that keep under control not only specific autoaggressive effectors, but also "bystander" ones. These highlight the importance of regarding chronic autoimmune disorders as pathological conditions in which it is relevant not only to identify the primary autoantigen or the primary event leading to the autoaggressive reaction, but also (and perhaps especially) to disclose the default of those protective processes that halt the progression toward stable autoimmunity. In this context heat shock proteins (Hsps) offer at least two starting points for investigation: they are ubiquitous, strong immunogens and may represent potential non-CNS-specific autoantigens; moreover the original definition of Hsps as chaperones and protective factors in stressful conditions is of obvious interest in disorders of a chronic inflammatory nature.

The rationale for investigating the Hsps as potential autoantigens in multiple sclerosis (MS) is common to other immune-mediated disorders. Hsps are the most conserved molecules known to date. Furthermore they are also relevant immune targets during infection. Hence, there is a theoretically high chance of cross-reactive responses to epitopes shared by host and microbe Hsps. This would suggest the possibility that Hsps trigger an autoaggressive response through a mechanism of molecular mimicry. Moreover, as stress-induced proteins, Hsps are upregulated antigens during inflammatory processes, being ideal candidates for sustaining the immunopathological loop through the so-called intra- and intermolecular spreading (an autoaggressive T lymphocyte response that initiates against one epitope on an autoantigen and then, as the tissue damage progresses and other epitopes/antigens become exposed in a pro-inflammatory context, spreads to other determinants on the same or on other autoantigens within the target organ).

The potential protective functions of Hsps have not been as extensively investigated as their antigenic role. However, interesting findings came from studies on Hsps as chaperones for myelin proteins or as protective factors against immune-mediated injury. In this context studies on Hsp gene polymorphisms (especially those of the 70-kDa family that are included in the human MHC class III genes) began to be carried out to define haplotypes that may confer protection or susceptibility to MS.

Finally, indirect evidence of the importance of Hsps in the pathogenesis of MS comes from studies on $\gamma\delta$ T cells. These lymphocytes are a subset (from 1% to 10% in the peripheral blood of healthy individuals) of T cells whose function and specificity are ill-defined in spite of more than a decade of research. They are thought to represent a first line of defense, at the boundaries between the innate (phylogenetically ancient, non-antigen-specific reactivity against pathogens carried out by macrophages, NK cells and other effectors) and the adaptive immune response (phylogenetically recent, based on the selection of antigen-specific T cells). Their antigen specificity seems to be directed primarily against non-proteinaceous moieties. Nonetheless, among canonical protein antigens, Hsps (particularly the 60-kDa family) may be target molecules of this elusive subset of T cells. If $\gamma\delta$ T cells are implicated in the pathogenesis of MS as suggested by various reports, Hsps may be involved as autoantigens recognized by these lymphocytes.

B. Hsp Expression in Inflammatory, Demyelinating Diseases of the Brain

The cellular distribution of Hsp expression in the brain is reviewed elsewhere in this book. Suffice here to say that it is compatible with Hsps being target antigens in a disease of the cerebral white matter. Nonetheless, given the complex events that regulate Hsp expression in different cell types, depending on the intensity and duration of various stress events, the induction of Hsp has been verified in glial cells following proinflammatory stimuli and during autoimmune inflammatory conditions of the brain.

I. Hsp Expression in Glial Cells

While there is consensus about the expression of Hsp60 on cultured oligodendrocytes and not on cultured astrocytes, data about the induction of this class of Hsp following cytokine exposure are not available yet. The role of cytokines in Hsp induction on glial cells has been investigated in detail with stress proteins of the 70-kDa family (D' Souza et al. 1994). Hsp70 is induced on mixed glial cultures following IL-1a, IFNγ and TNFα exposure. However, in purified oligodendrocytes, only IL-1a was able to induce Hsp70. An IL-1 receptor antagonist abrogated this effect as well as that due to IFNγ and TNFα on mixed glial cultures, suggesting that the final pathway for the induction of Hsp70 expression on glial cells in the presence of proinflammatory stimuli is an IL-1 mediated event. This was supported by the detection of IL-1 receptors on oligodendrocytes. With respect to these findings, it is interesting to notice that myelin basic protein-Hsc70 complexes exist in oligodendrocytes (AQUINO et al. 1996). This observation may have implications both for the failure of attempts to remyelinate in MS lesions as well as for myelin damage (see also below on the role of Hsp as molecule involved in antigen presentation).

IL-1 also induces Hsp27 on human fetal astrocytes (BROSNAN et al. 1996). Others (SATOH and KIM 1995) were not able to demonstrate this induction on human astrocytes following IL-1 or TNFα but noted a marked increase in the extent of phosphorylation (from the coexpression of the nonphosphorylated and monophosphorylated forms in unstressed astrocytes to the diphosphorylated form, without modifications of the cellular distribution of the protein content).

II. Experimental Allergic Encephalomyelitis

Much of the data available on Hsp expression in inflammatory, demyelinating diseases of the brain come from the extensive work of the group of Celia Brosnan and Cedric Raine. Hsp60, Hsp70 and Hsp27 are the best studied families of stress proteins. This choice reflects data from basic immunology indicating that Hsp60 and Hsp70 are major immunogens and data on the T cell response to αB-crystallin, Hsp60 and Hsp70 in EAE and in MS that provided evidence of a disregulated reactivity against these antigens (see below).

Experimental allergic encephalomyelitis (EAE) is the accepted animal model for MS. It can be induced following immunization with myelin antigens in complete Freund's adjuvant in various animals including mice, rats and non-human primates. In addition to immunization in adjuvant (so-called "active transfer" of the disease), EAE can be obtained by injecting syngeneic myelin-specific T cells ("passive transfer" of the disease).

Hsp expression in EAE has been investigated in mice and rats. In the mouse model, SJL mice have been studied for the expression of Hsp60 following passive transfer of MBP-specific T lymphocytes (Gao et al. 1995; Brosnan et al. 1996). In these animals the course of the disease mimics multiple sclerosis, being a relapsing-remitting illness that begins with an acute episode followed by exacerbations and remissions. During the acute onset, Hsp60 immunoreactivity defines an increased expression of the protein associated with inflammatory cells at lesion sites. In the more chronic phase that follows, Hsp60 expression is detected in oligodendrocytes and astrocytes in areas adjacent to the inflammatory infiltrates. The intracellular localization of the protein in the diseased animals is not exclusively mitochondrial. In lesioned areas Hsp60 is found in the cytosol as well as on the cell surface of inflammatory cells. These findings may have implications in terms of modulation of the inflammatory response. During immune responses to infectious agents, leukocytes can express Hsps that can become targets of cytotoxic cells (Koga et al. 1989). It is possible that a cytotoxic reactivity to Hsp60 present on the cell surface of infiltrating leukocytes helps limit the auto-aggressive process.

In the rat model, Lewis rats have been studied for the expression of the 70-kDa cognate (Hsc70) and stress inducible (Hsp70) proteins. EAE in these animals is an acute, monophasic event that develops 10–11 days after immunization with the autoantigen in complete adjuvant. The disease lasts on average 8 days. In a study by Aquino et al. (1993), Hsp70 mRNA levels and protein content did not vary throughout the disease course. Hsc70 mRNA increased approximately sixfold over control values but only a slight or no increase of protein production could be detected. In situ hybridization showed that Hsc70 mRNA was predominantly localized in neurons with some reactivity noted in glia (probably astrocytes). At variance with data on Hsp60 expression in mouse EAE, inflammatory cells were not responsible for the notable increase in Hsc70 message indicating that, in this case, a role in the downregulation of inflammation is unlikely.

III. Multiple Sclerosis

Hsp60 expression pattern in the human brain during MS is not dissimilar to the one already described in EAE (BROSNAN et al. 1996; Fig. 1). Hsp60 immuno-reactivity is present in mitochondria but also in the cytosol (RAINE 1994a, b). In active demyelinating lesions the protein expression is maximal. Cell types involved are inflammatory cells, oligodendrocytes (including proliferating oligodendrocytes at the lesion edge), reactive astrocytes, endothelial cells and microglia. In more chronic lesions Hsp60 immunoreactivity is confined to reactive astrocytes. In the white matter from other neurological diseases including AIDS, Alzheimer's disease and stroke, Hsp60 is also expressed by oligodendrocytes and astrocytes in lesioned areas of the brain though at lower levels compared to MS plaques.

At the beginning of studies on the expression of Hsp in MS, Hsp70 received less attention than Hsp60. Using an antibody that recognizes both Hsc70 and Hsp70, BROSNAN and RAINE (BROSNAN et al. 1996) have found strongly reactive astrocytes within the lesion and at the lesion edge. Also some

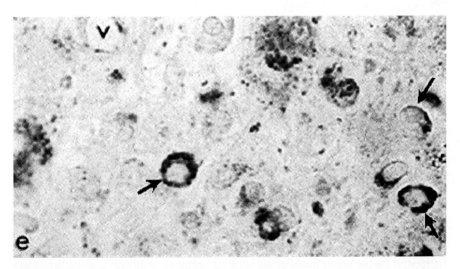

Fig. 1. Hsp60 expression in acute MS lesion. Intense punctate and cytosolic staining for Hsp60 in the oligodendrocytes (*arrows*) within the demyelinated parenchyma. Two hypertrophic astrocytes (*center left and top*), endothelial cells around a vessel (*v*) and foamy macrophages display elevated mitochondrial staining

macrophages in the perivascular infiltrate stained positively. More recently, Aquino et al. (1997) showed that even though a large proportion of Hsc70 was lost from lesioned white matter (30–50% less protein than the surrounding white matter), levels of myelin-associated Hsc70 were higher in active plaques, in the remaining myelin or in newly synthesized myelin. Hsp70 was less abundant than Hsc70. It was found in MS myelin but not in normal myelin (Fig. 2).

Much attention has been devoted to the study of Hsp27 (Fig. 3) given the important results on the T cell response to this antigen in MS (see below). Intense immunoreactivity for αB-crystallin was detected in oligodendrocytes in active lesions and in reactive astrocytes in older plaques (van Noort et al. 1995). Using a different antibody (against a 25-kDa protein isolated from breast cancer cells that showed complete sequence identity with Hsp27), others (Brosnan et al. 1996) could detect immunoreactivity in active MS lesions with reactive astrocytes, endothelial cells and axons at the edge of the lesions but no positive staining for oligodendrocytes. More recently, Aquino et al. (1997) reported a 2.5- to 4-fold enhanced Hsp27 expression in active MS lesions. Others (Bajramovic et al. 1997) have examined αB-crystallin expression at different stages of MS lesion development (using the human natural killer cell marker, HNK-1, as a marker for immature oligodendrocytes), showing that the percentage of oligodendrocytes that stained positively was 5–10% in active plaques and tenfold less in inactive lesions. There was no correlation between expression patterns of HNK-1 and αB-crystallin, indicating that this stress protein is expressed both in mature and in immature oligodendrocytes. Finally, in one study the expression of Hsp90 in MS lesions has been described (Wucherpfennig et al. 1992).

C. Immune Response to Hsps

As for other immune-mediated diseases, in MS we do not know which autoantigen(s) trigger and sustain the inflammatory process. In animal models of autoimmune disease, there is evidence of a T cell response that is initially focused on a single epitope of an autoantigen and then spreads to other epitopes on the same molecule, as well as to other determinants (Lehman et al. 1992; Kaufman et al. 1993). This paradigm may hold true also for MS: there are now a number of reports which confirm, by using different experimental approaches, a disregulated T cell response to myelin basic protein (MBP; Wucherpfennig et al. 1991; Allegretta et al. 1990; Olsson et al. 1992) and to other myelin antigens (Hohlfeld 1997). This broad T cell auto-reactivity may not be confined only to canonical myelin determinants as suggested by the encephalitogenic cellular immune response against the astrocyte-derived S-100β protein (Kojima et al. 1994) and by studies, which will be reviewed here, on the T cell response to Hsps in MS and their expression in MS lesions.

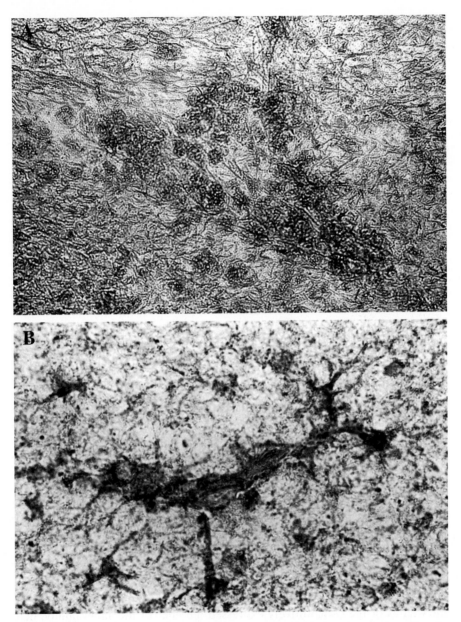

Fig. 2A,B. Hsc70 expression in MS and normal CNS tissue. **A** At the margin of a chronic active MS lesion a perivascular cuff of Hsc70+ macrophages is seen. Note also the twig-like background staining of fibrous astroglial cell processes, and in the adjacent white matter (*upper right*) some myelin staining, ×480. **B** Hsc70 expression in the white matter from a normal subject. Note the intense staining of the blood vessel (*center*) and many surrounding astrocytes. The background matrix shows a higher level of reactivity than in MS lesions

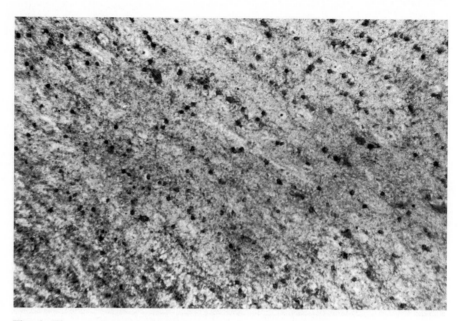

Fig. 3. The myelinated edge of another chronic active MS lesion shows intense immunoreactivity for Hsp27 on chains of interfascicular oligodendrocytes in numbers suggestive of hyperplasia. The lesion center is *to the lower left*

Most studies performed so far have focused on the 27-, 60- and 70-kDa families. For the small Hsps, αB-crystallin was investigated, demonstrating its peculiar expression in myelin preparation from MS patients (van Noort et al. 1995). Hsp60 has been extensively investigated in other organ-specific auto-immune illnesses where their pathogenetic role was studied in both animal models and human diseases (see previous chapters on insulin dependent diabetes and rheumatoid arthritis). Hsp27, Hsp60 and Hsp70 largely share the same rationale for being involved in autoimmunity. The risk of cross-reactivities may be particularly high in the case of Hsp70. Given the phyloge-netically ancient nature of this protein, especially in its N-terminal domain, microbes might have exerted an evolutionary pressure on the selection of T cells specific for conserved Hsp70 sequences (Gupta and Bhag 1992; Gupta and Golding 1993). Moreover, peptides representing sequences of self-Hsp70 (as well as of the constitutively expressed member of the same family, heat shock cognate, Hsc70) are among the motifs most frequently bound to an MHC class II molecule (HLA-DR11; Newcomb and Cresswell 1993) and it is reasonable to speculate that these antigenic sequences may contribute to the shaping of the T cell repertoire by promoting memory and viability (Williams et al. 1990; Gray and Matzinger 1991) of T cells with this specificity. Within the T cell response to Hsp the risk of autoreactivity is, therefore, exceedingly high.

I. Experimental Allergic Encephalomyelitis

At variance with numerous efforts to study the immune response to Hsps in MS (see below), few studies faced the matter in animal models. A study by MOR and COHEN (1992) showed Hsp-reactive T cells in the CNS of rats with EAE. The interesting point of the work was that the enrichment of Hsp-specific T lymphocytes was evident not only after actively-induced EAE, but also after passive transfer of the disease through an MBP-reactive T cell clone. The latter result suggests a role for the immune response to Hsp in sustaining the immunopathogenetic process. The mechanism underlying Hsp sensitization remains, however, obscure: intermolecular spreading or molecular mimicry between myelin basic protein and Hsp epitopes are possible explanations. Moreover, in the actively-induced EAE, CNS recruitment of T cells responding to Hsps contained in adjuvant (*Mycobacterium tuberculosis* is a component of Freund's adjuvant) preparation cannot be excluded.

The only example of Hsp-induced EAE is with an αB-crystallin peptide in Biozzi/ABH mice (AMOR et al. 1997 and J.M. VAN NOORT, personal communication). These observations represent further evidence for a relevant role of αB-crystallin as a potential autoantigen in MS (see below).

Two studies focused on the relevance of Hsps and $\gamma\delta$ T cells in EAE. The first one (GAO et al. submitted) showed that, in a model of passive disease in mice, there was a specific CNS accumulation of $\gamma\delta$ T cells in association with disease activity; colocalization of $\gamma\delta$ T cells and increased expression of Hsp60 suggested the pathogenetic relevance of the findings. Another study by the same group (RAJAN et al. 1996) was aimed at assessing the effect of depleting monoclonal antibodies specific for all TCR δ chain. mouse. The authors concluded that $\gamma\delta$ T cells play a pathogenetic role in EAE during both acute and chronic/progressive phases.

Another field of investigation on Hsps in EAE includes the studies on preimmunization to induce protection. In EAE, like in other animal models of autoimmune diseases, preimmunization with the autoantigen(s) confers protection. BIRNBAUM and colleagues (1996) recently identified a *M. leprae* Hsp65 peptide that is cross-reactive with the myelin protein 2',3' cyclic nucleotide 3' phosphodiesterase (CNP). Preimmunization with this peptide in incomplete Freund's adjuvant protected from EAE; a mechanism of immune deviation (i.e., a Th2 response to the cross-reactive peptide) mediated the protective effect. CNP was not encephalitogenic per se; however, the authors concluded for the relevance of the immune response to either Hsps or cross-reactive myelin proteins in the development of EAE. A study by FIORI et al. (1997) showed that preimmunization with diverse antigens (including autoantigens, foreign antigens and mitogens) suspended in IFA confers resistance to EAE induction. In this study the recombinant *M. tuberculosis* Hsp70 and *M. Bovis* Hsp65 were effective in protecting rats from EAE. The authors concluded that general mechanisms, independent and addictive to the specific tolerization of autoreactive effectors, contribute to the protection. They include not only immune deviation, but also downregulation of CD4 co-receptor and seem to be linked to the immunogenicity of the preimmunizing moiety rather than to its antigenic specificity.

II. Multiple Sclerosis

Early studies on the T cell response to heat shock proteins in MS dealt with mycobacterial antigens. We investigated the T cell proliferative primary responses to recombinant mycobacterial 65-(Hsp65) and 70 kDa Hsp in patients with MS, patients with other neurological diseases and healthy controls (Salvetti et al. 1992). Positive responses to Hsp70 but not to Hsp65 were significantly more frequent in patients with MS than in the other two groups. In the same study, long-term T cell lines specific for the purified protein derivative (PPD; which contains mycobacterial Hsps) and derived from the peripheral blood of patients and controls were screened for their proliferative response to Hsp65 and Hsp70. Again, the Hsp70-specific lines were significantly more common among MS patients than among healthy controls while the number of lines responding to Hsp65 was also increased though not significantly.

In another work by Birnbaum et al. (1993), T lymphocytes responsive to a sonicate of *M. tuberculosis* were found to be recruited or expanded in the cerebrospinal fluid (CSF) of patients with MS. In this study, responses to Hsp65 were minimal while Hsp70 was not investigated.

If a disregulated response to microbial Hsp70 exists in MS, the next step is to quantify the autoaggressive potential of T cells with this specificity by assessing their ability to cross-react with human Hsp70 sequences. Mapping experiments with recombinant proteins representing mycobacterial and human Hsp70 sequences, and with a panel of synthetic peptides encompassing the whole sequence of *Mycobacterium leprae* Hsp70, showed that the response to conserved epitopes of Hsp70 is a customary event in MS, often leading to the cross-recognition of microbial and human sequences (Salvetti et al. 1996). The same pattern of reactivity was observed in healthy controls and in patients with tuberculosis, thus supporting the hypothesis of a preformed response to motifs shared by microbial and human Hsps, which is useful during infection and detrimental – because quantitatively disregulated – in autoimmunity. The mechanisms that may keep this response under control and the reasons that may justify the exposure to such a risk constitute a matter of lively debate in immunology (Cohen 1992a,b). What may be relevant for the pathogenesis of immune-mediated diseases is that one can expect a considerable number of Hsp-specific, memory T lymphocytes to be present in the healthy immune repertoire. In normal conditions these cells are kept viable, under strict control and constitute a sort of preformed response that may be useful if exploited in the appropriate context (infection). In the wrong context (autoimmunity), the impact of a disregulated response to these antigens may constitute a potent promoter of the inflammatory process.

A disregulation of the T cell response to Hsp70 in MS is also suggested by a study by STINISSEN et al. (1995), who screened peripheral blood- and CSF-derived $\gamma\delta$ T cell clones for their proliferative response to recombinant mycobacterial Hsp65 and Hsp70. In this study, a considerable proportion of these clones responded to Hsp70 while only one clone responded to Hsp65. Furthermore, in spite of some bias given by two "high responder" MS patients, the responses to Hsp70 were more numerous in $\gamma\delta$ clones derived from patients with MS compared to patients with other neurological diseases and healthy controls. The Hsp70-specific clones originated preferentially from the peripheral blood. When tested, these clones proliferated also in response to two bacterial superantigens (staphylococcal enterotoxin B and toxic shock syndrome toxin-1).

During our screening of PPD-specific T lines for their response to mycobacterial Hsp, a weak association was noticed between Hsp70 reactivity and slight increases in the percentage of $\gamma\delta$ T cells within the lines assayed by flow cytometry. When PCR and direct sequencing analysis of the δ chain gene was performed, comparing freshly isolated PBLs, Hsp70-specific lines and PPD-specific lines from each donor (patients with MS, patients with tuberculosis and healthy donors were included in this study; BATTISTINI et al. 1995a), a clear link between Hsp70 specificity and a Vδ2-Jδ3 rearrangement was observed. To determine whether clonality occurred within the Hsp70-specific lines, the PCR products were subjected to direct sequence analysis. Direct sequence analysis showed oligoclonality of the response in all the Hsp70 specific lines while those specific for PPD alone were polyclonal. In 4 of 4 MS patients and in 2 of 3 TB patients, the dominant Vδ2-Jδ3 sequences were identical to one another and also identical to the one described as predominant in MS brain lesions (BATTISTINI et al. 1995b). Hence, Vδ2-Jδ3 TCR bearing cells may be responding to Hsp70 in MS plaques. Of interest, in this context, is that an oligoclonal expansion in a Vδ2-Jδ3 population was observed in one case of polymyositis (PLUSCHKE et al. 1992), further suggesting an autoreactive potential for these cells.

We are now trying to finely characterize the relationship between Hsp70, subclasses of $\gamma\delta$ T cells and MS. Preliminary results have been obtained investigating Vδ2+ T cells sorted (4G6 anti-Vδ2) from a healthy donor. They responded to whole Hsp70 and to residues 358–377, an exposed and conserved epitope of Hsp70. The proliferative responses did not require the presence of antigen presenting cells and the antigens could not be pulsed. Vδ2+ T cell responses to Hsp70, also in the absence of "canonical" APC, may be relevant in the CNS where professional APC are scanty. These results refer also to the possible role of Hsps as antigen-presenting molecules: several studies have already reported about the chaperoning activity of Hsps in both class I and class II pathways of antigen processing/presentation. Our work on Vδ2+ T cell may suggest a role of Hsp70, besides as an antigenic moiety, also as a direct presentation element, possibly of the non-peptide ligands for $\gamma\delta$ T cells recently reported in several studies (PORCELLI et al. 1996).

Another field of intensive work, involving $\gamma\delta$ T cells and Hsps, regards the oligodendrocyte lysis that may contribute to the mechanisms of myelin damage in MS. $\gamma\delta$ T cells have indeed potent cytotoxic activity and secrete significant amounts of proinflammatory cytokines and chemokines. Recent studies by Freedman and colleagues (1997) showed that Vδ2+ T cells were expanded in the presence of oligodendrocytes expressing Hsp, suggesting local expansion of $\gamma\delta$ T cells within MS plaques. The same group had demonstrated that freshly isolated $\gamma\delta$ T cells are capable of lysing oligodendrocytes in vitro (Freedman et al. 1991). This MHC-unrestricted activity could not be blocked by antibodies to Hsp70. Whether Hsps can be a target for $\gamma\delta$-mediated lysis of oligodendrocytes is still unclear. Another recent study by Freedman and colleagues (1997), aimed at addressing this possibility, supports the notion that more than one ligand, including Hsp, mediate $\gamma\delta$ T cell cytotoxic activity to oligodendrocytes.

A recent study by van Noort and colleagues (1995) showed the importance of the small Hsps as potential autoantigens in MS. In this study the authors, using reverse-phase HPLC, identified a low molecular weight fraction from MS-affected brain tissue which stimulated mononuclear peripheral blood cells of patients, but not of controls. The antigenic moiety present in the stimulating fraction was then further characterized as a protein of 23 kDa homologous to αB-crystallin. The potential role as an autoantigen of αB-crystallin was confirmed by immunohistochemical studies that showed the expression of this protein only on oligodendrocytes and astrocytes of the affected white matter, but not of normal brain areas (see above paragraphs). Current investigations by the same group have shown virus-induced expression and class II presentation of this candidate autoantigen by human peripheral blood lymphocytes: the possible priming of αB-crystallin-specific T cells by the virus-induced protein in the periphery and the subsequent response to the protein in the central nervous system are under scrutiny (van Noort, personal communication).

Few studies have been carried out so far on the humoral response to Hsp in MS. In one of these (Prabhakar et al. 1994), a significant correlation between antibody titers to human Hsp60 and the presence of oligoclonal bands in CSF was claimed. In another study (Gao et al. 1994) the presence of anti-Hsp60 was at low titer, prevalently in patients with severe disability, and non-specific for MS (being detectable also in patients with degenerative neurological diseases). Overall these data favor the hypothesis that immune responses to Hsp contribute to the chronicization of the immunopathological process.

D. Other Roles of Hsps in MS

Most of the studies on Hsps in EAE or MS dealt with their role as potential target of the disregulated immune response underlying the diseases. Nonetheless, the involvement of Hsps as chaperones and protective factors in disorders of a chronic inflammatory nature obviously deserve attention, being in accord with the role classically attributed to these proteins.

Several studies have already shown the nonspecific protective effect of Hsps in CNS injury of various kinds. Two protective actions of Hsps have been recently investigated with special interest, given their relevance in the mechanisms of myelin damage: the protection from TNF-induced cytotoxicity and the modulation of inducible nitric oxide synthase (iNOS) activity.

One group showed that Hsp70, but not Hsp27, protected from TNF injury by interfering with the activation of phospholipase A2 (JAATTELA et al. 1989). Another group, however, considered only Hsp27 and found that it had a protective role against TNF-mediated cytotoxicity as well (MEHLEN et al. 1995a,b). Despite the discordant result for Hsp27, which may be due to the different system used, these observations are of interest for the well known role of TNF in demyelinating diseases and raise the possibility that differential brain expression of Hsps may affect not only the autoaggressive response, but also the resistance to the consequences of such immune-mediated insult.

Other relevant results come from the work of FEINSTEIN et al. (1996) on the Hsp interference with iNOS. The levels of nitric oxide (NO) correlate with the severity of EAE and the inhibition of iNOS protects the animal against the disease. The expression of Hsp70 reduces iNOS activity and then limits the proinflammatory and neurotoxic effect of NO. These observations are of particular interest given the involvement of NO in mediating the TNF-induced cytotoxicity: Hsp70 appears then as a potential general modulator of different (and interplaying) pathways leading to myelin insult.

Indistinguishable from the role of Hsps as protective factors is that as chaperones, perhaps the best characterized biological activity of Hsps, especially of 60 and 70 kDa. The role of chaperones includes the participation in processes of protein folding and assembly into multimeric structures as well as the translocation or insertion across or into membranes. A study by AQUINO et al. (1996) showed that the reduction of constitutive Hsp70 levels (obtained by an antisense oligonucleotide) correlated with a decreased synthesis of MBP, but not of proteolipid protein (PLP). The findings confirmed their working hypothesis: the biochemical properties of the two myelin proteins differ for their requirement of chaperoning activity, which is much lower for PLP (an intrinsic membrane protein) compared to MBP (an extremely hydrophilic and cationic protein). These data suggest a role for Hsp70 in remyelination and once again are in favor of its protective potential in the demyelination process.

Another field of interest for the involvement of Hsp in the pathogenesis of MS is the search for polymorphisms of their genes. Hsp gene polymorphisms that confer protection or susceptibility to MS may have relevance for their role as both autoantigens and defensive factors. Alterations in the genetic control of the expression of these proteins in the brain could explain the disregulated immune response to these (auto-) antigens in an organ-specific disease, in spite of their ubiquity. Differential Hsp expression may also affect repair events and protection mechanisms.

For the Hsp70 family member encoded by the Hsp70–1 gene, the first study in this direction yielded a negative result showing the lack of association between polymorphisms in the promoter region of Hsp70–1 and MS (Cascino et al. 1994). Another study by Ramachandran and Bell (1995) also failed to reveal significant differences between patients and controls for restriction fragment length polymorphisms of Hsp70–2 and Hsp70-*hom* genes.

E. Conclusions and Future Work

The studies illustrated so far seem to satisfy some of the requirements for candidate Hsps as potential autoantigens in MS: (a) a disregulated T cell response to families of Hsp can be detected in MS; (b) this response can be autoreactive; and (c) these antigens are expressed in the target organ at the lesion level.

At present it is difficult to dissect this role as candidate autoantigens from the classical function as chaperones and protective factors. This will require further analyses in order to clarify the extent to which Hsps contribute to the defensive mechanisms against myelin injury. This is an interesting matter also in terms of therapeutic approaches possibly aimed at facilitating remyelination.

Points of interest in the near future include the role of Hsp-specific T cells as pathogenic or protective effectors. The possibility that the response to Hsp70 in MS is associated with regulatory functions rather than with disease progression needs to be further investigated: in juvenile chronic arthritis the T cell response to human Hsp60 is associated with a relatively mild and remitting subtype of the disease (de Graeff-Meeder 1995). Two main research directions of the studies on immune response to Hsps may help verify these possibilities: a stratification of the patients according to pathophysiological criteria (longitudinal investigations on the same patient in different phases of the disease; studies on patients with benign MS or monophasic acute encephalomyelitis; research on monozygotic twins discordant for the disease); and functional studies aimed at analyzing the phenotype of autoreactive effectors in terms of surface markers and cytokines produced.

Another direction for future studies is the relationship between $\gamma\delta$ T lymphocytes and Hsps; of special interest, in this context, are the particular conditions of antigen recognition by such T cells. Hopefully, the availability of Vδ2-Jδ3 TCR bearing T lymphocyte clones will help clarify the above relationship and the relevance of these cells to lesion development.

The widely accepted influence of infectious agents on the development and progression of the disease (Kurtzke 1992; Goverman 1993; Ragheb and Lisak 1993) makes Hsp-related events of molecular mimicry of obvious interest. Potential cross-reactivity between Hsps and microbial (but also myelin) antigens deserves specific investigation.

References

Allegretta M, Nicklas JA, Sriram S et al (1990) T cells responsive to myelin basic protein in patients with multiple scerosis. Science 247:718–721

Amor S, Baker D, Layward L et al (1997) Multiple sclerosis: variations on a theme. Immunol Today 18:368–371

Aquino DA, Klipfel AA, Brosnan CF, Norton WT (1993) The 70 kDa heat shock cognate protein is a major constituent of the central nervous system and is upregulated only at the mRNA level in acute experimental autoimmune encephalomyelitis. J Neurochem 61:1340–1348

Aquino DA, Lopez C, Farooq M (1996) Antisense oligonucleotide to the 70-kDa heat shock protein inhibits synthesis of myelin basic protein. Neurochem Res 21:417–422

Aquino DA, Capello E, Weisstein J et al (1997) Multiple sclerosis: altered expression of 70- and 27-kDa heat shock proteins in lesions and myelin. J Neuropathol Exp Neurol 56:664–672

Bajramovic JJ, Lassmann H, van Noort JM (1997) Expression of aB-crystallin in glia cells during lesional development in multiple sclerosis. J Neuroimmunol 78:143–151

Battistini L, Salvetti M, Ristori G et al. (1995a) HSP70 selects for an oligoclonal Vδ2-Jδ3 T cell receptor response in lymphocytes from healthy individuals, multiple sclerosis and tuberculosis patients. Mol Med 1:554–562

Battistini L, Selmaj K, Kowal C et al (1995b) Multiple sclerosis: limited diversity of the Vδ2-Jδ3 T-cell receptor in chronic-active lesions. Ann Neurol 37:198–205

Birnbaum G, Kotilinek L, Albrecht L (1993) Spinal fluid lymphocytes from a subgroup of multiple sclerosis patients respond to mycobacterial antigens. Ann Neurol 34:18–24

Birnbaum G, Kotilinek L, Schlievert P et al (1996) Heat shock proteins and experimental autoimmune encephalomyelitis (EAE): I. Immunization with a peptide of the myelin protein 2′, 3′ cyclic nucleotide 3′ phosphodiesterase that is cross-reactive with a heat shock protein alters the course of EAE. J Neurosci Res 44:381–396

Brosnan CF, Battistini L, Gao YL, Raine CS, Aquino DA (1996) Heat shock proteins and multiple sclerosis. J Neuropathol Exp Neurol 55:389–402

Cascino I, Galeazzi M, Salvetti M et al (1994) HSP70–1 promoter region polymorphisms tested in three autoimmune diseases. Immunogenetics 39:291–293

Cohen IR (1992a) The cognitive principle challenges clonal selection. Immunol Today 13:441–444

Cohen IR (1992b) The cognitive paradigm and the immunological homunculus. Immunol Today 13:490–494

de Graeff-Meeder ER, van Eden W, Rijkers GT et al (1995) Juvenile chronic arthritis: T cell reactivity to human HSP60 in patients with a favorable course of arthritis. J Clin Invest 95:934–940

D'Souza SD, Antel JP, Freedman MS (1994) Cytokine induction of heat shock protein expression in human oligodendrocytes: an interleukin-1 mediated mechanism. J Neuroimmunol 50:17–24

Feinstein DL, Galea E, Aquino DA et al (1996) Heat shock protein 70 suppresses astroglial-inducible nitric-oxide synthase expression by decreasing NFkappaB activation. J Biol Chem 271:17724–17732

Fiori P, Ristori G, Cacciani A et al (1997) Down-regulation of cell-surface CD4 co-receptor expression and modulation of experimental allergic encephalomyelitis. Int Immunol 9:541–545

Freedman MS, Ruijs TCG, Selin LK et al (1991) Peripheral blood gamma-delta T cells lyse fresh human brain-derived oligodendrocytes. Ann Neurol 30:794–800

Freedman MS, Buu NN, Ruijs TC, Williams K, Antel JP (1992) Differential expression of heat shock proteins by human glial cells. J Neuroimmunol 41:231–238

Freedman MS, D'Souza S, Antel JP (1997) Gamma delta T-cell-human glial cell interaction I. In vitro induction of gamma delta T cell expansion by human glial cells. J Neuroimmunol 74:135–142

Gao YL, Raine CS, Brosnan CF (1994) Humoral response to hsp 65 in multiple sclerosis and other neurological conditions. Neurology 44:941–946

Gao YL, Brosnan CF, Raine CS (1995) Experimental autoimmune encephalomyelitis. Qualitative and quantitative differences in heat shock protein 60 expression in the central nervous system. J Immunol 154:3448–3456

Goverman J, Woods A, Larson L et al (1993) Transgenic mice that express a myelin basic protein-specific T cell receptor develop spontaneous autoimmunity. Cell 72:551–560

Gray D, Matzinger P (1991) T cell memory is short lived in the absence of antigen. J Exp Med 174:969–974

Gupta RS, Bhag S (1992) Cloning of the HSP70 gene from *Halobacterium marismortui*: relatedness of archaebacterial HSP70 to its eubacterial homologs and a model for the evolution of the HSP70 gene. J Bacteriol 174:4594–4605

Gupta RS, Golding GB (1993) Evolution of HSP70 gene and its implications regarding relationships between archaebacteria, eubacteria and eukaryotes. J Mol Evol 37:573–582

Head MW, Corbin E, Goldman JE (1993) Overexpression and abnormal modification of the stress proteins alpha-B crystallin and HSP27 in Alexander disease. Am J Pathol 143:1743–1753

Holfeld R (1997) Biotechnological agents for the immunotherapy of multiple sclerosis. Principles, problems and perspectives. Brain 120:865–916

Jaattela M, Saksela K, Saksela E (1989) Heat shock protects WEHI-164 target cells from the cytolysis by tumor necrosis factor α and β. Eur J Immunol 19:1413–1417

Kaufman DL, Clare-Salzler M, Tian J (1993) Spontaneous loss of T cell tolerance to glutamic acid decarboxylase in murine insulin dependent diabetes. Nature 366:69–72

Kaufmann SH (1994) Heat shock proteins and autoimmunity. Intl Arch All Immunol 103:317–322

Koga T, Wand-Wurttenberger A, de Bruyn J, Munk ME, Schoel B, Kaufmann SH (1989) T cells against a bacterial heat shock protein recognize stressed macrophages. Sciences 246:1112–1115

Kojima K, Berger T, Lassmann H et al (1994) Experimental autoimmune panencephalitis and uveoretinitis transferred to the Lewis rat by T lymphocytes specific for the S100β molecule, a calcium binding protein of astroglia. J Exp Med 180:817–829

Kurtzke JF (1992) Epidemiologic evidence for multiple sclerosis as an infection. Clin Microbiol Rev 6:382–427

Lehmann PV, Forsthuber T, Miller A, Sercarz EE (1992) Spreading of T cell autoimmunity to cryptic determinants of an autoantigen. Nature 358:155–157

Lowe J, McDermott H, Pike I, Spendlove I, Landon M, Mayer RJ (1992) Alpha-B crystallin expression in non-lenticular tissues and selective presence in ubiquitinated inclusion bodies in human diseases. J Pathol 166:61–68

Mehlen P, Preville X, Chareyron P et al (1995a) Constitutive expression of human hsp27, Drosophila hasp27, or human alpha-B crystallin confers resistance to TNF- and oxidative stress-induced cytotoxicity in stably tranfected murine L929 fibroblasts. J Immunol 154:363–374

Mehlen P, Mehlen A, Guillet D et al (1995b) Tumor necrosis factor-alpha induced changes in the phosphorylation, cellular localization, and oligomerization of human hsp27, a stress protein that confers cellular resistance to this cytokine. J Cell Biochem 58:248–259

Mor F, Cohen IR (1992) T cells in the lesion of experimental autoimmune encephalomyelitis. Enrichment for reactivities to myelin basic protein and to heat shock protein. J Clin Invest 90:2447–2455

Newcomb JR, Cresswell P (1993) Characterization of endogenous peptides bound to purified HLA-DR molecules and their absence from invariant chain-associated $\alpha\beta$ dimers. J Immunol 150:499–507

Olsson T, Sun J, Hillert J et al (1992) Increased numbers of T cells recognizing multiple myelin basic protein-epitopes in multiple sclerosis. Eur J Immunol 22:1083–1087

Pluschke G, Ruegg D, Hohlfeld R et al (1992) Autoaggressive myocytotoxic T lymphocytes expressing an unusual gamma/delta T cell receptor. J Exp Med 176:1785–1793

Porcelli SA, Morita CT, Modlin PL (1996) T cell recognition of non-peptide antigens. Curr Opin Immunol 8:510–516

Prabhakar, S Kurien E, Gupta RS et al (1994) Heat shock protein immunoreactivity in CSF: correlation with oligoclonal banding and demyelinating disease. Neurology 44:1644–1648

Ragheb S, Lisak RP (1993) Multiple sclerosis: genetic background versus environment. Ann Neurol 34:509–510

Raine CS (1994a) Immune system molecule expression in the central nervous system. J Neuropath Exp Neurol 53:328–337

Raine CS (1994b) The Dale E. McFarlin memorial lecture: the immunology of the multiple sclerosis. Ann Neurol 36:S61–S72

Rajan AJ, Gao YL, Raine CS et al (1996) A pathogenic role for gamma delta T cells in relapsing-remitting experimental allergic encephalomyelitis in the SJL mouse. J Immunol 157:941–949

Ramachandran S, Bell RB (1995) Heavy shock protein 70 gene polymorphisms and multiple sclerosis. Tissue Antigens 46:140–141

Salvetti M, Buttinelli C, Ristori G et al (1992) T-lymphocyte reactivity to the recombinant mycobacterial 65- and 70-kD heat shock proteins in multiple sclerosis. J Autoimmun 5:691–702

Salvetti M, Ristori G, Buttinelli C et al (1996) The immune response to mycobacterial 70-kDa heat shock proteins frequently involves autoreactive T cells and is quantitatively disregulated in multiple sclerosis. J Neuroimmunol 65:143–153

Satoh J, Kim SU (1994) HSP induction by heat stress in human neurons and glial cells in culture. Brain Res 653:243–250

Satoh J, Kim SU (1995a) Constitutive and inducible expression of heat shock protein HSP72 in oligodendrocytes in culture. Neuroreport 6:1081–1084

Satoh J, Kim SU (1995b) Cytokines and growth factors induce HSP72 phosphorylation in human astrocytes. J Neuropath Exp Neurol 54:504–512

Satoh J, Nomaguchi H, Tabira T (1992a) Constitutive expression of 65-kDa heat shock protein-like immunoreactivity in cultured mouse oligodendrocytes. Brain Res 595:281–290

Satoh J, Yamamura T, Kunishita T, Tabira T (1992b) Heterogeneous induction of 72-kDa heat shock protein (HSP72) in cultured mouse oligodendrocytes and astrocytes. Brain Res 573:37–43

Satoh J, Tabira T, Yamamura T, Kim SU (1994) HSP72 induction by heat stress is not universal in mammalian neural cell lines. J Neurosci Res 37:44–53

Selmaj K, Brosnan CF, Raine CS (1991) Colocalization of lymphocytes bearing the $\gamma\delta$ T cell receptor and heat shock protein 65+ oligodendrocytes in multiple sclerosis. Proc Natl Acad Sci USA 88:6452–6456

Shinohara H, Inaguma Y, Goto S, Inagaki T, Kato K (1993) Alpha-B crystallin and HSP28 are enhanced in the cerebral cortex of patients with Alzheimer's disease. J Neurol Sci 119:203–208

Stinissen P, Vandevyver C, Medaer R et al (1995) Increased frequency of gd T cells in the cerebrospinal fluid and peripheral blood of patients with multiple sclerosis: reactivity, cytotoxity and T cell receptor V gene rearrangements. J Immunol 154:4883–4894 van Noort JM, van Sechel AC, Bajramovic JJ, el Ouagmiri M, Polman CH, Lassmann H, Ravid R (1995) The small heat shock protein alpha-B crystallin as a candidate autoantigen in multiple sclerosis. Nature 375:798–801

Williams GT, Smith TA, Spooncer E et al (1990) Haematopoietic colony stimulating factors promote cell survival by suppressing apoptosis. Nature 343:76–79

Wucherpfennig KW, Weiner HL, Hafler DA (1991) T cell recognition of myelin basic protein. Immunol Today 12:277–282

Wucherpfennig KW, Newcombe J, Li H, Keddy C, Cuzner ML, Hafler DA (1992) $\gamma\delta$ T-cell receptor repertoire in acute multiple sclerosis lesions. Proc Natl Acad Sci USA 89:4588–4592

CHAPTER 18
Heat Shock Proteins in Atherosclerosis

A.D. JOHNSON

A. Introduction

Atherosclerosis is a progressive pathological lesion of the arterial wall, arising as a consequence of chronic vascular injury. This review will briefly describe changes in heat shock protein (Hsp) distribution and expression levels that have been observed during atherosclerotic plaque evolution, then examine evidence that altered Hsp production and distribution within the arterial wall directly contributes to lesion formation and progression. Throughout the review, and in closing, some currently unanswered questions needing further investigation also will be outlined.

B. Pathogenesis of Primary Atherosclerotic Lesions

Development of the primary atherosclerotic lesion has been extensively reviewed previously (STARY et al. 1992, 1994, 1995). Major stages in primary plaque evolution that will be important for discussing contributions by Hsps are as follows:

1. Endothelial injury/activation by atherogenic risk factors.
2. Recruitment of macrophages and T-lymphocytes, and inflammation of the injured area.
3. Lipid and cholesterol deposition, with concurrent smooth muscle hyperplasia and migration.
4. Organization of newly synthesized intimal extracellular matrix into an eccentric intimal lesion.
5. Centralized cell death to form a thrombogenic necrotic core.
6. Destabilization of SMC-rich fibrous tissue overlying the necrotic core, allowing macroscopic plaque rupture.
7. Sudden partial or complete occlusion of flow, causing down-stream ischemia/infarction.

Briefly, primary atherosclerotic lesions develop when atherogenic risk factors injure and activate the endothelial cells (ECs) lining an artery. Activated ECs recruit T-lymphocytes and macrophages that initiate a localized inflammatory response to injury. If hypercholesterolemia is also present, cholesterol esters and triglycerides are extensively deposited within and around macrophages at the site of infiltration. Concurrently, smooth muscle cells (SMCs) proliferate and enter the intima from the underlying media. This further thickens the arterial wall, forming a mature atherosclerotic plaque. As plaque progression continues, sites of necrosis form in the deepest regions of the plaque, that can rupture into the vessel lumen, causing rapid thrombosis and occluding flow.

In addition to primary plaques, there are two other types of lesions that will be important for this discussion. The first type, graft arteriosclerosis, is a clinical complication of organ transplantation, and occurs when the graft arterial wall is attacked by the host immune system. Resultant SMC hyperplasia leads to vascular occlusion, and is a frequent cause of myocardial graft failure. Restenosis (or secondary atherosclerosis) is a distinct form of atherosclerosis that develops in about 30% of patients undergoing coronary angioplasty to relieve myocardial ischemia. Stretch injury of a primary lesion stimulates robust SMC hyperplasia and fibrotic reorganization, that can reduce flow as much or more than the original primary plaque.

C. Arterial Hsp Expression After Vascular Injury, and During Development of Atherosclerotic Lesions

Western blotting and immunohistochemical surveys indicate that uninjured adult vessels contain low basal levels of multiple Hsps (HIGHTOWER and WHITE 1982; BERBERIAN et al. 1990; UDELSMAN et al. 1991, 1993; JOHNSON et al. 1993; WILKINSON and POLLARD 1993). As in other tissues though, vascular injury stimulates a strong stress response, including increased Hsp production. HIGHTOWER and WHITE (1982) determined that explants of rat thoracic aortas expressed increasing amounts of Hsp70 and Grp78 as a function of time in vitro. Furthermore, SMCs underlying mechanically damaged ECs expressed more Hsp70 than did SMCs underlying intact endothelium. Similarly, mechanical or ischemic injury of arteries induced synthesis of Hsp70 (CURRIE et al. 1987; UDELSMAN et al. 1991). In vitro, cell culture was also a stimulus for expression of multiple Hsps by SMCs and ECs (BITAR et al. 1991; PIOTROWICZ et al. 1995; PORTIG et al. 1996).

Primary atherosclerotic plaque formation and progression is associated with dramatic changes in the distribution of multiple Hsps in vessels (BERBERIAN et al. 1990; JOHNSON et al. 1993). Serial sections of human carotid endarterectomy and thoracic aorta specimens, or aortas of cynomolgus macaques with diet-induced atherosclerosis, were immunostained for Hsp90, -70, -60, and -28. Compared to normal-appearing arterial wall, immunoreactivity against all four Hsps became more heterogeneously distributed as lesion severity increased. All four Hsps were detected both intracellularly, and in association with extracellular matrix of expanding plaques. Macrophages and lymphocytes in or near lipid-rich areas of plaques in particular exhibited strong Hsp70 immunoreactivity. In contrast, some SMC-rich areas exhibited increased Hsp immunoreactivity, while immediately adjacent areas of smooth muscle appeared to contain almost no immunoreactive Hsps. Histological analyses by other investigators have likewise demonstrated increased Hsp60 immunoreactivity in association with ECs, SMCs, and mononuclear cells of aortic or carotid atherosclerotic plaques (KLEINDIENST et al. 1993). Hsp27 was also elevated in the periluminal cells of atherosclerotic aortic segments obtained during cardiopulmonary bypass surgery (CAMPISI et al. 1993).

D. Association of Hsps with Specific Stages of Atherosclerotic Lesion Development

Present evidence indicates that Hsps contribute to atherosclerotic lesion development and progression at three specific stages. These are:

1. Predisposing risk factors for atherosclerotic lesion formation. Both hypertension and diabetes will stimulate or accelerate atherosclerotic lesion formation (for reviews, see ROSS 1995; STRONG et al. 1996). Hsp expression is dramatically altered during both environmentally and genetically induced forms of hypertension. Evidence will be presented which suggests that altered Hsp expression associated with hypertension may significantly accelerate arterial lesion formation. Since the roles of Hsps in diabetes have been reviewed by Altmann (see Chap. 17), they will not be reiterated here.
2. Lymphocyte activation during the inflammatory stage of primary and graft arteriosclerosis. Specific immunological responses against Hsps expressed by vascular cells promote both primary and graft atherosclerotic plaque progression, and so will be discussed in some detail. More general actions of Hsps during inflammation have been reviewed by Polla (see Chap. 13). Potential relationships between Hsps and cytokines derived from the recruited inflammatory cells have been discussed by Stephanou and Latchman (see Chap. 7).
3. Protection of plaque cells against toxic injury in mature and late-stage primary lesions. Evidence will be presented that suggests Hsps protect vascular cells from oxidized lipoprotein toxicity, ischemia, and oxygen free radical damage in advanced atherosclerosis, while loss of Hsp cytoprotection accelerates plaque necrotic core formation and rupture. Additional studies in other models of ischemia are reviewed by Nowak, Brown, and Yellon (see Chaps. 8, 11, and 12).

I. Hsps in Hypertension, a Risk Factor for Atherosclerosis

Essential hypertension is a complex, multifactorial disease, but in simplest terms, results from interactions between environmentally induced vasoconstriction, and genetic predisposing factors. Large arteries compensate for increased hemodynamic load and greater wall shear and tensile stress during hypertension through a series of physical and biochemical changes (for a review, see MULVANY 1996). Part of this compensatory reaction includes altered expression of Hsps.

Transiently increased aortic Hsp mRNA and/or protein production has been repeatedly observed in response to environmental hypertensive stimuli. In vitro, the vasoconstrictor angiotensin II (Ang-II) at 1 μM induced transient expression of Hsp60 and -70 within 24h in serum-free, quiescent SMCs in culture (PATTON et al. 1995). However, chronic stimulation with Ang-II over several days did not result in continuously elevated Hsp70 production (KOHANE et al. 1990). In vivo, intravenous injection of Ang-II, phenylephrine, or vasopressin elevated mean arterial pressure, and induced Hsp70 mRNA expression within 1–2h in rat aortas, but not in other tissues (MOALIC et al. 1989). UDELSMAN et al. (1993) similarly reported that acute hypertension induced by restraint stress of rats increased aortic Hsp27 and Hsp70 mRNA synthesis within 1h, and protein synthesis within 3–6h. However, more recent evidence obtained using the rat restraint model suggests that aortic smooth muscle Hsp synthesis during restraint is not specifically induced by activation of vasoconstrictor receptors. Rather, it is a generalized adaptive response to acute mechanical loading, that is independent of the particular hypertensive stimulus (BLAKE et al. 1995; UDELSMAN et al. 1995; XU et al. 1995). Moreover, chronic elevation of blood pressure in genetically normal animals does not lead to chronic overexpression of Hsps.

Unlike environmentally induced hypertension, genetically induced hypertension has been specifically linked to the hsp70 gene, and in particular, to inheritable defects in the regulatory elements or coding sequence. Results of cross-breeding studies between normotensive and spontaneously hypertensive rats (SHR) or mice (SHM) indicated that at least one genetic locus associated with hypertension co-assorted with the hsp70 gene in both models (HAMET et al. 1990a). Restriction fragment length polymorphism analysis also identified a variant 3.0-kB BamHI fragment within the hsp70 gene in SHR that was positively associated with a 15 mm Hg increase in mean blood pressure of hypertensive animals, compared to normotensive controls (HAMET et al. 1992). Hsp70 mRNA accumulation was higher in the kidneys of SHM after whole body hyperthermia, as well as in heat-shocked cultures of aortic SMCs derived from SHR, than in equivalent tissues or cells from normotensive controls (HAMET et al. 1990b; LUKASHEV et al. 1991; TREMBLAY et al. 1992). Multiple tissues from SHR and SHM also exhibited an accelerated rate of Hsp70 mRNA synthesis and decay post-heat shock, compared to normotensive controls (MALO et al. 1989; HAMET et al. 1990b,c; HASHIMOTO et al. 1991; TREMBLAY et al. 1992). Likewise, Hsp72 protein accumulated to a significantly greater level in the left ventricles of SHR after whole body hyperthermia than in ventricles of similarly treated, normotensive Wistar-Kyoto rats (BONGRAZIO et al. 1994). In humans, circulating lymphocytes obtained from patients diagnosed with essential hypertension constitutively expressed more Hsp70 than did lymphocytes of normotensive control patients (KUNES et al. 1992).

While they overexpress Hsp70 mRNA and protein, genetically hypertensive animals are markedly less resistant to thermal injury (HAMET et al. 1994; HAMET 1995). WRIGHT et al. (1977) noted that intact SHR were more sensitive to lethal whole body hyperthermia, and succumbed to heat stress 36–56% earlier than did normotensive Sprague-Dawley controls. Thermal hypersensitivity developed at an earlier age than did detectable hypertension, suggesting that poor heat tolerance resulted from a primary gene defect, rather than being a secondary consequence of hypertension (McMURTRY and WEXLER 1981, 1983). At the cellular level, cultured SMCs derived from aortas of SHR were similarly less thermotolerant than SMCs derived from normotensive control rats (TREMBLAY et al. 1992). SHR-derived SMCs also entered S-phase and proliferated more rapidly than did SMCs derived from Wistar-Kyoto normotensive controls, despite the fact that overexpression of Hsp70 typically inhibits SMC proliferation (SLEPIAN et al. 1996; NESCHIS et al. 1997). These observations suggest that the Hsp70 protein synthesized by tissues from genetically hypertensive animals is partially or completely nonfunctional, leaving these tissues more susceptible to injury. Defective Hsp70 protein could also alter feedback regulation, potentially causing both Hsp70 overexpression, and more rapid Hsp70 mRNA synthesis and degradation.

The described differences in Hsp70 protein expression in genetic versus environmental models of hypertension suggest a working hypothesis for future studies. First, it can be assumed that acute Hsp70 expression in response to hemodynamic loading (as occurs in environmentally induced hypertension) prevents or limits resultant arterial wall injury. If so, then genetically hypertensive animals and patients would be more susceptible to mechanical overload injury, because their aberrant stress response would make them vulnerable to benign environmental hypertensive factors. Genetically hypertensive individuals might also be more sensitive to other atherogenic risk factors as well. It is critical that future studies determine whether stable genetic defects in the stress response exacerbate atherosclerotic plaque formation.

II. Immunological Responses to Hsps in Primary Plaque Development

Early in plaque formation, atherogenic risk factors stimulate endothelial cells to recruit circulating macrophages and T-lymphocytes, that extravasate and initiate a localized inflammatory reaction at the site of the nascent plaque. The macrophages subsequently endocytose lipids to form foam cells, while cytokines released by both foam cells and T-lymphocytes stimulate SMC migration into and proliferation within the intima (for a review, see HANSSON 1994). Concurrently, complement antibody complexes are deposited in the expanding lesion as well. Proponents of immunological activation in atherosclerosis have suggested that Hsps are one of several antigenic targets in plaques (HANSSON and LIBBY 1996). Experimental evidence has now confirmed that Hsp60 expressed at sites of vascular injury can stimulate T-lymphocyte clonal expansion, and implicate Hsp60 in deposition of antibody/complement complexes that in turn stimulate necrotic core formation.

1. T-lymphocyte Activation by Hsp60 in Atherosclerotic Vessels

Initial evidence for T-lymphocyte activation by Hsp60 expressed by plaque cells arose fortuitously. In order to map immunodominant antigens in atherosclerotic lesions, WICK and colleagues (XU et al. 1992, 1993a) attempted to raise specific antibodies by immunizing New Zealand White rabbits with atherosclerotic plaque-derived proteins obtained from Watenabe heritable hyperlipidemic rabbits or humans, and mixed with Freund's complete adjuvant (FCA). Surprisingly, aortic intimal atherosclerotic lesions developed within 16 weeks after the first immunization, beginning at branch points, then spreading into adjacent areas. Moreover, this was observed even in control animals immunized with FCA alone. Subsequent mitogenic outgrowth assays revealed that inflammatory cells investing the lesions reacted strongly against the mycobacterial Hsp65 present in FCA adjuvant. In a longer study, rabbits were immunized once with purified Hsp65 minus adjuvant (XU et al. 1996). Eight of 10 rabbits developed aortic intimal lesions by 16 weeks; by 32 weeks though, only 3 of 10 immunized animals had discernible lesions, and average lesion area was not statistically different from that observed in non-immunized controls. Histologically, immunization with Hsp65 alone induced formation of simple lesions that consisted of intimal accumulations of T-lymphocytes, monocyte/macrophages, and some SMCs. However, if cholesterol feeding was combined with Hsp65 immunization, stable, complex lesions formed that did not resolve by 32 weeks, even if the high cholesterol diet was discontinued at 16 weeks. These complex lesions additionally contained foam cells and extracellular deposits of cholesterol, and exhibited some early necrotic changes. The authors concluded that Hsp65/60-reactive T-lymphocytes (initially stimulated by Hsp65 immunization) reversibly infiltrate the arterial wall and proliferate at sites where endogenous Hsp60 expression has been induced by non-laminar flow, altered shear stress, or endothelial injury by other atherogenic risk factors. If hypercholesterolemia is also present, then these transient inflammatory lesions progress to form stable atherosclerotic plaques.

Hsp65/60-reactive T-lymphocytes also invest injured and atherosclerotic human vessels. HENG and HENG (1994) examined 35 branches of gastroduodenenal, superior mesenteric, and inferior mesenteric arteries obtained after surgical resection; 22 had ligatures of 0.5–4 h duration. Histological sections of all 22 ligation sites cross-reacted more strongly with antibodies against mycobacterial Hsp65 than did sections from unligated control areas. Activated γ/δ T-lymphocytes invested the intima, adventitia, and occasionally the media, of all 22 ligated regions, and were consistently apposed to Hsp65-

immunopositive cells. Similarly, KLEINDIENST et al. (1993) obtained human arterial and venous segments from coronary bypass or endarterectomy surgery specimens, then immunostained them for mammalian Hsp60 plus cellular markers of T-lymphocyte subclasses, macrophages, endothelial cells, or smooth muscle. In atherosclerotic areas, Hsp60-immunoreactive ECs, SMCs, and T-lymphocytes were consistently observed, and again, α/β and γ/δ T-lymphocytes were routinely found in close association with Hsp60-expressing SMCs or ECs.

Despite the close association, T-lymphocytes are recruited into injured vessels or atherosclerotic lesions via the typical adhesion cascade (DUSTIN and SPRINGER 1991), rather than by binding directly to Hsp60. If intact rats were treated with lipopolysaccharide (LPS) for up to 24 h, activated ECs expressed both Hsp60 and intercellular adhesion molecule-1 (ICAM-1). ICAM-1 expression was elevated as early as 3 h post-LPS treatment, and coincided with T-lymphocyte margination; conversely, Hsp60 expression increased only after 16–18 h. T-lymphocyte adhesion could also be partially blocked by pre-incubating aortic specimens with monoclonal antibodies directed against ICAM-1 or other adhesion molecules, while blocking antibodies against Hsp65 were ineffective. Thus, T-lymphocyte recruitment did not require Hsp60 expression by ECs (SEITZ et al. 1996). Once T-lymphocytes enter the arterial wall though, they apparently undergo rapid clonal expansion in response to Hsp60 expressed by plaque cells (XU et al. 1993a).

The issue of how T-lymphocytes might gain access to Hsp60 expressed by plaque cells was addressed in a study by XU et al. (1994), in which resting ECs were activated with T-lymphocyte conditioned media or tumor necrosis factor alpha (TNF-α). When the stimulated ECs were immunostained using monoclonal antibodies against specific Hsp60 epitopes, immunoreactive Hsp60 protein was present both intracellularly, and as an integral membrane protein on the cell surface. The authors concluded that it is the superficially expressed pool of Hsp60 which induces T-lymphocyte clonal expansion. Surface expression of Hsp60 by ECs might also be stimulated by high shear or other mechanical forces during arterial ligation, or by balloon dilatation injury (XU et al. 1993a,b; KLEINDIENST et al. 1993), again causing T-lymphocyte proliferation. Macrophages and SMCs in the atherosclerotic lesion are also immunoreactive for Hsp60 after injury, and the close association of activated T-lymphocytes with Hsp60-immunopositive SMCs and macrophages is strong circumstantial evidence that these cells also express superficial Hsp60. If so, then Hsp60 expression may serve as the final common integration point for multiple atherogenic risk factors, that all stimulate T-lymphocyte proliferation within the injured arterial wall. In the presence of additional risk factors, Hsp60-stimulated T-lymphocytes would in turn induce smooth muscle migration and hyperplasia, through multiple paths that have been reviewed previously (HANSSON 1994; STEMME and HANSSON 1994; WATENABE et al. 1997).

2. Do Anti-Hsp60 Antibodies Contribute to Necrotic Core Formation?

The humoral arm of the immune system also reacts to Hsps expressed by cells of the atherosclerotic lesion. XU et al. (1993b) documented that serum titers of anti-mycobacterial Hsp65 antibodies were significantly higher in patients over 60 years old, and with sonographically detectable carotid atherosclerosis. IgG antibodies purified from the sera of 5 patients with high anti-Hsp65 titers reacted strongly with ECs and macrophages in histological sections of heterologous atherosclerotic plaques, and with the mammalian homologue Hsp60 in western blots. Subsequent peptide epitope mapping confirmed that purified anti-Hsp65 IgGs recognized three conserved epitopes that are also present in mammalian Hsp60 (METZLER et al. 1997). Moreover, circulating antibodies obtained from clinically healthy subjects also bound these same three epitopes. The presence of anti-Hsp65/60 antibodies in healthy individuals suggests that prior mycobacterial challenge may stimulate production of anti-Hsp65 antibodies, that subsequently could cross-react with mammalian Hsp60 expressed by ECs, SMCs, and macrophages in nascent plaques. Initial antibody binding would then stimulate further antibody synthesis by plasma cells, increasing the serum anti-Hsp65/60 titer.

Chronically elevated anti-Hsp65/60 serum titers are also associated with development of restenotic (i.e., secondary atherosclerotic) lesions. MUKHERJEE et al. (1996) determined the serum anti-Hsp65 titers of patients undergoing percutaneous transluminal coronary angioplasty (PTCA), pre-operatively through 6 months post-PTCA. Compared to normal, age-matched controls, anti-Hsp65 titers were higher at baseline for all patients, consistent with the presence of primary lesions. Patients with a serum anti-Hsp65 titer that was significantly reduced immediately post-procedure and at discharge (relative to their pre-operative titer) did not develop significant restenotic complications within 6 months after PTCA. In contrast, serum anti-Hsp65 titers remained unchanged in patients that developed restenosis within 6 months post-PTCA. The authors of the study suggested that chronically high levels of circulating anti-Hsp65 antibodies post-PTCA induce an antibody-mediated inflammatory reaction at the site of balloon injury, that initiates restenotic growth. As yet though, there has been no explanation of how or why anti-Hsp65 titers decline post-PTCA, or why this fails to occur prior to development of restenosis.

The pathological significance of anti-Hsp60 antibodies to lesion development remains unclear. One suggestion has been that anti-Hsp60 antibodies deposited in the plaque contribute to formation and expansion of the necrotic core (HANSSON 1994). XU et al. (1994), and SCHETT et al. (1995) both demonstrated that antibodies against Hsp65/60 caused ^{51}Cr release from stressed ECs, if they were subsequently incubated with either complement or peripheral blood mononuclear cells. In contrast, other antibodies, including anti-Hsp70 antibodies, did not cause cytolysis. Given that ECs, SMCs, and macrophages may all express surface Hsp60 (see Sect. D.II.1), it is plausible to suggest that anti-Hsp60 antibodies precipitate endothelial denudation and necrotic core expansion via cell-dependent or antibody-dependent cytolysis. At present though, this hypothesis remains untested.

3. Biphasic Effects of Hsp Expression in Organ Transplantation: Tissue Preservation Versus Graft Arteriosclerosis

Since cardiac transplantation began, initial allograft survival has risen dramatically as acute rejection declined. Consequently, occlusive arteriosclerosis within the coronary arteries of donor hearts has emerged as a distinct clinical threat to graft viability. Graft arteriosclerosis develops much more rapidly than primary lesions, but appears to include similar immunological recognition of Hsps, particularly by T-lymphocytes.

QIAN et al. (1995) divided the pathophysiological changes of the myocardial graft into three distinct stages: an initial ischemia/reperfusion response by the allograft tissue, subsequent infiltration and activation of recipient T-lymphocytes, and finally, overt inflammatory rejection of the allograft with tissue destruction. There is a similar cascade in the coronary arteries, except that concentric SMC hyperplasia and vascular occlusion follows T-lymphocytic infiltration (reviewed in GORDON 1996). Transplantation in and of itself induces rapid and transient Hsp expression by rat hearts (CURRIE et al. 1987), presumably in response to ischemia/reperfusion injury. It has also been shown that pre-induction of the stress response effectively increased survival and function of rat and pig kidney, or rat heart grafts (PERDRIZET et al. 1989, 1993; TAMAKI et al. 1997). To date though, no one has directly examined whether the stress response preserves coronary and renal vasomotor activity in allografts.

Post-ischemic recruitment and activation of T-lymphocytes in grafted myocardium and its associated coronary vessels begins as early as three days post-transplantation. During this period, a large number of CD4+/8-, CD4-/8+, and γ/δ T-lymphocytes infiltrate both the myocardium and coronary vessels (IZUTANI et al. 1995; LIU et al. 1996). Initially, it appeared that overtly donor-specific, alloreactive T-lymphocytes comprised a rather small proportion of the monocytic infiltrate (DUQUESNOY et al. 1995). However, results of in vitro mitogenic outgrowth analyses have subsequently shown that many supposedly "non-specific" T-lymphocytes recognize donor Hsp70 or Hsp60 (MOLITERNO et al. 1995a,b; LIU et al. 1996, 1997; TRIEB et al. 1996; LATIF et al. 1997). This is not an autoimmune response; rather, donor Hsps are recognized as "non-self" antigens. In retrospect, this observation is not surprising, since clonal deletion during the fetal period can only eliminate T-lymphocytes that recognize Hsp peptides complexed with "self" major histocompatibility complex (MHC) molecules (LAFFERTY and GAZDA 1997). The MHC repertoire of non-dedicated antigen presenting cells (SMCs, ECs, and macrophages) within the graft (SUTTLES et al. 1995; POBER et al. 1996) will very likely bind different Hsp-derived antigenic peptides, that will be recognized as "non-self" by the host lymphocyte population. Antigenic recognition is also aided by the fact

that many graft recipients will have memory T-lymphocytes from prior mycobacterial challenge, that recognize the immunodominant mycobacterial antigens, Hsp71 and Hsp65. These memory cells would rapidly proliferate after
cross-reacting with donor tissue homologues.

Preliminary evidence also suggests that γ/δ T-lymphocytes infiltrating the
grafted vessels are also directly activated by Hsps. While γ/δ cells typically
comprises only 1–3% of circulating T-lymphocytes, they make up 10–30% of
T-lymphocytes in endomyocardial biopsies, and in both primary and graft
arteriosclerotic lesions (KLEINDIENST et al. 1993; DUQUESNOY et al. 1995;
MOLITERNO et al. 1995a). These histological findings suggest γ/δ T-
lymphocytes are either preferentially recruited from the circulation, or proliferate in situ. The ligands responsible for activating γ/δ T-lymphocyte receptors
are still poorly defined, but appear to include cellular products generated
when tissue is abraded, overheated, exposed to toxic metals, or attacked by
invading organisms; in short, products of stress-induced genes (O'BRIEN et al.
1991). Furthermore, γ/δ T-lymphocytes readily proliferate in response to
soluble Hsp60 or -70; unlike α/β T-lymphocytes though, γ/δ T-lymphocytes do
not require that Hsps to be bound to MHCs for recognition (FU et al. 1994;
SRIVASTAVA 1994; MOLITERNO et al. 1995b). Therefore, increased Hsp
expression or release after ischemic or mechanical injury of the arterial
wall associated with grafting may directly stimulate clonal expansion of the $\gamma/$
δ T-lymphocyte subpopulation. Whether γ/δ T-lymphocytes uniquely contribute to progression of graft arteriosclerosis remains unknown, but is under
investigation.

As with primary atherosclerotic lesions, general effects of T-lymphocyte
activation upon subsequent SMC proliferation in graft arteriosclerosis have
been reviewed elsewhere (HANSSON et al. 1993; ADAMS et al. 1993; LIBBY and
TANAKA 1994).

III. Hsps in the Mature Atherosclerotic Plaque

As stated previously, intermediate and advanced, necrotic atherosclerotic lesions routinely contain discrete areas of increased as well as reduced Hsp
immunoreactivity, compared to normal vessels. This section outlines how
oxidized lipoprotein and cholesterol deposition, ischemia, and oxygen radical
injury associated with mature lesions combine to both stimulate and inhibit
Hsp production by arterial cells. Data from in vitro studies will be reviewed
which suggest that reduced Hsp expression may predispose certain areas to
necrotic core formation or expansion.

1. Induction of Hsp Expression by oxLDL

During chronic hypercholesterolemia, circulating low density lipoprotein particles (LDL) are oxidized by ECs, SMCs, and macrophages to form oxLDL.
The oxidized components are then deposited throughout the plaque, via mul-

tiple pathways (GREENSPAN et al. 1997). This concentrates mitogenic and toxic oxidation by-products in and around the plaque cells, that potently affect vascular Hsp synthesis.

If cultured EAhy-926 or HUVEC endothelial cells were exposed to 200 μg/ml oxLDL for 7 h, Hsp70 synthesis increased dramatically; equivalent results were obtained for human umbilical vein SMCs or the A-617 SMC line (ZHU et al. 1994, 1995). Murine peritoneal macrophages and U937 or HL60 promonocytic leukemia cells also responded to oxLDL challenge by expressing Hsp23 and Hsp60 (YAMAGUCHI et al. 1993; FROSTEGARD et al. 1996) However, higher concentrations of oxLDL, or longer exposure times, reduced Hsp expression and killed SMCs, ECs, and macrophages. In SMCs, both the early stress response and subsequent cytotoxicity in response to oxLDL treatment were greatest in subconfluent, actively proliferating cells, while quiescent or confluent SMCs were minimally affected (ZHU et al. 1995). Furthermore, highly de-differentiated A-617 SMCs expressed less Hsp70 in response to oxLDL than did more fully differentiated primary SMCs, and were more sensitive to oxLDL injury. OxLDL toxicity also increased dramatically if SMCs or ECs were treated with simvastatin, which blocked the endogenous stress response (PIRILLO et al. 1997). Cumulatively, these in vitro studies suggest that increased Hsp expression within the atherosclerotic plaque in vivo protects plaque cells against oxLDL-induced injury, while reduced Hsp expression sensitizes them.

How oxLDL can initially stimulate then later inhibit Hsp production, or only affect proliferating cells, remains incompletely defined. Different rates of oxLDL internalization are not responsible, because oxLDL uptake did not differ between subconfluent and confluent ECs after scrape-wounding (ZHU et al. 1996). It is more likely that individual components within the oxLDL moiety affect plaque cell Hsp expression differently. For example, ZHU et al. (1996) reported that extractable lipids from oxLDL rapidly and specifically stimulated Hsp production, but only in proliferating cells. Similarly, the oxidized lipid 4-hydroxynonenal, that is present in both oxLDL and atherosclerotic lesions, strongly induces heat shock factor activation in rapidly proliferating HeLa cells (CAJONE et al. 1989). Oxidized, lipid-soluble apolipoproteins or protein fragments might also trigger Hsp synthesis, if they are recognized as damaged or denatured proteins. In contrast, Hsp synthesis may be inhibited by one or more oxysterols (oxidized cholesterol derivatives) within oxLDL (PETTERSEN et al. 1991; MALAVASI et al. 1992; RAMASAMY et al. 1992). Some purified oxysterols do not significantly affect Hsp production by plaque cells, but others such as cholestane-3α,5β,6α-triol (C3ol) block Hsp synthesis by heat shocked SMCs in a dose-dependent manner (JOHNSON et al. 1995; ZHU et al. 1996).

The preceding observations in cultured vascular cells suggest oxLDL exposure affects Hsp expression in a complex manner. The stress response by plaque cells in vivo will be still more complex, because the extent of LDL oxidation can vary, which will affect its toxicity. Moreover, the total quantity of oxLDL present in and around plaque cells is variable with time, as is the relative proliferation rate and the extent of cellular de-differentiation (at least of SMCs). Finally, the length of time of oxLDL exposure will vary for cells in different areas within the plaque. The exact effect of all these interacting factors upon Hsp expression remains for future studies to define. Given current evidence though, it is plausible to speculate that oxLDL deposition contributes to formation of areas of high versus low Hsp expression in atherosclerotic plaques. Areas of reduced Hsp expression would be more sensitive to injury, either by oxLDL itself, or by other stresses associated with advanced necrotic lesions (see below).

2. The Stress Response and Plaque Cell Survival Versus Necrosis

As the necrotic core expands within an advanced atherosclerotic plaque, a dynamic equilibrium develops between fibrotic organization and rupture. SMCs continue to proliferate and lay down extracellular matrix, encapsulating the debris and stabilizing the plaque. In opposition, there is increased mechanical stress, altered shear flow, further cellular injury, and matrix proteolysis by activated macrophages, which promote plaque fissuring and rupture (FUSTER et al. 1992; STARY et al. 1995). Present evidence suggests that acute Hsp expression by plaque cells is affected by ischemia, oxygen free radical injury, and oxysterol toxicity, each of which may injure plaque cells, expand the necrotic core, and destabilize the plaque.

One deleterious consequence of plaque growth and fibrotic organization is that oxygen and nutrient deprived ischemic areas develop, that can serve as foci for necrosis. Studies by the reviewer (JOHNSON et al. 1990, 1995) examined whether endogenous Hsp synthesis protected aortic SMCs against ischemia. Cultured rabbit or macaque SMCs responded to up to 40 h of chronic serum/ O_2 deprivation (an in vitro model of plaque ischemic injury) with a classical transient stress response, that included cellular accumulation of Hsp70. Hsp70 levels in ischemic SMCs peaked at 5–10 h after onset of serum/O_2 deprivation in vitro, then steadily declined below baseline levels through 40 h. Cell viability (assessed by Trypan blue or propidium iodide exclusion) also declined after 10 h, paralleling the cell-associated Hsp70. Pre-treating SMCs for 30 min at 42°C, with 6 h of recovery at 37°C, dramatically delayed SMC death during subsequent serum/O_2 deprivation, and coincided with an approximately 3-fold higher level of cell-associated Hsp70 than was observed after serum/O_2 deprivation alone. Once again, overall SMC viability declined only after cell-associated Hsp70 levels fell below baseline. Thus, acutely elevated Hsps appear to protect SMCs against ischemic cell death, while their loss makes SMCs more susceptible.

There is also evidence that exogenous application of Hsp70 will transiently protect SMCs against ischemia. Hsp70 has been detected in the extracellular matrix of advanced atherosclerotic plaques, although the mechanism for its release remains undetermined (BERBERIAN et al. 1990; JOHNSON et al.

1993). Given this observation, cultured SMCs were treated with 10 μg/ml of purified Hsc70 (a mixture of 90–95% Hsc70, and 5–10% Hsp70, purified from bovine brain), then serum/O_2 deprived as before (JOHNSON et al. 1990, 1995). In multiple in vitro trials, ischemic SMCs treated with exogenous Hsc70 survived longer than either ischemic SMCs given vehicle only, or SMCs that were pre-treated with heat shock. SMCs treated with fluorescently tagged or ^{125}I-radiolabeled Hsc70 bound sufficient exogenous Hsc70 to increase total cell-associated Hsp70 approximately 3-fold over the level observed in untreated cells. Yet greater than 90% of radiolabeled Hsc70 remained trypsin sensitive, indicating that exogenous Hsc70 induced cytoprotection without being internalized by stressed SMCs (JOHNSON and TYTELL 1993, and unpublished data). No specific "Hsp receptor" has been reported, but there is mounting evidence that Hsps can bind directly to MHC molecules (possibly via the Vβ domain) of antigen presenting cells (YOUNG 1993; SCHLIEVERT 1997), a group that includes SMCs. Thus, Hsp70 released during a focal arterial injury may act through MHC molecules as a paracrine "alarm" signal that stimulates resistance of surrounding plaque cells to a subsequent injury. This model is highly speculative though, and additional studies both in culture and in intact vessels are required. These results also suggest there are possible therapeutic uses for exogenous Hsc70 (and perhaps other Hsps) in preventing rupture of advanced atherosclerotic lesions.

Like ischemia, oxygen free radical (oxyradical) injury occurs repeatedly as atherosclerotic lesions mature. In vitro studies suggest that reperfusion of ischemic zones by neovascularization is a significant source of oxyradicals in plaques (CLERGET and POLLA 1990). They might also be generated by metabolism of xanthine/hypoxanthine, that accumulates as a consequence of ATP depletion secondary to ischemia (McCORD et al. 1985). A third oxyradical source is the respiratory burst of macrophages that have been stimulated by lipoprotein uptake (VIRELLA et al. 1995). Regardless of their origin, oxyradicals can injure ECs, SMCs, and resident macrophage foam cells (SUSSMAN and BULKLEY 1990; MAZIERE et al. 1996), stimulating a stress response and increasing HSP synthesis (DONATI et al. 1990). Currently available evidence though, suggests that differences in the magnitude of Hsp expression by ECs versus macrophages in response to oxyradicals may dramatically affect their relative survival times in late-stage lesions.

Endothelial activation or injury is an essential step during in atherosclerotic lesion formation, but clinically, overt denudation does not occur until the later stages of plaque development (STARY et al. 1995). Therefore, later events such as oxyradical injury may be major contributors to endothelial loss. Compared to other plaque-associated stresses though, umbilical and coronary vein ECs in culture are exceptionally sensitive to oxyradical injury, either by peroxide or hypoxanthine/xanthine oxidase (JORNOT and JUNOD 1992; LU et al. 1993). Both oxyradical sources will stimulate endothelial Hsp70 production, but interestingly, total accumulated Hsp70 is considerably less than observed in heat shocked ECs (JORNOT et al. 1991; AUCOIN et al. 1995), suggesting that reduced Hsp production is in part responsible for endothelial sensitivity to oxyradicals. Why there is only a weak endothelial stress response to oxyradical injury remains undetermined, but WANG et al. (1997) recently reported that oxyradical injury followed by heat shock caused cultured bovine pulmonary arterial ECs to undergo apoptosis. This observation suggests that a robust stress response is somehow incompatible with other endothelial responses to oxyradicals. It remains for future studies to determine whether induction of apoptosis by an inappropriately timed stress response contributes to late en-dothelial loss over atherosclerotic plaques.

In contrast to ECs, differentiated monocytes and macrophages readily synthesize Hsps after oxidative injury. U937 cells, a nondifferentiated human promonocytic cell line, do not accumulate significant amounts of Hsps basally, or in response to peroxide injury (MINISINI et al. 1994). In contrast, mature, circulating human monocytes maintain higher levels of Hsp70 and Hsp27 than other cell types under non-stressed conditions, suggesting that monocytes are constitutively protected against respiratory burst injury (JAATTELA and WISSING 1993). In vitro, stimulation of the respiratory burst with macrophage colony stimulating factor (which is present in atherosclerotic plaques), vitamin D_3, or erythrophagocytosis activates heat shock factor, induces synthesis of additional Hsp90, -70, and -60, and dramatically increases the resistance of macrophages to both endogenous and exogenous peroxide injury (POLLA et al. 1987; CLERGET and POLLA 1990; TESHIMA et al. 1996; WANG et al. 1996). From these data, it can be postulated that strong Hsp expression by macrophages in the context of atherosclerotic lesions similarly protects them from oxyradical injury. Conversely, reduced Hsp expression (as is observed in immunostained atherosclerotic plaques) predisposes them to cytolysis, that in turn contributes to necrotic core formation. However, protecting macrophages by enhancing their endogenous stress response may also bolster their ability to erode the fibrous cap, again leading to plaque destabilization. Further studies will be necessary to determine which process is more important in plaque progression to rupture, and how the stress response is involved.

Along with ischemia and oxyradicals, oxysterols also contribute to ne-crotic core formation and expansion. Oxysterols are both deposited in lesions in conjunction with oxLDL, and are formed in situ by oxyradical attack on native cholesterol or cholesterol esters (GUYTON et al. 1990; RAMASAMY et al. 1992; AUPEIX et al. 1995; NISHIO and WATENABE 1996). However, unlike ischemia or oxyradicals, certain oxysterols inhibit cytoprotective Hsp synthe-sis. In cultured SMCs, sublethal and lethal concentrations of either C3ol or 25-hydroxycholesterol (25HC), two toxic oxysterols found in necrotic plaques, failed to induce Hsp70 expression (JOHNSON et al. 1995). Results of subsequent ^{35}S-methionine labeling assays indicated that C3ol blocked almost all protein production by SMCs, both before and after heat shock. As a result, heat shock pre-treatment in the presence of C3ol did not protect SMCs against oxysterol intoxication, and only exogenous Hsp70 prevented C3ol-induced cell death. In contrast, 25HC only slightly reduced Hsp synthesis after heat shock. Moreover, SMC death in response to 25HC exposure was reduced both by heat shock pretreatment, and by exogenous Hsc70. These results suggest that oxysterols present in plaques can kill SMCs directly, as well as by inhibiting the stress response that protects SMCs against other plaque-associated insults like ischemia, etc. Similar effects of oxysterol intoxication may also inhibit expression of Hsps by macrophages, accelerating necrotic core expansion by blocking Hsp production during the respiratory burst.

Finally, although the acute stress response protects plaque cells against injury, chronic or repetitive stimulation of the stress response may be deleterious to overall lesion stability. SLEPIAN et al. (1996) reported that inducing the stress response of cultured rat SMCs with heat, cadmium, or tributylin prior to scrape wounding reduced subsequent ^{3}H-thymidine uptake 75%. NESCHIS et al. (1997) confirmed that whole body hyperthermic preconditioning of intact rats significantly reduced intimal hyperplasia resulting from balloon catheter injury. The reduction in SMC growth appeared to result from both preservation of medial cell viability after stretch injury, and from direct inhibition of SMC hyperplasia. The authors of both studies suggested that "pre-modifying" SMCs by inducing an acute stress response might beneficially reduce aggressive SMC proliferation associated with balloon angioplasty. However, these observations also suggest that repeated activation of the stress response in primary lesions may inhibit smooth muscle proliferation required for fibrotic organization, which would indirectly accelerate plaque rupture. While it is only a speculation at present, appropriate transgenic animal models are already available for testing this hypothesis (see below).

E. Future Directions

Studies by many different investigators clearly indicate that Hsps have multiple roles in formation and progression of primary, secondary, and graft arteriosclerotic lesions. Yet this is still an emerging field, with a number of unanswered questions. For example, although hsp70 gene regulation is clearly altered in animals and human patients with an inherited predisposition to hypertension, no one has yet published the sequence of the mutated promoter or coding regions of a polymorphic hsp70 gene. Similarly, Hsp70 protein derived from hypertensive animals has never been directly assayed for altered protein binding, or for other biochemical characteristics that might promote hypertension or atherosclerotic lesion formation.

Possible functions of the low molecular weight (20–30 kDa) and 90 kDa Hsps during atherosclerosis have also remained virtually unexplored. For example, there is mounting evidence for both genomic and non-genomic effects of estrogens on atherosclerotic lesion progression (NATHAN and CHAUDHURI 1997). Given that the estrogen receptor complex requires Hsp90 to maintain its inactive conformation, altered Hsp90 expression in atherosclerotic plaques may radically change estrogen receptor activity in the arterial wall. Similarly, low molecular weight Hsps are involved with both SMC contractile "latch state", and mitogenic signal transduction by SMCs and ECs (Bitar et al. 1991; Santell et al. 1992; Piotrowicz et al. 1995; Yamada et al. 1995). Since expression of low molecular weight family members is also altered in the primary lesion (Johnson et al. 1993), contraction and mitogenic signaling are likely affected as well.

Finally, attempts to understand the pathophysiology of plaque rupture in general have been hampered by the lack of a robust whole animal model. Most of our knowledge of plaque disruption has come from histological surveys of advanced human lesions, along with a handful of experimental studies. Recently though, apolipoprotein E (apoE) knockout mice were described that develop atherosclerotic lesions which progress to necrosis in a manner similar to human vessels (Nakashima et al. 1994; Reddick et al. 1994). It is now possible to directly test many of the predictions made from in vitro studies (and described in this review) in a whole animal context. In particular, serial analyses are needed that establish the timing, distribution patterns, and consequences of acute versus chronic Hsp expression in early and late stage lesions. Another issue is, does the stress response cause endothelial apoptosis and denudation? Is SMC death consistently observed in areas with reduced Hsp expression? An especially interesting question is whether chronic stimulation causes sustained down-regulation of the stress response; in other words, can the stress response be "exhausted" in plaques, leaving the component cells more vulnerable to toxicity? If so, will constitutive, heterologous Hsp expression prevent or reduce plaque progression? Alternatively, is atherosclerotic plaque progression accelerated when apoE knockout mice are crossed with transgenic animals that constitutively overexpress Hsp70? It will also be possible to examine the roles of Hsps in vascular reparative responses after plaque rupture, or after balloon angioplastic injury of primary lesions.

Acknowledgements. The author gratefully acknowledges the contribution of Dr. Mike Tytell, Bowman Gray School of Medicine of Wake Forest University, Winston-Salem, NC, for helpful comments and critiques during preparation of this manuscript.

References

Adams DH, Wyner LR, Karnovsky MJ (1993) Cardiac graft arteriosclerosis in the rat. Transplant Proc 25:2071–2073

Aucoin MM, Barhoumi R, Kochevar DT, Granger HJ, Burghardt RC (1995) Oxidative injury of coronary venular endothelial cells depletes intracellular glutathione and induces HSP 70 mRNA. Am J Physiol 268:H1651–H1658

Aupeix K, Weltin D, Mejia JE, Christ M, Marchal J, Freyssinet JM, Bischoff P (1995) Oxysterol-induced apoptosis in human monocytic cell lines. Immunobiology 194:415–428

Berberian PA, Myers W, Tytell M, Challa V, Bond MG (1990) Immunohistochemical localization of heat shock protein-70 in normal-appearing and atherosclerotic specimens of human arteries. Am J Pathol 136:71–80

Bitar KN, Kaminski MS, Hailat N, Cease KB, Strahler JR (1991) Hsp27 is a mediator of sustained smooth muscle contraction in response to bombesin. Biochem Biophys Res Commun 181:1192–1200

Blake MJ, Klevay LM, Halas ES, Bode AM (1995) Blood pressure and heat shock protein expression in response to acute and chronic stress. Hypertension 25:539–544

Bongrazio M, Comini L, Gaia G, Bachetti T, Ferrari R (1994) Hypertension, aging, and myocardial synthesis of heat-shock protein 72. Hypertension 24:620–624

Cajone F, Salina M, Benelli-Zazzera A (1989) 4-Hydroxynonenal induces a DNA-binding protein similar to the heat-shock factor. Biochem J 262:977–979

Campisi D, Cutolo M, Carruba G, Lo Casto M, Comito L, Granata OM, Valentino B, King RJB, Castagnetta L (1993) Evidence for soluble and nuclear site I binding of estrogens in human aorta. Atherosclerosis 103:267–277

Clerget M, Polla BS (1990) Erythrophagocytosis induces heat shock protein synthesis by human monocytes-macrophages. Proc Natl Acad Sci USA 87:1081–1085

Currie RW, Sharma VK, Stepkowski SM, Payce RF (1987) Protein synthesis in heterotopically transplanted rat hearts. Exp Cell Biol 55:46–56

Donati YR, Slosman DO, Polla BS (1990) Oxidative injury and the heat shock response. Biochem Pharmacol 40:2571–2577

Duquesnoy RJ, Moliterno R, Qian J, Donovan-Peluso M, Pan F, Valdivia L (1995) Role of heat shock protein immunity in allograft rejection. Transplant Proc 27:468–470

Dustin ML, Springer TA (1991) Role of lymphocyte adhesion receptors in transient interactions and cell locomotion. Annu Rev Immunol 9:27–66

Frostegard J, Kjellman B, Gidlund M, Andersson B, Jindal S, Kiessling R (1996) Induction of heat shock protein in monocytic cells by oxidized low density lipoprotein. Atherosclerosis 121:93–103

Fu YX, Vollmer M, Kalataradi H, Heyborne K, Reardon C, Miles C, O'Brien R, Born W (1994) Structural requirements for peptides that stimulate a subset of γ/δ T cells. J Immunol 152:1578–1588

Fuster V, Badimon L, Badimon JJ, Chesebro JH (1992) The pathogenesis of coronary artery disease and the acute coronary syndromes, part 1. N Engl J Med 326:242–250

Gordon D (1996) Transplant arteriosclerosis. In: Fuster V, Ross R, Topol E (eds) Atherosclerosis and coronary artery disease. Lippencott-Raven, Philadelphia, pp 715–726

Greenspan P, Yu H, Mao F, Gutman RL (1997) Cholesterol deposition in macrophages: foam cell formation mediated by cholesterol-enriched oxidized low density lipoprotein. J Lipid Res 38:101–109

Guyton JR, Black BL, Seidel CL (1990) Focal toxicity of oxysterols in vascular smooth muscle cell culture. A model of the atherosclerotic core region. Am J Pathol 137:425–434

Hamet P, Malo D, Hashimoto T, Tremblay J (1990a) Heat stress genes in hypertension. J Hypertens 8:S47–S52

Hamet P, Malo D, Tremblay J (1990b) Increased transcription of a major stress gene in spontaneously hypertensive mice. Hypertension 15:904–908

Hamet P, Tremblay J, Malo D, Kunes J, Hashimoto T (1990c) Genetic hypertension is characterized by the abnormal expression of a gene localized in major histocompatibility complex HSP70. Transplant Proc 22:2566–2567

Hamet P, Kong D, Pravenec M, Kunes J, Kren V, Klir P, Sun YL, Tremblay J (1992) Restriction fragment length polymorphism of hsp70 gene, localized in the RT1 complex, is associated with hypertension in spontaneously hypertensive rats. Hypertension 19:611–614

Hamet P, Sun YL, Malo D, Kong D, Kren V, Pravenec M, Kunes J, Dumas P, Richard L, Gagnon F et al (1994) Genes of stress in experimental hypertension. Clin Exp Pharmacol Physiol 21:907–911

Hamet P (1995) Environmental stress and genes of hypertension. Clin Exp Pharmacol Physiol 22:S394–S398

Hansson GK, Geng YJ, Holm J, Stemme S (1993) Lymphocyte adhesion and cellular immune reactions in chronic rejection and graft arteriosclerosis. Transplant Proc 25:2050–2051

Hansson GK (1994) Immunological control mechanisms in plaque formation. Bas Res Cardiol 89 [Suppl 1]:41–46

Hansson GK, Libby P (1996) The role of the lymphocyte. In: Fuster V, Ross R, Topol E (eds) Atherosclerosis and coronary artery disease. Lippincott-Raven, Philadelphia, pp 557–568

Hashimoto T, Mosser RD, Tremblay J, Hamet P (1991) Increased accumulation of hsp70 messenger RNA due to enhanced activation of heat-shock transcription factor in spontaneously hypertensive rats. J Hypertens 9:S170–S171

Heng MK, Heng MC (1994) Heat-shock protein 65 and activated γ/δ T cells in injured arteries. Lancet 344:921–923

Hightower LE, White FP (1982) Preferential synthesis of rat heat-shock and glucose-regulated proteins in stressed cardiovascular cells. In: Schlesinger M (ed) Heat-shock: from bacteria to man. Cold Spring Harbor Press, Cold Spring Harbor, New York, pp 369–377

Izutani H, Miyagawa S, Matsumiya G, Nakata S, Shimazaki Y, Matsuda H, Shirakura R (1995) Graft coronary arteriosclerosis in rat heart transplant model with FK 506 short-term administration. Transplant Proc 27:579–581

Jaattela M, Wissing D (1993) Heat-shock proteins protect cells from monocyte cytotoxicity: possible mechanism of self-protection. J Exp Med 177:231–236

Johnson AD, Berberian PA, Bond MG (1990) Effect of heat shock proteins on survival of isolated aortic cells from normal and atherosclerotic cynomolgus macaques. Atherosclerosis 84:111–119

Johnson AD, Berberian PA, Tytell M, Bond MG (1993) Atherosclerosis alters the localization of HSP70 in human and macaque aortas. Exp Mol Pathol 58:155–168

Johnson AD, Tytell M (1993) Exogenous HSP70 becomes cell associated, but not internalized, by stressed arterial smooth muscle cells. In Vitro Cell Dev Biol 29A:807–812

Johnson AD, Berberian PA, Tytell M, Bond MG (1995) Differential distribution of 70-kD heat shock protein in atherosclerosis. Its potential role in arterial SMC survival. Arterioscler Thromb Vasc Biol 15:27–36

Johnson HM, Torres BA, Soos JM (1996) Superantigens: structure and relevance to human disease. Proc Soc Exp Biol Med 212:99–109

Jornot L, Mirault ME, Junod AF (1991) Differential expression of hsp70 stress proteins in human endothelial cells exposed to heat shock and hydrogen peroxide. Am J Respir Cell Mol Biol 5:265–275

Jornot L, Junod AF (1992) Response of human endothelial cell antioxidant enzymes to hyperoxia. Am J Respir Cell Mol Biol 6:107–115

Kleindienst R, Xu Q, Willeit J, Waldenberger FR, Weimann S, Wick G (1993) Immunology of atherosclerosis. Demonstration of heat shock protein 60 expression and T lymphocytes bearing α/β or γ/δ receptor in human atherosclerotic lesions. Am J Pathol 142:1927–1937

Kohane DS, Sarzani R, Schwartz JH, Chobanian AV, Brecher P (1990) Stress-induced proteins in aortic smooth muscle cells and aorta of hypertensive rats. Am J Physiol 258:H1699–H1705

Kunes J, Poirier M, Tremblay J, Hamet P (1992) Expression of hsp70 gene in lymphocytes from normotensive and hypertensive humans. Acta Physiol Scand 146:307–311

Lafferty KJ, Gazda LS (1997) Tolerance: a case of self/not-self discrimination maintained by clonal deletion? Hum Immunol 52:119–126

Latif N, Yacoub MH, Dunn MJ (1997) Association of pretransplant anti-heart antibodies against human heat shock protein 60 with clinical course following cardiac transplantation. Transplant Proc 29:1039–1040

Libby P, Tanaka H (1994) The pathogenesis of coronary arteriosclerosis ("chronic rejection") in transplanted hearts. Clin Transplant 8:313–318

Liu K, Moliterno RA, Qian J, Attfield D, Valdivia L, Duquesnoy RJ (1996) Role of heat shock proteins in heart transplant rejection. J Heart Lung Transplant 15:222–228

Liu K, Moliterno RA, Fu XF, Duquesnoy RJ (1997) Identification of two types of autoreactive T lymphocyte clones cultured from cardiac allograft-infiltrating cells incubated with recombinant mycobacterial heat shock protein 71. Transplant Immunol 5:57–65

Lu D, Maulik N, Moraru II, Kreutzer DL, Das DK (1993) Molecular adaptation of vascular endothelial cells to oxidative stress. Am J Physiol 264:C715–C722

Lukashev ME, Klimanskaya IV, Postnov YV (1991) Synthesis of heat-shock proteins in cultured fibroblasts from normotensive and spontaneously hypertensive rat embryos. J Hypertens 9:S182–S183

Malavasi B, Rasetti MF, Roma P, Fogliatto R, Allevi P, Catapano AL, Galli G (1992) Evidence for the presence of 7-hydroperoxycholest-5-en-3 beta-ol in oxidized human LDL. Chem Phys Lipids 62:209–214

Malo D, Schlager G, Tremblay J, Hamet P (1989) Thermosensitivity, a possible new locus involved in genetic hypertension. Hypertension 14:121–128

Maziere C, Auclair M, Djavaheri-Mergny M, Packer L, Maziere JC (1996) Oxidized low density lipoprotein induces activation of the transcription factor NF-κB in fibroblasts, endothelial and smooth muscle cells. Biochem Mol Biol Int 39:1201–1207

McCord JM, Roy RS, Schaffer SW (1985) Free radicals and myocardial ischemia. The role of xanthine oxidase. Adv Myocardiol 5:183–189

McMurtry JP, Wexler BC (1981) Hypersensitivity of spontaneously hypertensive rats (SHR) to heat, ether, and immobilization. Endocrinology 108:1730–1736

McMurtry JP, Wexler BC (1983) Hypersensitivity of spontaneously hypertensive rats to heat and ether before the onset of high blood pressure. Endocrinology 112:166–171

Metzler B, Schett G, Kleindienst R, van der Zee R, Ottenhoff T, Hajeer A, Bernstein R, Xu Q, Wick G (1997) Epitope specificity of anti-heat shock protein 65/60 serum antibodies in atherosclerosis. Arterioscler Thromb Vasc Biol 17:536–541

Minisini MP, Kantengwa S, Polla BS (1994) DNA damage and stress protein synthesis induced by oxidative stress proceed independently in the human premonocytic line U937. Mut Res 315:169–179

Moalic JM, Bauters C, Himbert D, Bercovici J, Mouas C, Guicheney P, Baudoin-Legros M, Rappaport L, Emanoil-Ravier R, Mezger V et al (1989) Phenylephrine, vasopressin and angiotensin II as determinants of proto-oncogene and heat-shock protein gene expression in adult rat heart and aorta. J Hypertens 7:195–201

Moliterno R, Valdivia L, Pan F, Duquesnoy RJ (1995a) Heat shock protein reactivity of lymphocytes isolated from heterotopic rat cardiac allografts. Transplantation 59:598–604

Moliterno R, Woan M, Bentlejewski C, Qian J, Zeevi A, Pham S, Griffith BP, Duquesnoy RJ (1995b) Heat shock protein-induced T-lymphocyte propagation from endomyocardial biopsies in heart transplantation. J Heart Lung Transplant 14:329–337

Mukherjee M, De Benedictis C, Jewitt D, Kakkar VV (1996) Association of antibodies to heat-shock protein-65 with percutaneous transluminal coronary angioplasty and subsequent restenosis. Thromb Haemost 75:258–260

Mulvany MJ (1996) Peripheral vasculature in essential hypertension. Clin Exp Pharmacol Physiol 23:S6–S10

Nakashima Y, Plump AS, Raines EW, Breslow JL, Ross R (1994) ApoE-deficient mice develop lesions of all phases of atherosclerosis throughout the arterial tree. Arterioscler Thromb Vasc Biol 14:133–140

Nathan L, Chaudhuri G (1997) Estrogens and atherosclerosis. Annu Rev Pharmacol Toxicol 37:477–515

Neschis DG, Barnathan ES, Safford SD, Kariko K, Langer DJ, Hanna AK, Golden MA (1997) Thermal preconditioning prior to arterial balloon injury: limitation of injury and sustained reduction of intimal thickening. FASEB J 11:A317 (abstract)

Nishio E, Watanabe Y (1996) Oxysterols induced apoptosis in cultured smooth muscle cells through CPP32 protease activation and bcl-2 protein downregulation. Biochem Biophys Res Commun 226:928–934

O'Brien RL, Happ MP, Dallas A, Cranfill R, Hall L, Lang J, Fu YX, Kubo R, Born W (1991) Recognition of a single hsp-60 epitope by an entire subset of gamma delta T lymphocytes. Immunol Rev 121:155–170

Patton WF, Erdjument-Bromage H, Marks AR, Tempst P, Taubman MB (1995) Components of the protein synthesis and folding machinery are induced in vascular smooth muscle cells by hypertrophic and hyperplastic agents. Identification by comparative protein phenotyping and microsequencing. J Biol Chem 270:21404–21410

Perdrizet GA, Heffron TG, Buckingham FC, Salciunas PJ, Gaber AO, Stuart FP, Thistlethwaite JR (1989) Stress conditioning: a novel approach to organ preservation. Curr Surg 46:23–26

Perdrizet GA, Kaneko H, Buckley TM, Fishman MS, Pleau M, Bow L, Schweizer RT (1993) Heat shock and recovery protects renal allografts from warm ischemic injury and enhances HSP72 production. Transplant Proc 25:1670–1673

Pettersen KS, Boberg KM, Stabursvik A, Prydz H (1991) Toxicity of oxygenated cholesterol derivatives toward cultured human umbilical vein endothelial cells. Arterioscler Thromb 11:423–428

Piotrowicz RS, Weber LA, Hickey E, Levin EG (1995) Accelerated growth and senescence of arterial endothelial cells expressing the small molecular weight heat-shock protein HSP27. FASEB J 9:1079–1084

Pirillo A, Jacoviello C, Longoni C, Radaelli A, Catapano AL (1997) Simvastatin modulates the heat shock response and cytotoxicity mediated by oxidized LDL in cultured human endothelial smooth muscle cells. Biochem Biophys Res Commun 231:437–441

Pober JS, Orosz CG, Rose ML, Savage CO (1996) Can graft endothelial cells initiate a host anti-graft immune response? Transplantation 61:343–349

Polla BS, Healy AM, Byrne M, Krane SM (1987) 1,25-Dihydroxyvitamin D_3 induces collagen binding to the human monocyte line U937. J Clin Invest 80:962–969

Portig I, Pankuweit S, Lottspeich F, Maisch B (1996) Identification of stress proteins in endothelial cells. Electrophoresis 17:803–808

Qian J, Moliterno R, Donovan-Peluso MA, Liu K, Suzow J, Valdivia L, Pan F, Duquesnoy RJ (1995) Expression of stress proteins and lymphocyte reactivity in heterotopic cardiac allografts undergoing cellular rejection. Transplant Immunol 3:114–123

Ramasamy S, Boissonneault GA, Hennig B (1992) Oxysterol-induced endothelial cell dysfunction in culture. J Am Coll Nutr 11:532–538

Reddick RL, Zhang SH, Maeda N (1994) Atherosclerosis in mice lacking apo E. Evaluation of lesional development and progression. Arterioscler Thromb 14:141–147

Ross R (1995) Cell biology of atherosclerosis. Annu Rev Physiol 57:791–804

Santell L, Bartfeld NS, Levin EG (1992) Identification of a protein transiently phosphorylated by activators of endothelial cell function as the heat-shock protein HSP27. Biochem J 284:705–710

Schett G, Xu Q, Amberger A, van der Zee R, Recheis H, Willeit J, Wick G (1995) Autoantibodies against heat shock protein 60 mediate endothelial cytotoxicity. J Clin Invest 96:2569–2577

Schlievert PM (1997) Searching for superantigens. Immunol Invest 26:283–290

Seitz CS, Kleindienst R, Xu Q, Wick G (1996) Coexpression of heat-shock protein 60 and intercellular-adhesion molecule-1 is related to increased adhesion of monocytes and T cells to aortic endothelium of rats in response to endotoxin. Lab Invest 74:241–252

Slepian MJ, Massia SP, Whitesell L (1996) Pre-conditioning of smooth muscle cells via induction of the heat shock response limits proliferation following mechanical injury. Biochem Biophys Res Commun 225:600–605

Srivastava PK (1994) Heat shock proteins in immune response to cancer: the Fourth Paradigm. Experientia 50:1054–1060

Stary HC, Blankenhorn DH, Chandler AB, Glagov S, Insull W Jr, Richardson M, Rosenfeld ME, Schaffer SA, Schwartz CJ, Wagner WD et al (1992) A definition of the intima of human arteries and of its atherosclerosis-prone regions. A report from the Committee on Vascular Lesions of the Council on Arteriosclerosis, American Heart Association. Arterioscler Thromb 12:120–134

Stary HC, Chandler AB, Glagov S, Guyton JR, Insull W,Jr., Rosenfeld ME, Schaffer SA, Schwartz CJ, Wagner WD, Wissler RW (1994) A definition of initial, fatty streak, and intermediate lesions of atherosclerosis. A report from the Committee on Vascular Lesions of the Council on Arteriosclerosis, American Heart Association. Circulation 89:2462–2478

Stary HC, Chandler AB, Dinsmore RE, Fuster V, Glagov S, Insull W Jr, Rosenfeld ME, Schwartz CJ, Wagner WD, Wissler RW (1995) A definition of advanced types of atherosclerotic lesions and a histological classification of atherosclerosis. A report from the Committee on Vascular Lesions of the Council on Arteriosclerosis, American Heart Association. Arterioscler Thromb Vasc Biol 15:1512–1531

Stemme S, Hansson GK (1994) Immune mechanisms in atherogenesis. Ann Med 26:141–146

Strong JP, Malcom GT, Oalmann MC (1995) Environmental and genetic risk factors in early human atherogenesis: lessons from the PDAY study. Pathobiological Determinants of Atherosclerosis in Youth. Pathol Intl 45:403–408

Sussman MS, Bulkley GB (1990) Oxygen-derived free radicals in reperfusion injury. Methods Enzymol 186:711–723

Suttles J, Miller RW, Moyer CF (1995) T cell-vascular smooth muscle cell interactions: antigen-specific activation and cell cycle blockade of T helper clones by cloned vascular smooth muscle cells. Exp Cell Res 218:331–338

Tamaki T, Tanaka M, Konoeda Y, Koizumi H, Yamaguchi H, Kawamura A (1997) Cytoprotective effects of the 60kD heat-shock protein enhanced by heat stress in rat cardiac grafts with warm ischemic injuries. Transplant Proc 29:1340–1341

Teshima S, Rokutan K, Takahashi M, Nikawa T, Kishi K (1996) Induction of heat shock proteins and their possible roles in macrophages during activation by macrophage colony-stimulating factor. Biochem J 315:497–504

Tremblay J, Hadrava V, Kruppa U, Hashimoto T, Hamet P (1992) Enhanced growth-dependent expression of TGF-β 1 and hsp70 genes in aortic smooth muscle cells from spontaneously hypertensive rats. Can J Physiol Pharmacol 70:565–572

Trieb K, Grubeck-Loebenstein B, Eberl T, Margreiter R (1996) T cells from rejected human kidney allografts respond to heat shock protein 72. Transplant Immunol 4:43–45

Udelsman R, Blake MJ, Holbrook NJ (1991) Molecular response to surgical stress: specific and simultaneous heat shock protein induction in the adrenal cortex, aorta, and vena cava. Surgery 110:1125–1131

Udelsman R, Blake MJ, Stagg CA, Li DG, Putney DJ, Holbrook NJ (1993) Vascular heat shock protein expression in response to stress. Endocrine and autonomic regulation of this age-dependent response. J Clin Invest 91:465–473

Udelsman R, Li DG, Stagg CA, Holbrook NJ (1995) Aortic crosstransplantation between young and old rats: effect upon the heat shock protein 70 stress response. J Gerontol 50:B187–B192

Virella G, Munoz JF, Galbraith GM, Gissinger C, Chassereau C, Lopes-Virella MF (1995) Activation of human monocyte-derived macrophages by immune complexes containing low-density lipoprotein. Clin Immunol Immunopathol 75:179–189

Wang JH, Redmond HP, Watson RW, Bouchier-Hayes D (1997) Induction of human endothelial cell apoptosis requires both heat shock and oxidative stress responses. Am J Physiol 272:C1543–C1551

Wang YR, Xiao XZ, Huang SN, Luo FJ, You JL, Luo H, Luo ZY (1996) Heat shock pretreatment prevents hydrogen peroxide injury of pulmonary endothelial cells and macrophages in culture. Shock 6:134–141

Watanabe T, Haraoka S, Shimokama T (1997) Inflammatory and immunological nature of atherosclerosis. Intl J Cardiol 54[Suppl]:S51–S60

Wilkinson JM, Pollard I (1993) Immunohistochemical localisation of the 25 kDa heat shock protein in unstressed rats: possible functional implications. Anat Rec 237:453–457

Wright G, Iams S, Knecht E (1977) Resistance to heat stress in the spontaneously hypertensive rat. Can J Physiol Pharmacol 55:975–982

Xu Q, Dietrich H, Steiner HJ, Gown AM, Schoel B, Mikuz G, Kaufmann SH, Wick G (1992) Induction of arteriosclerosis in normocholesterolemic rabbits by immunization with heat shock protein 65. Arterioscler Thromb 12:789–799

Xu Q, Kleindienst R, Waitz W, Dietrich H, Wick G (1993a) Increased expression of heat shock protein 65 coincides with a population of infiltrating T lymphocytes in atherosclerotic lesions of rabbits specifically responding to heat shock protein 65. J Clin Invest 91:2693–2702

Xu Q, Luef G, Weimann S, Gupta RS, Wolf H, Wick G (1993b) Staining of endothelial cells and macrophages in atherosclerotic lesions with human heat-shock protein-reactive antisera. Arterioscler Thromb 13:1763–1769

Xu Q, Schett G, Seitz CS, Hu Y, Gupta RS, Wick G (1994) Surface staining and cytotoxic activity of heat-shock protein 60 antibody in stressed aortic endothelial cells. Circ Res 75:1078–1085

Xu Q, Li DG, Holbrook NJ, Udelsman R (1995) Acute hypertension induces heat-shock protein 70 gene expression in rat aorta. Circulation 92:1223–1229

Xu Q, Kleindienst R, Schett G, Waitz W, Jindal S, Gupta RS, Dietrich H, Wick G (1996) Regression of arteriosclerotic lesions induced by immunization with heat shock protein 65-containing material in normocholesterolemic, but not hypercholesterolemic, rabbits. Atherosclerosis 123:145–155

Yamada H, Strahler J, Welsh MJ, Bitar KN (1995) Activation of MAP kinase and translocation with HSP27 in bombesin-induced contraction of rectosigmoid smooth muscle. Am J Physiol 32:G683–G691

Yamaguchi M, Sato H, Bannai S (1993) Induction of stress proteins in mouse peritoneal macrophages by oxidized low-density lipoprotein. Biochem Biophys Res Commun 193:1198–1201

Young D, Roman E, Moreno C, O'Brien R, Born W (1993) Molecular chaperones and the immune response. Phil Transact R Soc Lond Biol Sci 339:363–367

Young RA (1990) Stress proteins and immunology. Annu Rev Immunol 8:401–420

Zhu W, Roma P, Pellegatta F, Catapano AL (1994) Oxidized-LDL induce the expression of heat shock protein 70 in human endothelial cells. Biochem Biophys Res Commun 200:389–394

Zhu WM, Roma P, Pirillo A, Pellegatta F, Catapano AL (1995) Oxidized LDL induce hsp70 expression in human smooth muscle cells. FEBS Lett 372:1–5

Zhu W, Roma P, Pirillo A, Pellegatta F, Catapano AL (1996) Human endothelial cells exposed to oxidized LDL express hsp70 only when proliferating. Arterioscler Thromb Vasc Biol 16:1104–1111

CHAPTER 19

Heat Shock Protein-Peptide Interaction: Basis for a New Generation of Vaccines Against Cancers and Intracellular Infections

P.K. Srivastava

A. Introduction

The observations that inbred mice and rats can be immunized against their own tumors or against tumors of the same genetic background were convincingly made between 1943 and 1962 (Gross 1943; Foley 1953; Prehn and Main 1957; Klein et al. 1960; Old et al. 1962, for review, see Srivastava and Old 1988). They provided the underpinnings for the idea of immunogenicity of cancers and by deduction, of the existence of tumor-specific antigens. These studies showed that mice vaccinated with inactivated cancer cells are immune to subsequent challenges of live cancer cells. The phenomenon was shown to be individually tumor-specific, in that mice were resistant specifically to the tumors which were used to immunize them and not to other tumors (Basombrío 1970; Globerson and Feldman 1964), hence the name *individually distinct* tumor rejection antigens. The demonstration of immunogenicity of cancer cells led to a search for the cancer-derived molecules, which elicit resistance to tumor challenges. The general structure of these experiments was to fractionate cancer-derived proteins and test them individually for their ability to immunize mice against the cancers from which the fractions were prepared (see Srivastava and Old 1988; Old 1981, and Boon 1992 for other approaches to identification of tumor-specific antigens). The proteins of 96 kDa, 90 kDa and 70 kDa size identified by the method (Table 1; Srivastava and Das 1984; Srivastava et al. 1986; Ullrich et al. 1986; Udono and Srivastava 1993) turned out to be related to a class of proteins known as heat shock proteins (hsps) or stress-induced proteins (see Lindquist and Craig 1988). Similar to the immunogenicity of intact tumor cells, it turned out that hsps gp96, hsp90 and hsp70 isolated from tumors were able to immunize and elicit protective immunity specifically against the tumors from which the hsps were isolated. Hsps isolated from normal tissues were found to be unable to elicit immunity to any cancers tested.

Heat shock proteins (hsps) or stress-induced proteins are expressed in all living cells and are among the most abundant proteins in living cells. They are expressed at high levels constitutively; however, the expression of some members of the hsp family is induced by stress-causing conditions such as heat shock, glucose deprivation etc. Hsps are generally classified by their molecular weights; thus, hsp110, hsp90, hsp70, hsp60 and hsp28 represent hsps of

Table 1. Cancer-derived proteins which have been shown to elicit protective immunity to cancers

Cancer	Induced by	Immunogenicity	Host	Molecules	Assay	Reference
Zajdela hepatocarcinoma	Chemical	++	Rat	gp100 (gp96)	Prophylaxis	SRIVASTAVA and DAS 1984
Meth A fibrosarcoma	Chemical	++	BALB/c mice	gp96 gp96 hsp90 hsp70	Prophylaxis Therapy Prophylaxis Prophylaxis	SRIVASTAVA et al. 1986 TAMURA et al. 1997 ULLRICH et al. 1986 UDONO and SRIVASTAVA 1993
CMS5	Chemical	+	BALB/c mice	gp96	Prophylaxis	SRIVASTAVA et al. 1986
CMS13	Chemical	++	BALB/c mice	gp96	Prophylaxis	PALLADINO et al. 1987
Dunning G prostate cancer	Chemical	−−	Rat	gp96 gp96	Prophylaxis Therapy	TIWARI et al. 1996 TIWARI et al. 1996
Lewis Lung carcinoma	Spontaneous	−−	C57BL/6 mice	gp96 hsp70	Therapy Therapy	TAMURA et al. 1997 TAMURA et al. 1997
B16 melanoma	Spontaneous	−−	C57BL/6 mice	gp96	Therapy	TAMURA et al. 1997
CT26 colon carcinoma	Chemical	++	BALB/c mice	gp96	Therapy	TAMURA et al. 1997
UV6138	UV	+++	C3H mice	gp96	Prophylaxis	JANETZKI et al. 1998
UV6139	UV	++	C3H mice	gp96 gp96	Prophylaxis Therapy	JANETZKI et al. 1998 TAMURA et al. 1997

110 kDa, 90 kDa, 70 kDa, 60 kDa and 28 kDa sizes respectively. Each such hsp family may have one or more members which are highly homologous to each other; however, there is little homology among different families. Thus, hsp70 and hsp90 have no homology among them. Further, each specific hsp family is highly conserved phylogenetically. Thus, hsp70 molecules in bacteria and humans are very similar to each other.

The identification of hsps as eliciting individually tumor-specific tumor immunity is a curious paradox. Hsps are among the most highly conserved and abundant proteins in living systems; they are found across the phylogenetic ladder from archaebacteria to primates and differ modestly among different species, let alone within an inbred strain. All in all, they are the most unlikely candidates for tumor-specific antigens.

The observations which have helped define and substantiate this paradox are described here, as are the recent results which now explain it. The specificity of immunogenicity of hsps derived from tumor cells and the lack of tumor immunogenicity of hsp preparations derived from normal tissues suggested that hsps might be hot spots for mutations during malignant transformation such that the hsp genes will show variation between normal tissues and tumors, and among tumors. However, sequencing of gp96 cDNAs from tumors and normal tissues did not reveal any tumor-specific, individually distinct polymorphisms (SRIVASTAVA and MAKI 1991). This directed our attention to the role of N-linked sugars of gp96 (there are no O-linked sugars in gp96). However, the following observations ruled out their role in tumor-specific immunogenicity of gp96: (i) tumor cells cultured in presence of tunicamycin can be used to specifically immunize mice against the tumor. (ii) Gp96 derived from tumor cells grown in presence of tunicamycin did not bind to Con A-Sepharose and successfully immunized mice specifically against the tumor (unpublished observations). In the case of hsp90 and hsp70, the question of a role for sugars in the specific immunogenicity does not arise, as they are not glycosylated. It is conceivable that specificity might reside in other post-translational modifications; however, we consider this unlikely. Thus, there appears little reason to believe that specific immunogenicity of hsps lies in hsps per se.

B. Hsps Chaperone Antigenic Peptides

The lack of diversity in hsps led us to consider molecules associated with them. As the gp96 used to immunize mice is homogeneous by all criteria [(a) single band on overloaded silver-stained gels; (b) single amino terminus during Edman degradation; (c) anti-peptide antibody to gp96 depletes a preparation of its immunogenic activity), our attention was focussed on small moieties. A number of hsps bind to a wide array of molecules, including peptides (FLYNN et al. 1989, 1991); In this light, we proposed that gp96 molecules are not immunogenic per se, but act as carriers of antigenic peptides (SRIVASTAVA and MAKI 1991; SRIVASTAVA and HEIKE 1991). In view of the predominant localization of gp96 in the endoplasmic reticulum (ER; Booth and Koch 1989), we further suggested that gp96 acts as peptide-acceptor for peptides transported to ER and may be accessory to loading of MHC class I molecules (Srivastava and Maki 1991; Srivastava et al. 1994). These ideas have now found a solid degree of experimental support from our and other laboratories (Li and Srivastava 1993; Udono and Srivatsava 1993; Suto and Srivastava 1995; Peng et al. 1997; Arnold et al. 1995, 1997; Lammert et al. 1997; Neefjes 1997; Nieland et al. 1996). It is now clear from studies from several laboratories and in a variety of experimental systems that the hsps are not antigenic per se, but act as carriers of antigenic peptides. The idea that hsps chaperone not only tumor-specific peptides but indeed all peptides generated within a cell, including viral peptides generated during infection, has also been developed and successfully tested (Blachere et al. 1993; Suto and Srivastava 1995; Nieland et al. 1996; Heikema et al. 1997). These results have led to the development of hsp-peptide complexes as the basis of a new generation of specific vaccines against cancers and intracellular infections.

C. Unique Advantages of Hsp-Peptide Vaccines

Hsp-peptide complexes offer unique and unprecedented advantages over other types of vaccines against cancers, infectious and transforming viruses, intracellular bacteria and protozoa:

1. Knowledge of the antigenic epitopes which elicit immunity is a prerequisite for all forms of vaccination. Hsp-peptide based vaccination circumvents this necessity, as hsps are naturally complexed with the repertoire of peptides generated in a cell. For this reason it is an ideal means for vaccination against infections in which the protective epitopes are yet undefined, or where a single epitope may not be sufficient for eliciting immunity, or where the infectious agent is so highly variable (in a population, season or individual-specific manner) that the prospect of identifying the immunogenic epitopes for each variant is simply impractical.

2. Even in the case of infectious diseases where the relevant antigenic epitopes *have been* defined, hsp-based vaccination offers a unique advantage: hsp-peptide complexes can be readily stripped of their natural peptides and these "empty" hsps can be reconstituted with known, synthetic peptide epitope(s) (Blachere et al. 1997). The non-covalent hsp-peptide complexes reconstituted in vitro elicit potent *T cell response* to the complexed peptide. Further, if the peptide is conjugated covalently to the hsp, the hsp-peptide complexes elicit potent *antibody responses* to the complexed peptide (Lussow et al. 1991; Barrios et al. 1992).

3. One of the major conceptual difficulties in cancer immunotherapy has been the possibility that human cancers, like cancers of experimental animals, are antigenically distinct (Globerson and Feldman 1964; Basombrío 1970). The prospect of identification of immunogenic antigens of individual

tumors from cancer patients is daunting to the extent of being impractical. The ability of hsps to chaperone the entire repertoire of antigenic peptides of the cells from which they are derived circumvents this extraordinary hurdle. Thus, patients can be vaccinated with hsp-peptide complexes derived from their own tumors, or cell lines derived from the tumors, without any need for identification of the antigenic epitopes of the patient's tumor. In light of the emerging evidence for existence of "shared" tumor antigens (PARDOLL 1994; HOUGHTON 1994), this particular point of advantage for hsp-based vaccines may at first sight appear less profound than originally imagined; however, closer scrutiny suggests than even if human cancer antigens are cross-reactive rather than individually distinct, hsp-peptide complexes offer a uniquely effective method of vaccination (see next section).

4. As hsps are non-polymorphic (i.e., show no allelic diversity, although there are several hsp families), they bind the entire spectrum of peptides regardless of the MHC haplotype of a cell. Thus, an hsp-peptide complex isolated from cells of a given haplotype may be used to vaccinate individuals of other haplotypes (SUTO and SRIVASTAVA 1995; ARNOLD et al. 1995; NIELAND et al. 1996).

5. Vaccination with hsp-peptide complexes elicits CD8+ T cells without the use of live (attenuated or otherwise) agents and in spite of exogenous administration (UDONO et al. 1994; BLACHERE et al. 1997; TAMURA et al. 1997).

6. Hsp-peptide based vaccines are inherently multivalent because Hsps chaperone, not one or a few but the entire repertoire of epitopes generated in a cell.

7. As hsp-peptide complexes can be purified easily to apparent homogeneity, vaccination with such preparations circumvents the risks associated with vaccination with attenuated organisms or undefined biological extracts which contain transforming DNA and immunosuppressive factors such as TGFβ.

D. Use of Hsp-Peptide Complexes as Cancer Vaccines

One of the advantages of hsp-peptide complexes for vaccination against human cancers lies in the possibility that human cancers, like their murine counterparts, are antigenically diverse and individually distinct (GLOBERSON and FELDMAN 1964; BASOMBRÍO 1970). In that scenario, it would be practically impossible to identify the antigenic epitopes of individual cancer patients and autologous tumor-derived hsp-peptide vaccines would present an attractive opportunity.

If, however, human tumors are antigenically cross-reactive (as is argued by some on the basis of the "experience" with human melanomas), anti-tumor vaccines could be designed simply on the basis of peptide epitopes of known cross-reactive melanoma antigens instead of hsp-peptide complexes isolated from individual tumors. It is the premise of this article that hsp-peptide complexes provide a uniquely effective method of vaccination, regardless of antigenic individuality or cross-reactivity of human tumor antigens. This premise is elaborated as follows.

Firstly, the antigenic cross-reactivity among human melanomas suggests that once a number of shared melanoma antigens are identified, patients can be immunized with synthetic peptides corresponding to the relevant epitopes and that the vaccinated patients will elicit a CD8+ CTL response. Vaccination with peptides under suitable conditions has indeed been shown to elicit CD8+ CTLs in a number of systems (NOGUCHI et al. 1994; SCHULZ et al. 1991). Such conditions usually include the use of incomplete Freund's adjuvant along with large quantities (~100 µg peptide for a 20-g mouse) of peptide. This is clearly incompatible with human use. Alternative approaches, such as addition of a lipophilic tail to the peptides, have been employed successfully (DERES et al. 1989) and could be potentially suited for human applications. In this context, vaccination with in vitro reconstituted complexes of human hsps with the relevant antigenic peptides offers an economical and technologically simple method of vaccination. The ability of hsp-peptide complexes, *administered in saline and without any adjuvants*, to prime naive CD8+ CTLs in vivo has been reported recently (BLACHERE et al. 1997). The use of human hsps and synthetic peptides is also attractive from the point of view of circumventing hazards associated with vaccination of patients with chimeric molecular constructs of unknown toxicity.

Secondly, vaccination with a given peptide will be effective only for patients with a given HLA allele. If different epitopes from a single molecule are recognized by different HLA alleles (as appears to be the case in the case of tyrosinase, ref. BRICHARD et al. 1993; WOLFEL et al. 1994), a cocktail of peptides will have to be used for vaccination of a general population. Even for a given patient, a cocktail may have to be used, as humans are outbred and possess several restriction elements. A far more effective and simpler alternative will be to isolate hsp-peptide complexes from human cell lines transfected with the relevant gene under the control of a high expression promoter. The hsp-peptide complexes purified from such transfectants will consist of the entire repertoire of antigenic peptides derived from that particular protein. As hsp-peptide binding is proximal to HLA-peptide binding during antigen processing, there is no HLA restriction in the hsp-bound peptides. Peptides capable of binding to all possible HLA alleles will be represented among the hsp-peptide complexes.

Thirdly, the methodology of identification of CTL epitopes of human cancers suggests that these epitopes may represent a significantly biased sample of the antigenic repertoire of human cancers. Generation of cell lines is an essential prerequisite for isolation of CTLs and only a very small proportion of human cancers (less than 2% for breast cancers to about 30% for melanomas) lend themselves to it. Of the tumors from which cell lines are developed, only a small proportion permit generation of CTLs. Thus, the CTL epitopes being identified may represent an atypical and sparse sampling of the cancer antigenic repertoire. Immunogenicity of cancers represents, in all likelihood, the sum total of immunogenicity of a large number of immunogenic epitopes and effective anti-cancer vaccines should include this antigenic multiplicity. Hsp-peptide complexes are such a multi-component, multivalent vaccine. If cancer antigens are shared and not individually indistinct (examined in detail in the next section), the use of hsp-peptide complexes isolated from tumors becomes even simpler, such that these complexes need not be isolated from cancers of individual patients. Instead, they could be purified from a mixture of human melanomas and be used to vaccinate allogeneic melanomas. The lack of HLA-restriction of hsp-bound peptides (discussed in the preceding section) is a key advantage in this regard. Vaccination with multi-component, multivalent vaccines rather than single or oligo-component vaccines is also necessary for protection against antigenic escape or preexisting antigenic heterogeneity of human cancers.

These three major considerations indicate that regardless of the cross-reactive or individually distinct nature of human cancer antigens, hsp-peptide complexes offer unique and unprecedented advantages over other existing methods, in vaccination against human cancer.

E. Protective Human Cancer Antigens: Unique to Each Individual Cancer or Shared Between Cancers?

It has been demonstrated convincingly in most systems analyzed that the CD8+ T cells are a crucial element of the protective immune response to cancer (SHU et al. 1986; YOSHIZAWA et al. 1992; UDONO et al. 1994). However, there is little evidence that the crucial tumor-protective CD8+ cells correspond to the CTLs generated, developed into lines and clones, in vitro. In a number of tumor systems, protective immunity can be shown to be dependent on a CD8+ response, but no CTLs against these tumors can be isolated from hyperimmunized mice (unpublished). At the same time, CTLs generated against tumors can be shown, in a number of systems, to have little protective value (RAMARATHINAM et al. 1995; LEVRAUD et al. 1996). At the heart of this discussion is the process by which anti-tumor CTLs are generally isolated. The process involves co-culturing of splenic or peripheral blood cells with cancer cells in vitro for one to several weeks, followed by testing of the CTLs for activity. In all instances, the T cells isolated from an animal or a patient have no primary cytotoxic effect against the tumor; in vitro stimulation with the cognate cancer cell is necessary. The requirement for in vitro stimulation raises the question if the CTLs obtained after stimulation in vitro were necessarily active in vivo. The stimulation in vitro is generally necessary because of low precursor CTL frequencies, which must be brought into a measurable range. While this is reasonable, it also indicates that the CTLs generated in vitro might simply represent activation of preexisting memory populations which are not active in vivo for a number of reasons, such as the regulatory interactions. To the extent this is so, the CTL response and the CTL epitopes identified on the basis of such a response are misleading as to their role in tumor protection in vivo. It is of interest in this regard that CTLs against melanoma antigens are detected not only in melanoma patients but also in healthy volunteers; further, in melanoma patients, there appears to be little correlation between the CTL response to these antigens and the course of disease (SALGALLER et al. 1995). If one views the anti-tumor CTL response, as it is traditionally measured, it appears reasonable that strong responses to shared antigens would be detected. Antigens such as differentiation antigens of the melanocytic lineage or antigens expressed in sites of immunological privilege are not expressed in the thymus and CTLs recognizing them are not expected to be centrally deleted. However, CTLs against such antigens must be kept from activation by peripheral mechanisms. Once a response is freed from such constraints in vitro, one would expect to see a strong response to such antigens, but this phenomenon in vitro is unrelated to the anti-tumor response in vivo. Thus, the dissonance between the view of cancer antigens as obtained by tumor transplantation and the view of cancer antigens developed on the basis of CTL epitopology results from the erroneous extrapolation of CTL epitopes into tumor protective antigens. The root of this error can be traced to the fact that contemporary methods of isolation, propagation and cloning of CTLs do not permit a selective identification of CTLs which are active in vivo; instead, they provide a much broader view of all potential reactivities.

The idea that the identified CTL epitopes are not necessarily tumor-protective is strengthened by recent studies, which have directly examined the tumor-protective ability of such epitopes. CTL epitopes of three murine tumors have been identified so far. These are the P815 mastocytoma, the Lewis Lung carcinoma and CT26 colon carcinoma. The CTL epitopes of the three tumors are derived respectively from the P1 A (VAN DEN EYNDE et al.1991), connexin 37 (MANDELBOIM et al. 1995) and the gp70 murine leukemia virus antigens (HUANG et al. 1996). RAMARATHINAM et al. have studied the tumor protective activity of the P1 A antigen, which is the mouse prototype for the MAGE type of human melanoma antigens. These authors find that the P1 A molecule is expressed in a number of tumors and that all the P1A+ tumors elicit P1 A specific CTL response. However, the tumors are not cross-reactive in tumor protection assays. Further, immunization with P1 A peptide does not elicit protective immunity against any of the P1A+ tumors. In earlier studies with the P1 A epitope, BOON et al. had shown that the P815 variants which grew out in immunized mice had lost the P1 A epitope, thus implicating the epitope in protection from tumor growth (UTTENHOVE et al. 1983). Those data are difficult to interpret in this manner because one cannot determine what other epitope(s) in addition to P1 A may have been lost. In studies with the CT26 colon carcinoma, HUANG et al. have identified a dominant CTL epitope derived from the gp70 molecule of the endogenous murine leukemia virus. The CTL epitope in CT26 is unmutated with respect to the endogenous gene. These authors find that adoptive transfer of gp70-specific CTLs into mice provides protection against the CT26 tumor. However, vaccination of mice with the gp70 epitope by itself, or peptide-pulsed dendritic cells does not confer protective immunity against CT26. The results suggest that there are immunological constraints on generation of CTL response against the gp70 epitope although a preexisting response (such as CTLs generated in vitro) can confer protection from tumor growth. In the case of the Lewis lung carcinoma, the CTL epitope is mutated with respect to the normal sequence of connexin 37 and vaccination with the epitope-pulsed antigen presenting cells, but not with the epitope alone, confers protection against micrometastatic disease (MANDELBOIM et al. 1995). This observation is hard to interpret as the nature of mutations in the connexin 37 epitope in the tumor with respect to its normal counterpart has raised questions about genetic drift between the tumor and the mice in which this tumor first arose. Further, the original observations now appear to have been retracted (MANDELBOIM et al. 1997; ROSEN et al., submitted).

The results with these three CTL epitopes support the contention that CTL epitopes defined so far may have little to do with tumor protection (as in the case of P815), and the possibilities of the use of such epitopes for vaccination against cancers remain to be realized, possibly because of tolerance to epitopes defined by the CTLs (as in the case of CT26 and P815).

F. Hsp-Peptide Complexes as Vaccines Against Intracellular Infectious Agents

Although the ideas discussed here were arrived at through pursuit of cancer immunity, it soon became clear that the peptide-chaperoning role hypothesized for the hsps is a general one, i.e., valid for any cellular protein and not only for tumor antigens. In that scenario, Hsp-peptide complexes isolated from virus-infected cells would be expected to elicit T cell response and immunity against the cognate virus. We and others have tested this idea in a number of viral systems including the influenza virus, simian virus 40, and vesicular stromatitis virus. Gp96 preparations were obtained from flu (PR8)-infected BALB/c fibroblasts and used to immunize BALB/c mice. T cells generated from these mice and stimulated in vitro with flu-infected cells showed significant antigen-specific, MHC class I restricted CTL activity (BLACHERE et al. 1993; HEIKEMA et al. 1997). In other studies, mice were immunized with gp96 derived from SV40 transformed PSC3H cells, or with non-SV40-transformed C3H fibroblasts. Mixed lymphocyte-stimulator cultures generated from these mice and stimulated with PSC3H cells were tested on PSC3H and non-SV40 transformed fibroblasts. Class I restricted cytotoxicity against PSC3H was seen in MLTCs generated from mice immunized with

PSC3H gp96 but not the fibroblast-derived gp96 (Blachere and Srivastava, unpublished). Similarly, mice immunized with gp96 purified from VSV-infected cells showed antigen-specific, MHC I-restricted CTL response to VSV (Suto and Srivastava 1995). As further proof of these ideas, we have recently demonstrated that one can complex hsps to synthetic peptides in vitro and use such complexes to elicit antigen-specific T cell responses (Blachere et al. 1997). These studies indicate that the ability of hsps to chaperone a broad array of peptides can be used effectively not only against cancers but against infectious agents as well. These should include intracellular agents such as viruses, rickettsiae, intracellular bacteria such as mycobacteria, and certain protozoan parasites.

References

Arnold D, Faath S, Rammensee H, Schild H (1995) Cross-priming of minor histocompatibility antigen-specific cytotoxic T cells upon immunization with the heat shock protein gp96. J Exp Med 182(3):885–889

Arnold D, Wahl C, Faath S, Rammensee HG, Schild H (1997) Influences of transporter associated with antigen processing (TAP) on the repertoire of peptides associated with the endoplasmic reticulum-resident stress protein gp96. J Exp Med 186(3):461–466

Barrios C, Lussow AR, van Embden J, Van der Zee R, Rappuoli R, Costantino P, Louis JA, Lambert PH, Del Giudice G (1992) Mycobacterial heat shock proteins as carrier molecules II: the use of the 70 kda mycobacterial heat shock protein as carrier for conjugated vaccines can circumvent the need for adjuvants and BCG priming. Eur J Immunol 22:1365

Basombrío MA (1970) Search for common antigenicities among 25 sarcomas induced by methylcholanthrene. Cancer Res 30:2458

Blachere NE, Li Z, Chandawarkar RY, Suto R, Jaikaria NS, Basu S, Udono H, Srivastava PK (1997) Heat shock protein-peptide complexes, reconstituted in vitro, elicit peptide-specific cytotoxic T lymphocyte response and tumor immunity. J Exp Med 186(8):1315–1322

Blachere NE, Udono H, Janetzki S, Li Z, Heike M, Srivastava PK (1993) Heat shock protein vaccines against cancer. J Immunother 14:352

Boon T (1992) Toward a genetic analysis of tumor rejection antigens. Adv Cancer Res 58:177–210

Booth C, Koch GL (1989) Perturbation of cellular calcium induces secretion of luminal ER proteins. Cell 59(4):729–737

Brichard V, van Pel A, Wolfel T, DePlaen E, Lethe B, Coulie P, Boon T (1993) The tyrosinase gene codes for an antigen recognized by autologous cytolytic T lymphocytes on HLA-A2 melanomas. J Exp Med 178(2):489–495

Deres K, Schild H, Wiesmuller KH, Jung G, Rammensee HG (1989) In vivo priming of virus-specific cytotoxic T lymphocytes with synthetic lipopeptide vaccines. Nature 342:561

Flynn GC, Chappell TG, Rothman JE (1989) Peptide binding and release by proteins implicated as catalysts of protein assembly. Science 245(4916):385–390

Flynn GC, Pohl J, Flocco MT, Rothman JE (1991) Peptide-binding specificity of the molecular chaperone BiP. Nature 353(6346):726–730

Foley EJ (1953) Cancer Res 13:835–837

Globerson A, Feldman M (1964) Antigenic specificity of benzo(a)pyrene induced sarcomas. J Natl Cancer Inst 32:1229

Gross L (1943) Cancer Res 3:323–326

Heikema A, Agsteribbe E, Wiscjut J, Huckriede A (1997) Generation of heat shock protein-based vaccines by intracellular loading of gp96 with antigenic peptides. Immunol Lett 57(1–3):69–74

Houghton AN (1994) Cancer antigens: immune recognition of self and altered self. J Exp Med 180(1):1–4

Huang AYC, Gulden PH, Woods AS, Thomas MC, Tong CD, Wang W, Engelhard VH, Pasternack G, Cotter R, Hunt D, Pardoll D, Jaffe E (1996) The immunodominant MHC class I-restricted antigen of a murine colon tumor derives from an endogenous retroviral gene product. Proc Natl Acad Sci USA 93(18):9730–9735

Klein G, Sjorgen HO, Klein E, Hellstrom KE (1960) Cancer Res 20:1561–1572

Levraud J, Pannetier C, Langlade-Demoyen , Brichard V, Kourilsky P (1996) Recurrent T cell receptor rearrangements in the cytotoxic T lymphocyte response in vivo against the p815 murine tumor. J Exp Med 183:439–449

Li Z, Srivastava PK (1993) Tumor rejection antigen Gp96/ Grp94 is an ATPase: implications for protein folding and antigen presentation. EMBO J 12:3143

Lindquist S, Craig EA (1988) The heat-shock proteins. Annu Rev Genet 22:631–677

Lussow AR, Barrios C, van Embden J, Van der Zee R, Verdini AS, Pessi A, Louis JA, Lambert PH, Del Giudice G (1991) Mycobacterial heat shock proteins as carrier molecules. Eur J Immunol 21:2297

Mandelboim O, Berke G, Fridkin M, Feldman M, Eisenstein M, Eisenbach L (1997) CTL induction by a tumor-associated antigen octapeptide derived from a murine lung carcinoma. Nature 390:643

Mandelboim O, Vadai E, Fridkin M, Katz-Hillel A, Feldman M, Berke G, Eisenbach L (1995) Regression of established murine carcinoma metastases following vaccination with tumour-associated antigen peptides. Nature Med 1:1179–1183

Nieland TJ, Tan MC, Monne-van Muijen M, Koning F, Kruisbeek AM, van Bleek GM (1996) Isolation of an immunodominant viral peptide that is endogenously bound to the stress protein GP96/GRP94. Proc Natl Acad Sci USA 93(12):6135–6139

Noguchi Y, Chen YT, Old LJ (1994) A mouse mutant p53 product recognized by CD4+ and CD8+ T cells. Proc Natl Acad Sci USA 91(8):3171–3175

Old LJ, Boyse EA, Clarke DA, Carswell EA (1962) Antigenic properties of chemically induced tumors. Ann NY Acad Sci 101:80

Old LJ (1981) The search for specificity: GHA Clowes Memorial Lecture. Cancer Res 41:361–375

Pardoll DM (1994) Tumour antigens. A new look for the 1990 s. Nature 369(6479):357

Peng P, Menoret A, Srivastava PK (1997) Purification of immunogenic heat shock protein 70-peptide complexes by ADP-affinity chromatography. J Immunol Methods 204(1):13–21

Prehn RT, Main JM (1957) Immunity to methylcholanthrene-induced sarcomas. J Natl Cancer Inst 18:769

Ramarathinam L, Sarma S, Maric M, Zhao M, Yang G, Chen L, Liu Y (1995) Multiple lineages of tumors express a common tumor antigen, P1 A, but they are not cross protected. J Immunol 155:5323

Rosen D, Brookenthal K, Berke G A reappraisal of the tumor immunoprotective and therapeutic activities of the octapeptide Mut 1 and 2 from mouse Lewis lung carcinoma 3LL. J Immunol (submitted)

Salgaller ML, Afshar A, Marincola FM, Rivoltini L, Rosenberg YA (1995) Recognition of multiple epitopes in the human melanoma antigen gp100 by peripheral blood lymphocytes stimulated in vitro with synthetic peptides. Cancer Res 21:4972–4979

Schulz M, Zinkernagel RM, Hengartner H (1991) Peptide-induced antiviral protection by cytotoxic T cells. Proc Natl Acad Sci USA 88(3):991–993

Shu S, Chou T, Rosenberg SA (1986) In vitro differentiation of T-cells capable of mediating the regression of established syngeneic tumors in mice. Cancer Res 47:1354–1360

Spee P, Neefjes J (1997) TAP-translocated peptides specifically bind proteins in the endoplasmic reticulum, including gp96, protein disulfide isomerase and calreticulin. Eur J Immunol 27(9):2441–2449

Srivastava PK, Maki RG (1991) Stress-induced proteins in immune response to cancer. In: Capron A, Compans RW, Cooper M et al (eds) Current topics in microbiology and immunology, vol 167. Springer, Berlin Heidelberg New York, pp 109–123

Srivastava PK, Heike M (1991) Tumor-specific immunogenicity of stress-induced proteins: convergence of two evolutionary pathways of antigen presentation? Semin Immunol 3(1):57–64

Srivastava PK, Old LJ (1988) Individually distinct transplantation antigens of chemically induced mouse tumors. Immunol Today 9:78

Srivastava PK, DeLeo AB, Old LJ (1986) Tumor rejection antigens of chemically induced sarcomas of inbred mice. Proc Natl Acad Sci USA 83:3407

Srivastava PK, Udono H, Blachere NE, Li Z (1994) Heat shock proteins transfer peptides during antigen processing and CTL priming. Immunogenetics 39:93

Suto R, Srivastava P (1995) A mechanism for the specific immunogenicity of heat shock protein-chaperoned peptides. Science 269(5230):1585–1588

Tamura Y, Peng P, Kang L, Daou M, Srivastava P (1997) Immunotherapy of tumors with autologous tumor-derived heat shock protein preparations. Science 278:117–120

Udono H, Srivastava PK (1993) Heat shock protein 70-associated peptides elicit specific cancer immunity. J Exp Med 178:1391

Udono H, Levey DL, Srivastava PK (1994) Definition of T cell sub-sets mediating tumor-specific immunogenicity of cognate heat shock protein gp96. Proc Natl Acad Sci USA 91:3077

Ullrich SJ, Robinson EA, Law LW, Willingham M, Appella E (1986) A mouse tumor-specific transplantation antigen is a heat shock-related protein. Proc Natl Acad Sci USA 83:3121–3125

Uttenhove C, Maryanski J, Boon T (1983) Escape of mouse mastocytoma p815 after nearly complete rejection is due to antigen-loss variants rather than immunosupression. J Exp Med 157:1040–1052

Van den Eynde B, Lethe B, Van Pel A, DePlaen E, Boon T (1991) The gene coding for a major tumor rejection antigen of tumor p815 is identical to the normal gene of syngeneic DBA/2. J Exp Med 173:1373–1384

Wolfel T, Van Pel A, Brichard V, Schneider J, Seliger B, Meyer zum Buschenfelde KH, Boon T (1994) Two tyrosinase nonapeptides recognized on HLA-A2 melanomas by autologous cytolytic T lymphocytes. Eur J Immunol 24(3):759–764

Yoshizawa H, Chang AE, Shu S (1992) Cellular interactions in effector cell generation and tumor regression mediated by anti-CD3/interleukin 2-activated tumor-draining lymph node cells. Cancer Res 52:1129–1136

Subject Index

CHESTER COLLEGE LIBRARY

EXETER COLLEGE LIBRARY

Springer
and the
environment

At Springer we firmly believe that an
international science publisher has a
special obligation to the environment,
and our corporate policies consistently
reflect this conviction.
We also expect our business partners –
paper mills, printers, packaging
manufacturers, etc. – to commit
themselves to using materials and
production processes that do not harm
the environment. The paper in this
book is made from low- or no-chlorine
pulp and is acid free, in conformance
with international standards for paper
permanency.

CHESTER COLLEGE
ACC. No.
CLASS No.
LIBRARY

 Springer

Printing: Saladruck, Berlin
Binding: Buchbinderei Saladruck, Berlin